Progress in Combustion Diagnostics, Science and Technology

Progress in Combustion Diagnostics, Science and Technology

Special Issue Editors

Paul Medwell
Michael Evans
Shaun Chan

MDPI • Basel • Beijing • Wuhan • Barcelona • Belgrade • Manchester • Tokyo • Cluj • Tianjin

Special Issue Editors

Paul Medwell
School of Mechanical
Engineering, The University
of Adelaide
Australia

Michael Evans
School of Mechanical
Engineering, The University
of Adelaide
Australia

Shaun Chan
School of Mechanical and
Manufacturing Engineering,
University of New South Wales
Australia

Editorial Office
MDPI
St. Alban-Anlage 66
4052 Basel, Switzerland

This is a reprint of articles from the Special Issue published online in the open access journal *Applied Sciences* (ISSN 2076-3417) (available at: https://www.mdpi.com/journal/applsci/special_issues/Combustion_Diagnostics).

For citation purposes, cite each article independently as indicated on the article page online and as indicated below:

LastName, A.A.; LastName, B.B.; LastName, C.C. Article Title. *Journal Name* **Year**, *Article Number*, Page Range.

ISBN 978-3-03928-510-5 (Pbk)
ISBN 978-3-03928-511-2 (PDF)

Contents

About the Special Issue Editors

Paul Medwell is an Associate Professor at the School of Mechanical Engineering at the University of Adelaide. He has many years of combustion research experience, initially in the field of moderate or intense low oxygen dilution (MILD) combustion and the development and application of advanced laser diagnostic techniques for the measurement of flames and harsh reacting environments. Over the years, he has expanded his research interests to other areas of combustion and humanitarian engineering. Dr. Medwell is an Australian Research Council Future Fellow and was a Discovery Early Career Researcher Award recipient.

Michael Evans is a Research Associate at the School of Mechanical Engineering at the University of Adelaide, having completed his Ph.D. in 2017 and B.Eng. and B.S. in 2013. In 2015, Dr. Evans was awarded a prestigious Endeavour Postgraduate Scholarship to fund an extended stay at Stanford University. Dr Evans' research involves a combination of numerical simulations and experimental laser diagnostics predominantly focussing on both the combustion of alternative, carbon-free, and liquid fuels and the differences in and transition between conventional flames and low-emissions flameless or moderate or intense low oxygen dilution (MILD) combustion.

Qing Nian (Shaun) Chan is a Senior Lecturer at the School of Mechanical and Manufacturing Engineering at UNSW Sydney. He holds a B.Eng. in chemical engineering, a Ph.D. in combustion and has extensive experience in developing and implementing optical- and laser-based diagnostics techniques for the detailed analysis of combustion systems. Dr. Chan's Ph.D. research primarily focused on the development and the innovative use of advanced laser-based measurements to investigate reacting and non-reacting flows, and spray and combustion diagnostics in the engine environment as well as in fire scenarios. He contributed toward the securing of $18 million of new competitive research funding and research contracts. Dr. Chan has published 60+ journal articles and 40+ conference papers in the areas of combustion and engine diagnostics, as well as fire.

Preface to "Progress in Combustion Diagnostics, Science and Technology"

The role played by combustion in energy systems remains crucial in supplying the world's ever-increasing power demands. In the quest for improving efficiency, additional knowledge is essential for developing new combustion technologies and appliances. Focus is increasing on the conservation of energy and addressing environmental concerns, which together necessitate cleaner and more efficient combustion processes using a range of fuel sources. This is essential to respond to global challenges in energy supply and to continue to address the decarbonization of the sector. In addition to power production, understanding combustion also plays a critical role in both managing fires and the materials synthesis sectors. To meet the objectives of progress and innovation in combustion science, new experimental measurements are needed and complemented by computational approaches. This book includes a series of 17 research studies that reveal new knowledge about combustion and its application. The topics covered span diverse areas associated with combustion including: fires, engines and applications, and acoustics. In combination, these complementary contributions provide a substantial body of knowledge in the field of progress in combustion diagnostics, science and technology, hence the apt name of this exciting publication..

Paul Medwell, Michael Evans, Shaun Chan
Special Issue Editors

Article

Numerical Study of the Comparison of Symmetrical and Asymmetrical Eddy-Generation Scheme on the Fire Whirl Formulation and Evolution

Cheng Wang [1], Anthony Chun Yin Yuen [1,*], Qing Nian Chan [1], Timothy Bo Yuan Chen [1], Ho Lung Yip [1], Sherman Chi-Pok Cheung [2], Sanghoon Kook [1] and Guan Heng Yeoh [1,3]

[1]   School of Mechanical and Manufacturing Engineering, University of New South Wales, Sydney, NSW 2052, Australia; c.wang@unsw.edu.au (C.W.); qing.chan@unsw.edu.au (Q.N.C.); timothy.chen@unsw.edu.au (T.B.Y.C.); h.l.yip@unsw.edu.au (H.L.Y.); s.kook@unsw.edu.au (S.K.); g.yeoh@unsw.edu.au (G.H.Y.)

[2]   School of Mechanical and Automotive Engineering, RMIT University, Melbourne, VIC 3000, Australia; chipok.cheung@rmit.edu.au

[3]   Australian Nuclear Science and Technology Organisation (ANSTO), Locked Bag 2001, Kirrawee DC, NSW 2232, Australia

*   Correspondence: c.y.yuen@unsw.edu.au; Tel.: +61-2-9385-5697

Received: 30 November 2019; Accepted: 28 December 2019; Published: 1 January 2020

Abstract: A numerical study of the fire whirl formation under symmetrical and asymmetrical entraining configuration is presented. This work aims to assess the effect of eddy-generation configuration on the evolution of the intriguing phenomenon coupled with both flow dynamics and combustion. The numerical framework implements large-eddy simulation, detailed chemistry to capture the sophisticated turbulence-chemistry interaction under reasonable computational cost. It also adopts liquid-based clean fuel with fixed injection rate and uniformed discretisation scheme to eliminate potential interference introduced by various aspects of uncertainties. The result reveals that the nascent fire whirl formulates significantly rapidly under the symmetrical two-slit configuration, with extended flame height and constrained vortex structure, compared with the asymmetrical baseline. However, its revolution orbit gradually diverges from domain centreline and eventually stabilises with a large radius of rotation, whereas the revolution pattern of that from the baseline case is relatively unchanged from the inception of nascent fire whirl. Through the analysis, the observed difference in evaluation pathway could be explained using the concept of circular motion with constant centripetal force. This methodology showcases its feasibility to reveal and visualise the fundamental insight and facilitate profound understanding of the flaming behaviour to benefit both research and industrial sectors.

Keywords: fire whirl; computational fluid dynamics; eddy-generation mechanism; combustion modelling; detailed chemistry; large eddy simulation

---

1. Introduction

Fire whirl is a unique combustion behaviour with a twisting flame structure which significantly intensifies combustion and fluid mechanics, that is often observed in urban and wildland fire scenarios [1]. The swirling reacting flow limits the dispersion rate of the radial flame and stretches the hot plume to progress up to an elevated vertical angular path. Compared with a free-standing diffusion flame, the formulation of the fire whirl often results in a significant increase in rate of combustion, visible flame height, peak flame temperature and the intensity of radiation emitted from flames towards the surrounding Ref [2]. Owing to the unique features associated with this very combustion behaviour, the inception of the fire whirl in a fire scenario could lead to catastrophic effects. In particular, the intensification of heat energy contained within the stretched flame plume and the enhancement in radiation heat transfer to the neighbouring points resulted from a fire whirl could significantly promote the fire spreading towards the surroundings in an accelerated rate. The spatial movement of the swirling plume may also aid the mixing of the reacting gas mixture with the oxidiser to increase the rate of combustion and impose more spatial uncertainties [3]. The presence of the fire whirl has been reported in many notorious fire incidents and been identified as the root causes to aggravate the fire scenario to an unmanageable and untenable manner [4,5]. The formation and evolution of fire whirl, therefore, is a topic of great interest by both industrial communities and industrial sectors.

Previous investigations have identified the three essential criteria that are essential for the formation of this particular swirling flame, namely a thermally driven fluid sink, a radial boundary layer created by a surface drag force, and an eddy-generation source [6–8]. In a typical buoyant diffusion flame, the flame structure is acting as the fluid sink which generates hot plumes that naturally drives horizontal flow radically towards the vortex column, which fulfil the first two requirements [9,10]. The eddy-generation mechanism, as the only remaining criteria, therefore, is the most critical element in the transition from free-standing flame towards fire whirl.

The generation of an eddy could naturally occur in a wildland fire scenario that is triggered by topological obstacles, leeward slope, unpredicted weather conditions, etc. [11–13]. Such an eddy-generation source could also be observed in compartment fire situation [14–16], particularly with the current trend of urban development and the need to construct high-rise building with complex geometry and interior design. A typical example of source of eddy-generation that may trigger the formation of fire whirl in a high-rise building could be an enclosed structure with openings for flow entrainment that induces the circulation, which includes atrium, lift pit, spiral staircase, etc. Those features have been widely implanted as common features in modern building design, such as the Macquarie's global headquarters dispatched in Figure 1 [17]. Such building geometrical configuration could potentially act as to induce the eddy generation and hence pose a risk to trigger the formulation of fire whirl in a fire incident. For instance, despite the recently developed bio-based flame-retardant materials that effectively restrain the fire from spreading [18–22], the occurrence of the swirling flame has been observed in some recent high profile skyscraper fire incidents including the Beijing Television Cultural Center, Plasco Building, Grenfell Tower, etc. [23]. Understanding the formation of fire whirl and flaming behaviour in an enclosed configuration that resembles the high-rise buildings, is of great benefits to ensure the safety of the resident and occupant as well as to prevent property lost [24–26], and therefore is identified as the topic of this study.

Figure 1. Example of the internal structure of a high-rise building features atrium and spiral staircase [17].

Fire whirl has been systematically investigated using experimental means during the past decades to characterise the flame structure and gain the understanding of the transformation of fire whirl from buoyancy-driven free-standing diffusion [7,27–33]. However, there are some inevitable limitations associated with the experimental method. For instance, the parameters that could be quantified by an experimental approach are limited to temperature, velocity and exhaust gas concentration, which may not be sufficient to reveal the fundamental physical-chemical behaviour and probe the root causes of the observed flaming behaviour. In addition, if the intrusive measuring technique is implemented, e.g., thermocouple for temperature and bidirectional venturi tube for velocity, the experimental data obtained from the literature are often limited to a few points of interest, due to the constraint of the set locations of the measuring device and its associated spatial resolution. Nevertheless, as mentioned previously, the fire whirl is swirling in an unpredictable manner and the flaming structure may not reside in a fixed position. Furthermore, the existence of the measuring devices could inevitably be perturbative, hence potentially alter the very flow dynamics and combustion behaviour [34]. On the other hand, the optical-based diagnostic technique improves the data collection to planar scale without disturbing the measured conditions. It is, however, widely agreed that the optical properties that are critical for the reliability and accuracy of the optical-based measuring result, is often challenging to acquire [35]. The numerical approach hence can serve as a great aid to provide a completed set of data that is often difficult to assess by experimental means alone. The comprehensive data set generated by numerical simulation could work concurrently with experimental measurement to shed light on the fundamental understanding of this topic of interest.

Through the literature review, it has been revealed that the sophisticated flaming behaviour of the fire whirl is governed by the entangled coupling between flow dynamics and combustion kinetics, which could be inevitably affected by various aspects, such as the temperature, buoyancy, vortex, the combustion reaction, etc. which are interrelated and interact during the formulation and evolution of the fire whirl. Such intriguing interaction makes the understanding of the not-well-understood fundamental of the swirling reaction flow more unfathomable. For example, the inception of fire whirl within the enclosure intensifies the combustion process and subsequently varies the energy transfer from the flame structure to the surroundings [36–39]. If a pool fire configuration is used, the burning rate, which depends on the rate of the liquid fuel, convert into combustible gas mixture via evaporation will be inevitably altered, compared with non-swirled flame counterpart. As a result, ensuring combustion behaviour and heat release rate could also be affected. Similarly, the residence of the swirling reacting flow promotes circulation within the enclosure and enhances the mixing of the reactant with oxidiser and result in a more completed combustion event. The soot formation mainly due to the incomplete combustion within such flame should theoretically be hinder [40–43]. Nevertheless, the aggravation of the combustion due to the enhance in mixing increase the flame temperature to that often excesses the threshold for soot nucleation, therefore facilitate the inception of in-flame generated soot species [44,45]. The arguably intensified or suppressed generation of soot species is directly correlated to the radiative heat transfer and varies the ensuing combustion process. In addition to the intricate physical and chemical coupling involved, the uncertainties introduced via the numerical modelling could also make the assessment of the fire whirl more cumbersome. For example, the region where combustion occurs is expected to be unfixed, as the hot plume is spinning with the formation of fire whirl as well as with the associated fire tilting, stretching, converging, etc. If the numerical domain is discretised in a non-uniformed manner, the variation in the spatial resolution of each discretised volume may significantly vary the prediction of the parameter of interest, particularly for the evaluation of the sensitive flame temperature [46].

The abovementioned complex coupling and interaction between various aspects have nevertheless not been carefully taken into consideration or restrained in may of the previous studies. It is often noted that heavy sooty flame, pool-based configuration and unevenly distributed mesh strategy that applies finer mesh at near fuel pan regions and coarse mesh at the fringes, has often been implemented in the studies found in the literature. Such implementation makes the quantitative analysis of the fire whirl behaviour unattainable. It is therefore needed to formulate a numerical framework that considers and constraints the preceding coupling processes, and enables the establishment of the correlation between observed changes in flaming behaviour and the proposed parameter of interest.

It also needs to highlight that, the nature of the fire whirl evolution is highly irregular. A fire whirl with typical vortex shape spinning structure could be observed located right above the fuel source at one instant of time, and it could be transformed back to a randomly flickering flame at the next monitoring time instant. The fire whirl rotation pattern and evolution pathway are also highly unpredictable. Such an unpredictable nature of the evolvement and randomly spatial movement of the fire structure and plume poses significant difficulties in fire control, fire prevention, and evacuation planning, if the fire whirl occurs in a high rise building. The studies on fire whirl conducted in the past decades, however, is mainly focused on the characterisation and quantification of the formulated fire whirl from flow dynamic and combustion perspective, upon its formation, or at one very particular time instant that is ideal for the analysis. The qualitative investigation of the evolution of the fire whirl with a detailed description of its various stages, including the ignition, buoyant fire development, formulation of the nascent fire whirl, and the ensuing developments and evolution, has not been thoroughly investigated.

In light of the abovementioned gap in knowledge. This paper will present the first time, using advanced numerical approaches to describe the evolution pathway of two fire whirls formulated under different entrainment configurations from 0.00 s to 50.00 s. This numerical investigation is conducted in a fully controlled numerical environment that isolates all possible variation introduced by intriguing flow dynamic and combustion coupling as well as by variance in numerical modelling aspects, and focuses to establishing the comparison of two fire whirl scenario solely attributed to the different eddy-generation scheme. The enhanced understanding of the evolution pathway of the fire whirl may largely benefit in various aspects including architecture design, fire evacuation planning, as well as fire prevention [47] and extinguishing planning.

2. Numerical Details

A numerical domain of the baseline model replicating the test rig of one previous experimental study of the fire whirl [48], as well as that with an additional flow channelling slit, was constructed accordingly. The geometric features of the two domains are shown in Figure 2. The detailed description of the numerical configuration of the model has been detailed in our previous work of the characterisation of the fire whirl formulated in an enclosed chamber with different entrainment schemes [49], for the sake of brevity, such information would not be repeated herein. In summary, the geometrical features of the model include a fuel pan located in the domain centreline. The fuel pan is converted from circular to square configuration with the same cross-section area to match the experimental setup to achieve a fully structured mesh, as it significantly enhances the numerical accuracy and computational efficiency.

The domain is discretised to about $800,000$ elements using a uniform division algorithm, to eliminate the numerical uncertainty associated with the spatial resolution of the discretised control volume. A doubling of the number of elements to $1,600,000$ resulted in only a 5% difference in the centreline temperature, thus achieving a mesh-independent solution.

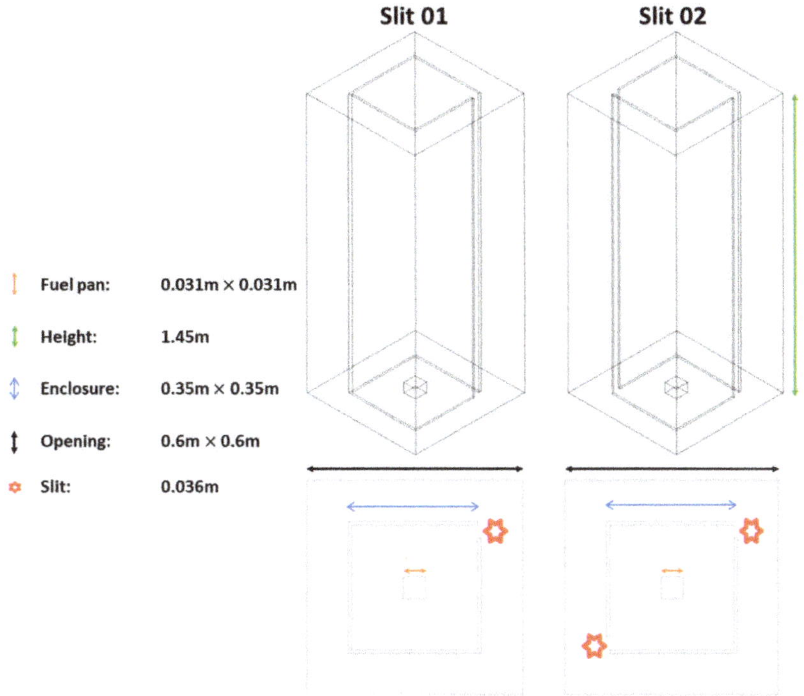

Figure 2. Geometric features of the computational domain.

The fluid flow and heat transfer within the compartment is described through the conservation equations of continuity, Navier–Stokes, and scalar quantities. A general form of transport equation can be expressed as:

$$\frac{\partial}{\partial t}(\overline{\rho}\Phi) + \frac{\partial}{\partial x_i}\left(\overline{\rho}\Phi\widetilde{U}_i\right) = (\Gamma_\Phi)\frac{\partial^2 \overline{\rho}\Phi}{\partial x_i \partial x_i} + \widetilde{S(\Phi)} \tag{1}$$

where Φ is the general field variable dependent on space and time, $\overline{\rho}$ is the mean density, $\widetilde{U}_i$ is the fluid velocity, Γ_Φ is the diffusion term, and $\widetilde{S(\Phi)}$ is the source term of the general variable. According to the general formula, the transport equations are tabulated in Table 1 as:

Table 1. General transport equation variable, diffusion and source terms.

Φ	Γ_Φ	$\widetilde{S(\Phi)}$
1	0	0
$\widetilde{U}_j$	$\mu + \mu_T$	$\frac{\partial \overline{p}}{\partial x_j} + \frac{\partial}{\partial x_j}\left(\mu \frac{\partial \widetilde{U}_i}{\partial x_j} - \frac{2}{3}\mu \frac{\partial \widetilde{U}_k}{\partial x_k}\right) - \frac{\partial \tau_{ij}}{\partial x_j} + \left(\overline{\rho} - \rho_{ref}\right)g_j + \widetilde{S(U)}$
$c_p\widetilde{T}$	$\frac{\mu}{Pr} + \frac{\mu_T}{Pr_T}$	$\widetilde{\omega}_T + \widetilde{S(rad)}$
Φ	$\frac{\mu}{Sc} + \frac{\mu_T}{Sc_T}$	$\widetilde{S(\Phi)}$

Large eddy simulation (LES) and the wall-adapting local eddy-viscosity (WALE) function, is adopted as the turbulence model to describe the turbulent reaction flow behaviour. This turbulence model has been extensively validated and proved to be a valid approach for resolving various wall-bounded turbulence flow applications with a reasonable computational course [8,43,50–54].

The chemical reaction source term in the transport equations of the involved reacting scalars is determined by strained laminar flamelet approach, in which the chemistry is a pre-assumed probability density function (pre-PDF) of two quantities including mixture fraction (f) and the scalar dissipation (χ). In essence, the mixture fraction governs the amount of the fuel mixture in each control volume element in the simulation domain. The scalar dissipation is a term introduced to describe the strain and extinction of the flame, in which the larger this quantity depicts its departure from its chemical equilibrium [55]. It should be noted that, in the present work, the GRI-MECH 3.0 detailed chemical reaction mechanisms, with includes 325 reaction steps and 53 chemical species [56], was implemented to formulate the flamelet library for the strained laminar flamelet model, with ethylene (C_2H_4) selected as parental fuel. The presented approach of the modelling turbulence–chemistry interaction has been validated in previous studies to provide a reasonable result with moderate computational burden [50,56,57]. This approach has been validated in previous numerical studies for modelling turbulence chemistry interaction, and has been proved to provide a reasonable result with moderate computational consumption [50,57,58]. It should be highlighted that alcohol-based fuel methanol (CH_3OH), is deliberately selected as the parent fuel to constrain the coupling between radiative energy feedback and the combustion process, due to its feature of clean and soot-free burning behaviour [45]. For the same reason, the concentration of the key building structure unit of soot formation, acetylene (C_2H_2), in the gas mixture produced by this very parent fuel, is proved to be negligible. The most commonly adopted soot models formulated on C_2H_2 precursor-based inception and surface growth mechanism are not applicable in this study [59–62]. As a result, a primitive and computational lightweight two-step soot model is integrated, for concept verification purpose only.

With respect to the boundary conditions, a set flow rate of 0.0216 ms^{-1} in the direction normal to the fuel surface is applied on the fuel pan. The applied injection velocity is determined on the basis of the cross-sectional area of the fire pan, the density of the parent fuel at a reference temperature, and the heat of combustion of the parent fuel in order to match the targeted heat release rate reported in the experimental study. The constant injection rate ensures an evenly distributed burning profile, regardless of the intensity of flow circulation and energy feedback. The fuel surface is set at an elevated height according to the experimental setup. The top and the periphery of the domain is set as the opening to allow naturally convected air entrainment in and out of the system. The base of the domain is set as a non-slip adiabatic wall, across which no heat or matter is allowed to pass. The simulation is initiated with standard temperature and pressure replicating the ambient conditions for the combustion process to proceed. For convenience, the case with one side entrainment slit is referred to as the Slit 01 case, case 1, or the baseline case/model, and that with two side entrainment slits is referred as the Slit 02 case, case 2, or the comparison case/model, in the following sections.

It should be noted that the simulation result is validated against experimental data and achieve good comparison in temperature profile as well as the incoming velocity at the entraining slits at various HABs (height above burners), which is explained in detail in the co-published work [49]. For the sake of brevity, the validation process is not presented in this work.

3. Results

3.1. The Formation and Evolution of Fire Whirl

The flame temperature profile at the domain centreline of all monitoring HABs, from the start of the simulation to 50.00 *s* for both Slit 01 case and Slit 02 case, are presented in Figures 3 and 4. In general, the evolution of this very flaming phenomenon, based on its combustion behaviours, can be categorised into three main stages. Those include Stage A: flame development; Stage B: fire whirl development and formation; Stage C: fire whirl evolution (the cross symbol indicates the particular time instant the core of the fire whirl is centre in the domain centerline). The characterisation of the formation pathway and evolution of the fire whirl is discussed in detail in the following sections.

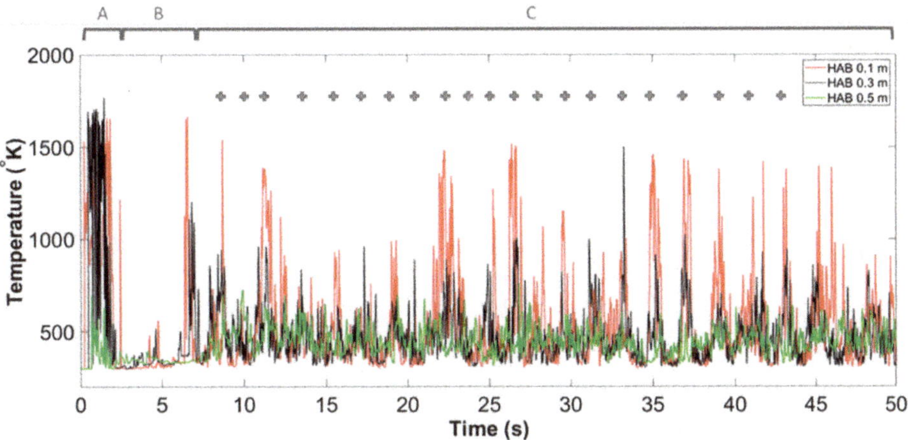

Figure 3. Centerline temperature of Slit 01 case at all monitoring HABs, from 00.00 s to 50.00 s, consist of three main stages of the development and evolution of fire whirl, namely Stage A: flame development, Stage B: fire whirl development and Stage C: fire whirl formation and evolution.

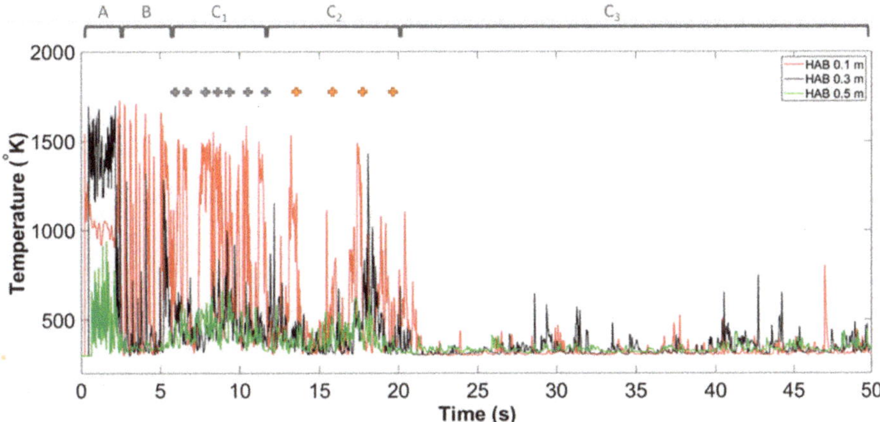

Figure 4. Centerline temperature of Slit 02 case at all monitoring height above burners (HABs), from 00.00 s to 50.00 s, consist of three main stages of the development and evolution of fire whirl, namely Stage A: flame development, Stage B: fire whirl development and Stage C: fire whirl formation and evolution.

3.1.1. Stage A: Flame Development

Stage A: flame development represents the period from starting the ignition until the full development of the buoyancy-driven diffusion flame. The flame temperature in the domain centreline of all HABs of both Slit 01 case and Slit 02 case during Stage A of the combustion process are presented in Figure 5 and Figure 8. The temperature iso-surface at representative instants of time are illustrated in Figures 6 and 7, and Figures 9 and 10.

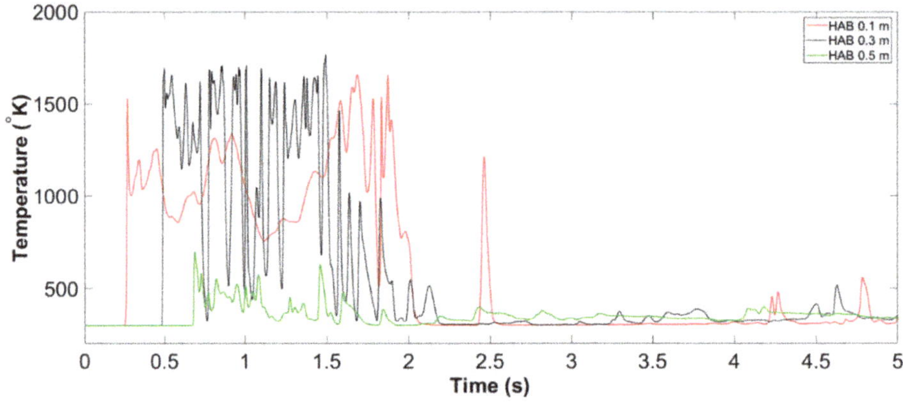

Figure 5. Centerline temperature of Slit 01 case at all monitoring HABs, from 00.00 s to 05.00 s (example of Stage A: flame development).

The flame temperature at domain centerline during Stage A of Slit 01 case is shown in Figure 5. Combined with the temperature iso-surface presented in Figures 6 and 7, the time duration for the stage of flame development can be approximately defined as from $t = 0.00$ s to $t = 2.55$ s. The stage A could be further splitted into two periods, the period flame developing upwards from the time of ignition, demonstrated in Figure 6, and the period flame fluctuation propagates from flame tip downwards to near fuel pan surface, presented in Figure 7.

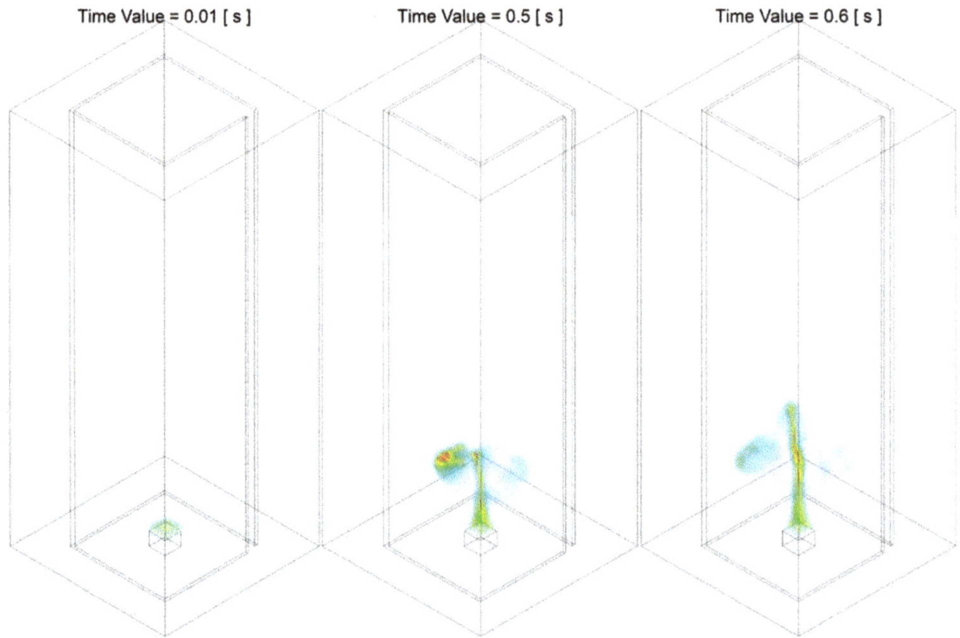

Figure 6. Temperature iso-surface at representative instant of time during Stage A of the combustion process: from the ignition to the fully developed buoyant diffusion flame, of Slit 01 case.

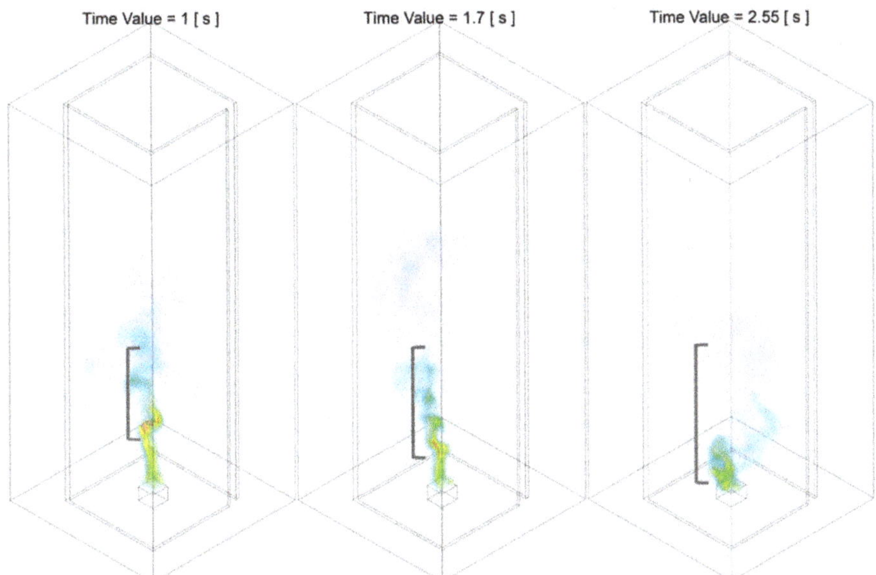

Figure 7. Temperature iso-surface at representative instant of time during Stage A of the combustion process: from fully developed flame to flame tilting and flickering in a random manner, of the Slit 01 case. The region highlighted in grey indicates that flame starts to flicker, whereas the region further upstream persists a relatively straight profile.

Figure 6 demonstrates a typical development of a buoyant diffusion flame during the period of $t = 0.00$ s to $t = 0.60$ s. The flame proporgates in the vertical direction, due to the dominant of momentum at the initialisation stage of the combustion. The flame front reaches the three monitoring HABs, 0.1 m, 0.3 m and 0.5 m, at approximately $t = 0.25$ s, $t = 0.50$ s and $t = 0.60$ s.

Upon the extension of the flame height to its maximum, the flame with the straight profile starts to flicker, from flame tip and gradually propagated to the further upstream region, as demonstrated in Figure 7. It can be seen from the figure that, the region where flame starts to fluctuate starting from the flame tip at approximately $t = 1.00$ s, reaches the low-intermediate region at approximately $t = 1.70$ s and finally proporgates to near fuel pan surface at approximately $t = 2.55$ s. The flucturation of the flame is developing towards upstream along the domain centreline axis, as the figure demonstrated that, the flame structure that beneath the flickering region remains straight and in a relatively regulated shape. The transition of the flaming behaviour could be contributed to momentum-driven upward motion is overwhelmed by the thermo-dynamic of the buoyant diffusion upon the completion of its development. It should be highlighted that the fluctuation of the flame pattern at this particular stage is highly randomised but relatively centred with respect to the location of fuel pan, i.e., differing from the structure of a formulated fire whirl, no tendency of tilting, rotation can be observed explicitly at this instant of time.

The flame temperature at domain centerline during Stage A of Slit 02 case is shown in Figure 8. Alongside with the temperature iso-surface presented in Figures 9 and 10, the time duration for this stage of flame development can be approximately defined as from $t = 0.00$ s to $t = 2.55$ s. Similar to what observed in Slit 01 case, the Stage A of Slit 02 could be further split into the flame vertical development period and flame flickering period.

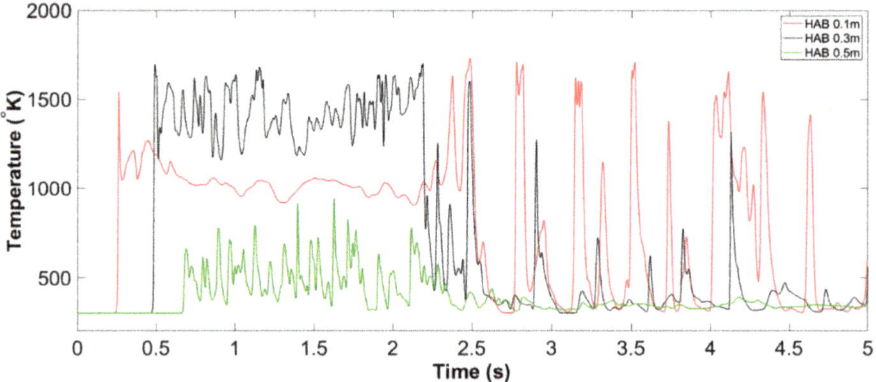

Figure 8. Centerline temperature of Slit 02 case at all monitoring HABs, from 00.00 s to 05.00 s (example of Stage A: flame development).

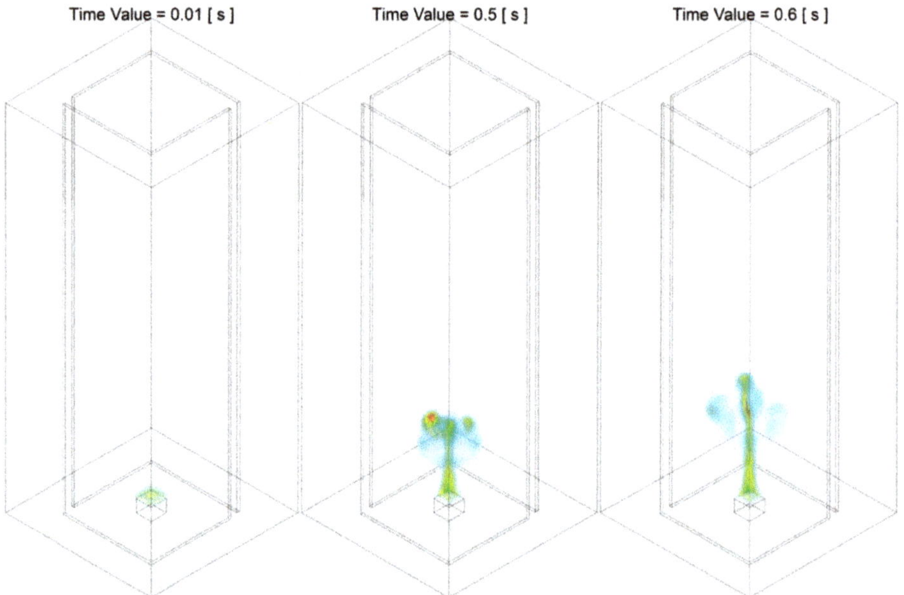

Figure 9. Temperature iso-surface at representative instant of time during Stage A of the combustion process: from the ignition to the fully developed buoyant diffusion flame, of Slit 02 case.

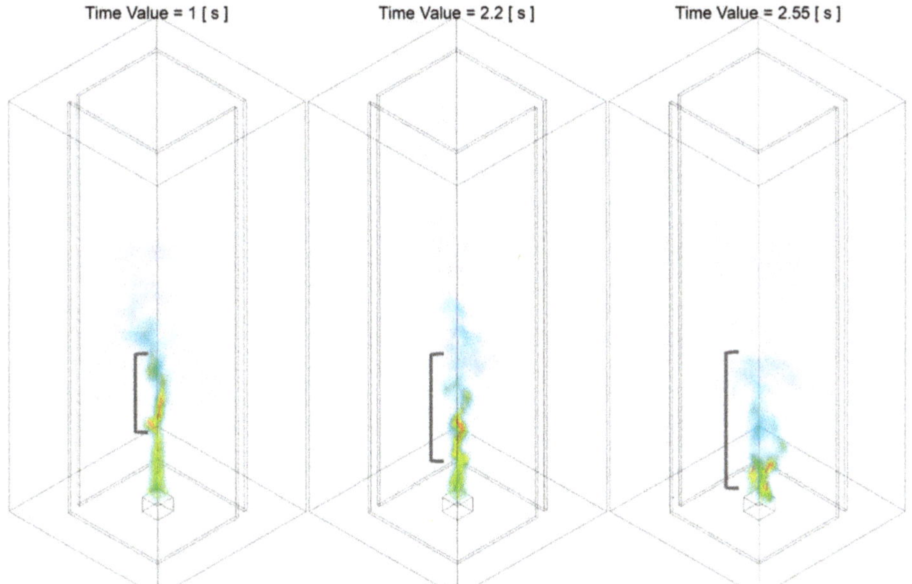

Figure 10. Temperature iso-surface at representative instant of time during Stage A of the combustion process: from fully developed flame to flame tilting and flickering in a random manner, of Slit 02 case. The region highlighted in grey indicates that flame starts to flicker, whereas the region further upstream persists a relatively straight profile.

As demonstrated in Figure 9, the flame develops upwards in the vertical direction during the period of $t = 0.00$ s to $t = 0.60$ s, due to the dominant of momentum at the initialisation stage of the combustion. The flame front approaches the three monitoring HABs, 0.1 m, 0.3 m and 0.5 m, at approximately $t = 0.25\ s$, $t = 0.50\ s$ and $t = 0.60\ s$.

Nearly identical to the process observed in the Slit 01 case, as the flame puffs up to its maximum height, a region at the downstream starts to flicker and such region gradually develops downwards towards the fuel source. As illustrated in Figure 10, the region where flame starts to fluctuate starting from the flame tip at approximately $t = 1.00\ s$, expands to the low-intermediate region at approximately $t = 2.20\ s$ and finally proporgates to near fuel pan surface at approximately $t = 2.55\ s$. A noticeable difference of the flickering pattern observed, compared with that of the Slit 01 case, is the flame structure is relatively more symmetric. For example, by comparing the flame structure of both cases at $t = 2.55\ s$ where the region of the fluctuation reaches near fuel source surface, the flame generated from Slit 02 case is symmetrically split into two streams, with each of the streams slightly expands towards the air entrainment slit. Such well-formed symmetrical flame structure could be potentially attributed to the symmetrical air entrainment due to the two slits configuration.

In general, during the stage of flame development, the flame developing pattern of both Slit 01 case and Slit 02 case are nearly identical and resembles what expected for a typical free-standing diffusion flame in an enclosed configuration. At this stage, the entrainment air introduced from the slit(s) has not started to influence and alter the general flame propagation process.

3.1.2. Stage B: Fire Whirl Development and Formation

The transition period from the flame flickering in a randomised manner to the formation of nascent fire whirl is categorised as Stage B: fire whirl development and formation. The flame temperature in the domain centreline of all HABs of both Slit 01 case and Slit 02 case during Stage B of the combustion process are presented in Figure 11 and Figure 15. The temperature iso-surface at representative instant of time are illustrated in from Figures 12–14 and from Figures 16–18.

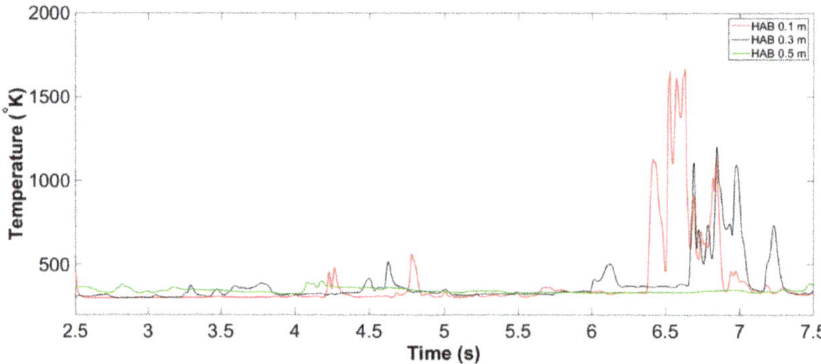

Figure 11. Centerline temperature of Slit 01 case at all monitoring HABs, from 02.50 s to 07.50 s (example of Stage B: fire whirl development and formation).

The flame temperature at domain centerline during Stage B of Slit 01 case is shown in Figure 11. Combined with the temperature iso-surface presented in from Figures 12–14, the time duration for the stage of fire whirl development and formation can be approximately defined as from $t = 2.55$ s to $t = 8.50$ s. Stage B could be further splitted into two periods, the period of transition from randomly flickering flame to the emerging of rotating reacting flow, presented in Figures 12 and 13, and the period of formation of nascent fire whirl, shown in Figure 14.

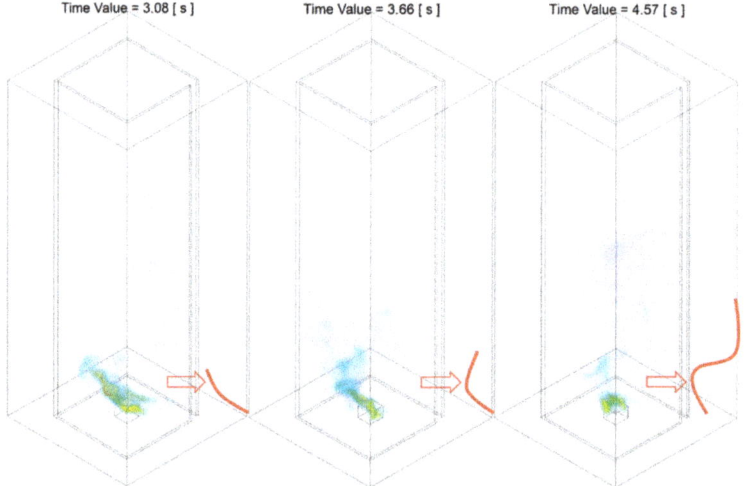

Figure 12. Temperature iso-surface at representative instant of time during Stage B of the combustion process: from randomly flickering flame to the emerging of rotating reacting flow, of the Slit 01 case. The red solid line approximately illustrates the shape of the flame core region structure.

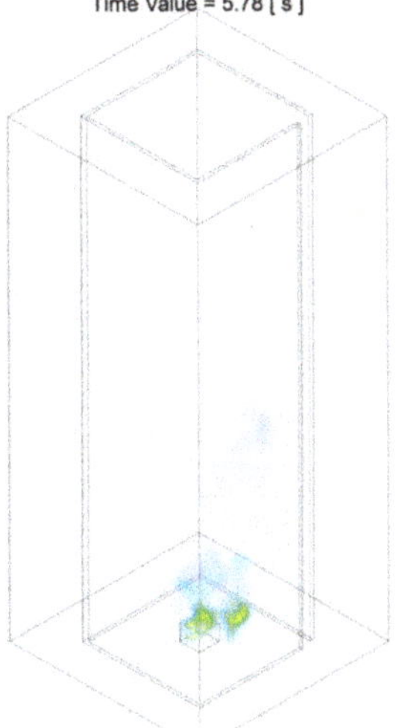

Figure 13. Temperature iso-surface at representative instant of time during Stage B of the combustion process: flame restoring from emerging of the swirling reacting flow back to that flickering randomly, of the Slit 01 case.

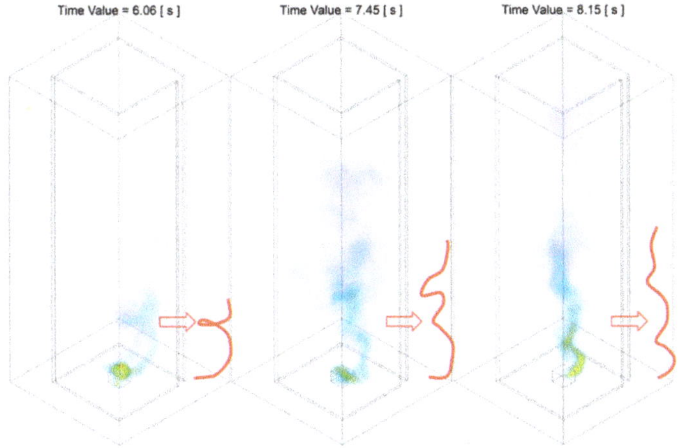

Figure 14. Temperature iso-surface at representative instant of time during Stage B of the combustion process: the formation of nascent fire whirl, of the Slit 01 case. The red solid line approximately illustrates the shape of the flame core region structure.

The first transition period of Stage B from a flickering flame to the emerging of rotating reacting flow is can be approximately defined from $t = 2.55$ s to $t = 6.00$ s. As demonstrated in Figure 12, during the transition period, the flame started from flame tilting towards one side in a relatively straight line form, i.e., at $t = 3.08$ s. It follows by the transformation into that flame tilting towards one direction in the low-intermediate region and bending toward the opposite direction in the downstream region, i.e., at $t = 3.66$ s. It should be noted that, up to this instant of time, the profile of the lower-intermediate flame core structure as well the downstream plume tilting in the reverse direction all resembles a straight line in an L-shape, with no rational or twisting motion observed. It finally developed into the flame that tilting in one direction in the region close to fuel source, and bending towards the opposite direction in the intermediate region and rotating back to vertically straight in the down streaming plume region, i.e., at $t = 4.57$ s. At this time instant, the profile presented in a Z-shape format and indicates the formation of incipient fire whirl.

Nevertheless, it should be noted that such transition from tilting flame towards emerging of rotating reacting flow as demonstrated in Figure 12, is not perpetual, and could be revolved back to the previous stage at any instant of time during this period. For example, as presented in Figure 13, after the emerging of rotating reacting flow at $t = 4.57$ s, the flame shape becomes irregular and flucturate randomly, as a typical flame observed in Stage A during the second period of flame flickering.

The emerging of the incipient rotating reacting flow indicates the starting of the second period of Stage B, the formation of nascent fire whirl, defined from $t = 6.00$ s to $t = 8.50$ s. During this period, the flame structure rapidly transformed from whirling flow rotating in a relatively large radius, i.e., at $t = 6.06$ s, into the nascent fire whirl that is shifting towards domain centreline with reduces rotating radius and increased flame height, i.e., at $t = 7.45$ s, and eventually developed into a fire whirl that is centred with respect to fuel source and confined with a relatively small rotating radius, i.e., at $t = 8.15$ s. Unlike of that observed in the first period of Stage B, during this period, the evolution of the fire whirl will not be revolved back to the previous stage, and the formulated fire whirl remains in a quasi-steady state.

It should be highlighted that the domain centreline temperature of all monitoring HABs during the first period, is relatively low, i.e., near-ambient condition, compared with that of Stage A. This agrees well with the abovementioned description made based on temperature iso-surface, as the flame at this stage are likely to be tilting away from the domain centreline, thus resulting the observed low-temperature profile at all HABs. The raise in centreline temperature starts as the transition moves to the second period, which is again consistent with the tendency that the core structure of the flame is shifting towards domain centreline with reduced rotating radius and extended flame height.

The flame temperature at domain centerline during Stage B of the Slit 02 case is shown in Figure 15. Alongside with the temperature iso-surface presented in from Figures 16–18, the time duration for this stage of flame development can be approximately defined as from $t = 2.50$ s to $t = 5.50$ s. Similar to what observed in Slit 01 case, the Stage B of the Slit 02 could also split into two period, the period of transition from randomly flickering flame to the emerging of rotating reacting flow, presented in Figures 16 and 17, and the period of formation of nascent fire whirl, shown in Figure 18.

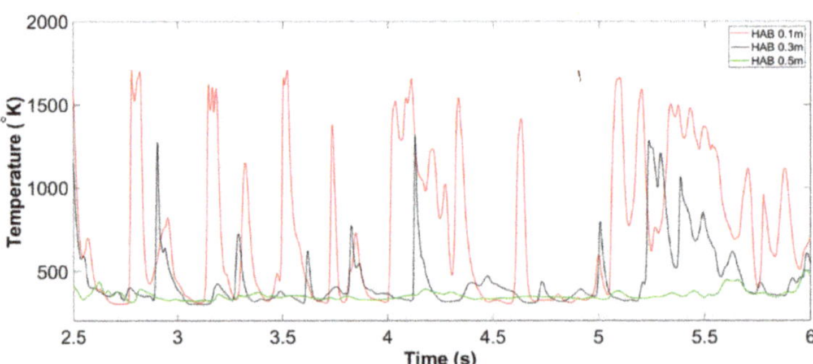

Figure 15. Centerline temperature of Slit 02 case at all monitoring HABs, from 02.50 s to 06.00 s (example of Stage B: fire whirl development and formation).

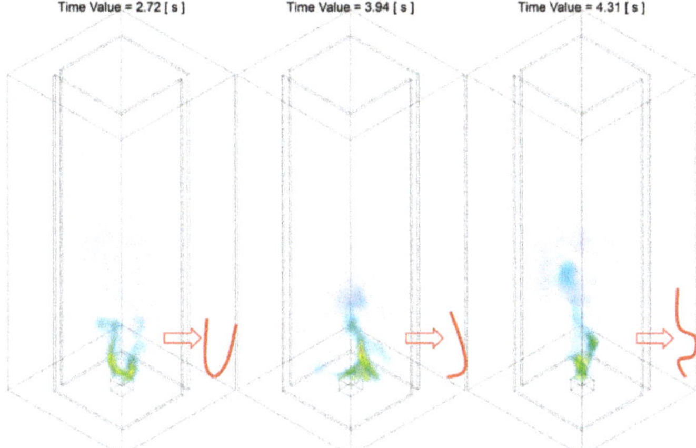

Figure 16. Temperature iso-surface at representative instant of time during Stage B of the combustion process: from randomly flickering flame to the emerging of rotating reacting flow, of the Slit 02 case. The red solid line approximately illustrates the shape of the flame core region structure.

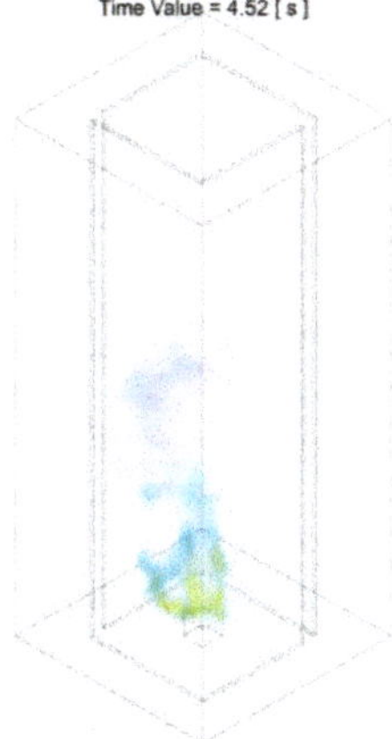

Figure 17. Temperature iso-surface at representative instant of time during Stage B of the combustion process: flame restoring from emerging of the swirling reacting flow back to that flickering randomly, of Slit 02 case.

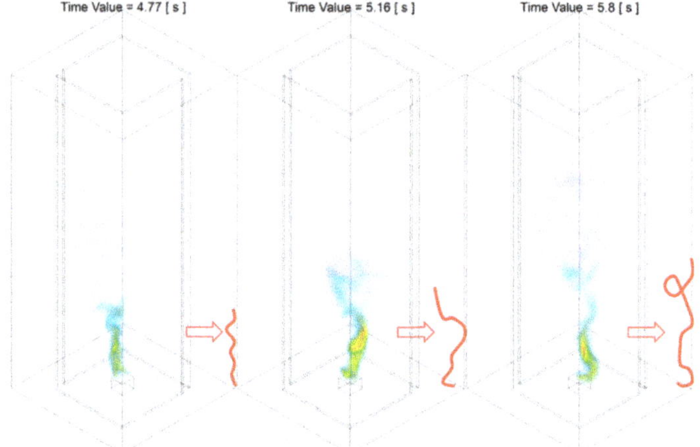

Figure 18. Temperature iso-surface at representative instant of time during Stage B of the combustion process: the formation of nascent fire whirl, of Slit 02 case. The red solid line approximately illustrates the shape of the flame core region structure.

As illustrated in Figure 16, the three-phase transition pattern previously demonstrated in the Slit 01 case, is observed in Slit 02 case. The three-phase transition pattern can be identified as: flame tilting towards one side with relatively straight line format, i.e., at $t = 2.72$ s; flame tilting towards one direction in the low-intermediate height and bending towards the opposite direction in the downstream, i.e., at $t = 3.94$ s; and the emerging of the rotating reacting flow with zigzag flame structure in the low-intermediate region and straight in vertical direction in the plume region, i.e., at $t = 4.31$ s.

Despite the general agreement with that observed in the Slit 01 case during this transition period, there are some noticeable difference need to be highlighted. Firstly is the difference regarding the flame tilting in a straight line yield at $t = 2.72$s compared with that of Slit 01 case. It can be seen that the flame is split into two streaming and tilting towards the respective air entrainment slit, compared with that tilting towards one side demonstrated in Slit 01 case. This symmetrical tilting pattern again could be potentially attributed to the symmetric geometrical configuration of the Slit 02 case. The second noticeable difference is the reduction in the first transition period in Stage B, which has a duration of approximately 2.00 s, which is reduced by 46.67% compared with that of Slit 01 case.

Similarly, during the first transition period, the transition has not been settled. In other words, the evolvement towards the regulated swirling motion could revolve back to the previous stage at any time instant. As demonstrated in Figure 17, the flame transformed from that resemble the initial inception of the reacting flow back to the flickering flame as typically observed in Stage A.

As the flame evolves into the second period of Stage B, the formation of nascent fire whirl, defined from $t = 4.55$ s to $t = 6.00$ s, the flame structure swiftly transformed from a swilling flow with relatively large rotating radius into a semi-stabilised fire whirl that spinning with the respect of the centre of fuel source within a confined cylindrical region with extended flame height. Likewise, the duration of the second period is calculated as 1.45 s, reduced by 42.00% compared with that of the Slit 01 case.

In regard to the characterisation of the domain centreline profile, it presents a significant difference compared with that of the Slit 01 case. Unlike the flat temperature profile reported in the baseline case, the rise of temperature in domain centreline appears at a much earlier instant of time. The temperature profile agrees well with the aforementioned reduction of duration for the first and second period. This indicates that with the introduction of the additional air-entraining slit and the resulted symmetrical side co-flow profile, the transition from a free-standing flame into a nascent fire is accelerated. The flame structures during the transition period are more likely to be positioned around domain centerline, due to the symmetrical geometrical configuration.

Generally speaking, during the stage of fire whirl development and formation, both cases share the same transmission patterns. The most significant differences observed in Slit 02 case can be summarised as the 49.58% reduction in the overall duration of Stage B, and the relatively symmetrical flame shape compared with that of the Slit 01 case.

3.1.3. Stage C: Fire Whirl Evolution

The instant of time that post-formation of the nascent fire whirl is herein defined as Stage C: fire whirl evolution. Prior to the detailed analysis of the evolution of the fire whirl, the characterisation of the formulated nascent fire whirl formulated in both Slit 01 case and Slit 02 case is firstly presented and discussed.

Characterisation of the Formulated Fire Whirl

With the formulation of the nascent fire whirl after Stage B, assorted key parameters have been generated to assessed to compare the characteristic of the swirling reacting flow generated by the two cases. A quantitative assessment of key parameters of both flow dynamic and combustion perspectives of the formulated fire whirl has been described in detail in the co-published work [49], and hence for the sake of brevity, would not repeat herein. In summary, it could be concluded that, the nascent fire whirl formulated under stronger eddy-generation mechanisms, i.e., domain with two slit configuration, has elongated, constrained swirling combustion region, compared with that generated in the single slit baseline model. As illustrated in Figure 19, the visible flame height with cut off flame tip temperature of 600 °K, as well as the peak flame temperature of the core structure of the swirling reacting flame formulated in Slit 02 case increased from 0.32 m to 0.59 m by 84.38% and from 1380 K to 1510 K by 9.42%, respectively, when compared with the baseline. Meanwhile, the axial velocity, axial velocity dominant region of the fire whirl core structure of the Slit 02 case increased by 6.81% and by 46.14% when compared with the single slit counterpart. In addition, the vortex core structure, determined by flame temperature as well as evaluated based on velocity field, of the fire whirl generated in Slit 02 case remains relatively unchanged through all monitoring HABs, compared with that increased by 153.33% and 136.91% observed in the Slit 01 case.

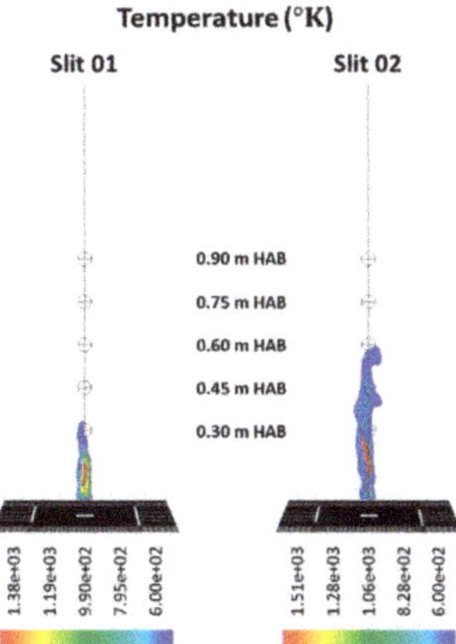

Figure 19. The visible flame height of the fire whirl formulated by the Slit 01 case and the Slit 02 case.

Evolution of Fire Whirl

The flame temperature at domain centerline during Stage C: evolution of fire whirl of Slit 01 case, can be defined as from the inception of nascent fire whirl till the end of the simulation, i.e., from $t = 8.50$ s to $t = 50.00$ s. It can be seen from Figure 3 that, the flame temperature at the domain centreline appears to fluctuate and repeat in a periodic manner, indicating a quasi-steady state of the status of the formulated fire whirl. Such status could be confirmed by inspecting the temperature iso-surface of the four representative instant of time during Stage C, as illustrated in Figure 21.

A selective section of the period of Stage C, i.e., from $t = 10.00$ s to $t = 20.00$ s is presented in Figure 20. Unlike the flat temperature profile reported in Stage B due to the flame tilting, the centreline flame temperature in Stage C appears to fluctuate periodically. To be more specific, there are approximately four peaks in every five seconds can be observed to occur repeatedly. The pattern of the peak in centreline flame temperature could be correlated to the frequency of the revolution of the spinning motion associated with the fire whirl.

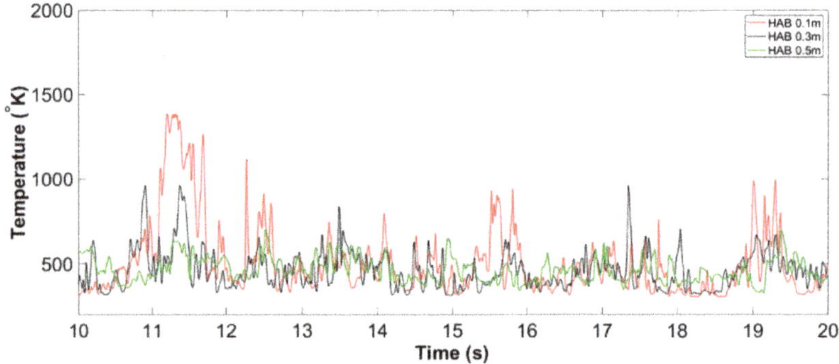

Figure 20. Centerline temperature of the Slit 01 case at all monitoring HABs, from 10.00 s to 20.00 s (example of Stage C: fire whirl evolution).

The characterisation of the fire whirls formulated presented in Figure 21 are almost identical in terms of visible flame height, flame temperature, core radius, and the frequency of the revolution, etc. Such similarity in the characterisation of fire whirls persists during the entire Stage C till the end of the simulation. It could be, therefore, conclude that, the fire whirl formulated by Slit 01 case has reached semi-steady state during Stage C.

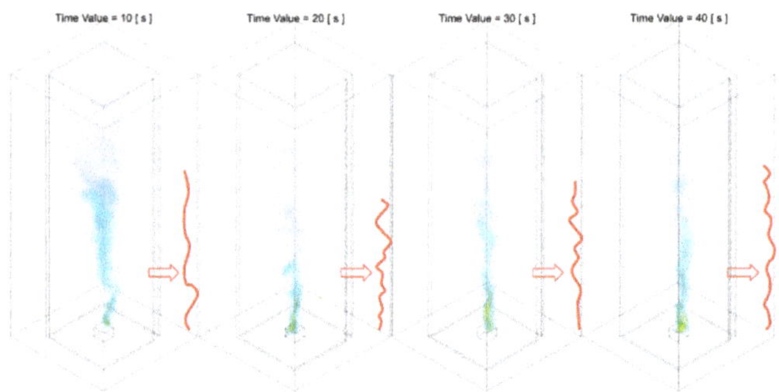

Figure 21. Temperature iso-surface at representative instant of time during Stage C of the combustion process: the evolution of fire whirl, of the Slit 01 case. The red solid line approximately illustrates the shape of the flame core region structure.

Differ from the flame centreline appears in a repeated pattern of the baseline case, the centerline flame temperature of Slit 02 case varies significantly during Stage C. As demonstrated in Figure 4, the occurrence pattern of relative high flame temperature at the domain centerline transformed from appearing intensively to occasionally and the temperature profile eventually becomes flat. It is, therefore, reasonable to subdivide the Stage C of Slit 02 case into separate periods and investigate the characterisation of each period individually.

The first period of the fire whirl evolution, Stage C_1, of the Slit 02 case is characterised as the period when the raise of domain centerline occurs frequently. From the observation of Figure 4, such period can be identified as from $t = 6.00$ s to $t = 12.00$ s, and the centerline temperature profile at all monitoring HABs is plotted in Figure 22. It is apparent that, by compared with that of Stage C of the Slit 01 case report previously, the occurrence of peak temperature at the domain centerline is observed to be more recurrent. For example, compared with peaks three or four times in every five seconds in the baseline case, the pattern of temperature rise at centerline demonstrated a two to three peak occurs in every second. This trend is more predominant during the period between $t = 7.50$ s to $t = 9.00$ s, where the centerline flame temperature at HAB 0.1 m remains roughly constant at a maximum value. This observation agrees well with the previous analysis in comparing the nascent fire whirl formulated in both cases. Due to the intensified eddy-generation mechanism from the second entrainment slit, the nascent fire whirl generated in the Slit 02 case is more spatially centralised in a confined space in the vicinity of domain centerline as well as with a more extended flame height, and thus persists with a relatively smaller radius and leads to higher centerline flame temperature, when compared with that of the baseline case.

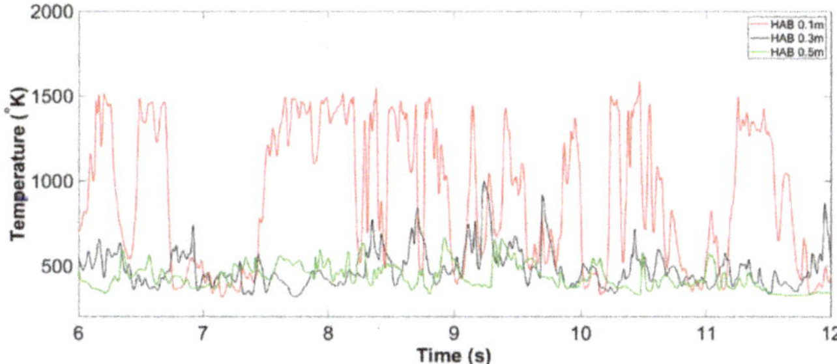

Figure 22. Centerline temperature of the Slit 02 case at all monitoring HABs, from 06.00 s to 12.00 s (example of Stage C_1: fire whirl evolution).

At this stage, the fire whirl in a fairly stable status, as the temperature iso-surface at representative instant of time during Stage C_1 is presented in Figure 23 is nearly identical in terms of all primary parameters of interest during this period.

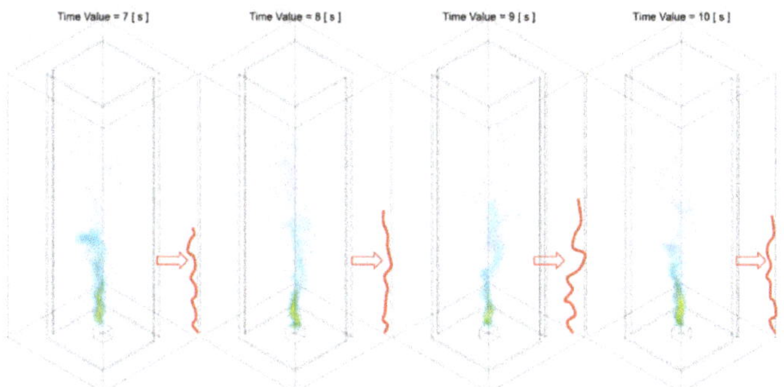

Figure 23. Temperature iso-surface at representative instant of time during Stage C_1 of the combustion process: the first phase of the evolution of fire whirl, of the Slit 02 case. The red solid line approximately illustrates the shape of the flame core region structure.

As the fire whirls revolving under the symmetrical and intensified entrainment configuration, it gradually moves to the second stage of fire whirl evolution, Stage C_2: the fire whirl gradually deviates from the previously observed confined region and departs from domain centerline in an outbound direction. With a close inspection of the centerline flame temperature profile as well as the temperature iso-surface at representative instant of time, presented in Figures 24 and 25 respectively, the time duration of the Stage C_2 can be defined as from $t = 12.00\ s$ to $t = 20.00\ s$. From the observation of flame centerline temperature, it can be concluded that the occurrence of a temperature peak at domain centerline are tending to become scatted during this period, compared with that reported in the Stage C_1, i.e., the frequency of the flame temperature peak at the domain centerline reduced from two to three appearances every second into approximately three appearances in every five seconds, which resembles the nascent fire whirl formulated in the baseline case.

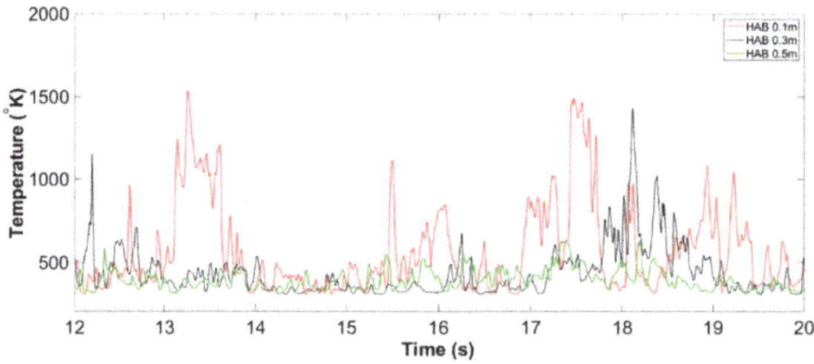

Figure 24. Centerline temperature of the Slit 02 case at all monitoring HABs, from 12.00 s to 20.00 s (example of Stage C_2: fire whirl evolution).

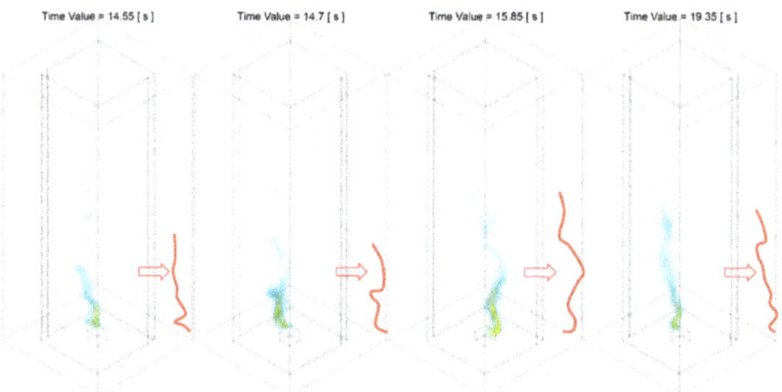

Figure 25. Temperature iso-surface at representative instant of time during Stage C_2 of the combustion process: the second phase of the evolution of fire whirl, of the Slit 02 case. The red solid line approximately illustrates the shape of the flame core region structure.

The reduction in the frequency that high flame temperature arises at the domain centerline could be explained by assessing the iso-surface of representative instant of time during this very period. It can be seen from Figure 25 that, over time, the lower-intermediate region, as well as the downstream plume region of the fire whirl, starts to deviate from the confined space that observed in previous period and departs from domain centerline towards the enclosure. The flame core region where high temperature persists is less likely to sweep across the domain centerline and result in the temperature rise compared with that of Stage C_1. It should be noted that there are two unique features of the flame structure of the fire whirl generated in this very period. Firstly, despite deviating from the domain centerline, it increases in cone radius, and the tilting or bending occurs limited at the elevated height close to the fuel source, i.e., near the bottom boundary, with the downstream region of the flame remains relatively straight. Secondly, despite the fact that the flame core structure is tilting towards the domain enclosure, the entire region of combustion, including the flame core as well as the plume is observed to be spinning and remains a fairly regulated cylindrical shape, which is fundamentally different from that observed in Stage B, i.e., tilting file during fire whirl development and formation.

The flame core region keeps departing from the domain centerline as the fire evolution progresses into the final period, Stage C_3: the core structure fire whirl continued to deviate from domain centerline and eventually settles as it swilling in a relatively large radius at regions closed the domain enclosure. The duration of this period can be defined as from the completion of Stage C_2 till the completion of the simulation, i.e., from $t = 20.00\ s$ to $t = 50.00\ s$.

The flame centerline temperature profile, as well as the temperature iso-surface at representative instant of time, is plotted in Figures 26 and 27. From the observation of the figures, there is evidence that the flame core region continues to deviate from the domain centerline and rotates with respect to the fuel source in a relatively large radius. The departing from the domain centerline of the fire whirl core structure leads to a flat centerline temperature profile at all monitoring HABs, resembles what was observed in Stage B as the flame is tilting. Similar to that noticed in Stage C_2, the intermediate and plume region of the fire whirl remains relatively upright straight, and the tilting and bending region are limited to low HAB where close to the bottom boundary. The flame structure is also different from tilting flame as strong rotational motion at any part of the flame can be observed.

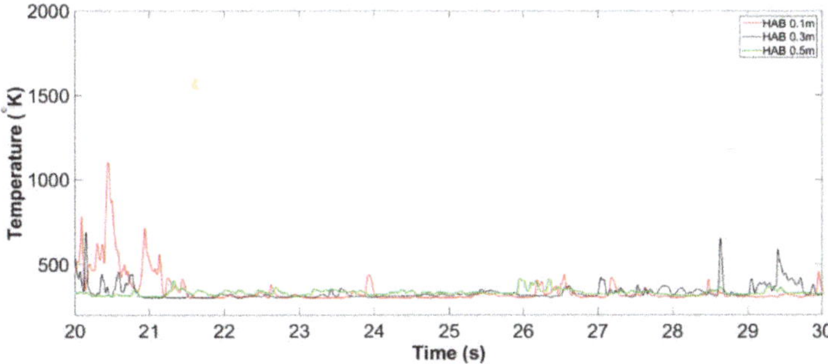

Figure 26. Centerline temperature of the Slit 02 case at all monitoring HABs, from 20.00 s to 30.00 s (example of Stage C_3: fire whirl evolution).

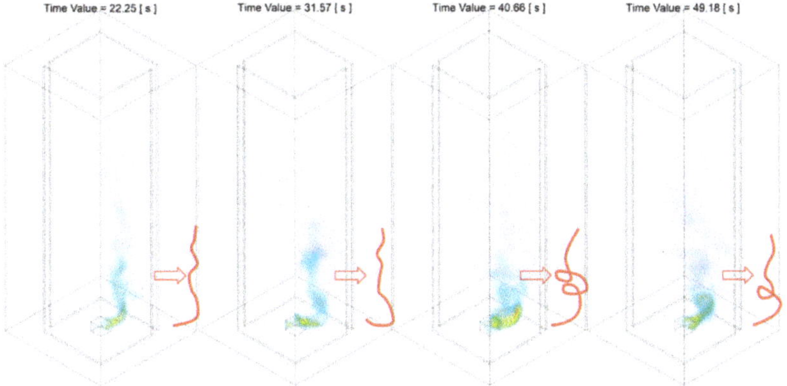

Figure 27. Temperature iso-surface at representative instant of time during Stage C_3 of the combustion process: the final phase of the evolution of fire whirl, of the Slit 02 case. The red solid line approximately illustrates the shape of the flame core region structure.

It should be highlighted that the departure from the domain centerline or increases in core radius will ease as the core structure of the fire whirl approaches the domain boundary and is constrained by the enclosure. Up to this stage, the fire whirl is reaching a quasi-steady state and such revolution behaviour is observed to remain till the end of the simulation.

It should also be noted that the results presented herein are limited to describing the characterisation of the formulated fire whirl at various time instances based on their development stage. The potential causes for such behaviour, including that such a flame remained relatively steady in Slit 01 case but departed away from the centerline and eventually rotated in a relatively large radius in Slit 02 case, is discussed in detail in the following section.

3.2. The Potential Causes of the Observations

It is also commonly agreed upon that the pressure deficit between the environment and cyclone eye is expected in a typical air spinning scenario. Similarly, the velocity vector field gives an indication of the magnitude as well as the approximate size of the vortex region, i.e., that with strong rational motion. Pressure contour and velocity vector field of both cases are therefore generated and presented to further assess flaming behaviour and transition during fire whirl evolution and as a probe for potential causes. Based on the observation from the previous section, it appears that the region where fire whirl tilting and deviating from the domain centerline occurred was in the low HABs region, i.e., near the bottom boundary, whereas the downstream of the fire structure remained relatively upright straight. The pressure contour and velocity vector field at the selected instant of time, therefore, is limited to the lower monitoring HAB, i.e., HAB 0.1 m, for both the Slit 01 case and Slit 02 case. It should also be noted that the representative time instant is selected based on when the incoming velocity of the air-entraining slit, at its maximum, minimum and at one time instant in between. The location of the vortex core of the fire whirl, determined collectively from assessing the pressure and velocity, is subsequently recorded and transferred to generate the revolution orbit during the revolution.

In addition, for this particular geometrical configuration of both cases, it is generally known that the fire whirl is induced by the air entrainment introduced from the slit(s). For this reason, the incoming velocity profile at HAB 0.1 m, is also presented to establish the correlation with the observed flaming behaviour and key parameters of interest.

The information of pressure, velocity at the boundary layer as well at the slit, and the pathway of orbiting, of the two cases, are collectively assessed to probe the potential causes of the observed evolving flaming behaviour.

3.2.1. The Potential Causes of Observations of Case 01

The pressure contour and velocity vector field of the Slit 01 case at HAB 0.1 m at the selected instant of time, are presented in Figures 28 and 29.

The figures show that the air at the monitoring HAB is flowing from high pressure to low pressure region, i.e., drawn from surrounding to the enclosed space via entraining slit. This observation generally agrees well with the commonly agreed flaming behaviour in the fire scenario. To be more specific, a fire in the compartment is expected to grow in direct proportion to available oxygen. In a typical compartment fire, air carrying oxygen is often observed to be drawn from outside into the base of the fire through openings in the building, e.g., doors, windows, roof. The opening of the slit in the current case acting as the source of entraining air promotes the combustion process, and at the same time facilitates the movement of combustible gas mixture, and the subsequently involved flame structure, heat and smoke, from higher pressure areas towards lower pressure areas accessible openings.

The pressure contour combined with velocity vector field also indicates the relative vortex core location and the approximate size of the spinning reacting flow. It can be seen from the figures that, the size of the swirling reacting flow, as well as the rotating radius, of the fire whirl of the Slit 01 case is relatively small when compared with that of the Slit 02 case shown in from Figures 32–35. It is suggested that the core structure of the formulated fire whirl is constrained in a confined space, and is rotating in a relatively small radius with respect to the fuel pan.

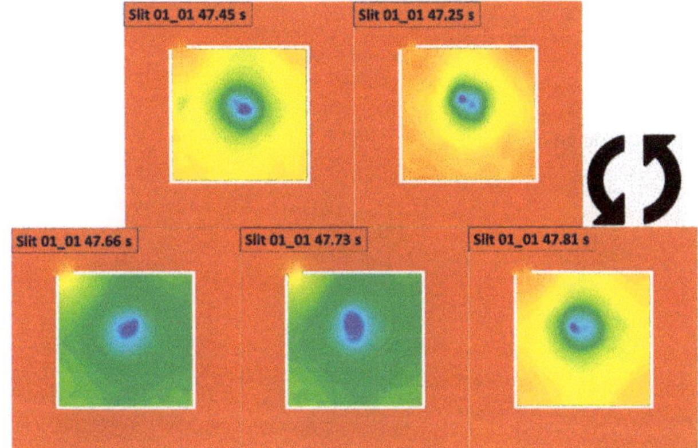

Figure 28. Pressure contour of the domain at 0.1 m HAB, at five timesteps of the Slit 01 case selected based on representative time instants associated with Slit 01 (anti-clockwise from top-right with increase in time). The contours indicate the fire whirl core location and collectively illustrate a full circle of orbital revolution of the whirling reacting flow with respect to fuel source.

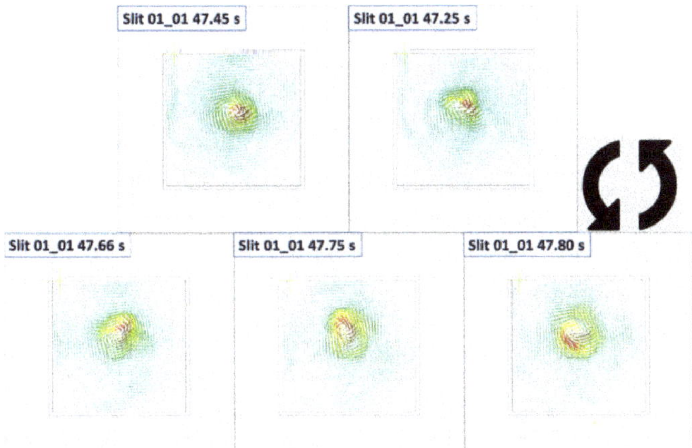

Figure 29. Vector field plot of the domain at 0.1 m HAB, at five timesteps of the Slit 01 case selected based on representative time instants associated with Slit 01 (anti-clockwise from top-right with increase in time). The plots indicate the fire whirl core location and collectively illustrate a full circle of orbital revolution of the whirling reacting flow with respect to fuel source.

The revolution orbit of the fire whirl with respect to the fuel source is obtained by transferring the location of the vortex of all time instant of interest and plotting into Figure 30. In addition, the incoming velocity of the slit of the monitoring HAB, at the corresponding time instant, is presented in Figure 31. The location of the fire whirl core is indicated by the ✪ in various colour, in both figures.

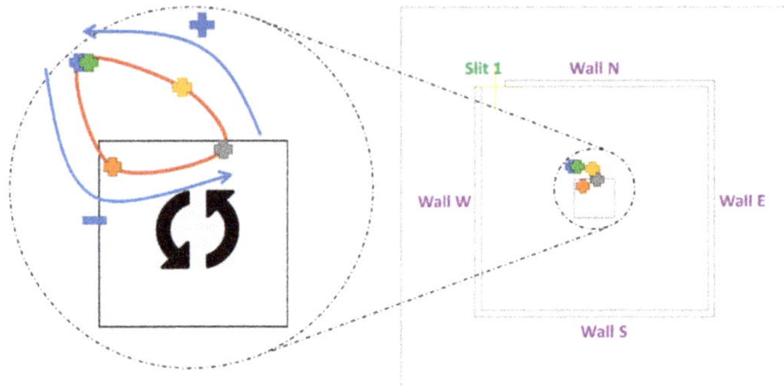

Figure 30. Orbit of fire whirl core centre of the Slit 01 case, indicating the fire whirl's revolution around the fuel source, starring from ●, and end with ● (with approximately one complete circle of revolution in anti-clockwise direction). The blue arrow indicates the tendency of changing of slit incoming velocity.

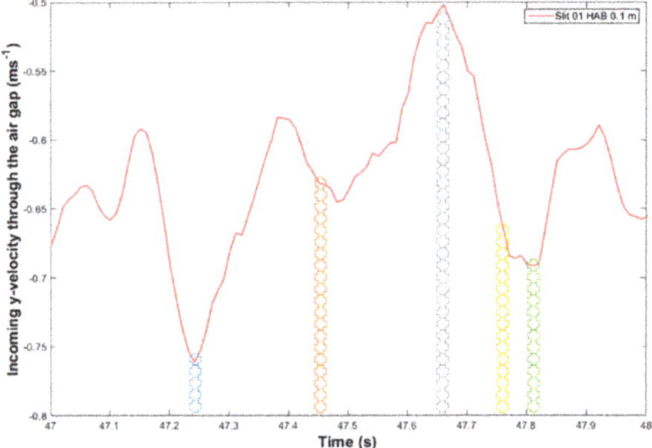

Figure 31. Plot of incoming velocity of the Slit 01 for Slit 01 case, with the marks indicating the representative time instant corresponding to Figure 30.

From Figure 30, it can be seen the anti-clockwise revolution orbit of the fire whirl of the Slit 01 case, starting from ●, and ending with ●, is not in a shape of a regular circle. It appears the vortex core location deviates from domain centerline as it approaches the entrainment slit, and restores to the near fuel source location as it moves away from entrainment slit. To be more specific, for the fire whirl originated from the fuel pan centre, the rotating radius can be defined as the distance between the vortex core and the fuel pan centre. It can be observed that, from time instant of $t = 47.25\ s$ (●) to $t = 47.45\ s$ (●), the fire whirl is rotating anti-clockwise from northern wall region to western wall region and moves away from the air entrainment slit. The rotating radius decreases significantly. While from time instant of $t = 47.66\ s$ (●) followed by $t = 47.73\ s$ (●) and to $t = 47.81\ s$ (●), the fire while sweep from near eastern wall region towards near northern wall region and moves towards the air entrainment slit, the rotating radius increases significantly.

This observation could potentially be explained using the concept of the circular motion with a constant centripetal force, which can be expressed as:

$$\vec{F}_C = \frac{mv^2}{r} \tag{2}$$

where $\vec{F}_C$ is the centripetal force towards the centre of the circle, m is the mass, v is the velocity and r is the radius of the rotation.

For a circular motion with a constant centripetal force, the radius is proportional to the square of the velocity. For the current fire whirl scenario spinning around the fuel pan in a single slit, the centripetal resembles the surface drag force to create the radial boundary layer, which can be considered as a constant, due to the fixed burning rate introduced to the domain, i.e., constant injection velocity of the parent fuel. Such a centripetal force is acting to ensure the reacting flow originated from the domain centre is spinning around the fire source.

Due to the pressure gradient, the air is entrained from the surrounding to the chamber. As the fire whirl core structure approaching the slit, the air entrainment acted as a supply source to intensify the velocity field of the flame core structure. As a result, the increase in the velocity field of the rotation reaction flow may increase its rotation radius to achieve the balance with the relatively fixed centripetal force acting perpendicular to the circle, for example, for time instant of 47.80 s (●). On the other hand, as the flame structure departing from near slit region towards where away from entrainment sources, i.e., region between northern and eastern walls such as indicated in 47.66 s (●), the intensification of the velocity field of the fire whirl eases, therefore the location of the fire whirl core structure restores back to the original near fuel pan region.

The incoming velocity of the air entrainment source, the slit, against the time, is also plotted in Figure 31, with the representative time instant denoted as ●.

It is obvious that the amount/rate of the air drawn into the chamber via the slit is proportional to the pressure gradient between the two sides of the slit, which is as expected. For example, with the region of the vortex with larger pressure gradient approaching the slit, the rate of air drawn into the domain is increasing, i.e., from time instant 47.66 s (●) to that in 47.80 (●), whereas the rate of air entraining rate decreases as the core of the fire moves away from the slit, such as from time instant 47.25 s (●) to that in 47.66 s (●).

3.2.2. The Potential Causes of Observations of Case 02

Similarly, the pressure contour and velocity vector field at HAB 0.1 m at the selected instant of time, is presented in from Figures 32–35, to reveal the information of the size and radius of rotation of the fire whirl of the Slit 02 case. Herein, the time instant denoted in ● are corresponding to those the peak, bottom and median velocity instant associated with slit 01, and those denoted as ○ are representative for the same time instant set associated with Slit 02.

Through the close review of the pressure contours and velocity vector field, and compared with those generated by the Slit 01 case, some similar trends that have been seen in the baseline model can be again observed in here. For example, the air is again drawn from the surrounding to the enclosure through the slits, and the reacting flow is swirling around, with respect to the fire source in an anti-clockwise direction.

However, there is some distinct difference that can be noticed. The size of the core structure is significantly larger, and the flame structure is rotating with a large radius, with respect to the geometry domain centerline, when compared with that of the baseline case. To be more specific, the core structure of the baseline case, is in a narrow form and its circulation movement is constrained in the vicinity of fuel pan regions. On the other hand, the structure of the fire whirl formulated in Slit 02 case are circulating around the fuel source in a significantly large radius, and the size of the structure is significantly larger, i.e., it can be even observed that the shape of the swirling reacting flow is constrained and limited by the enclosed wall boundaries.

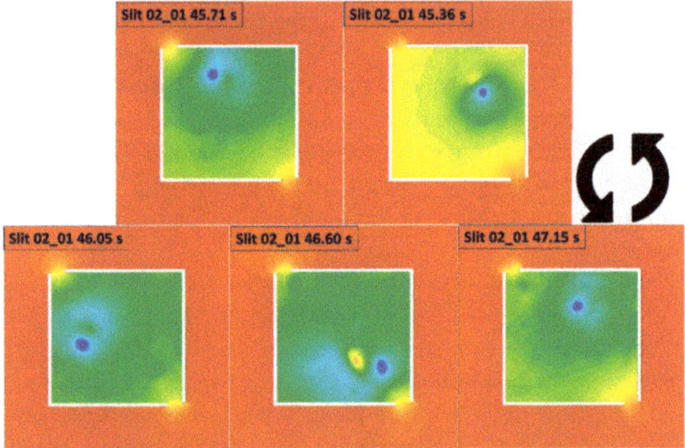

Figure 32. Pressure contour of the domain at 0.1 m HAB, at five timesteps of Slit 02 case selected based on representative time instants associated with Slit 1 (anti-clockwise from top-right with increase in time). The contours indicate the fire whirl core location and collectively illustrate a full circle of orbital revolution of the whirling reacting flow with respect to fuel source.

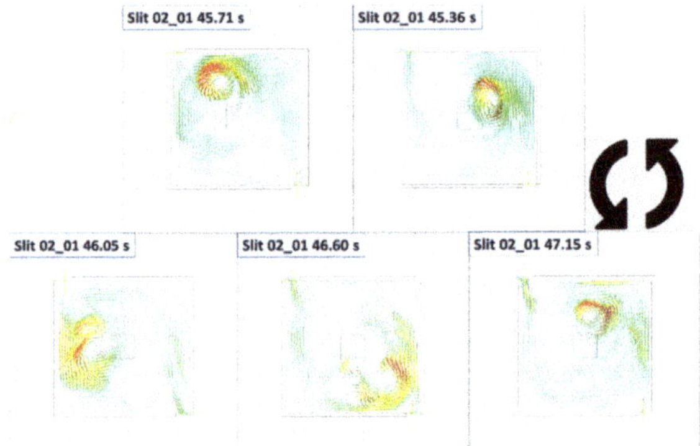

Figure 33. Vector field plot of the domain at 0.1 m HAB, at five timesteps of Slit 02 case selected based on representative time instants associated with Slit 1 (anti-clockwise from top-right with increase in time). The plots indicate the fire whirl core location and collectively illustrate a full circle of orbital revolution of the whirling reacting flow with respect to fuel source.

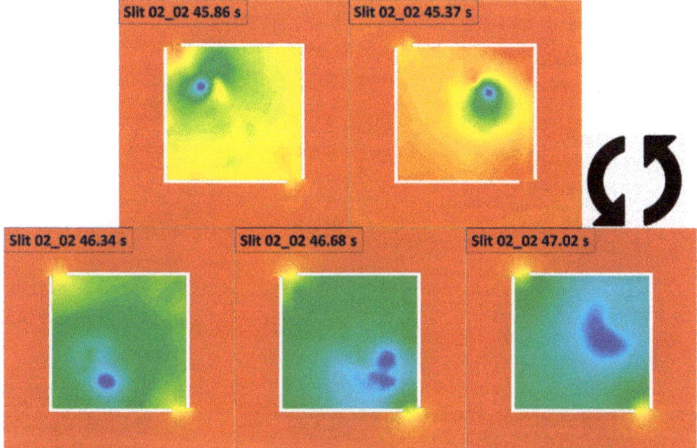

Figure 34. Pressure contour of the domain at 0.1 m HAB, at five timesteps of Slit 02 case selected based on representative time instants associated with Slit 2 (anti-clockwise from top-right with increase in time). The contours indicate the fire whirl core location and collectively illustrate a full circle of orbital revolution of the whirling reacting flow with respect to fuel source.

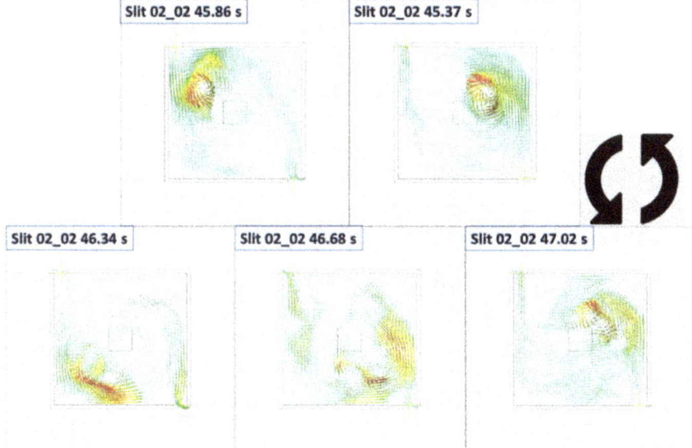

Figure 35. Vector field plot of the domain at 0.1 m HAB, at five timesteps of Slit 02 case selected based on representative time instants associated with Slit 2 (anti-clockwise from top-right with increase in time). The plots indicate the fire whirl core location and collectively illustrate a full circle of orbital revolution of the whirling reacting flow with respect to fuel source.

This speculation can be further justified when transferring the location of the vortex core and plotting the orbit of revolution of the reacting flow, as showing in Figure 36. The radius is further away from the fuel pan when compared with the baseline, especially when the core structure is approaching to the slits. Please note that the representative time instant associated with Slit 1 and Slit 2 denoted as ✿ and ◯ respectively.

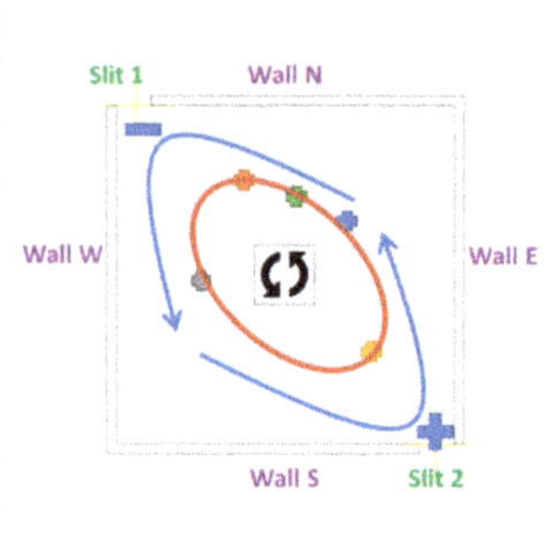
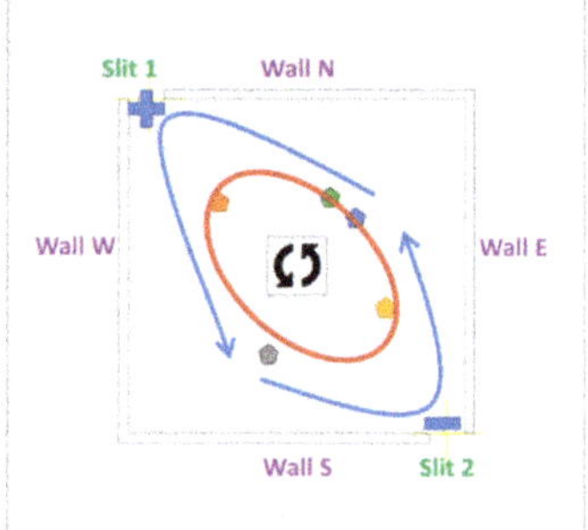

Figure 36. Orbit of fire whirl core centre of the Slit 02 case, indicating the fire whirl's revolution around the fuel source, with time instant associated with Slit 1 and Slit 2 denoted as ⬤ and ◯ respectively. The blue arrow indicates the tendency of changing of slit incoming velocity.

The revolution orbit demonstrated in Figure 36 also agrees well with pseudo-centripetal and associated circulation motion concept proposed previously. With the vortex approaching the air entrainment supplier source, i.e., slit and the resulted intensification of the velocity field of the structure of the swirling reacting flow, the radius of the revolution increases, based on a near-constant centripetal force resembled surface drag force that creates the radial boundary layer, and vice versa. For example, when the fire whirl sweep from enclosed corner towards slits, i.e., from time instant 45.36 s (⬤) to 45.71 s (⬤), from 46.05 s (⬤) to 46.60 s (⬤), from 45.37 s (⬤) to 45.86 s (⬤), and from 46.34 s (⬤) to 46.68 s (◯), a significant increase in rotation radius is observed. On the other hand, the centripetal force restores the orbit of the revolution by dragging the vortex structure back to the near fuel pan regions, when the fire whirl spinning from near slit region towards enclosed corner, i.e., from time instant 45.71 s (⬤) to 46.05 s (⬤), from 46.60 s (⬤) to 47.15 s (⬤), from 45.86 s (⬤) to 46.34 s (⬤), and from 46.68 s (◯) to 47.02 s (⬤).

The incoming velocity of the air entrainment source, both Slit 1 and Slit 2, against the time, is also plotted in Figure 37. It is surprisingly that the correlation of the distance between the fire whirl core to the slit and the incoming velocity of the slit demonstrated in the Slit 01 case can not be observed in the Slit 02 case. In contrary, the minimum incoming velocity is observed when the fire whirl core is at the closest location with respect to the slit. For example, for incoming velocity associated with Slit 01, it can be seen that the incoming velocity decreases as the fire whirl approaches and sweep through the Slit 01, i.e., from time instant of 45.36 s (⬤) to 46.05 s (⬤), and the incoming velocity increases as it shifted away from the slit, for example from time instant of 46.05 s (⬤) to 47.15 s (⬤). The similar tendency can be observed with relative fire whirl location with respect to the incoming velocity associated with Slit 02.

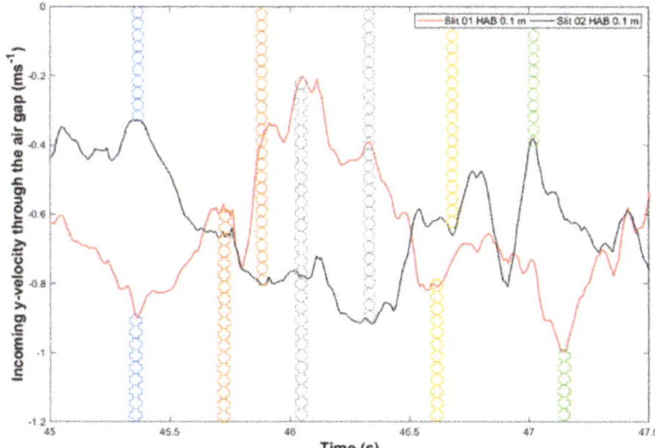

Figure 37. Plot of incoming velocity of the Slit 01 and Slit 02 cases, with the marks indicating the representative time instant corresponding to Figure 36.

It is reasonable to expect that the increased amount of air is drawn from the surrounding to the enclosures through the slit due to the intensified pressure gradient between the two side of the slit as the fire whirl approaching, as demonstrated in the Slit 01 case. However, it is also true that the pattern of the incoming flow via the opening could be inevitably intrusive by the presence and the interaction of the large plume of swirling structure at near slit regions as observed in the Slit 02 case, and as a result, disturb the pattern of incoming flow via the slit and lead to a reduced incoming velocity.

Prior to the conduction of the simulation and comparing the fire whirl evolution of the Slit 01 case and Slit 02 case. It is generally agreed and expected that the model with a symmetrical geometry configuration, i.e., Slit 02 case, should formulate a relatively centralised and stabilised fire whirl, when compared with the baseline model that has an asymmetrical eddy generation mechanism. However, the results of the assessment demonstrated a contradictory observation with the previous speculation.

It appears that the for the current burning configuration, i.e., constant fuel injection rate hence fixed burning rate, the effect of the surface drag force to create a radial boundary layer and act as the pseudo-centripetal force is limited. For single slit configuration, the vortex structure diverges from the domain centreline when it is approaching the opening can be restored and regulated to the semi-stable status once it is spinning through the enclosed wall regions, within the one circle of the revolution. On the contrary, the vortex core structure is drifted towards slit two times in one circle of revolution, and the restoration period of the vortex to drag back to near-source pan region is significantly shorten compared with baseline case. The orbit of the revolution of the Slit 02 case is regulated and balanced due to the constraint of the enclosure wall that stops the vortex from further drifting away from the fuel centreline, and this explains the observed development of the fire whirl rotation behaviour from Stage C_1, that centralised around fuel pan region towards Stage C_3 that circulation around with a relatively large radius of rotation.

4. Conclusions

A numerical modelling of the fire whirl formulated and evolved in an enclosed configuration with different entrainment schemes was performed to evaluate the effect of additional eddy generation to the fire whirl formation and development pathway. The model adopted detailed chemistry, WALES large eddy simulation turbulence and combustion coupling approach to capture the intricate whirling flame flow behaviour.

The modelling is conducted in a controlled numerical environment, for example, soot-free alcohol-based fuel injected with fixed inlet velocity and resolved numerically based on a uniformed fully structured discretisation scheme. This approach isolates and constrains the complicated coupling effect between the combustion process and the flow dynamics due to theoretically and numerically introduced uncertainties such as radiation feedback from soot particulates, variance in burning rate due to pool-based fuel configuration and numerical variation in spatial resolution due to locally refined mesh that typically implemented in previous studies in the literature.

The present numerical simulation of the baseline case consists of a single side flow channelling slit, replicate the experimental setup from the literature. The result yielded from the baseline model compares well with the experimental data, and hence the same numerical methodologies are applied to construct the comparison group with two identical symmetrical entrainment slits, and the results of the two models, from the time of ignition to 50.00 s is presented and compared to reveal the effect of the two eddy generation scheme on the fire whirl formation and development. In conclusion, the following observations can be made:

- With the existence of the eddy generation sources, i.e., slit(s) on the side of the enclosure, both Slit 01 and Slit 02 case observed the formulation and evolvement of the fire whirl from a buoyancy-driven diffusion flame that flickering randomly into a swirling reacting flow that spanning around the chamber with respect to domain centreline;
- Three-stage of the fire whirl formulation and development pathway can be observed in both cases, namely Stage A as the flame development, Stage B as the fire whirl development and the formation and Stage C as the fire whirl evolution;
- Compared with the baseline model, the Slit 02 case formulated the fire whirl much faster, i.e., 49.58% reduction of the duration in Stage B which transforms from the free-standing buoyant flame into nascent fire whirl;
- The nascent fire whirl formulated in Slit 02 is more intensified and spatially extended compared with baseline case, with the visible height is increased by 84.38%, from 0.32 m to 0.59 m, peak flame temperature increased 9.42%, from 1380 K to 1510 K and relatively consistent vortex core radius compared with that increase of the monitoring flame height;
- Once the nascent fire whirl is formulated, the fire whirl for the baseline model is spinning around the centreline with a relatively small radius of revolution in a semi-steady pattern, for the rest of the simulation duration up to 50 s. On the other hand, the highly centralised fire whirl formulated in the Slit 02 case may gradually diverge via swirling with an increasing radius of revolution. It will eventually achieve an internal balanced semi-stable status that the revolution radius is intensified by the introduction of the additional eddy via the slits and at the same time constrained by the enclosures boundary walls;
- The revolution obit of the fire whirl could be potentially explained based on the theory of circular motion with constant surface drag force to create a radial boundary layer, acting as the centripetal force that balances the velocity field of the vortex and the radius of revolution. It has been observed in both cases that increased radius of revolution is observed as the fire whirl core structure approaches the slit and hence intensified its velocity field, and vice verse decreased as it departs from the near slit region, to balance the constant burning rate that fixed the surface drag force.
- The incoming velocity of the slit is observed to be proportional with the distance between the vortex core centre and the slit in the baseline case, which agrees well with the flow dynamic driven by pressure gradient. However, the incoming velocity is observed to decrease as the swirling plume approaches the slit and increase as it departs, which may be attributable to the disturbance of and potential interaction between of the swirling reacting flow and naturally ventilated flow pattern.

In summary, the results collectively compare the complete formation and evolution pathway of the fire whirl under both under symmetrical and asymmetrical entraining scheme. The result demonstrates that the fire whirl formulated in the single asymmetrical configuration is relatively smaller in size, and it swirling with respect to the fuel pan in a relatively smaller radius as the centripetal drag force corrected its obit as it rotates through the enclosed wall regions. Comparatively, the fire whirl generated by symmetrical slits configuration is observed to have higher visible height and relatively more regulated share once it has been formulated. Its revolution orbit gradually evolves as its radius of rotation constantly increased by the intensification of the vortex core velocity field, and eventually reaches a semi-stable status as its move pattern is constrained by the enclosed wall. The result presented in the current work provides a visual of the entire development of the combustion event that aids the understanding of the complexed phenomenon couples with flow dynamics and combustion. It also demonstrates that the proposed numerical framework is feasible to reveal the fundamental information and probe the possible correlation between the geometrical configuration and the development of the combustion behaviour. The information delivered could be beneficial for both research and industrial communities and could be further implemented to enhance in building design, hazard prevention and control as well as evacuation planning.

Author Contributions: Conceptualization, A.C.Y.Y., Q.N.C., S.K. and G.H.Y.; data curation, C.W. and H.L.Y.; formal analysis, C.W., T.B.Y.C., S.C.-P.C. and H.L.Y.; funding acquisition, A.C.Y.Y. and G.H.Y.; methodology, A.C.Y.Y., Q.N.C., S.K. and G.H.Y.; resources, S.K.; supervision, A.C.Y.Y., Q.N.C., S.K. and G.H.Y.; validation, S.C.-P.C.; visualization, C.W.; writing—original draft, C.W.; writing—review and editing, T.B.Y.C. and S.C.-P.C. All authors have read and agreed to the published version of the manuscript.

Funding: This research was funded by the Australian Research Council (ARC Industrial Transformation Training Centre IC170100032), and Wuhan Shuanglian-Xingxin Machinery & Equipment Co. Ltd., China.

Acknowledgments: The authors wish to acknowledge the financial support of Australian Research Council (ARC Industrial Transformation Training Centre IC170100032) and the Australian Government Research Training Program Scholarship. It is also sponsored by Wuhan Shuanglian-Xingxin Machinery and Equipment Co. Ltd., China. The authors deeply appreciate all financial and technical supports.

Conflicts of Interest: The authors declare no conflicts of interest.

References

1. Tohidi, A.; Gollner, M.J.; Xiao, H. Fire Whirls. *Annu. Rev. Fluid Mech.* **2018**, *50*, 187–213. [CrossRef]
2. Wang, P.; Liu, N.; Bai, Y.; Zhang, L.; Satoh, K.; Liu, X. An experimental study on thermal radiation of fire whirl. *Int. J. Wildland Fire* **2017**, *26*, 693. [CrossRef]
3. Medwell, P.R.; Chan, Q.N.; Kalt, P.A.M.; Alwahabi, Z.T.; Dally, B.B.; Nathan, G.J. Instantaneous Temperature Imaging of Diffusion Flames Using Two-Line Atomic Fluorescence. *Appl. Spectrosc.* **2010**, *64*, 173–176. [CrossRef] [PubMed]
4. Soma, S.; Saito, K. Reconstruction of fire whirls using scale models. *Combust. Flame* **1991**, *86*, 269–284. [CrossRef]
5. Zhou, K.; Liu, N.; Yin, P.; Yuan, X.; Jiang, J. Fire Whirl due to Interaction between Line Fire and Cross Wind. *Fire Saf. Sci.* **2014**, *11*, 1420–1429. [CrossRef]
6. Byram, G.; Martin, R. The Modeling of Fire Whirlwinds. *For. Sci.* **1970**, *16*, 386–399.
7. Lei, J.; Liu, N.; Tu, R. Flame height of turbulent fire whirls: A model study by concept of turbulence suppression. *Proc. Combust. Inst.* **2017**, *36*, 3131–3138. [CrossRef]
8. Yuen, A.C.Y.; Yeoh, G.H.; Cheung, S.C.P.; Chan, Q.N.; Chen, T.B.Y.; Yang, W.; Lu, H. Numerical study of the development and angular speed of a small-scale fire whirl. *J. Comput. Sci.* **2018**, *27*, 21–34. [CrossRef]
9. Church, C.R.; Snow, J.T.; Dessens, J. Intense Atmospheric Vortices Associated with a 1000 MW Fire. *Bull. Am. Meteorol. Soc.* **1980**, *61*, 682–694. [CrossRef]
10. Forthofer, J.M.; Goodrick, S.L. Review of Vortices in Wildland Fire. *J. Combust.* **2011**, *2011*, 984363. [CrossRef]
11. Liu, N.; Liu, Q.; Deng, Z.; Kohyu, S.; Zhu, J. Burn-out time data analysis on interaction effects among multiple fires in fire arrays. *Proc. Combust. Inst.* **2007**, *31*, 2589–2597. [CrossRef]
12. Emori, R.I.; Saito, K. Model experiment of hazardous forest fire whirl. *Fire Technol.* **1982**, *18*, 319–327. [CrossRef]
13. Kuwana, K.; Morishita, S.; Dobashi, R.; Chuah, K.H.; Saito, K. The burning rate's effect on the flame length of weak fire whirls. *Proc. Combust. Inst.* **2011**, *33*, 2425–2432. [CrossRef]
14. Chuah, K.H.; Kuwana, K.; Saito, K.; Williams, F.A. Inclined fire whirls. *Proc. Combust. Inst.* **2011**, *33*, 2417–2424. [CrossRef]

15. Lei, J.; Liu, N.; Zhang, L.; Chen, H.; Shu, L.; Chen, P.; Deng, Z.; Zhu, J.; Satoh, K.; De Ris, J.L. Experimental research on combustion dynamics of medium-scale fire whirl. *Proc. Combust. Inst.* **2011**, *33*, 2407–2415. [CrossRef]

16. Zhou, K.; Liu, N.; Lozano, J.S.; Shan, Y.; Yao, B.; Satoh, K. Effect of flow circulation on combustion dynamics of fire whirl. *Proc. Combust. Inst.* **2013**, *34*, 2617–2624. [CrossRef]

17. Taylor Thomson Whitting 50 Martin Place. Available online: https://www.ttw.com.au/projects/50-martin-place/ (accessed on 29 November 2019).

18. Lin, B.; Yuen, A.C.Y.; Li, A.; Zhang, Y.; Chen, T.B.Y.; Yu, B.; Lee, E.W.M.; Peng, S.; Yang, W.; Lu, H.-D.; et al. MXene/chitosan nanocoating for flexible polyurethane foam towards remarkable fire hazards reductions. *J. Hazard. Mater.* **2020**, *381*, 120952. [CrossRef]

19. Yang, W.-J.; Yuen, A.C.Y.; Li, A.; Lin, B.; Chen, T.B.Y.; Yang, W.; Lu, H.-D.; Yeoh, G.H. Recent progress in bio-based aerogel absorbents for oil/water separation. *Cellulose* **2019**, *26*, 6449–6476. [CrossRef]

20. Si, J.-Y.; Tawiah, B.; Sun, W.-L.; Lin, B.; Wang, C.; Yuen, A.C.Y.; Yu, B.; Li, A.; Yang, W.; Lu, H.-D.; et al. Functionalization of MXene Nanosheets for Polystyrene towards High Thermal Stability and Flame Retardant Properties. *Polymers* **2019**, *11*, 976. [CrossRef]

21. Yu, B.; Tawiah, B.; Wang, L.-Q.; Yuen, A.C.Y.; Zhang, Z.-C.; Shen, L.-L.; Lin, B.; Fei, B.; Yang, W.; Li, A.; et al. Interface decoration of exfoliated MXene ultra-thin nanosheets for fire and smoke suppressions of thermoplastic polyurethane elastomer. *J. Hazard Mater.* **2019**, *374*, 110–119. [CrossRef]

22. Yuen, A.C.Y.; Chen, T.B.Y.; Wang, C.; Wei, W.; Kabir, I.; Vargas, J.B.; Chan, Q.N.; Kook, S.; Yeoh, G.H. Utilising genetic algorithm to optimise pyrolysis kinetics for fire modelling and characterisation of chitosan/graphene oxide polyurethane composites. *Compos. Part B Eng.* **2020**, *182*, 107619. [CrossRef]

23. Chen, T.; Yuen, A.; Yeoh, G.; Yang, W.; Chan, Q. Fire Risk Assessment of Combustible Exterior Cladding Using a Collective Numerical Database. *Fire* **2019**, *2*, 11. [CrossRef]

24. Li, A.; Yuen, A.C.Y.; Chen, T.B.Y.; Wang, C.; Liu, H.; Cao, R.; Yang, W.; Yeoh, G.H.; Timchenko, V. Timchenko Computational Study of Wet Steam Flow to Optimize Steam Ejector Efficiency for Potential Fire Suppression Application. *Appl. Sci.* **2019**, *9*, 1486. [CrossRef]

25. Yuen, A.C.Y.; Yeoh, G.H.; Alexander, B.; Cook, M. Fire scene investigation of an arson fire incident using computational fluid dynamics based fire simulation. *Build. Simul.* **2014**, *7*, 477–487. [CrossRef]

26. Yuen, A.C.Y.; Yeoh, G.H.; Alexander, R.; Cook, M. Fire scene reconstruction of a furnished compartment room in a house fire. *Case Stud. Fire Saf.* **2014**, *1*, 29–35. [CrossRef]

27. Dobashi, R.; Okura, T.; Nagaoka, R.; Hayashi, Y.; Mogi, T. Experimental Study on Flame Height and Radiant Heat of Fire Whirls. *Fire Technol.* **2016**, *52*, 1069–1080. [CrossRef]

28. Hartl, K.A.; Smits, A.J. Scaling of a small scale burner fire whirl. *Combust. Flame* **2016**, *163*, 202–208. [CrossRef]

29. Wang, P.; Liu, N.; Hartl, K.; Smits, A. Measurement of the Flow Field of Fire Whirl. *Fire Technol.* **2016**, *52*, 263–272. [CrossRef]

30. Xiao, H.; Gollner, M.J.; Oran, E.S. From fire whirls to blue whirls and combustion with reduced pollution. *Proc. Natl. Acad. Sci. USA* **2016**, *113*, 9457–9462. [CrossRef]

31. Muraszew, A.; Fedele, J.B.; Kuby, W.C. The fire whirl phenomenon. *Combust. Flame* **1979**, *34*, 29–45. [CrossRef]

32. Chuah, K.H.; Kushida, G. The prediction of flame heights and flame shapes of small fire whirls. *Proc. Combust. Inst.* **2007**, *31*, 2599–2606. [CrossRef]

33. Parente, R.M.; Pereira, J.M.C.; Pereira, J.C.F. On the influence of circulation on fire whirl height. *Fire Saf. J.* **2019**, *106*, 146–154. [CrossRef]

34. Wang, C.; Chan, Q.N.; Kook, S.; Hawkes, E.R.; Medwell, P.R.; Lee, J. Development of an in-flame thermophoretic soot sampling device. In Proceedings of the Australian Combustion Symposium, Melbourne, Australia, 7–9 December 2015; pp. 360–363.

35. Wang, C.; Chan, Q.N.; Kook, S.; Hawkes, E.R.; Medwell, P.R.; Lee, J. External Irradiation Effect on the Evolution of In-flame Soot Species. In Proceedings of the Australian Combustion Symposium, Melbourne, Australia, 7–9 December 2015; pp. 364–367.

36. Lei, J.; Liu, N.; Satoh, K. Buoyant pool fires under imposed circulations before the formation of fire whirls. *Proc. Combust. Inst.* **2015**, *35*, 2503–2510. [CrossRef]

37. Lei, J.; Liu, N.; Zhang, L.; Satoh, K. Temperature, velocity and air entrainment of fire whirl plume: A comprehensive experimental investigation. *Combust. Flame* **2015**, *162*, 745–758. [CrossRef]

38. Sikanen, T.; Hostikka, S. Modeling and simulation of liquid pool fires with in-depth radiation absorption and heat transfer. *Fire Saf. J.* **2016**, *80*, 95–109. [CrossRef]
39. Yao, W.; Yin, J.; Hu, X.; Wang, J.; Zhang, H. Numerical modeling of liquid n-heptane pool fires based on heat feedback equilibrium. *Procedia Eng.* **2013**, *62*, 377–388. [CrossRef]
40. Wang, C.; Chan, Q.N.; Zhang, R.; Kook, S.; Hawkes, E.R.; Yeoh, G.H.; Medwell, P.R. Automated determination of size and morphology information from soot transmission electron microscope (TEM)-generated images. *J. Nanopart. Res.* **2016**, *18*, 127. [CrossRef]
41. Wang, C.; Chan, Q.N.; Kook, S.; Hawkes, E.R.; Lee, J.; Medwell, P.R. External irradiation effect on the growth and evolution of in-flame soot species. *Carbon N. Y.* **2016**, *102*, 161–171. [CrossRef]
42. Yuen, A.C.Y.; Yeoh, G.H.; Timchenko, V.; Cheung, S.C.P.; Chan, Q.N.; Chen, T. On the influences of key modelling constants of large eddy simulations for large-scale compartment fires predictions. *Int. J. Comut. Fluid Dyn.* **2017**, *31*, 324–337. [CrossRef]
43. Yuen, A.C.Y.; Yeoh, G.H.; Timchenko, V.; Chen, T.B.Y.; Chan, Q.N.; Wang, C.; Li, D.D. Comparison of detailed soot formation models for sooty and non-sooty flames in an under-ventilated ISO room. *Int. J. Heat Mass Transf.* **2017**, *115*, 717–729. [CrossRef]
44. Tree, D.R.; Svensson, K.I. Soot processes in compression ignition engines. *Prog. Energy Combust. Sci.* **2007**, *33*, 272–309. [CrossRef]
45. Wang, H. Formation of nascent soot and other condensed-phase materials in flames. *Proc. Combust. Inst.* **2011**, *33*, 41–67. [CrossRef]
46. Jones, W.P.; Whitelaw, J.H. Calculation methods for reacting turbulent flows: A review. *Combust. Flame* **1982**, *48*, 1–26. [CrossRef]
47. Liu, H.; Wang, C.; De Cachinho Cordeiro, I.M.; Yuen, A.C.Y.; Chen, Q.; Chan, Q.N.; Kook, S.; Yeoh, G.H. Critical assessment on operating water droplet sizes for fire sprinkler and water mist systems. *J. Build. Eng.* **2020**, *28*, 100999. [CrossRef]
48. Chow, W.; Han, S.S. Experimental Data on Scale Modeling Studies on Internal Fire Whirls. *Int. J. Eng. Perform. Based Fire Codes* **2011**, *10*, 63–74.
49. Wang, C.; Chun, A.; Yuen, Y.; Chan, Q.N.; Bo, T.; Chen, Y.; Chen, Q.; Cao, R.; Yip, H.L.; Kook, S.; et al. Influence of Eddy-Generation Mechanism on the Characteristic of On-Source Fire Whirl. *Appl. Sci.* **2019**, *9*, 3989. [CrossRef]
50. Chen, T.B.Y.; Yuen, A.C.Y.; Wang, C.; Yeoh, G.H.; Timchenko, V.; Cheung, S.C.P.; Chan, Q.N.; Yang, W. Predicting the fire spread rate of a sloped pine needle board utilizing pyrolysis modelling with detailed gas-phase combustion. *Int. J. Heat Mass Transf.* **2018**, *125*, 310–322. [CrossRef]
51. Yuen, A.; Chen, T.; Yang, W.; Wang, C.; Li, A.; Yeoh, G.; Chan, Q.; Chan, M. Natural Ventilated Smoke Control Simulation Case Study Using Different Settings of Smoke Vents and Curtains in a Large Atrium. *Fire* **2019**, *2*, 7. [CrossRef]
52. Yuen, A.C.Y.; Yeoh, G.H.; Timchenko, V.; Cheung, S.C.P.; Chen, T. Study of three LES subgrid-scale turbulence models for predictions of heat and mass transfer in large-scale compartment fires. *Numer. Heat Transf. Part A Appl.* **2016**, *69*, 1223–1241. [CrossRef]
53. Yuen, A.C.Y. On the Prediction of Combustion Products and Soot Particles in Compartment Fires. Ph.D. Thesis, University of New South Wales, Sydney, Australia, 2014.
54. Wang, C. External Irradiation Effect on the Evolution of In-flame Soot Species. M.E. Thesis, University of New South Wales, Sydney, Australia, 2016.
55. Yuen, A.C.Y.; Yeoh, G.H.; Timchenko, V.; Barber, T. LES and multi-step chemical reaction in compartment fires. *Numer. Heat Transf. Part A Appl.* **2015**, *68*, 711–736. [CrossRef]
56. Kee, R.J.; Rupley, F.M.; Miller, J.A.; Coltrin, M.E.; Grcar, J.F.; Meeks, E.; Moffat, H.K.; Lutz, A.E.; Dixon-Lewis, G.; Smooke, M.D.; et al. CHEMKIN collection Release 3.6. In *CHEMKIN Collection Release 3.6*; Reaction Design. Inc.: San Diego, CA, USA, 2000.
57. Yuen, A.C.Y.; Yeoh, G.H.; Timchenko, V.; Cheung, S.C.P.; Barber, T.J. Importance of detailed chemical kinetics on combustion and soot modelling of ventilated and under-ventilated fires in compartment. *Int. J. Heat Mass Transf.* **2016**, *96*, 171–188. [CrossRef]
58. Yuen, A.C.Y.; Chen, T.B.Y.; Yeoh, G.H.; Yang, W.; Cheung, S.C.P.; Cook, M.; Yu, B.; Chan, Q.N.; Yip, H.L. Establishing pyrolysis kinetics for the modelling of the flammability and burning characteristics of solid combustible materials. *J. Fire Sci.* **2018**, *36*, 494–517. [CrossRef]

59. Chen, T.B.Y.; Yuen, A.C.Y.; Yeoh, G.H.; Timchenko, V.; Cheung, S.C.P.; Chan, Q.N.; Yang, W.; Lu, H. Numerical study of fire spread using the level-set method with large eddy simulation incorporating detailed chemical kinetics gas-phase combustion model. *J. Comput. Sci.* **2018**, *24*, 8–23. [CrossRef]
60. Wang, C.; Yuen, A.C.Y.; Chan, Q.N.; Chen, T.B.Y.; Yang, W.; Cheung, S.C.-P.; Yeoh, G.H. Sensitivity Analysis of Key Parameters for Population Balance Based Soot Model for Low-Speed Diffusion Flames. *Energies* **2019**, *12*, 910. [CrossRef]
61. Mueller, M.E.; Chan, Q.N.; Qamar, N.H.; Dally, B.B.; Pitsch, H.; Alwahabi, Z.T.; Nathan, G.J. Experimental and computational study of soot evolution in a turbulent nonpremixed bluff body ethylene flame. *Combust. Flame* **2013**, *160*, 1298–1309. [CrossRef]
62. Wang, C.; Yuen, A.C.Y.; Chan, Q.N.; Chen, T.B.Y.; Yang, W.; Cheung, S.C.P.; Yeoh, G.H. Characterisation of soot particle size distribution through population balance approach and soot diagnostic techniques for a buoyant non-premixed flame. *J. Energy Inst.* **2019**, *93*, 112–128. [CrossRef]

Article

Computational Study of Wet Steam Flow to Optimize Steam Ejector Efficiency for Potential Fire Suppression Application

Ao Li [1], Anthony Chun Yin Yuen [1], Timothy Bo Yuan Chen [1], Cheng Wang [1], Hengrui Liu [1], Ruifeng Cao [1], Wei Yang [1,2,*], Guan Heng Yeoh [1,3] and Victoria Timchenko [1]

[1] School of Mechanical and Manufacturing Engineering, University of New South Wales, Sydney, NSW 2052, Australia; ao.li@unsw.edu.au (A.L.); c.y.yuen@unsw.edu.au (A.C.Y.Y.); timothy.chen@unsw.edu.au (T.B.Y.C.); c.wang@unsw.edu.au (C.W.); h.liu@unsw.edu.au (H.L.); ruifeng.cao@unsw.edu.au (R.C.); g.yeoh@unsw.edu.au (G.H.Y.); v.timchenko@unsw.edu.au (V.T.)

[2] Department of Chemical and Materials Engineering, Hefei University, Hefei 230601, China

[3] Australian Nuclear Science and Technology Organization (ANSTO), Locked Bag 2001, Kirrawee DC, NSW 2232, Australia

[*] Correspondence: yangwei@hfuu.edu.cn; Tel.: +61-2-9385-5697

Received: 2 March 2019; Accepted: 4 April 2019; Published: 9 April 2019

Abstract: The steam ejector is a core component of an ejector-based refrigeration system. Additionally, steam ejectors can also be potentially applied for a fire suppression system by using pressurized steam droplets to rapidly quench and extinguish the fire. The use of steam will significantly reduce the amount of water consumption and pipe flow rate compared to conventional sprinklers. However, the efficiency of the steam ejector nozzle is one of major factors that can influence the extinguishing mechanisms and the performance of pressurized steam for fire suppression. In this article, to formulate an assessment tool for studying the ideal entrainment ratio and initial flow wetness, a wet steam model has been proposed to enhance our understanding of the condensation and evaporation effects of water droplets from a numerical perspective. The entire steam-ejector system including the nozzle, mixing chamber, throat and diffuser were modeled to study the profiles in axial pressure and temperature across the system, and were compared with self-measured experimental data. In addition, the flow and heat transfer interactions between the fluid mixture and nucleating water droplets were numerically examined by comparing initial conditions with different liquid fractions, as opposed to the ideal gas assumption. With the application of the proposed wet-steam model, the numerical model showed vast improvement in the axial pressure distribution over the ideal gas model. Through numerical conditions, it was found that reducing the wetness of the secondary inlet flow will potentially optimize the system performance with a significant increase of the entrainment ratio from 0.38 to 0.47 (i.e., improvement of around 23%).

Keywords: steam ejector; wet steam model; computational fluid dynamics; condensation effect; fire suppression system

---

1. Introduction

A steam ejector is a fluid phase-change device, which can utilize low-grade energy, such as industrial waste heat, to generate vacuum environments. The first steam ejector was designed by Flugel in 1939 [1]. Ever since it has been developed, the steam ejector has been widely applied in multiple industrial sectors, including refrigeration, chemistry, metallurgy, petroleum, and food processing [2–5]. Steam ejectors can also be applied to a fire suppression system in place of water mist to provide greater displacement of oxygen from a fire plume due to the evaporation of water droplets. Based on nozzles and water mist, numerous studies on different characteristics of the system such as pressure, injection rate, cone angles, and droplet dimensions were conducted for different working conditions and were performed both experimentally and numerically in recent years [6–8].

A steam ejector utilizes high-pressure steam to pump low-pressure fluids. It has many advantages over other commonly used devices, such as no moving parts, less-required maintenance and low cost in manufacturing and energy consumption. Although the structure of a steam ejector is relatively simple, the flow behavior within the enclosure can be extremely complex, involving phenomena such as transonic flow, shockwaves, phase change etc.

The configuration of a typical steam ejector [9] along with the corresponding pressure changes and velocity distribution is shown in Figure 1. In general, a steam ejector consists of four main components: (i) primary nozzle, (ii) mixing chamber, (iii) throat, and (iv) subsonic diffuser. The steam ejector work cycle starts with the acceleration of highly pressurized primary fluid to supersonic after passing through the nozzle. Approaching the end of the nozzle, a relative low-pressure region is generated due to the high velocity, which entrains the secondary fluid stream from the evaporator into the mixing chamber. The velocity of the secondary fluid increases within the mixing chamber due to the energy and momentum exchange between the two fluid streams. Subsequently, the speed of the mixed flow is reduced to subsonic range, due to the normal shock wave occurring in the throat. The mixed flow is further compressed after exiting via the subsonic diffuser. This entire entrainment process is very complex, and analysis requires a deep understanding of the underlying physical fluid behaviors and the change in water droplets within the steam ejector. One of the key parameters to assess the performance of the steam ejector is the entrainment ratio or mass flow ratio [10], which is defined as:

$$Entrainment\,Ratio = \frac{Mass\ flow\ rate\ of\ primary\ flow}{Mass\ flow\ rate\ of\ secondary\ flow} \tag{1}$$

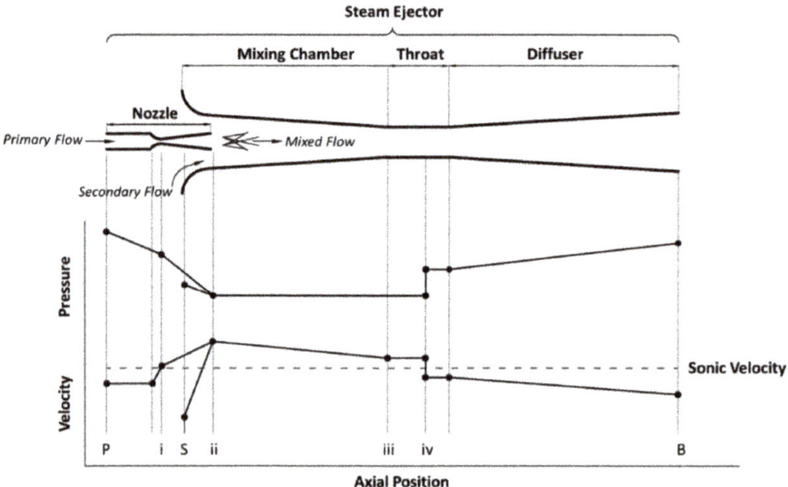

Figure 1. Configuration of a typical steam ejector and flow characteristic along the axis of the steam ejector.

As a characteristic index of efficiency, a stream ejector with a higher entrainment ratio is usually preferable. In addition, the performance of the steam ejector can also be described by the characteristic curve (see Figure 1) and herein the critical back pressure plays an important role. According to the critical back pressure, the steam ejector entrainment ratio can be separated into three modes [11]: (i) critical mode (double-choking), (ii) subcritical mode (single-choking) and (iii) back-flow mode (malfunction) as indicated in Figure 2. In the double-choking mode, the entrainment ratio of the stream ejector remains as a constant with the increase of back pressure up to a critical point. Afterwards, the entrainment ratio dramatically decreases to zero, and the ejector loses its function. Previous works have been carried out using both experimental and numerical approaches to evaluate the effect of key parameters on the performance of a steam ejector. These studies include the optimization of key components such as the mixing chamber, nozzle, throat etc. as well as the assessment of the system under various operating conditions have also been performed. For instance, Chunnanond et al. [9] conducted an experimental investigation on the pressure profile along the ejector axial direction and concluded that the performance of the ejector was predominately governed by the amount of secondary fluid as well as the momentum of the mixed stream. Sankarlal et al. [12] studied the effects of non-dimensional parameters including the compression and expansion ratios, on the system performance. The results showed that the coefficient of performance is proportional to the expansion ratio and is inversely proportional to the compression ratio. Aidoun et al. [13] and Wang et al. [14] applied various boundary conditions for inlet flows and investigated their effect on the entrainment ratio. The results suggested that, with the implementation of superheat inlet flow, improvement of entrainment ratio could be achieved, whereas the pressure at the exit remains unchanged. The performance of a steam ejector can be improved by optimizing the geometric configuration of the key components. Varga et al. [15–17] and Ma et al. [18] developed the relationship between the entrainment ratio and the area ratio. The area ratio is a critical non-dimensional factor that has a significant impact on the performance of the system which is defined as the ratio between the cross-section area of primary nozzle and constant area section. Other experimental [9] and numerical studies [19–21] have also been conducted on the nozzle exit position. It was reported that positioning the nozzle exit upstream of the mixing chamber entrance gave better performance than pushing it into the mixing chamber. Furthermore, Sun et al. [22] and Cizungu et al. [23] investigated the correlation between the primary nozzle diameter and system performance. Chaiwongsa et al. [24] and Ruangtrakoon et al. [25,26] performed parametric studies on the influence of different configurations of the primary nozzle, and suggested that the expansion angle of the nozzle is also critical to the performance of the steam ejector. Most experimental studies focused on the optimization of geometric parameters and operating conditions as well as the explanation of experimental phenomena. The running cost for the experiment is also too costly to generate a lot of data for comparison. However, the numerical simulation can perform a comprehensive analysis for the flow behavior based on the thermal dynamics and fluid dynamics with the advantages of reducing cost and increasing the efficiency.

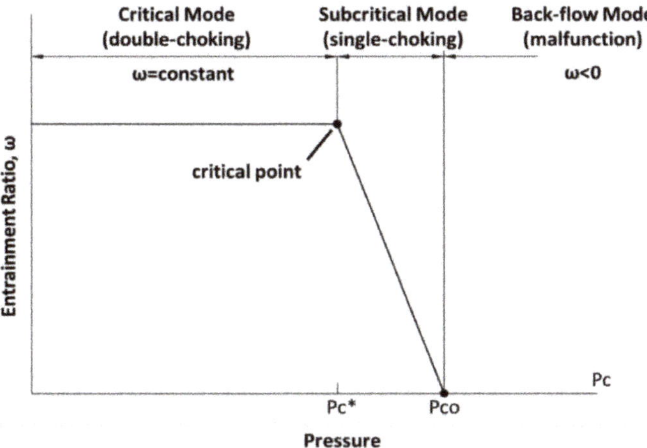

Figure 2. Characteristic curve of a steam ejector at a fixed temperature [10].

Computational fluid dynamics (CFD) simulation has become an effective tool to analyze the understanding of complex flow behavior due to the rapid advancement of numerical methods and computational power. Previous numerical studies demonstrated that, with the cautiously selected sub-molecular modeling components, CFD simulation could provide detailed information about the flow field that often difficult to assess by experimental means such as gaining more detail about nucleating small particles in the main flow field [27,28]. During the past few decades, extensive numerical research works based on ideal gas assumption have been performed to analyze the flow field development of the steam ejector. However, the spontaneous condensation phenomena occurring in the nozzle during operation [29] have not been captured under the ideal gas assumption. When the steam passes through the nozzle, the temperature of steam would be much lower than the stagnation temperature due to the inner energy partly transforming into kinetic energy caused by the occurrence of the condensation phenomena. Fluid thermodynamics characteristics of the inlet flows are described in Figure 3a. The state of primary fluid moves from the original state (point A) to point C, which is an isentropic expansion process. However, the real expansion process is line AC', due to energy loss existing in the actual process. The secondary fluid started from Point B is mixed with the primary fluid along the isobar line CB and arrives in point M at the end of mixing process. Subsequently, as for the shock wave, the mixed flow reaches a subsonic condition (Point E) and it arrives in condition F after it eventually passes through the diffuser. The steam jet refrigeration cycle is applied in the ejector refrigeration system [30]. Figure 3b,d illustrates a typical one stage ejector refrigeration system, where the steam ejector is shown in Figure 3c. Based on this system, numerical studies with the consideration of spontaneous condensation of the steam ejector have also been performed to analyze the development of the primary fluid and the complicated phase change process. For example, Yang et al. [31] simulated the condensation process during the nozzle, and Mazzelli et al. [32] compared the CFD modeling with an experimental test-case for the condensation. The spontaneous condensation in the nozzle of the steam ejector was studied in recent decades [33–35]. Sharifi et al. [34] found that the steam condensation increases the flow Mach number and the ejector's performance comparison with a dry gas. Wang et al. [33,35] investigated the influences of superheating operation of the primary fluid. It was concluded that 0 K–20 K superheating for the primary fluid has a positive impact on the entrainment ratio, while superheating over 20 K was meaningless and led to energy waste. Wang et al. [36] showed that the primary pseudo-shock flow gives a further downstream choking position and a higher secondary fluid flow velocity at the choking position with the wet steam model. Furthermore, Bakhtar et al. [37] applied a time-marching technique to obtain more accurate numerical predictions. This technique was proven to be effective and feasible for the two-dimensional flows of the nucleating and wet steam model in nozzles. Those studies, nevertheless, are limited to mainly focus on the individual parts of the system, i.e., the nozzle. Detailed analysis of the flow field development as well as the information about phase change within the entire steam ejector including all four major components, however, has so far not been performed. Furthermore, it is critical to have a clear study of the condensation effect inside the whole steam ejector, especially in the application of the wet steam model.

Considering the above-mentioned research gap, in this article, the working process of the entire steam ejector consisting of the nozzle, mixing chamber, throat and diffuser will be considered entirely utilizing CFD simulations. In essence, key objectives of this study can be summarized by the following statements:

1. The simulation will take advantage of a wet steam model to investigate the condensation effect on the flow behavior and steam ejector performance.
2. The numerical model will be validated against experimental data and the results will also be compared with the ideal gas model.
3. Different wall functions and turbulences models will be tested to determine the most optimized model in terms of prediction accuracy.

4. A series of simulations with different wetness operating conditions will be performed to investigate the relationship between the wetness of working flows and pump efficiency.

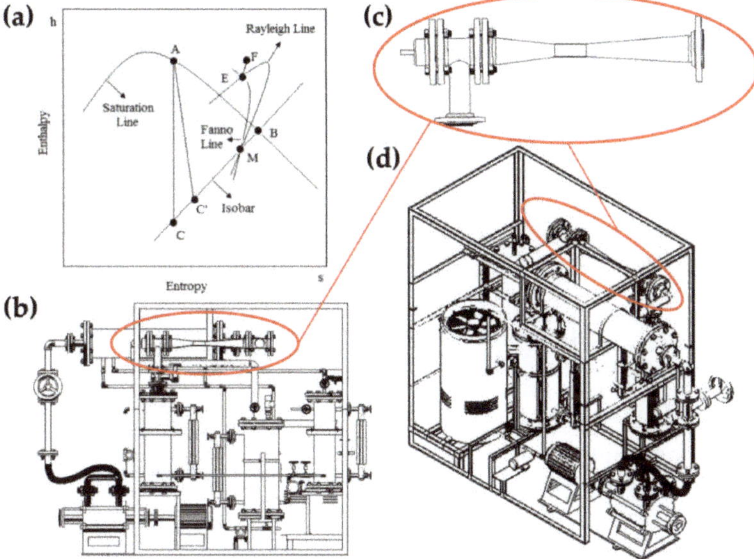

Figure 3. (**a**) Enthalpy-entropy diagram of the working process in the nozzle of the steam ejector. (**b**) Main view of the ejector refrigeration system (**c**) Steam ejector. (**d**) Three-dimensional isometric side view of the system.

2. Mathematical Model

In order to model the complex, non-linear and fully-coupled turbulent and steam interacting fluid mixture involved in a steam-ejector, the conserved properties including velocity, temperature, and mass fraction of water droplets were solved based on the governing equations of fluid dynamics and transport equations, in associated with the wet-steam model. The governing equations are mathematical statements of the physical and chemical phenomena of the considered fluid mixture, which can be numerically resolved in the form of computational codes to obtain solutions for fluid problems, whilst the heat exchange led by the phase-changes of steam droplets were considered by the additional transport properties for steam.

2.1. Wet Steam Modeling

The wet steam flow was modeled by the Eulerian-Eulerian approach. Compressible Navier-Stokes equations were adopted for modeling of the condensation process of the two-phase flow. In addition, two more transport equations accounting for the liquid-phase mass-fraction (β) and the number of liquid-droplets per unit volume (η) were used to capture and describe the distribution of water droplets within the field [38]. The two-equation model is a viable approach which governs the proportion and particle numbers by resolving two conservative transporting properties, and it has been proven to be effective in applications such as turbulence and smoke particles modeling [36,39,40]. Based on the classical non-isothermal nucleation theory, the formation of liquid-droplets in a homogeneous nonequilibrium condensation process is described in the phase change model.

As for the non-equilibrium state in supersonic flows, the nucleation and condensation of condensable gases are overwhelmingly complicated [41,42]. Based on the single-fluid Eulerian model [41], several assumptions were implemented to ensure the simulation was numerically stable and efficient. Since the mass fraction of the condensed phase is quite small (<0.2), the effect of droplets size to the main flow stream is negligible and the interactions between droplets are not considered. Moreover, since the droplets sizes are sufficiently small, the volume of the condensed phase and thermal capacity are negligible [27]. A non-slip condition was also assumed, which means that the velocity difference between the droplets and gaseous-phase is negligible. Within the phase change model, the droplet was assumed to be spherical and surrounded by infinite vapor space due to the small liquid volume fraction. The droplet growth was based on average representative mean radii.

With the aforementioned assumptions, the Euler Equations [43] can be written as:

$$\frac{\partial U}{\partial t} + \frac{\partial F}{\partial x} + \frac{\partial G}{\partial y} = 0 \tag{2}$$

where x and y represent the special coordinates while t depicts the physical time. The unsteady term U, flux matrices F and G respectively in x-, y-directions can be expressed as:

$$U = \begin{bmatrix} \rho \\ \rho u \\ \rho v \\ \rho E \end{bmatrix}, F = \begin{bmatrix} \rho u \\ \rho u^2 + P \\ \rho u v \\ (\rho E + P)u \end{bmatrix}, G = \begin{bmatrix} \rho v \\ \rho u v \\ \rho v^2 + P \\ (\rho E + P)v \end{bmatrix} \tag{3}$$

The mixture density (ρ) is given by:

$$\rho = \frac{\rho_v}{(1 - \beta)} \tag{4}$$

where (ρ_v) is the vapor density and (β) is the liquid-phase mass fraction. And in Equation (3), the internal energy (E) and the specific enthalpy (h) are given by:

$$E = h_0 - \frac{P}{\rho} \tag{5}$$

$$h_0 = h + \frac{\left(u^2 + v^2\right)}{2} \tag{6}$$

$$h = (1 - \beta)h_v + h_l \tag{7}$$

In order to describe the distribution of nucleating droplets within the computational field, two transport properties were introduced and solved. The first being the mass fraction of fluid condensed liquid phase (β), where the transport equation can be expressed as:

$$\frac{\partial \rho \beta}{\partial t} + \nabla \cdot \left(\rho \vec{v} \beta\right) = \Gamma \tag{8}$$

The second additional transport equation governs the number density of the droplets per unit volume, which is given by:

$$\frac{\partial \rho \eta}{\partial t} + \nabla \cdot \left(\rho \vec{v} \eta\right) = \rho I \tag{9}$$

In Equation (8), Γ is the mass generation rate, related to the nucleation rate (I). It is defined by the sum of a mass increase due to nucleation and growth/demise of these droplets during the nonequilibrium condensation process in the classical nucleation theory [38]:

$$\Gamma = \frac{4}{3}\pi \rho_l I r_*^{\,3} + 4\pi \rho_l \eta \bar{r}^2 \frac{\partial \bar{r}}{\partial t} \tag{10}$$

where r_* is the Kelvin-Helmholtz critical droplet radius above which the droplet will grow and below which the droplet will evaporate [44] and it is defined as:

$$r_* = \frac{2\sigma}{\rho_l RT \ln S} \tag{11}$$

where σ is the liquid surface tension evaluated at temperature T, and S is the supersaturation ratio defined as the ratio of vapor pressure to the equilibrium saturation pressure:

$$S = \frac{P}{P_{sat}(T)} \tag{12}$$

In Equation (9), the number of liquid-droplets per unit volume (η) is given by:

$$\eta = \frac{\beta}{(1-\beta)V_d(\rho_l/\rho_v)} \tag{13}$$

where V_d is the average droplet volume, which is calculated by the average droplet radius ($\bar{r}_d$) using the volume formula of a circle.

Also in Equation (8), the nucleation rate I is described by the steady-state classical homogeneous nucleation theory [45]:

$$I = \frac{q_c}{1+\theta}\left(\frac{\rho_v^2}{\rho_l}\right)\sqrt{\frac{2\sigma}{M_m^3\pi}}\exp\left(-\frac{4\pi r_*^2\sigma}{3K_bT}\right) \tag{14}$$

where θ is a non-isothermal correction factor and is determined by the following relation:

$$\theta = \frac{2(\gamma-1)}{(\gamma+1)}\left(\frac{h_{lv}}{RT}\right)\left(\frac{h_{lv}}{RT}\quad 0.5\right) \tag{15}$$

Based on the nucleation model, the condensation process involves two mechanisms: (i) Mass transfer from gas phase to droplets and (ii) the transfer of heat from the droplets to the vapor in the form of latent heat [46]. This energy transfer relation can be written as:

$$\frac{\partial \bar{r}}{\partial t} = \frac{P}{h_{lv}\rho_l\sqrt{2\pi RT}}\frac{(\gamma+1)}{2\gamma}C_p(T_0-T) \tag{16}$$

2.2. Wet Steam Equation of State

In wet steam flow calculations, it is more convenient to relate steam properties using a simpler form of the thermodynamic state equations [45]. The equation of state is given by:

$$P = \rho_v RT\left(1 + B\rho_v + C\rho_v^2\right) \tag{17}$$

where B and C are the second and the third virial coefficients are given by the following empirical expressions:

$$B = a_1\left(1+\frac{\tau}{\alpha}\right)^{-1} + a_2 e^\tau (1-e^{-\tau})^{\frac{5}{2}} + a_3\tau \tag{18}$$

where $\tau = 1500/T$, $\alpha = 10{,}000.0$, $a_1 = 0.0015$, $a_2 = -0.000942$, and $a_3 = -0.0004882$.

$$C = a(\tau-\tau_0)e^{-\alpha\tau} + b \tag{19}$$

where $\tau = T/647.286$, $\tau_0 = 0.8978$, $\alpha = 11.16$, $a = 1.772$, and $b = 1.5 \times 10^{-6}$.

The vapor isobaric specific heat capacity (Cp_v), specific enthalpy (h_v), and specific entropy (s_v), which is applied for calculating the state of vapor, are also given below.

$$Cp_v = Cp_0(T) + R\left\{[(1 - \alpha_v T)(B - B_1) - B_2]\rho_v + \left[(1 - 2\alpha_v T)C + \alpha_v T C_1 - \frac{C_2}{2}\right]\rho_v^2\right\} \tag{20}$$

$$h_v = h_0(T) + RT\left[(B - B_1)\rho_v + \left(C - \frac{C_1}{2}\right)\rho_v^2\right] \tag{21}$$

$$s_v = s_0(T) - R\left[\ln \rho_v + (B + B_1)\rho_v + \frac{(C + C_1)}{2}\rho_v^2\right] \tag{22}$$

where Cp_0, h_0, s_0 are the standard state isobaric specific heat capacity, enthalpy, entropy respectively.

3. Modeling Setup and Boundary Conditions

The dimensions of the steam ejector geometry were based on the experiments conducted by Sriveerakul et al. [47], which are summarized in Table 1. A two-dimensional axisymmetric geometry was applied in this study to improve the simulation efficiency [48]. Structured quadrilateral meshes were applied as displayed in Figure 4, and some refinements were performed around the primary nozzle region due to the occurrence of condensation and other sharp-changing properties. This section may be divided by subheadings. It should provide a concise and precise description of the experimental results, their interpretation as well as the experimental conclusions that can be drawn. A mesh sensitivity analysis was carried out to investigate the independence of mesh refinement towards the computational results. Three mesh sizes were used to perform the analysis, with coarse, medium and fine meshes comprising of 8117, 77,676 and 175,522 grid elements respectively. Comparing the test cases, shown in Figure 5, the medium mesh showed a better result than the coarse mesh. Nevertheless, the deviation of results between the medium mesh and the fine mesh was within 4.2%, which is relatively small. Considering simulation efficiency, while retaining the accuracy of further refined mesh, the medium mesh was chosen for all simulations carried out in this study.

Table 1. Detailed dimensions for the inlet and outlet diameter sizes, lengths of the components for the steam ejector.

Geometry	Value
Diameter of Nozzle inlet	7.75 mm
Diameter of Nozzle outlet	8 mm
Diameter of Mixing chamber inlet	24 mm
Diameter of Mixing throat	19 mm
Length of Nozzle	60 mm
Length of Mixing chamber	130 mm
Length of Throat	95 mm
Length of Diffuse	180 mm
Expanded angle of nozzle	10°
Nozzle exit position	0 mm

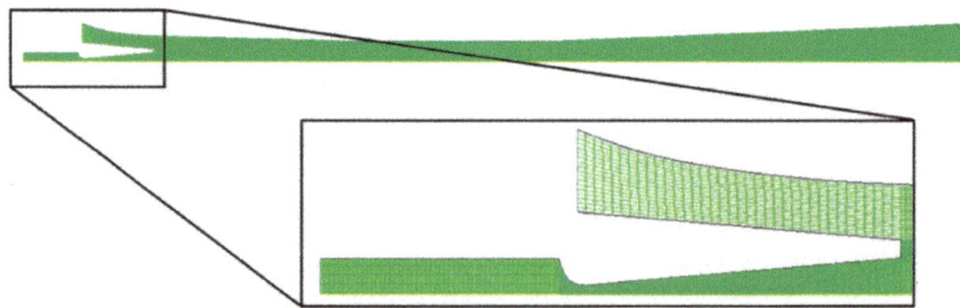

Figure 4. An enlarged view of the structural mesh system applied in the steam ejector fluid domain.

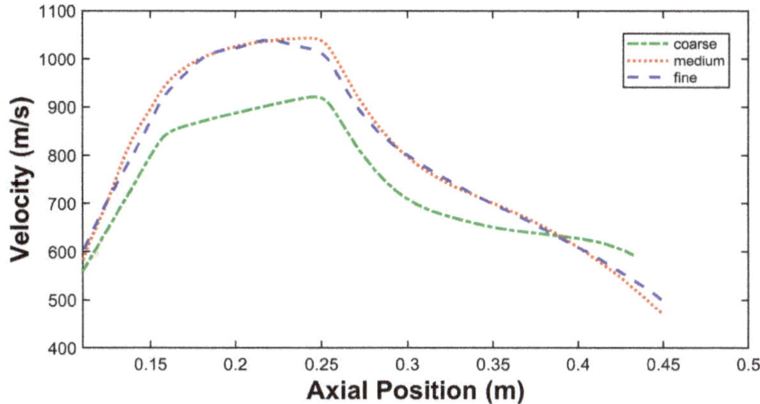

Figure 5. Comparison of velocity distribution along the axial position for different meshes.

The numerical simulation is performed by commercial CFD software package, ANSYS-Fluent with in-house attachment UDF. The inlet boundary conditions were summarized in Table 2. As the flow inside the nozzle region is transonic, the outlet pressure was then set to an arbitrary low number. The realizable k–ε turbulence model was adopted for the turbulence flow and enhanced wall functions were applied to provide a more accurate representation of the flow in the near-wall regions. The finite-volume approach was implemented for solving the governing equations. The discretized system was solved by the Green-Gauss Node based method and the convection terms were discretized with a second-order upwind scheme.

Table 2. Temperature and pressure configurations for the primary and secondary inlets.

Inlet Fluid	Temperature (K)	Pressure (Pa)
Primary fluid	393.15	200,000
Secondary fluid	283.15	1200

4. Validation of the Numerical Model

The static pressure distribution along the steam ejector upper wall was validated against the experimental measurements reported by Sriveerakul et al. [47]. As shown in Figure 6, the overall static pressure profile showed good agreement with the experimental data, and the position of the shockwave was also well aligned. Within the mixing chamber section (i.e., axial position from 0.06 m to 0.15 m), the predicted static pressure distribution was slightly lower than the experimental data. One possible reason for this error might be the over-predicted condensation effect in the simulation. Furthermore, the remaining static pressure distribution profile of the numerical result was over- predicted against experimental data. Considering deviations shown in Figure 6, there are limitations in the experimental data collection, and it is extremely difficult to obtain measurements in the steam ejector. In this study, the experimental data was obtained from the wall of the steam ejector, where the solution variables have large gradients and the momentum and other scalar transports occur most vigorously. Also, the steam of the primary flow was generated by the boiler, where the state is not the pure dry saturated water steam, and this has a negative effect on the experimental testing data. Additionally, before the primary flow passed through the nozzle, part of steam condensed inside the boiler, which makes the experimental pressure lower than the numerical simulation results. Apart from the deviations in static pressure at axial positions from 0.15–0.25 (which are clarified in the response above), the numerical predictions were in good agreement with experimental results. The trends between the experimental and numerical results are the same, which is that the pressure increased due to the shock wave happened at the throat. Also, the value of pressure was the same around 1200 Pa at the mixing chamber and eventually increased to 3000 Pa at the exit of diffuser. Therefore, the numerical results are validated against experimental results. Figure 7 illustrates the axial pressure distribution results from the wet steam model and ideal gas model against the experimental data. Overall, the numerical results from the wet steam model are in better agreement with the experimental data in comparison to the ideal gas case. With the consideration of the condensation effect in the wet steam model, there was a significant improvement in the pressure predictions within the mixing chamber (the axial position from 0.06 m to 0.15 m). Otherwise, the difference is small between the ideal gas and wet steam models for the rest of the profile.

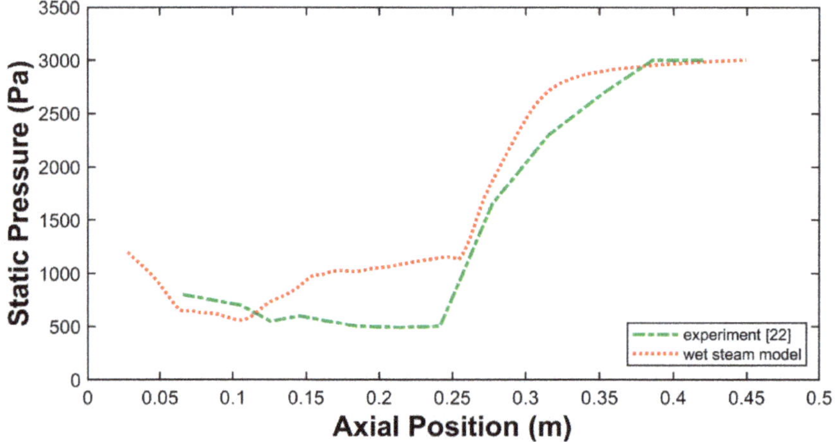

Figure 6. Comparison of static pressure distribution along the ejector wall between wet steam model and experimental data.

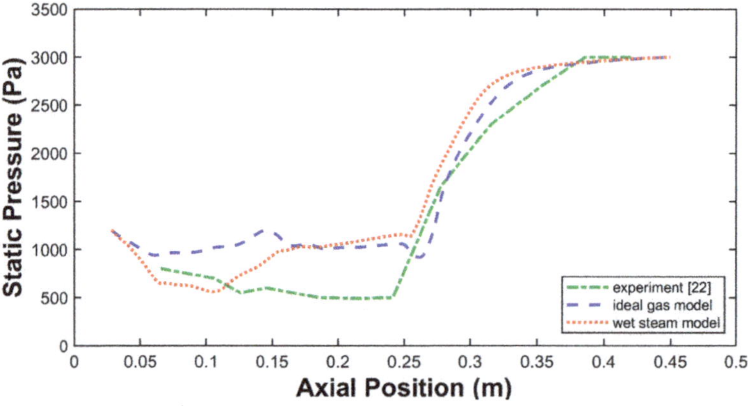

Figure 7. Static pressure distribution along ejector wall between ideal gas model and wet steam model against the experimental data.

5. Results and Discussion

5.1. Turbulence Model and Wall Functions for Steam Ejector Modeling

The flow state during the entire working process involves both high pressure and high-velocity flows and shockwaves. Therefore, a suitable turbulence model is essential for accurate numerical predictions. Figure 8 illustrates the static pressure results from three turbulence models against the experimental data. Three different turbulence models (i.e., standard $k-\varepsilon$, realizable $k-\varepsilon$ and the $k-\omega$ *SST* model) were applied in this study. As shown in Figure 8, the results of the realizable $k-\varepsilon$ model were the most accurate compared to experimental results. The only section in which the $k-\omega$ *SST* model showed slightly improved prediction over the realizable $k-\varepsilon$ model was during first half of the throat section (i.e., axial position from 0.15 m to 0.2 m). However, after the shock, the $k-\omega$ *SST* model shows a bigger deviation from experimental data than the $k-\varepsilon$ models. In addition to turbulence models, wall functions also need to be carefully considered because the turbulent flows near the wall region have large gradients, and the experimental data was collected from the wall of the steam ejector. Therefore, implementing wall functions should improve the accuracy of the wall-bounded turbulent flows. Figure 9 illustrates the axial pressure distribution for the realizable $k-\varepsilon$ model with different wall functions. As can be seen in Figure 9, the enhanced wall function showed a better performance compared to the non-equilibrium wall function. Taking into consideration the aforementioned results, the realizable $k-\varepsilon$ model with the enhanced wall functions was chosen to further investigate the condensation effect on the performance of the steam ejector.

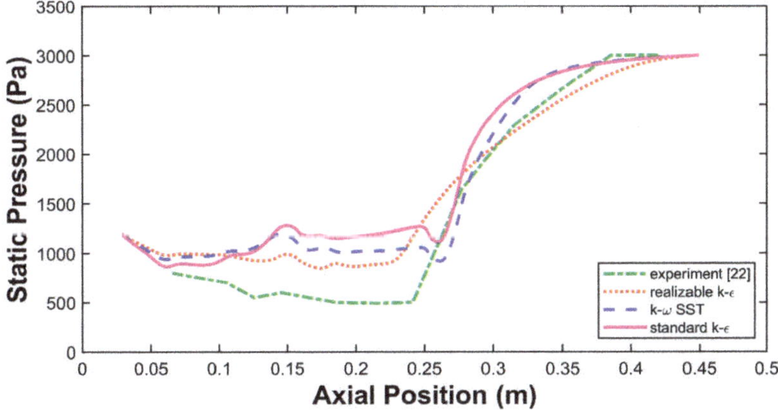

Figure 8. Axial static pressure distribution of different turbulence models against the experimental data.

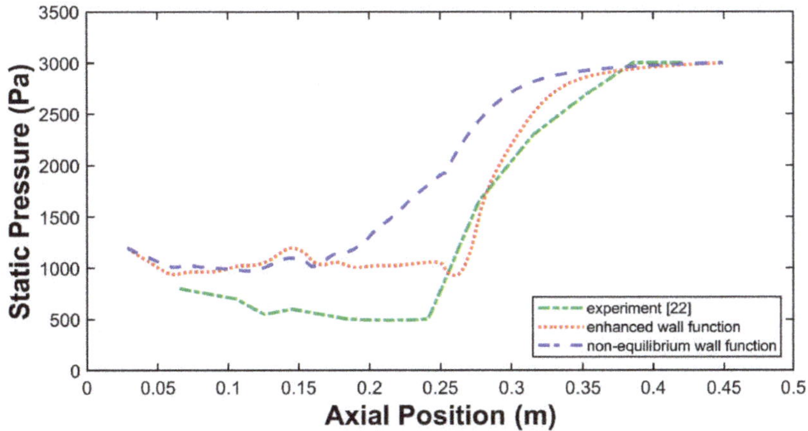

Figure 9. Static pressure distribution of different wall functions against the experimental data.

5.2. Wetness Influences of Inlets on Entrainment Ratio

Condensation occurs in the nozzle of the steam ejector and results in the formation of water droplets. These water droplets pass through the steam ejector in the primary flow and have significant influence on the mixing process and the flow behavior. The wet steam is a mixture of saturated vapor and fine liquid droplets, and the wetness is defined by the liquid mass fraction. In this article, four working conditions with different wetness settings on the primary and secondary flows were chosen to investigate the effects of wetness on the condensation effect. The case details are listed in Table 3.

Table 3. Different wetness settings for inlet flows.

Case Number	Wetness of Primary Flow (Inlet 1)	Wetness of Secondary Flow (Inlet 2)
Case 1	0	0
Case 2	0	0.1
Case 3	0.1	0
Case 4	0.1	0.1

Case 1 is a virtue case which tests a totally dry fluid stream, which means both the liquid mass fraction of primary and secondary flow are set as zero. *Case 4* is more like the actual scenario since in reality there is humidity and the steam should be somewhat wet. The liquid mass fraction was set to 0.1 on both inlet flows, which corresponds to the fact that most nozzle and turbine flows will have a wetness factor less than 0.1. In addition, large values of the wetness fraction lead to instability in the solution due to the large source terms in the transport equations.

The axial static temperature distribution for the 4 case studies are illustrated in Figure 10. *Case 1* and 2 shows a slightly different trend to *Case 3* and 4 during the nozzle section. It is proven that the wetness of primary flow has a more significant effect on the condensation in comparison to that of secondary flow. The gap in temperature distribution along the axis position between *Case 1 & 2* and *Case 3 & 4* is due to the heat realized from the condensation effect. From Figure 10, it is also concluded that: (i) For *Case 1*, since both the primary and secondary flow are totally dry, the condensation effect is the most significant and can be reflected from the figure as it has the highest temperature overall. (ii) On the contrary, *Case 4* has the least condensation effect since the incoming fluid has the greatest moisture content before it approaches the mixing chamber. (iii) Compared with *Case 4*, the temperature of *Case 2* in the nozzle part is higher since the condensation happened in the nozzle and some latent heat was released due to the condensing effect, which leads to a delay of the temperature drop. (iv) *Case 3* shows the same trend of *Case 4* during the nozzle part because the wetness of primary flow is the same. For the mixing chamber, with the different wetness of secondary flow, the condensation of secondary fluid for *Case 3* occurred and the process released heat to the mixed flow, leading to an increased temperature. (v) *Case 2* and 3 illustrates that the wetness of the primary flow has a more significant effect on the condensation in the whole working process of the steam ejector compared to the secondary flow.

From Figure 11, it is concluded that during the mixing process, the nucleation rate is high, and droplets were generated, which means that condensation is still happening in the mixing chamber. Compared to *Case 4*, *Case 3* shows a wider area of droplets nucleation, which leads to a higher entrainment ratio. Theoretically, reducing the wetness of secondary flow has a positive impact on improving the entrainment ratio and the performance of the steam ejector. With less wetness of the secondary flow, the nucleation can last from mixing chamber to throat, which increases the range of condensation effect and improves the entrainment ratio of the steam ejector. The pumping efficiency of the steam ejector was calculated based on the Equation (1) for all four cases, which was summarized in Figure 12. Compared to the ideal gas model, the entrainment ratio of *Case 1* and *3* was improved, due to the application of the wet steam. The condensation effect was taken into consideration during the pumping progress by using the wet steam model, and the primary flow entrained more secondary flow, which improves the pumping efficiency. It is also evident that reducing the wetness of the secondary fluid, the entrainment ratio is the highest to 0.47, which significantly improves the performance of the steam ejector.

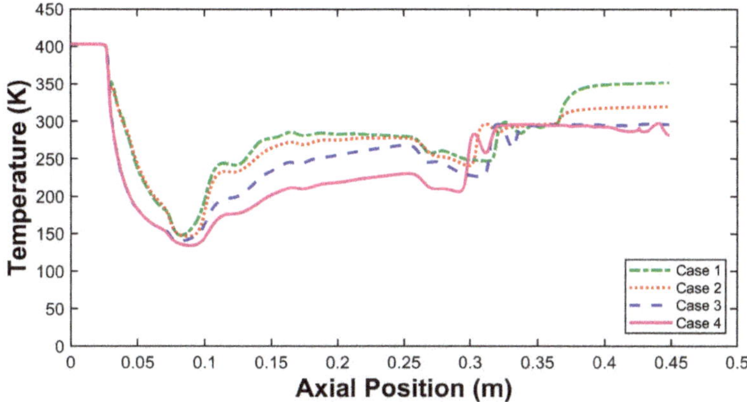

Figure 10. Axial static temperature distribution of different wetness of inlet flows based on the wet steam model.

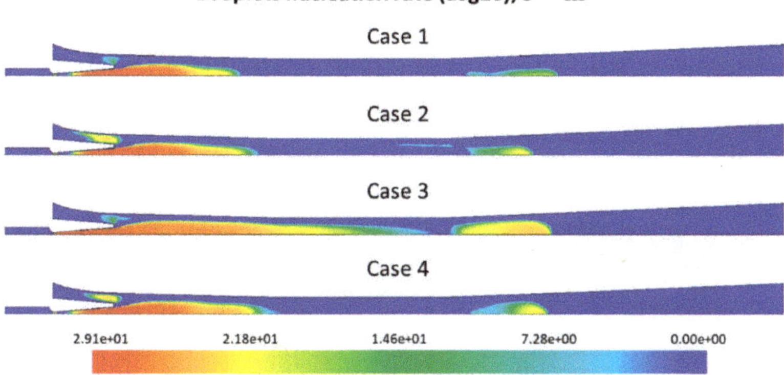

Figure 11. Contours of droplets nucleation rate with different wetness of inlet flows.

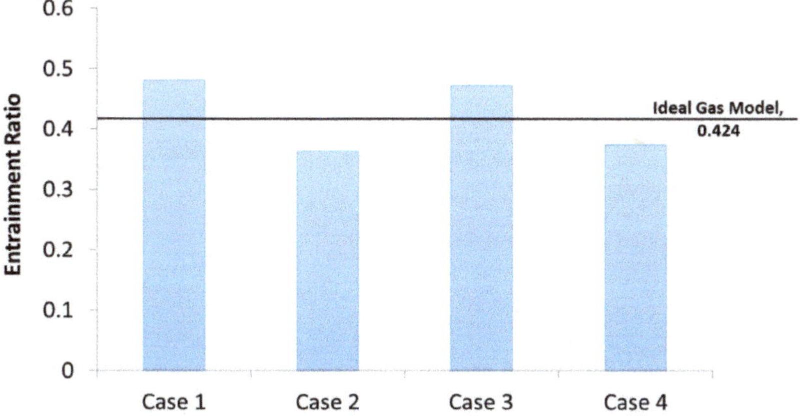

Figure 12. Entrainment Ratio of different wetness of inlet flows.

6. Conclusions

In this study, numerical simulations have been performed to study the entire steam ejector system. A wet steam model was implemented to describe the condensation and evaporation processes of the water droplets within the system. The numerical model was validated against experimental results which were found to agree well with the experimental data in terms of the static pressure distribution along the steam ejector. In addition, the model was able to provide an accurate prediction of the changes in water droplets, axial pressure, temperature, and pressure distribution within the entire system. The investigation into different wetness of inlet flows showed that in the whole working process, the wetness of secondary flow influences the pumping efficiency more than the primary flow. Within a certain range (from 0 to 0.1), the wetness of the secondary flow has an inverse proportion relationship with the entrainment ratio. The nucleation rate of water droplets is improved by the decrease of the wetness of the secondary flow at the front half of the mixing chamber, leading to better performance of the steam ejector. On the other hand, the wetness of the primary flow plays a domain role on the spontaneous condensation at the diffuser part. In summary, the following are the key findings from this study:

1. Based on the ideal gas model, the realizable $k-\varepsilon$ turbulence model with enhanced wall functions showed a better performance of simulating the working process of the steam ejector.
2. Compared with the ideal gas model, the wet steam model showed a better agreement against the experimental data, especially in the mixing chamber section with approximately 23% improvement.
3. The spontaneous condensation happened in the nozzle and during the mixing chamber, the condensation effect had an influence on the entrainment ratio.
4. By reducing the wetness of the secondary fluid or improving the wetness of the primary fluid, the condensation effect lasted longer and improved the entrainment ratio, which means the performance of the steam ejector is enhanced.

Author Contributions: Conceptualization, G.H.Y. and A.L.; methodology, A.L. and G.H.Y.; software, A.C.Y.Y. and T.B.Y.C.; data curation, A.L. and H.L.; formal analysis, A.C.Y.Y. and C.W.; numerical simulations, A.L. and A.C.Y.Y.; validation, T.B.Y.C. and C.W.; writing—original draft preparation, A.L. and A.C.Y.Y.; writing—review and editing, C.W. and W.Y.; visualization, R.C. and T.B.Y.C.; supervision, G.H.Y. and V.T.; project administration, A.L. and W.Y.; funding acquisition, G.H.Y.

Funding: This research was funded by the Australian Research Council (ARC Industrial Transformation Training Centre IC170,100,032) and the Australian Government Research Training Program. All financial and technical support are deeply appreciated by the authors.

Conflicts of Interest: The authors declare no conflict of interest.

Nomenclature

B, C	virial coefficients (m^3/kg, m^6/kg^2)
Cp, Cp_0	isobaric heat capacity, standard state heat capacity (J/kg·K)
E_m	entrainment ratio
h, h_0, h_l, h_v, h_{lv}	specific enthalpy, standard state enthalpy, liquid specific enthalpy, vapor specific enthalpy, specific enthalpy between phases (J/kg)
I	nucleation rate (1/s)
K_b	Boltzmann constant (J/mol·K)
k	turbulent kinetic energy (m^2/s^2)
M	molecular mass (kg)
P	pressure (Pa)
P_b	back pressure (Pa)
P_b^*	critical back pressure (Pa)

P_b^r	reversing back pressure (Pa)
P_{sat}	saturation pressure
q_c	evaporation coefficient
R	gas-law constant (J/mol·k)
$\bar{r}$	droplet average radius (m)
r_*	critical droplet radius (m)
S	super saturation ratio
s, s_0, s_v	specific entropy, standard state entropy, vapor specific entropy (J/kg·mol·K)
T	thermal temperature (K)
T_0	droplet temperature (K)
t	time (s)
u	velocity (m/s)
V_d	average droplet volume (m^3)
α_v	volume fraction
β	mass fraction
γ	specific heat capacities ratio
Γ	mass generation rate (kg/s)
ε	turbulent energy dissipation rate (m^2/s^3)
η	droplet number density (1/m^3)
θ	non-isothermal correction factor
μ, μ_t	dynamic viscosity, turbulent viscosity (Pa·s)
ν	kinetic viscosity (m^2/s),
ρ, ρ_l, ρ_v	mixture density, liquid density, vapor density (kg/m^3)
σ	droplet surface tension (N/m)
E	internal energy

References

1. Riffat, S.; Jiang, L.; Gan, G. Recent development in ejector technology—A review. *Int. J. Ambient Energy* **2005**, *26*, 13–26. [CrossRef]
2. Lucas, C.; Koehler, J. Experimental investigation of the COP improvement of a refrigeration cycle by use of an ejector. *Int. J. Refrig.* **2012**, *35*, 1595–1603. [CrossRef]
3. Dopazo, J.A.; Fernandez-Seara, J. Experimental evaluation of an ejector as liquid re-circulator in an overfeed NH3 system with a plate evaporator. *Int. J. Refrig.* **2011**, *34*, 1676–1683. [CrossRef]
4. Han, B.; Liu, Z.L.; Wu, H.Q.; Li, Y.X. Experimental study on a new method for improving the performance of thermal vapor compressors for multi-effect distillation desalination systems. *Desalination* **2014**, *344*, 391–395. [CrossRef]
5. Wang, X.; Dong, J.; Lei, H.; Li, A.; Wang, R.; Chen, W.; Tu, J. Modeling and simulation of steam-jet vacuum pump. *Zhenkong Kexue yu Jishu Xuebao/J. Vac. Sci. Technol.* **2013**, *33*, 1069–1073.
6. Yinshui, L.; Zhuo, J.; Dan, W.; Xiaohui, L. Experimental research on the water mist fire suppression performance in an enclosed space by changing the characteristics of nozzles. *Exp. Thermal Fluid Sci.* **2014**, *52*, 174–181. [CrossRef]
7. Yang, P.Z.; Liu, T.; Qin, X.A. Experimental and numerical study on water mist suppression system on room fire. *Build Environ.* **2010**, *45*, 2309–2316. [CrossRef]
8. Beji, T.; Zadeh, S.E.; Maragkos, G.; Merci, B. Influence of the particle injection rate, droplet size distribution and volume flux angular distribution on the results and computational time of water spray CFD simulations. *Fire Saf. J.* **2017**, *91*, 586–595. [CrossRef]
9. Chunnanond, K.; Aphornratana, S. An experimental investigation of a steam ejector refrigerator: The analysis of the pressure profile along the ejector. *Appl. Therm. Eng.* **2004**, *24*, 311–322. [CrossRef]
10. Huang, B.J.; Jiang, C.B.; Hu, F.L. Ejector Performance-Characteristics and Design Analysis of Jet Refrigeration System. *J. Eng. Gas Turbines Power* **1985**, *107*, 792–802. [CrossRef]
11. Huang, B.J.; Chang, J.M.; Wang, C.P.; Petrenko, V.A. A 1-D analysis of ejector performance. *Int. J. Refrig.* **1999**, *22*, 354–364. [CrossRef]

12. Sankarlal, T.; Mani, A. Experimental investigations on ejector refrigeration system with ammonia. *Renew. Energy* **2007**, *32*, 1403–1413. [CrossRef]

13. Aidoun, Z.; Ouzzane, A. The effect of operating conditions on the performance of a supersonic ejector for refrigeration. *Int. J. Refrig.* **2004**, *27*, 974–984. [CrossRef]

14. Wang, X.D.; Dong, J.L. Numerical study on the performances of steam-jet vacuum pump at different operating conditions. *Vacuum* **2010**, *84*, 1341–1346. [CrossRef]

15. Varga, S.; Oliveira, A.C.; Diaconu, B. Influence of geometrical factors on steam ejector performance—A numerical assessment. *Int. J. Refrig.* **2009**, *32*, 1694–1701. [CrossRef]

16. Varga, S.; Oliveira, A.C.; Diaconu, B. Numerical assessment of steam ejector efficiencies using CFD. *Int. J. Refrig.* **2009**, *32*, 1203–1211. [CrossRef]

17. Varga, S.; Oliveira, A.C.; Ma, X.L.; Omer, S.A.; Zhang, W.; Riffat, S.B. Experimental and numerical analysis of a variable area ratio steam ejector. *Int. J. Refrig.* **2011**, *34*, 1668–1675. [CrossRef]

18. Ma, X.L.; Zhang, W.; Omer, S.A.; Riffat, S.B. Experimental investigation of a novel steam ejector refrigerator suitable for solar energy applications. *Appl. Therm. Eng.* **2010**, *30*, 1320–1325. [CrossRef]

19. Yadav, R.L.; Patwardhan, A.W. Design aspects of ejectors: Effects of suction chamber geometry. *Chem. Eng. Sci.* **2008**, *63*, 3886–3897. [CrossRef]

20. Riffat, S.B.; Omer, S.A. CFD modelling and experimental investigation of an ejector refrigeration system using methanol as the working fluid. *Int. J. Energy Res.* **2001**, *25*, 115–128. [CrossRef]

21. Bartosiewicz, Y.; Aidoun, Z.; Desevaux, P.; Mercadier, Y. Numerical and experimental investigations on supersonic ejectors. *Int. J. Heat Fluid Flow* **2005**, *26*, 56–70. [CrossRef]

22. Sun, D.W. Variable geometry ejectors and their applications in ejector refrigeration systems. *Energy* **1996**, *21*, 919–929. [CrossRef]

23. Cizungu, K.; Groll, M.; Ling, Z.G. Modelling and optimization of two-phase ejectors for cooling systems. *Appl. Therm. Eng.* **2005**, *25*, 1979–1994. [CrossRef]

24. Chaiwongsa, P.; Wongwises, S. Experimental study on R-134a refrigeration system using a two-phase ejector as an expansion device. *Appl. Therm. Eng.* **2008**, *28*, 467–477. [CrossRef]

25. Ruangtrakoon, N.; Aphornratana, S.; Sriveerakul, T. Experimental studies of a steam jet refrigeration cycle: Effect of the primary nozzle geometries to system performance. *Exp. Thermal Fluid Sci.* **2011**, *35*, 676–683. [CrossRef]

26. Ruangtrakoon, N.; Thongtip, T.; Aphornratana, S.; Sriveerakul, T. CFD simulation on the effect of primary nozzle geometries for a steam ejector in refrigeration cycle. *Int. J. Therm. Sci.* **2013**, *63*, 133–145. [CrossRef]

27. Gerber, A.; Kermani, M. A pressure based Eulerian–Eulerian multi-phase model for non-equilibrium condensation in transonic steam flow. *Int. J. Heat Mass Transf.* **2004**, *47*, 2217–2231. [CrossRef]

28. Yuen, A.C.Y.; Yeoh, G.H.; Timchenko, V.; Chen, T.B.Y.; Chan, Q.N.; Wang, C.; Li, D.D. Comparison of detailed soot formation models for sooty and non-sooty flames in an under-ventilated ISO room. *Int. J. Heat Mass Transf.* **2017**, *115*, 717–729. [CrossRef]

29. Wang, X.D.; Dong, J.L.; Wang, T.; Tu, J.Y. Numerical analysis of spontaneously condensing phenomena in nozzle of steam-jet vacuum pump. *Vacuum* **2012**, *86*, 861–866. [CrossRef]

30. Munday, J.T.; Bagster, D.F. A new ejector theory applied to steam jet refrigeration. *Ind. Eng. Chem. Process Des. Dev.* **1977**, *16*, 442–449. [CrossRef]

31. Yang, Y.; Walther, J.H.; Yan, Y.Y.; Wen, C. CFD modeling of condensation process of water vapor in supersonic flows. *Appl. Therm. Eng.* **2017**, *115*, 1357–1362. [CrossRef]

32. Mazzelli, F.; Giacomelli, F.; Milazzo, A. CFD modeling of condensing steam ejectors: Comparison with an experimental test-case. *Int. J. Therm. Sci.* **2018**, *127*, 7–18. [CrossRef]

33. Wang, X.D.; Lei, H.J.; Dong, J.L.; Tu, J.Y. The spontaneously condensing phenomena in a steam-jet pump and its influence on the numerical simulation accuracy. *Int. J. Heat Mass Transf.* **2012**, *55*, 4682–4687. [CrossRef]

34. Sharifi, N.; Boroomand, M.; Sharifi, M. Numerical assessment of steam nucleation on thermodynamic performance of steam ejectors. *Appl. Therm. Eng.* **2013**, *52*, 449–459. [CrossRef]

35. Wang, X.D.; Dong, J.L.; Li, A.; Lei, H.J.; Tu, J.Y. Numerical study of primary steam superheating effects on steam ejector flow and its pumping performance. *Energy* **2014**, *78*, 205–211. [CrossRef]

36. Wang, X.D.; Dong, J.L.; Zhang, G.L.; Fu, Q.; Li, H.; Han, Y.; Tu, J.Y. The primary pseudo-shock pattern of steam ejector and its influence on pumping efficiency based on CFD approach. *Energy* **2019**, *167*, 224–234. [CrossRef]

37. Bakhtar, F.; Tochai, M.M. An investigation of two-dimensional flows of nucleating and wet steam by the time-marching method. *Int. J. Heat Fluid Flow* **1980**, *2*, 5–18. [CrossRef]

38. Ishizaka, K. A high-resolution numerical method for transonic non-equilibrium condensation flow through a steam turbine cascade. *Proc. of the 6th ISCFD* **1995**, *1*, 479–484.

39. Yuen, A.C.Y.; Yeoh, G.H.; Timchenko, V.; Cheung, S.C.P.; Barber, T.J. Importance of detailed chemical kinetics on combustion and soot modelling of ventilated and under-ventilated fires in compartment. *Int. J. Heat Mass Transf.* **2016**, *96*, 171–188. [CrossRef]

40. Chen, T.B.Y.; Yuen, A.C.Y.; Yeoh, G.H.; Timchenko, V.; Cheung, S.C.P.; Chan, Q.N.; Yang, W.; Lu, H. Numerical study of fire spread using the level-set method with large eddy simulation incorporating detailed chemical kinetics gas-phase combustion model. *J. Comput. Sci.* **2018**, *24*, 8–23. [CrossRef]

41. Yang, Y.; Zhu, X.; Yan, Y.; Ding, H.; Wen, C. Performance of supersonic steam ejectors considering the nonequilibrium condensation phenomenon for efficient energy utilisation. *Appl. Energy* **2019**, *242*, 157–167. [CrossRef]

42. Wen, C.; Karvounis, N.; Walther, J.H.; Yan, Y.; Feng, Y.; Yang, Y. An efficient approach to separate CO_2 using supersonic flows for carbon capture and storage. *Appl. Energy* **2019**, *238*, 311–319. [CrossRef]

43. Yang, Y.; Shen, S.Q. Numerical simulation on non-equilibrium spontaneous condensation in supersonic steam flow. *Int. Commun. Heat Mass Transf.* **2009**, *36*, 902–907. [CrossRef]

44. Young, J. Two-dimensional, nonequilibrium, wet-steam calculations for nozzles and turbine cascades. *J. Turbomach.* **1992**, *114*, 569–579. [CrossRef]

45. Young, J.B. An Equation of State for Steam for Turbomachinery and Other Flow Calculations. *J. Eng. Gas Turbines Power* **1988**, *110*, 1–7. [CrossRef]

46. Young, J. Spontaneous condensation of steam in supersonic nozzles. *Physicochem. Hydrodyn.* **1982**, *3*, 57–82.

47. Sriveerakul, T.; Aphornratana, S.; Chunnanond, K. Performance prediction of steam ejector using computational fluid dynamics: Part 1. Validation of the CFD results. *Int. J. Therm. Sci.* **2007**, *46*, 812–822. [CrossRef]

48. Pianthong, K.; SeehanaM, W.; Behnia, M.; SriveerakUl, T.; Aphornratana, S. Investigation and improvement of ejector refrigeration system using computational fluid dynamics technique. *Energy Convers. Manag.* **2007**, *48*, 2556–2564. [CrossRef]

Article

Influence of Eddy-Generation Mechanism on the Characteristic of On-Source Fire Whirl

Cheng Wang [1], Anthony Chun Yin Yuen [1,*], Qing Nian Chan [1], Timothy Bo Yuan Chen [1], Qian Chen [1], Ruifeng Cao [1], Ho Lung Yip [1], Sanghoon Kook [1] and Guan Heng Yeoh [1,2]

[1] School of Mechanical and Manufacturing Engineering, University of New South Wales, Sydney, NSW 2052, Australia; c.wang@unsw.edu.au (C.W.); qing.chan@unsw.edu.au (Q.N.C.); timothy.chen@unsw.edu.au (T.B.Y.C.); enzo.chen@student.unsw.edu.au (Q.C.); ruifeng.cao@unsw.edu.au (R.C.); h.l.yip@unsw.edu.au (H.L.Y.); s.kook@unsw.edu.au (S.K.); g.yeoh@unsw.edu.au (G.H.Y.)
[2] Australian Nuclear Science and Technology Organisation (ANSTO), Locked Bag 2001, Kirrawee DC, NSW 2232, Australia
* Correspondence: c.y.yuen@unsw.edu.au; Tel.: +61-2-9385-5697

Received: 9 August 2019; Accepted: 13 September 2019; Published: 24 September 2019

Abstract: This paper numerically examines the characterisation of fire whirl formulated under various entrainment conditions in an enclosed configuration. The numerical framework, integrating large eddy simulation and detailed chemistry, is constructed to assess the whirling flame behaviours. The proposed model constraints the convoluted coupling effects, e.g., the interrelation between combustion, flow dynamics and radiative feedback, thus focuses on assessing the impact on flame structure and flow behaviour solely attribute to the eddy-generation mechanisms. The baseline model is validated well against the experimental data. The data of the comparison case, with the introduction of additional flow channelling slit, is subsequently generated for comparison. The result suggests that, with the intensified circulation, the generated fire whirl increased by 9.42% in peak flame temperature, 84.38% in visible flame height, 6.81% in axial velocity, and 46.14% in velocity dominant region. The fire whirl core radius of the comparison case was well constrained within all monitored heights, whereas that of the baseline tended to disperse at 0.5 m height-above-burner. This study demonstrates that amplified eddy generation via the additional flow channelling slit enhances the mixing of all reactant species and intensifies the combustion process, resulting in an elongated and converging whirling core of the reacting flow.

Keywords: fire whirl; computational fluid dynamics; eddy-generation mechanism; combustion modelling; detailed chemistry; large eddy simulation

---

1. Introduction

Fire whirl is a unique swirling diffusion flame that often occurs in urban and wildland fires with disastrous effects [1]. The whirling combustion flow limits the radial flame dispersion and stretches the hot plume to rise upon an elevated vertical angular path, and as a result significantly increases the burning rate, flame height, and flame temperature and radiation flux to the surroundings [2]. Owing to the unique features associated with this particular combustion dynamic behaviour, the formation of the fire whirl in a fire scenario can be devastated. For instance, the intense heat energy contained within the elongated fire whirl core combined with the radiative heat transfer to the surrounding, promote the fire propagation to the neighbouring points. The swirling motion associated with the fire whirl could also intensify the combustion process by enhancing the mixing of the combustible gas mixture with the oxidiser, while at the same time making the unsteady movement of the hot plume region even more spatially unpredictable [3]. The presence of such phenomena associated with fire whirl has been observed and reported in many notorious fire cases [4,5], and it has been identified as the core factor in intensifying the combustion process and making the scenario unmanageable and untenable. For this very reason, fire whirls, in terms of their formation mechanisms, characteristic features, and physical behaviours, are of great interest to both the research and industrial sectors, and hence have been identified as the topic of interest for this study.

Previous experimental and theoretical studies have characterised and identified the three most critical criteria for the formation of a fire whirl, namely, a thermally driven fluid sink, a surface drag force to create a radial boundary layer, which facilitates air entrainment to the generated vortex column, and an eddy-generation mechanism [6–8]. In a typical buoyancy-driven diffusion flame, the flame can be considered as the fluid sink and the generated hot plume naturally drives horizontal flow radically towards the vortex column [9,10], which fulfils the first two requirements. Therefore, the most critical element in fire whirl formation is the presence of a vortex generation mechanism. The generation of vortex could occur in a wildland fire situation, induced, for example, by topological obstacles, leeward slope, unpredicted weather conditions, etc. [11–13]. The formation of such circulation can also be observed in a compartment fire scenario [14–16], particular with the current trend of needing to construct high-density, high-rise buildings with complex geometric configurations. A typical example of a high-risk region for fire whirl formation is an enclosed space with entrainment from the flow channelling slit to introduce the circulation, for example, atrium, lift pit, spiral staircase etc., which are commonly featured in modern constructions. Despite the recently developed bio-based flame-retardant materials that effectively restrain the fire from spreading [17–20], the occurrence of the spinning flaming flow in skyscraper fires, such as the recent incidents in Beijing Television Cultural Center, Plasco Building, Grenfell Tower, Marina Torch, FR Tower, etc., could aggressively promote an alarming fire propagation within the confined space and cause catastrophic damages [21]. This poses a severe risk to the occupants and makes the fire suppression process extremely challenging [22–24]. Therefore, the characterisation of fire whirl formulated for enclosed configurations was targeted as the investigation case.

Fire whirl formulated in enclosed configurations has been intensively studied, both experimentally and numerically [7,25–31]. The literature has revealed an entangled coupling between the combustion process and the dynamic flow, i.e., the temperature, buoyancy, vortex, the combustion reaction, etc., are interrelated and interact during the formation of the fire whirl. Such interrelation between the involved physical and chemical aspects makes the understanding of the not-well-understood fundamentals of the swirling reacting flow more unfathomable. For instance, the existence of the whirling flame within the enclosure alters the heat feedback from the flame sheet to the fuel surface, and results in an unevenly distributed heating profile over the fuel surface. In a pool fire configuration, such ununiformed heat feedback could affect the burning rate as it influences the evaporation of combustible gas mixture converted from liquid-based parent fuel, and hence varies the ensuing combustion process and heat release rate [32–35]. Similarly, the residence of the whirling flame escalates the circulation within the enclosure, enhances the mixing of the oxidiser and reactant, and promotes a more completed combustion process. Theoretically speaking, the formation of soot species, the fine particles resulting from incomplete combustion, and the coupled radiative heat transfer to and from the surrounding, should be suppressed, when compared with non-swirling freestanding flame [36–39]. Nevertheless, the aggravation of the combustion process increases the flame temperature, which often excesses the threshold of soot nucleation, therefore aid the inception of in-flame generated soot species [40,41]. The excess generated soot species could arguably intensify the radiative heat transfer to the fuel surface and further boost the heat release rate, which contradicts with the previous speculation. In addition to the intricate physical and chemical coupling process, the uncertainty caused by the numerical modelling configuration could make the characterisation of fire whirl more cumbersome. For example, the combustion region of the fire whirl is likely to swirl around the domain which may involve tilting, stretching, converging etc. in a highly unpredicted manner. If the numerical domain is discretised non-uniformly, the variation in spatial resolution of the discretised region may significantly affect the evaluated result, particularly for the prediction of the sensitive flame temperature [42].

The ambiguous correlation, involving various aspects, has nevertheless not been decoupled or restrained in many of the previous studies. Strategies, e.g., heavy sooty flame, pool-based flame generation mechanism where the burning rate depends on the evaporation rate, locally refined meshing algorithm with a denser grid at the domain centre, where a freestanding fire is expected to occur and coarse mesh at the fringes, etc., have commonly been adopted in previous studies, both experimental and numerical. Those implementations make quantitative analysis of the fire whirl behaviour unattainable. It is, therefore, critical to construct a numerical framework with the isolation and consideration of the preceding coupling parties, and focus on establishing the correlation between the eddy-generation mechanism and the characteristic of the formulated fire whirl.

In light of the abovementioned gap, the current study aims to characterise fire whirl formulated under various entrainment conditions via a numerical approach. The proposed numerical model, elaborated in more detail in the next section, is deliberately designed to isolate and restrain the coupling effect between some primary physical, chemical and numerical factors. The proposed model adopts soot-free and clean-burning alcohol-based fuel with a set injection speed, and aims to eliminate the soot radiative heat transfer to and from the surrounding, as well as the potential variation in the burning rate. In addition, the numerical domain of the entire enclosure, where both combustion and dynamic flow motion are expected to occur, is discretised with uniformed grid size to minimise the impact of the spatial resolution of computational control volume on the numerical result. The baseline case with one flow channelling slit is built and validated against experimental data. A modified case with the introduction of an additional entrainment slit is also constructed. The results generated by the two are compared and assessed.

2. Numerical Details

A numerical domain of the baseline model replicating the test rig of the previous experimental study [43], as well as that with an additional flow channelling slit, was constructed accordingly. The geometric features of the two models are shown in Figure 1. The fuel pan is converted from circular to square configuration with the same cross-section area as in the experimental setup to achieve a fully structured mesh, as this significantly improves the numerical accuracy and computational efficiency.

Despite the inevitable "flame wander" phenomenon, the domain is discretised to about 800,000 elements based on a uniform division algorithm, to limit the numerical uncertainty associated with the spatial resolution of the discretised control volume. A doubling of the number of elements to 1,600,000 resulted in only a 5% difference in the centreline temperature, thus achieving a mesh-independent solution.

The fluid flow and heat transfer within the compartment is described through the conservation equations of continuity, Navier-Stokes, and scalar quantities:

$$\frac{\partial \overline{\rho}}{\partial t} + \frac{\partial}{\partial x_i} \overline{\rho} \widetilde{U}_i = 0 \tag{1}$$

$$\frac{\partial}{\partial} \overline{\rho} \widetilde{U}_i + \frac{\partial}{\partial} \overline{\rho} \widetilde{U}_i \widetilde{U}_j + \frac{\partial}{\partial} \overline{\rho} \widetilde{u_i u_j} = -\frac{\partial \overline{p}}{\partial x_i} + (v + v_t) \frac{\partial^2 \overline{\rho} \widetilde{U}_i}{\partial x_j \partial x_j} \tag{2}$$

$$\frac{\partial}{\partial} \overline{\rho} \widetilde{\Phi}_\alpha + \frac{\partial}{\partial} \overline{\rho} \widetilde{\Phi}_\alpha \widetilde{U}_i + \frac{\partial}{\partial} \overline{\rho} \widetilde{\phi_\alpha u_i} = (\Gamma + \Gamma_t) \frac{\partial^2 \overline{\rho} \widetilde{\Phi}_\alpha}{\partial x_i \partial x_i} + \widetilde{S(\Phi)} \tag{3}$$

where $\overline{\rho}$ and $\overline{p}$ are the mean density and pressure, v and v_t are the kinematic and turbulent viscosity, $\widetilde{U}_i$ and u_i are the Favre-averaged and fluctuation value of the mean fluid velocity, $\widetilde{\Phi}_\alpha$ and ϕ_α are the mean and fluctuation value of the αth scalar, Γ and Γ_t are molecular and turbulent diffusivity, and $\widetilde{S(\Phi)}$ is the Favre-averaged chemical reaction source term.

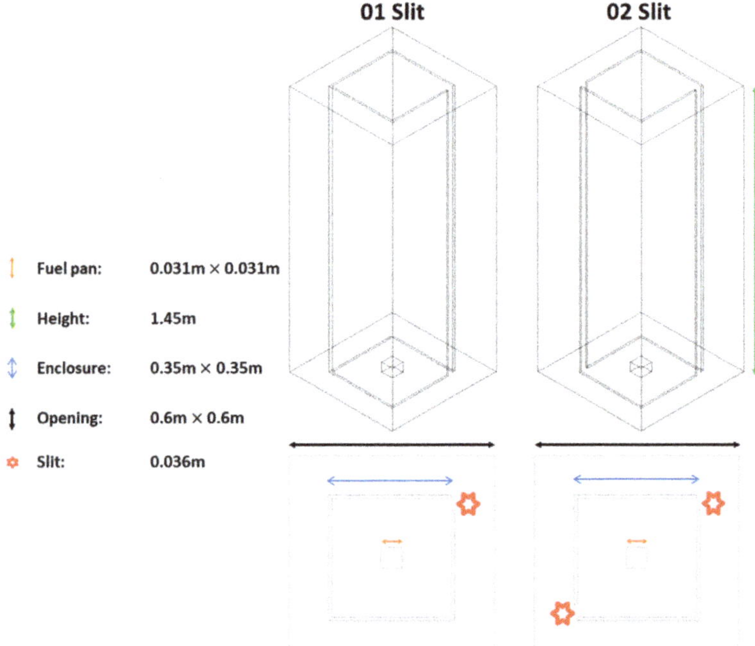

Figure 1. Geometric features of the computational domain.

Large eddy simulation (LES) with Wall-Adapting Local Eddy-viscosity (WALE) function, which accounts for both the strain and rotation rate of the turbulent structure is employed as the turbulence model to describe the turbulent reaction flow behaviour. This turbulence model has been extensively validated and proved to be a valid approach for resolving various wall-bounded turbulence flow applications [8,39,44–48].

The chemical reaction source term in the transport equations of the involved reacting scalars is evaluated via the strained laminar flamelet approach. The involved chemical phenomenon is described as the pre-assumed probability density function (pre-PDF) of two quantities, namely the mixture fraction (f) and the scalar dissipation (χ). Essentially, the mixture fraction governs the proportion of the fuel mixture in each control volume element in the discretised domain, and the first two moments of the mixture fraction, namely the mean mixture fraction, $\bar{f}$, and the mixture fraction variance, $\bar{g}$, are evaluated by resolving their conservation equations, respectively. Scalar dissipation is introduced to describe the strain and extinction of the flame, in which larger values of this quantity depict its departure from chemical equilibrium [49]. The scalar dissipation across the flamelet is determined based on the mean scalar dissipation of the flamelets embedded in turbulent flames at the position where $f = f_{st}$, as well as the ratio of the function of the mixture fraction and density at the point of interest and at the position where f is stoichiometric. A detailed GRI-MECH 3.0 chemical reaction mechanism, with the inclusion of 325 reaction steps and 53 chemical species [50], is implemented to establish the flamelet library of the strained laminar flamelet model considering parental fuel. This approach has been validated in previous numerical studies for modelling turbulence chemistry interaction, and has been proved to provide a reasonable result with moderate computational consumption [44,51]. It should be highlighted that methanol (CH_3OH), an alcohol-based fuel, was deliberately selected as the parent fuel to minimise the coupling between radiative energy feedback and the combustion process, due to its feature of clean and soot-free burning behaviour [41]. For the same reason, the concentration of the key intermediate species of soot formation, acetylene (C_2H_2), in the gas mixture produced by this very parent fuel, is proved to be negligible. The most commonly adopted soot models formulated on C_2H_2 precursor-based inception and surface growth mechanism are not applicable in this study [52–55]. As a result, a primitive and computational lightweight two-step soot model is integrated, for concept verification purpose only.

With respect to the boundary conditions, a set flow rate of 0.0216 ms^{-1} in the direction normal to the fuel surface is applied on the fuel pan. The applied injection velocity is determined on the basis of the cross-sectional area of the fire pan, the density of the parent fuel at reference temperature, and the heat of combustion of the parent fuel in order to match the targeted heat release rate reported in the experimental study. The constant injection rate ensures an evenly distributed burning profile, regardless of the intensity of flow circulation and energy feedback. The fuel surface is set at an elevated height according to the experimental setup. The top and the periphery of the domain is set as the opening to allow naturally convected air entrainment in and out of the system. The base of the domain is set as a non-slip adiabatic wall, across which no heat or matter is allowed to pass. The simulation is initiated with a standard temperature and pressure replicating the ambient conditions for the combustion process to proceed. For convenience, the case with one side entrainment slit is referred to as the 01 slit case, case 1, or the baseline case/model, and that with two side entrainment slits is referred as the 02 slit case, case 2, or the comparison case/model, in the following sections.

3. Results

The numerical result generated via the baseline case is firstly validated against the experimental data [43], and the characterisation of the fire whirl evaluated via the two cases is subsequently assessed and discussed in the following section.

3.1. Validation

An accurate description of the fire whirl's inner structure is intimately tied with a proper description of the temperature and the velocity field of the swirling reacting flow. For this reason, the temperature and velocity profile generated by the baseline model is used to compare with the that reported in the literature for validation purpose.

Due to the constraints related to the physical set up and experimental design, limited information is provided in prior studies. For clarity, herein, temperature measurement at domain centreline at various HABs is selected for comparison. Furthermore, the incoming y-velocity profile through the air gap is also validated against a prior numerical investigation of the similar experimental setup [8]. The result of the validation procedure is presented in Figures 2 and 3.

It can be seen from Figure 2 that the centreline temperature peaks at about HAB 0.15 m and sharply decreases to ambient as the combustion process develops towards downstream. Similarly, high incoming velocity at the air gap is observed in the low-intermediate region, and its quantity gradually drops with the increase of flame height. Despite the fact that the flame temperature in the intermittent region is slightly overestimated, potentially attributed to the absence of radiation correction, as well as the inevitable flame flickering, the comparison shows good agreement between the numerical simulation and experimental data. As presented in Figure 3, a similar agreement could be reached in the comparison of incoming y-velocity through the air gap with the prior study.

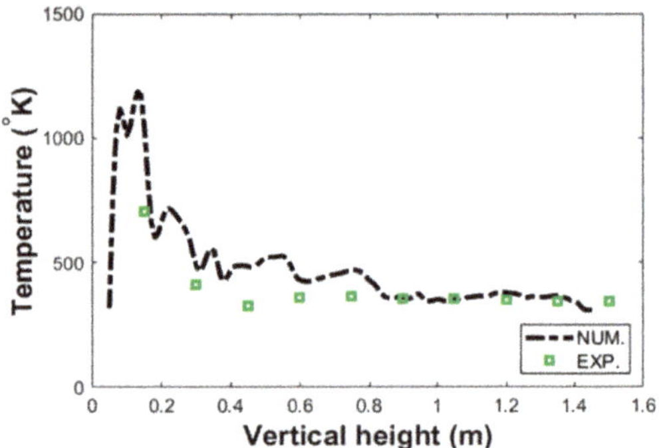

Figure 2. Numerical prediction of temperature at domain centreline, compared with experimental measurements.

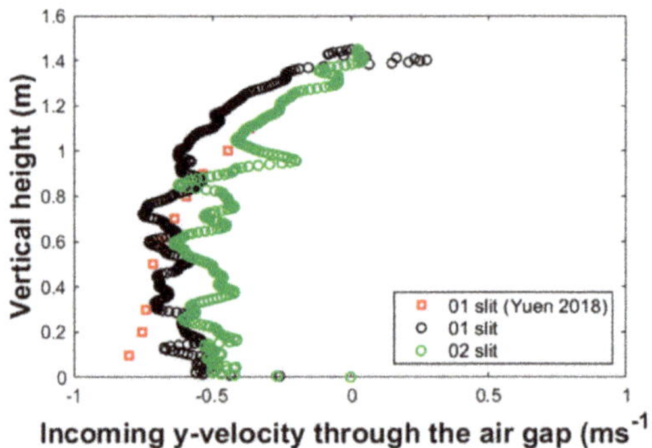

Figure 3. Numerical prediction of incoming y-velocity through the air gap, compared with that generated from prior study.

In addition, the soot volume fraction evaluated from both cases is proved to be negligible, i.e., on an order of magnitude of −21 and −18 in the baseline and comparison cases, respectively. It was further ascertained that with the adoption of the alcohol-based parent fuel, the in-flame soot formation and the ensuring radiation feedback can be safely neglected in this study.

With the numerical model of the baseline case validated, the model of the two slits cases is constructed accordingly, by implementing the identical numerical setup on the geometry with the introduction of an additional flow channelling slit. The result generated by the two slits model is therefore assumed to be justified. Temperature and velocity profiles, as mentioned, are the primary parameters for characterising the swirling reacting flow involved in the fire whirl. The assorted temperature and velocity data generated numerically by the baseline model are presented alongside those of the comparison model, to assess the effect of the geometrical configuration on the formation of fire whirl.

3.2. Temperature

The thermal component of the flame is undeniably the most critical aspect to be investigated in fire modelling. It is more critical in this study of fire whirl, as the temperature and velocity are interrelated, and such coupling is essential for the understanding of the ensuing formation and sustenance of the swirling combustion phenomenon. The qualitative analysis of the fire whirl therefore starts by assessing the temperature profile of the formulated swirling reacting flow field.

The temperature iso-surface of the flame generated by the two cases, presented in Figure 4, provides a glance of the geometric characteristics of the fire whirl formulated under different entrainment boundary conditions. The figure shows that, with the side flow channelling slit(s) acting as the eddy-generation mechanism, a highly twisted flow field with relatively high temperature can clearly be observed in both geometric configurations. This temperature profile exhibits a helical form, implying the persistence of strong swirl motion within the core structure of both flames. To be more specific, the fire whirl generated from the baseline model is exhibited in the format of a tortuous and highly twisted string. The hot plume accumulates and exits up near exit region in the vertical orientation. Whereas the fire whirl of that generated via the two-slit configuration resembles a convoluted cylinder centred on the lower-mediate region of the computational domain.

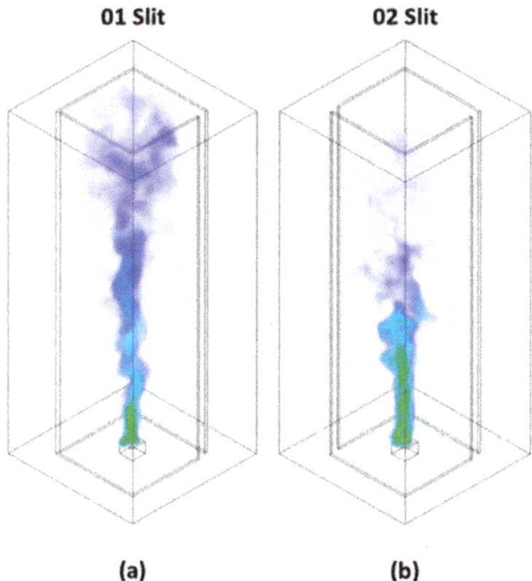

Figure 4. Temperature iso-surface of the fire whirl formulated in 01 slit case (**a**) and the 02 slit case (**b**).

The temperature contour of a cross-sectional plane of the computational domains of the two cases is illustrated in Figure 5a,b. The contour plot shows that the maximum flame temperature of the case with the two-slit configuration is higher, i.e., increased by about 9.42%, when compared with the baseline. Similar to the temperature iso-surface presented above, the cloud of the hot plume, accumulated at the downstream region observed in the baseline case, is cleared out when the additional slit is introduced.

The visible flame height of the two cases determined based on applying the cut the cut-off visible flame tip temperature of 600 °K is presented in Figure 5c,d. From the comparison, it is evident that the visible flame height for two-slit configuration grows significantly compared with that of the baseline model, i.e., it is increased by 84.38 % from 0.32 m to 0.59 m.

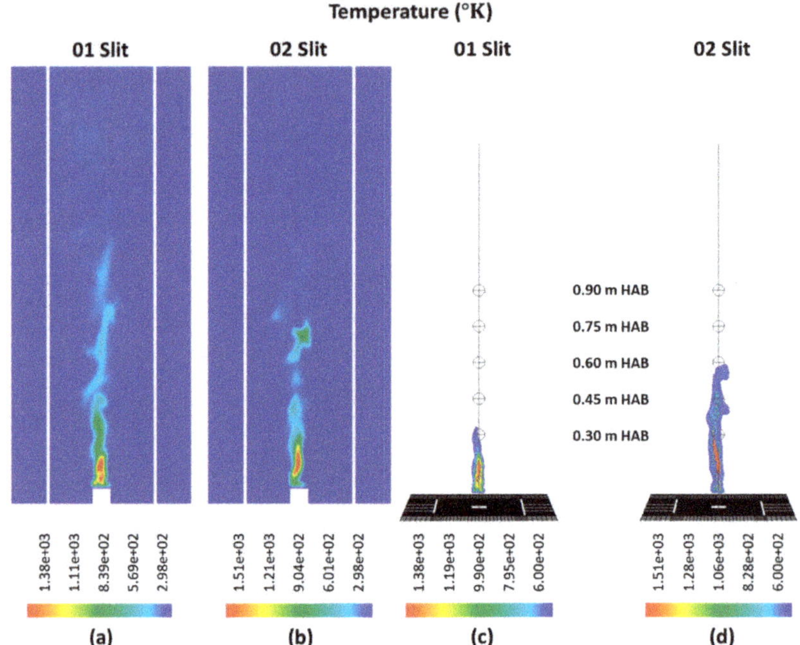

Figure 5. Temperature contour of a cross-sectional plane of the 01 slit case (**a**) and the 02 slit case (**b**), and the visible flame height of the fire whirl formulated by the 01 slit case (**c**) and the 02 slit case (**d**).

The extension of the flame height of the comparison model is further verified by comparing the spatial temperature distribution, along with the lateral direction, as presented in Figure 6. Theoretically, the peak flame temperature of a typical diffusion flame should occur at the reaction layers, i.e., the intersection where the fuel and the oxidiser encounter each other and react due to diffusion. The reaction layer in the continuous and intermittent region of a diffusion flame is typically away from the flame centreline, and as a result, the spatial distribution of the flame temperature should resemble a hump-like profile. The hump-like profile tends to shift to a Gaussian distribution, with its value maximised at the centreline, as the flame flows further to the downstream, i.e., enters the plume region.

The transition from a hump-like profile to a Gaussian distribution can be observed in the baseline model, as the HAB increased from 0.3 m to 0.5 m. This implies that the domain from HAB 0.5 m and beyond of this very flame can be identified as the plume region. On the other hand, the temperature profile generated by the 02 slit case features a hump-like profile for all monitoring HABs, means that the flaming structure up to HAB 0.5 m of this fire whirl can still be considered to be in the continuous and intermittent region. This therefore indicates that the flame height of the swirling reacting flow triggered by the intensified circulation is significantly higher compared with the baseline model.

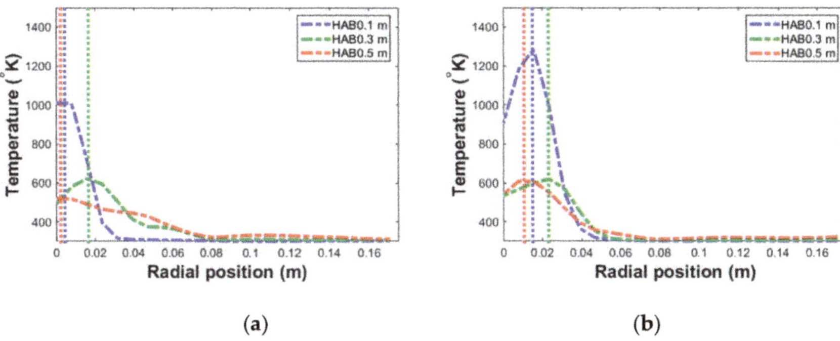

Figure 6. Plot of temperature along with the radial direction at three monitoring HABs: 01 slit case (**a**) and 02 slit case (**b**).

It is also worthwhile assessing the temperature core radius b_T, defined as the radial location where the excess temperature declines to half of the maximum recorded value, at the monitoring HABs of the two cases. Pioneer studies of the fire whirl suggested that the flame pulsation behaviour is suppressed in the continuous flame region due to the cyclostrophic balance effect, resulting in a near-constant temperature core radius. As the flame enters the plume region, the temperature core radius increases proportionally with the flame height [33].

As reported in Table 1, the temperature core radius of the baseline model expands significantly; for example, it is increased by 153.33% from 0.0660 m at HAB 0.3 m to 0.1672 m at HAB 0.5 m, while that of the 02 slit case is reported to be nearly identical at all monitoring HABs. The results suggest an elongated continuous flame region of the 02 slit model when compared with that of the baseline.

Table 1. Temperature core radius at three monitoring HABs: 01 slit case and 02 slit case.

HAB (m)	Temperature Core Radius (cm)	
	01 Slit	**02 Slit**
0.1	1.76	1.36
0.3	6.60	3.64
0.5	16.72	3.64

The preceding comparison of the spatial temperature distribution demonstrated that, as a buoyant driven diffusion flame, the intensification of the eddy generation promotes the air entrainment from the surroundings, enhances the mixing of fuel and oxidizer. This facilitates the combustion process, hence resulting in the formation of the flame field with higher peak temperature. The additional entrainment induced via the extra slit also amplifies the generation of vorticity, aggravates the swirl motion, and escalates the formulation of fire whirl, which leads to intensified combustion reactions and elongated combustion region. In addition, for this very reason, the 02 slit model promotes natural ventilation at the near exit region, and as a result, the hot plume is entrained out of the computational domain more effectively compared with 01 slit configuration.

3.3. Velocity

To characterise the fire whirl formulated under various entrainment conditions, it is imperative to evaluate and assess the related velocity field. In essence, the inner structure of an on-source fire whirl can be approximated as a 3-dimensional flow. The following vector plot, as well as the peak value of the velocity of U, V and W at all monitoring HABs, presented in Figures 7–12, illustrates the key features of the flow behaviour of the swirling flame structure of both cases. This provides compelling evidence that, with the introduction of the slit to trigger eddy generation, the presence of the highly swirled reacting flow field, i.e., the formation of the fire whirl, can be confirmed. Such whirling motion within the flame structure can be observed at all HABs of interest for both geometric configurations.

The peak value of the associated velocity at three monitoring HABs is plotted in Figures 8, 10 and 12. The comparison indicates that the peak values of U, V, and W velocity of the baseline model tend to firstly increase in the region between HAB 0.1 m and HAB 0.3 m, and then decrease sharply as the combustion spreads to downstream HABs. Meanwhile, for the 02 slit configuration, the magnitude of the drop of the assorted velocities from HAB 0.3 m to HAB 0.5 m is observed to be moderated compared with the 01 slit configuration. For example, the results report a −1.06%, +36.65% and −3.83% variation in U, V, and W velocity between HAB 0.3 m and HAB 0.5 m of the 02 slit case, compared with that of −9.25%, −26.34% and −28.34% of the baseline.

The dampened reduction in U and W velocity and the significant increase in V velocity from HAB 0.3 m to HAB 0.5 m demonstrate that, with the introduction of additional air entrainment slit, the formulated flame is still in the developing phase, i.e., in the continuous and intermittent regions at HAB of 0.5 m. Meanwhile, the flame yield for the baseline model is well-developed, and reached the plume region at the same elevated height; hence, it possesses a shorter flame length, which is in good agreement with the speculation deduced from the preceding analysis of the temperature profile.

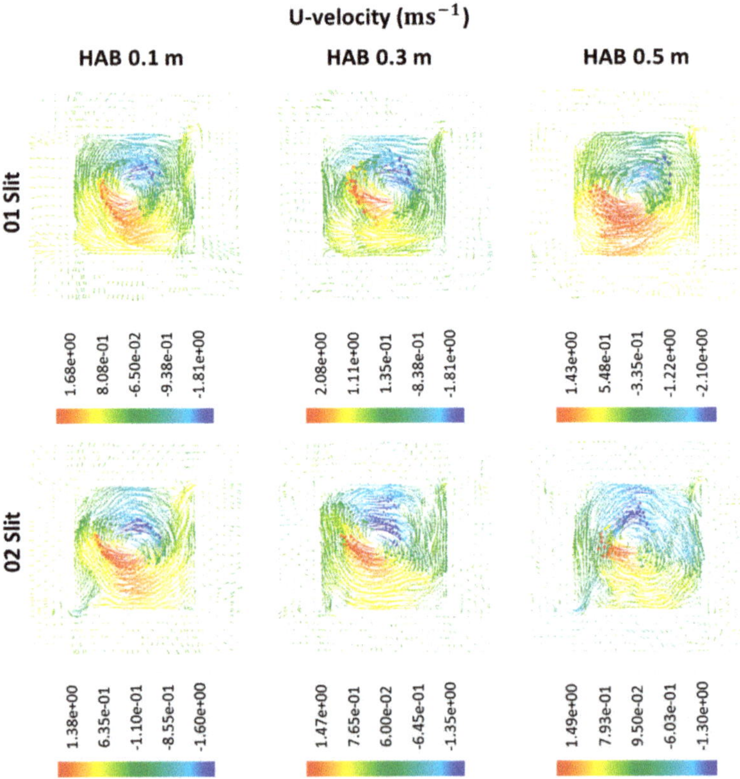

Figure 7. Plot of vector field of U-velocity of the two cases, at all monitoring HABs.

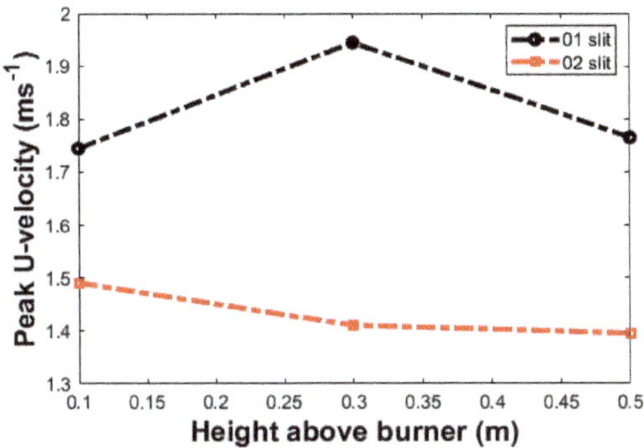

Figure 8. Peak value of U-velocity of the two cases, at all monitoring HABs.

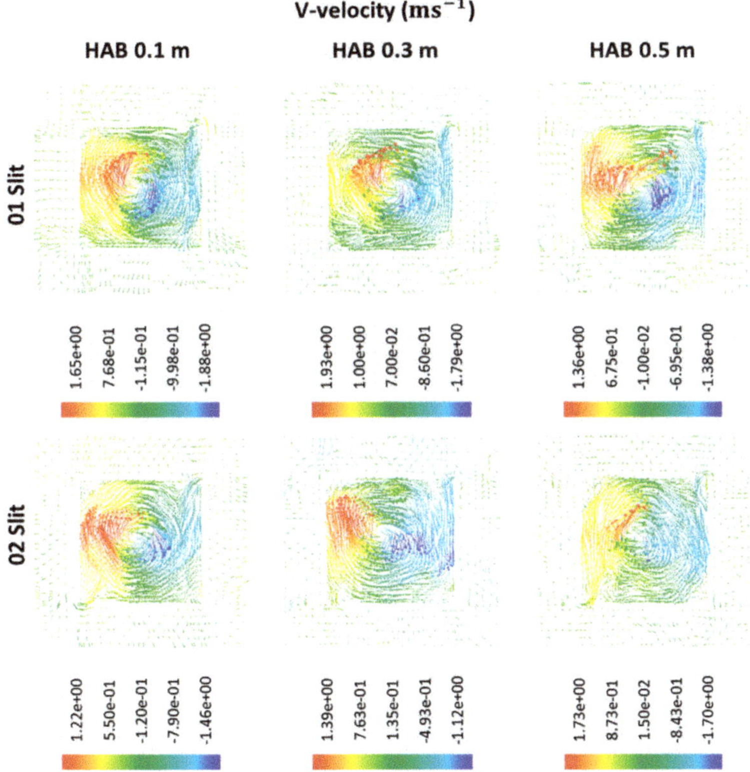

Figure 9. Plot of the vector field of V-velocity of the two cases, at all monitoring HABs.

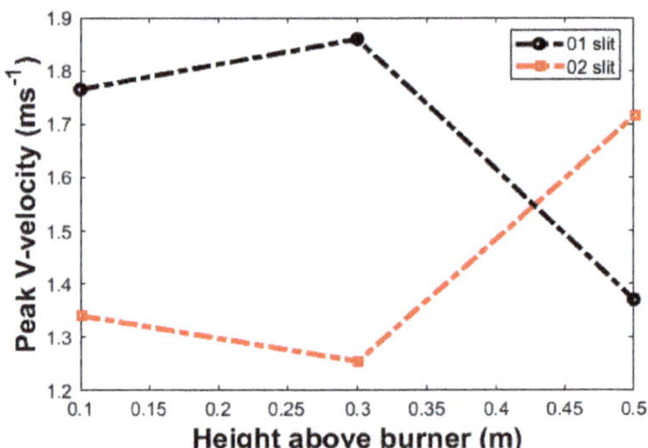

Figure 10. Peak value of V-velocity of the two cases, at all monitoring HABs.

W-velocity (ms⁻¹)

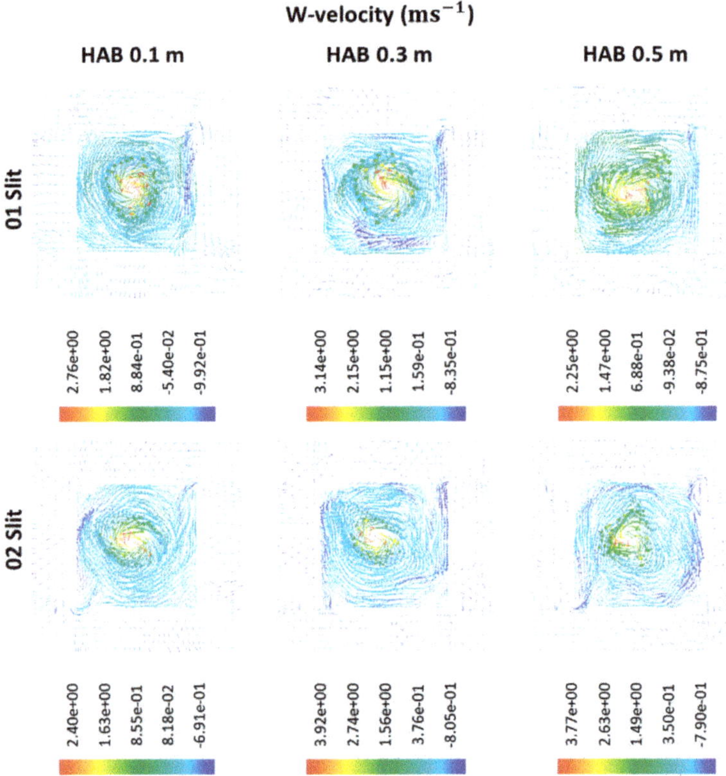

Figure 11. Plot of vector field of W-velocity of the two cases, at all monitoring HABs.

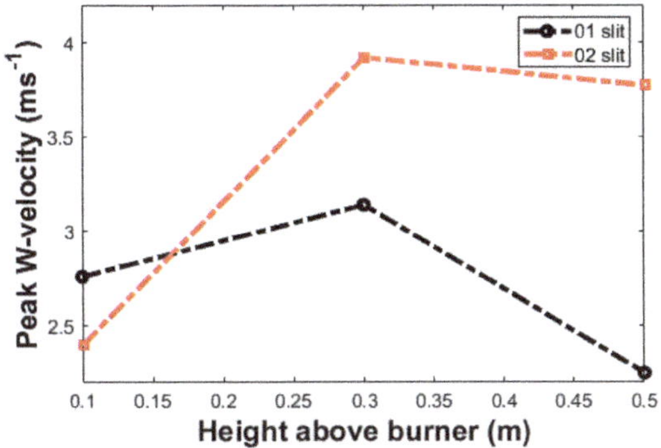

Figure 12. Peak value of W-velocity of the two cases, at all monitoring HABs.

3.3.1. Axial Velocity

The axial velocity contour of a cross-sectional plane generated by the two cases is presented in Figure 13 for comparison. The contour plot depicts that the maximum axial velocity of the case with 02 slit configuration is higher, i.e., increased by 6.81%, when compared with that of the baseline model. The region in which axial velocity is dominant was also extended with the introduction of additional air entrainment slit. The axial velocity dominant region, determined as the vertical location at which the excess axial velocity decreases to half of the maximum recorded value, increased by 46.14% to 0.5739 m, compared with 0.3927 m for the baseline model.

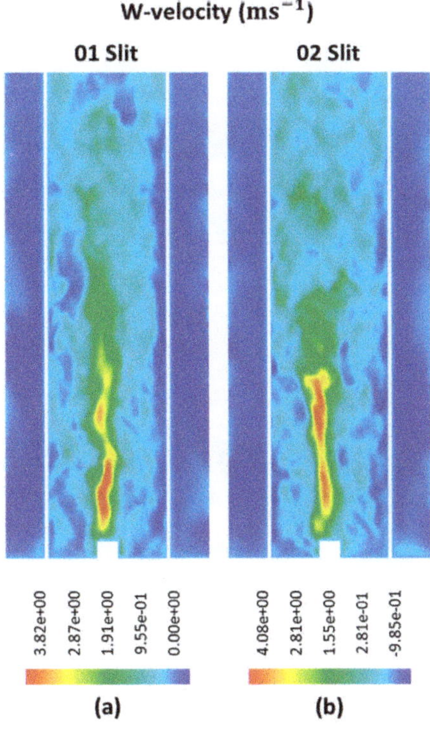

Figure 13. Axial velocity contour of a cross-sectional plane of the 01 slit case (**a**), and the 02 slit case (**b**).

The promotion of axial velocity in the 02 slit case is further justified by assessing the spatial distribution of the axial velocity plotted against the lateral direction at the monitoring HABs, presented in Figure 14. The literature reveals that the axial centreline velocity accelerates upwards to maximum in the continuous flame region. Due to the velocity fluctuation, the axial centreline velocity decreases sharply as the HAB increases to intermittent regions, and further decelerates to ambient conditions as it enters the plume region [56,57]. Such a transition can be clearly seen in the velocity profile yield from the baseline model, indicating that the continuous and intermittent region of the fire whirl generated in the baseline case should be well below HAB 0.5 m. Meanwhile, the peak axial centreline velocity reported in the comparison case is presented in an incremental pattern as the flame developing in the downstream direction, suggesting an extended continuous region when compared with the baseline case.

In addition, experimental data of PIV measurements confirmed that within the continuous flame region, the axial velocity reaches its maximum value in the vicinity of the reaction layer, i.e., away from the flame centreline, and its profile resembles a hump-like distribution along the radial direction. With increasing height from the intermittent flame region and beyond, the location of maximum excess temperature shifts toward the whirl's centreline and forms a Gaussian profile [27]. The transformation of the axial velocity profile from hump-like to Gaussian form, evidently observed in the baseline model as the HAB increased from 0.3 m to 0.5 m, however, had not yet occurred in the 02 slit model. The hump-like profile at all monitoring HABs up to 0.5 m in the comparison case again confirms the extension of the continuous flame region, i.e., to and beyond HAB 0.5 m, when compared with single slit configuration.

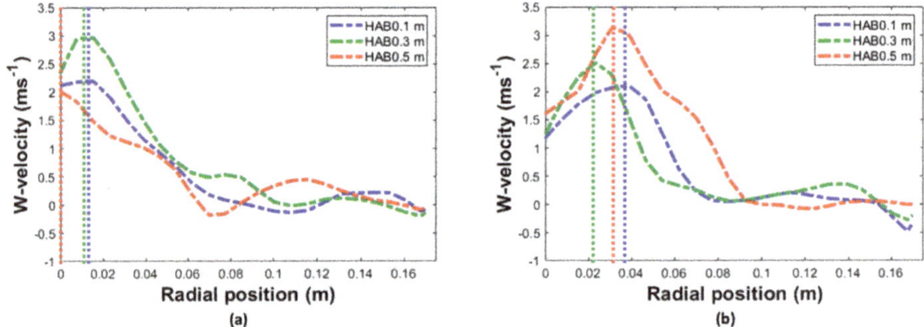

Figure 14. Plot of axial velocity, along with the radial direction at three monitoring HABs: 01 slit case (**a**), and 02 slit case (**b**).

3.3.2. Radial Velocity and Tangential Velocity

The profile of U, V and radial velocity, plotted as a function of the radial position at three monitoring HABs, is presented in Figure 15. The transformation of the profile of U, V and radial velocity from hump-like to a plateau with the increase of flame height was observed in the baseline model, whereas that hump-like velocity profile is inherently persistent in all HABs in the 02 slit case. Furthermore, the radial location at which the velocity of interest peaks shifts away from the centreline axis with an increase in the flame height in the baseline case, whereas this is nearly identical at all HABs in the comparison case. It has been described in prior studies that the maximum radial velocity occurs adjacent to the surface and reverses direction close to the upper parts of the boundary layer, indicating the presence of the circulation zone [26]. The departure of the boundary layer of the circulation zone away from the centreline, observed in the baseline case, implies the dissipation of the fire whirl core region. In other words, the flame departs from the continuous and intermittent region and enters the plume zone at the elevated height of 0.5 m. The near-constant radial position of the peak radial velocity observed in the 02 slit case depicts a well-maintained fire whirl core structure, i.e., the flame at high elevations can still be considered to be within the continuous region. The combustion region of the comparison case, therefore, is demonstrated to be well-extended compared with that of the baseline.

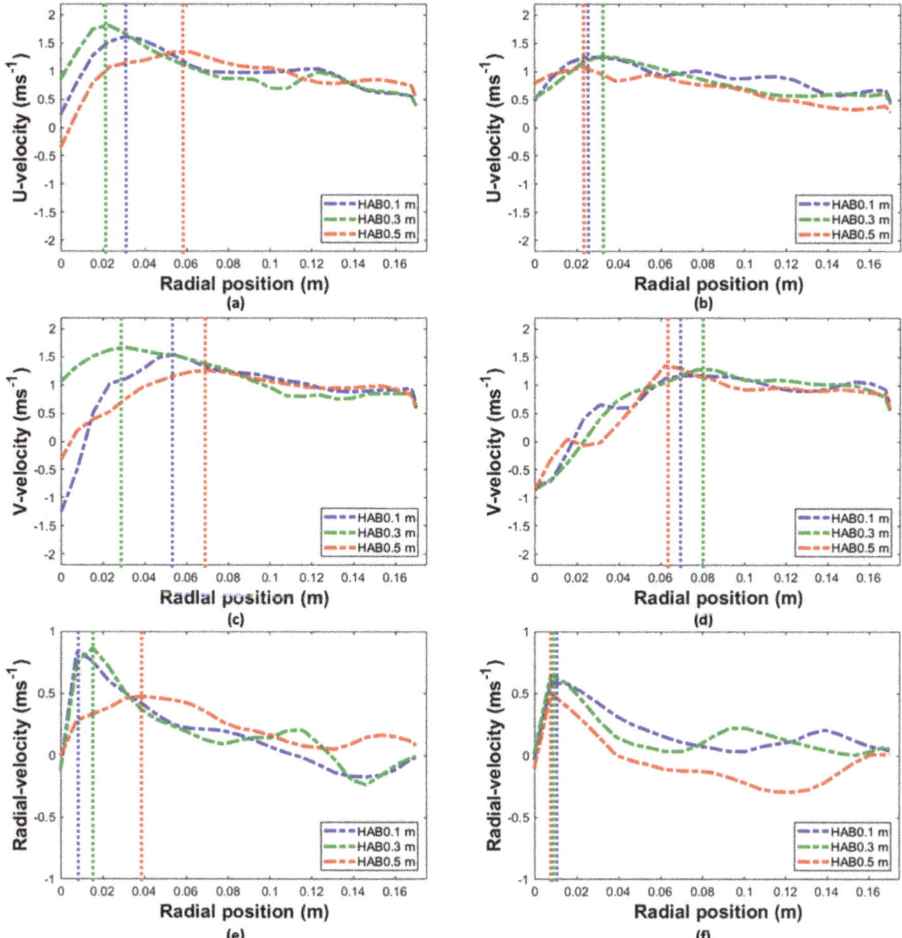

Figure 15. Plot of U, V and radial velocity along with the radial direction at three monitoring HABs: U-velocity of the 01 slit case (**a**), U-velocity of the 02 slit case (**b**), V-velocity of the 01 slit case (**c**), V-velocity of the 02 slit case (**d**), radial velocity of the 01 slit case (**e**), radial velocity of the 02 slit case (**f**).

Circulation is indeed the unique feature distinguishing fire whirls from non-swirling fires. The presence of the circulation disturbs the flow field and alters the ensuing combustion process, as a result significantly affecting assorted properties, such as flame temperature, combustion dominant region, as well as the velocity profile. The circulation is characterized by assessing the azimuthal velocity, or tangential velocity, in this study. Herein, the tangential velocity against the radial position at the selected HABs of the cases of interest is plotted in Figure 16.

Similar to the profile of radial velocity, The humped-like profile turns to flat and the radial location where the tangential velocity peaks departs away from domain centreline with the increase of flame height in the baseline model, while those of the comparison case remains relatively unchanged at all monitored heights.

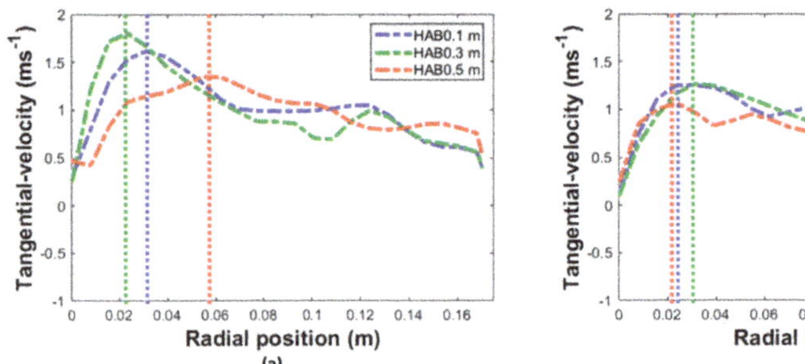

Figure 16. Plot of tangential velocity along with the radial direction at three monitoring HABs: 01 slit case (**a**), and 02 slit case (**b**).

To further elaborate the transition, the mean whirl vortex core radius at the elevated heights of the two cases are evaluated and listed in Table 2. The mean whirl core radius is determined based on the azimuthal velocity profile, i.e., the radial location at which the tangential velocity is at its maximum, and beyond which the circulation becomes nearly constant. The results reveal that the mean whirl core radius of the baseline case departs significantly from the centreline axis, i.e., it increases by 136.91% from 2.33 cm at HAB 0.3 m to 5.52 cm at HAB 0.5 m, while that reported for the 02 slit case is nearly identical at all monitoring HABs. Previous experimental study examining the correlation between the fire whirl radius and the stage of combustion suggests a reasonably constant whirl core radius in the continuous flame region, which increases with flame height in the intermittent and plume regions [33]. The results presented in Table 2, therefore, again verify the abovementioned speculation that the combustion region is well maintained and is extended by means of the introduction of the additional flow channelling slit.

In addition, by comparing the results presented in Table 2 with those in Table 1, it can be concluded that the radius determined based on the temperature is generally larger than that based on circulation, which again is in good agreement with the experimental observations [1].

Table 2. Mean vortex core radius at three monitoring HABs: 01 slit case and 02 slit case.

HAB (m)	Mean Vortex Core Radius (cm)	
	01 Slit	02 Slit
0.1	3.13	3.01
0.3	2.33	3.13
0.5	5.52	2.33

Collectively, the preceding assessment of the axial, radial and tangential velocity demonstrates that, with the additional vorticity induced by the entrainment via the additional slit, the swirling motion within the domain is intensified, and the circulation of the dominant region is elongated. The intensified circulation enhances the mixing of the oxidiser with combustible gas mixtures and facilitates the ensuing combustion process. As a result, a quasi-steady, on-source fire whirl with a unique flow core structure that extended to a higher elevated HAB was formulated and well maintained

4. Conclusions

A numerical simulation of a unique combustion process occurring in an enclosed configuration with one or two side entrainment slit(s) was performed to assess the character of a fire whirl formulated by means of various eddy-generation mechanisms. The numerical framework of the proposed model consists of the WALE large eddy simulation turbulence model coupled with reduced detailed chemistry to describe the intricate whirling flaming flow behaviour. The numerical model adopts a uniformed structured discretisation scheme, a fixed fuel injection rate, and a soot-free alcohol-based parent fuel, to isolate and constrain the complicated coupling effect between the combustion process and the flow dynamics.

The numerical simulation of the baseline model consists of one flow channelling slit, replicating the test rig layout in the literature. This was firstly validated against experimental data and achieved satisfactory agreement. A modified model built with the introduction of an additional slit was subsequently constructed, and the results generated were compared with the baseline for assessment. The analysis concluded that, with the intensified eddy-generation mechanism,

- The peak flame temperature of the formulated fire whirl increased by 9.42%, from 1380 K to 1510 K;
- The visible flame height increased by 84.38%, from 0.32 m to 0.59 m;
- The axial velocity and the axial velocity dominant region increased by 6.81% and by 46.14%, respectively;
- The flame structure was well-maintained as the temperature core radius remained relatively unchanged at all monitoring HABs, compared with the increase by 153.33% in the baseline case;
- The circulation core structure was well-sustained, and the mean vortex core radius was observed to be nearly identical at the elevated height of up to 0.5 m, whereas there was an increase of 136.91% in the baseline case.

In summary, the results generated from this numerical study demonstrate that the addition of an introduced slit intensifies the circulation, enhances the mixing of the oxidiser with the reactants, and hence facilitates the ensuing combustion process. As a result, the numerical model with stronger eddy-generation mechanisms, i.e., featuring two flow channelling slits, formulates and sustains a fire whirl possessing an elongated, constrained and swirling combustion region, compared with those generated from the single-slit configuration. In addition, this numerical study showcases the capability of the proposed numerical framework to capture various features of the whirling reacting flow, which are often difficult to acquire and assess by experimental means. The model also provides visualised information on the combustion process and flow field, thus delivering theoretical insight. The numerical model presented in this study, is therefore proved to be a feasible tool for assisting the understanding of the fundamentals of the formation and evolution of this intriguing combustion phenomenon.

Author Contributions: Conceptualization, A.C.Y.Y., Q.N.C., S.K. and G.H.Y.; Data curation, C.W. and H.L.Y.; Formal analysis, C.W., T.B.Y.C., R.C. and H.L.Y.; Funding acquisition, A.C.Y.Y. and G.H.Y.; Methodology, A.C.Y.Y., Q.N.C., S.K. and G.H.Y.; Resources, S.K.; Supervision, A.C.Y.Y., Q.N.C., S.K. and G.H.Y.; Validation, Q.C.; Visualization, Q.C.; Writing—original draft, C.W.; Writing—review & editing, T.B.Y.C. and R.C.

Funding: This research was funded by Australian Research Council (ARC Industrial Transformation Training Centre IC170100032), and Wuhan Shuanglian-Xingxin Machinery & Equipment Co. Ltd., China.

Acknowledgments: The authors wish to acknowledge the financial support of Australian Research Council (ARC Industrial Transformation Training Centre IC170100032) and the Australian Government Research Training Program Scholarship. It is also sponsored by Wuhan Shuanglian-Xingxin Machinery & Equipment Co. Ltd., China. The authors deeply appreciate all financial and technical supports.

References

1. Tohidi, A.; Gollner, M.J.; Xiao, H. Fire Whirls. *Annu. Rev. Fluid Mech.* **2018**, *50*, 187–213. [CrossRef]
2. Wang, P.; Liu, N.; Bai, Y.; Zhang, L.; Satoh, K.; Liu, X. An experimental study on thermal radiation of fire whirl. *Int. J. Wildl. Fire* **2017**, *26*, 693. [CrossRef]
3. Medwell, P.R.; Chan, Q.N.; Kalt, P.A.M.; Alwahabi, Z.T.; Dally, B.B.; Nathan, G.J. Instantaneous Temperature Imaging of Diffusion Flames Using Two-Line Atomic Fluorescence. *Appl. Spectrosc.* **2010**, *64*, 173–176. [CrossRef] [PubMed]

4.   Soma, S.; Saito, K. Reconstruction of fire whirls using scale models. *Combust. Flame* **1991**, *86*, 269–284. [CrossRef]

5.   Zhou, K.; Liu, N.; Yin, P.; Yuan, X.; Jiang, J. Fire Whirl due to Interaction between Line Fire and Cross Wind. *Fire Saf. Sci.* **2014**, *11*, 1420–1429. [CrossRef]

6.   Byram, G.; Martin, R. The Modeling of Fire Whirlwinds. *For. Sci.* **1970**, *16*, 386–399.

7.   Lei, J.; Liu, N.; Tu, R. Flame height of turbulent fire whirls: A model study by concept of turbulence suppression. *Proc. Combust. Inst.* **2017**, *36*, 3131–3138. [CrossRef]

8.   Yuen, A.C.Y.; Yeoh, G.H.; Cheung, S.C.P.; Chan, Q.N.; Chen, T.B.Y.; Yang, W.; Lu, H. Numerical study of the development and angular speed of a small-scale fire whirl. *J. Comput. Sci.* **2018**, *27*, 21–34. [CrossRef]

9.   Church, C.R.; Snow, J.T.; Dessens, J. Intense Atmospheric Vortices Associated with a 1000 MW Fire. *Bull. Am. Meteorol. Soc.* **1980**, *61*, 682–694. [CrossRef]

10. Forthofer, J.M.; Goodrick, S.L. Review of Vortices in Wildland Fire. *J. Combust.* **2011**, *2011*, 1–14. [CrossRef]

11. Liu, N.; Liu, Q.; Deng, Z.; Kohyu, S.; Zhu, J. Burn-out time data analysis on interaction effects among multiple fires in fire arrays. *Proc. Combust. Inst.* **2007**, *31*, 2589–2597. [CrossRef]

12. Emori, R.I.; Saito, K. Model experiment of hazardous forest fire whirl. *Fire Technol.* **1982**, *18*, 319–327. [CrossRef]

13. Kuwana, K.; Morishita, S.; Dobashi, R.; Chuah, K.H.; Saito, K. The burning rate's effect on the flame length of weak fire whirls. *Proc. Combust. Inst.* **2011**, *33*, 2425–2432. [CrossRef]

14. Chuah, K.H.; Kuwana, K.; Saito, K.; Williams, F.A. Inclined fire whirls. *Proc. Combust. Inst.* **2011**, *33*, 2417–2424. [CrossRef]

15. Lei, J.; Liu, N.; Zhang, L.; Chen, H.; Shu, L.; Chen, P.; Deng, Z.; Zhu, J.; Satoh, K.; De Ris, J.L. Experimental research on combustion dynamics of medium-scale fire whirl. *Proc. Combust. Inst.* **2011**, *33*, 2407–2415. [CrossRef]

16. Zhou, K.; Liu, N.; Lozano, J.S.; Shan, Y.; Yao, B.; Satoh, K. Effect of flow circulation on combustion dynamics of fire whirl. *Proc. Combust. Inst.* **2013**, *34*, 2617–2624. [CrossRef]

17. Lin, B.; Yuen, A.C.Y.; Li, A.; Zhang, Y.; Chen, T.B.Y.; Yu, B.; Lee, E.W.M.; Peng, S.; Yang, W.; Lu, H.-D.; et al. MXene/chitosan nanocoating for flexible polyurethane foam towards remarkable fire hazards reductions. *J. Hazard. Mater.* **2020**, *381*, 120952. [CrossRef] [PubMed]

18. Yang, W.-J.; Yuen, A.C.Y.; Li, A.; Lin, B.; Chen, T.B.Y.; Yang, W.; Lu, H.-D.; Yeoh, G.H. Recent progress in bio-based aerogel absorbents for oil/water separation. *Cellulose* **2019**, *26*, 6449–6476. [CrossRef]

19. Si, J.-Y.; Tawiah, B.; Sun, W.-L.; Lin, B.; Wang, C.; Yuen, A.C.Y.; Yu, B.; Li, A.; Yang, W.; Lu, H.-D.; et al. Functionalization of MXene Nanosheets for Polystyrene towards High Thermal Stability and Flame Retardant Properties. *Polymers* **2019**, *11*, 976. [CrossRef]

20. Yu, B.; Tawiah, B.; Wang, L.-Q.; Yin Yuen, A.C.; Zhang, Z.-C.; Shen, L.-L.; Lin, B.; Fei, B.; Yang, W.; Li, A.; et al. Interface decoration of exfoliated MXene ultra-thin nanosheets for fire and smoke suppressions of thermoplastic polyurethane elastomer. *J. Hazard. Mater.* **2019**, *374*, 110–119. [CrossRef]

21. Chen, T.; Yuen, A.; Yeoh, G.; Yang, W.; Chan, Q. Fire Risk Assessment of Combustible Exterior Cladding Using a Collective Numerical Database. *Fire* **2019**, *2*, 11. [CrossRef]

22. Li, A.; Yuen, A.C.Y.; Chen, T.B.Y.; Wang, C.; Liu, H.; Cao, R.; Yang, W.; Yeoh, G.H.; Timchenko, V. Computational Study of Wet Steam Flow to Optimize Steam Ejector Efficiency for Potential Fire Suppression Application. *Appl. Sci.* **2019**, *9*, 1486. [CrossRef]

23. Yuen, A.C.Y.; Yeoh, G.H.; Alexander, B.; Cook, M. Fire scene investigation of an arson fire incident using computational fluid dynamics based fire simulation. *Build. Simul.* **2014**, *7*, 477–487. [CrossRef]

24. Yuen, A.C.Y.; Yeoh, G.H.; Alexander, R.; Cook, M. Fire scene reconstruction of a furnished compartment room in a house fire. *Case Stud. Fire Saf.* **2014**, *1*, 29–35. [CrossRef]

25. Dobashi, R.; Okura, T.; Nagaoka, R.; Hayashi, Y.; Mogi, T. Experimental Study on Flame Height and Radiant Heat of Fire Whirls. *Fire Technol.* **2016**, *52*, 1069–1080. [CrossRef]

26. Hartl, K.A.; Smits, A.J. Scaling of a small scale burner fire whirl. *Combust. Flame* **2016**, *163*, 202–208. [CrossRef]

27. Wang, P.; Liu, N.; Hartl, K.; Smits, A. Measurement of the Flow Field of Fire Whirl. *Fire Technol.* **2016**, *52*, 263–272. [CrossRef]

28. Xiao, H.; Gollner, M.J.; Oran, E.S. From fire whirls to blue whirls and combustion with reduced pollution. *Proc. Natl. Acad. Sci. USA* **2016**, *113*, 9457–9462. [CrossRef] [PubMed]

29. Muraszew, A.; Fedele, J.B.; Kuby, W.C. The fire whirl phenomenon. *Combust. Flame* **1979**, *34*, 29–45. [CrossRef]

30. Chuah, K.H.; Kushida, G. The prediction of flame heights and flame shapes of small fire whirls. *Proc. Combust. Inst.* **2007**, *31*, 2599–2606. [CrossRef]

31. Parente, R.M.; Pereira, J.M.C.; Pereira, J.C.F. On the influence of circulation on fire whirl height. *Fire Saf. J.* **2019**, *106*, 146–154. [CrossRef]

32. Lei, J.; Liu, N.; Satoh, K. Buoyant pool fires under imposed circulations before the formation of fire whirls. *Proc. Combust. Inst.* **2015**, *35*, 2503–2510. [CrossRef]

33. Lei, J.; Liu, N.; Zhang, L.; Satoh, K. Temperature, velocity and air entrainment of fire whirl plume: A comprehensive experimental investigation. *Combust. Flame* **2015**, *162*, 745–758. [CrossRef]

34. Sikanen, T.; Hostikka, S. Modeling and simulation of liquid pool fires with in-depth radiation absorption and heat transfer. *Fire Saf. J.* **2016**, *80*, 95–109. [CrossRef]

35. Yao, W.; Yin, J.; Hu, X.; Wang, J.; Zhang, H. Numerical modeling of liquid n-heptane pool fires based on heat feedback equilibrium. *Procedia Eng.* **2013**, *62*, 377–388. [CrossRef]

36. Wang, C.; Chan, Q.N.; Zhang, R.; Kook, S.; Hawkes, E.R.; Yeoh, G.H.; Medwell, P.R. Automated determination of size and morphology information from soot transmission electron microscope (TEM)-generated images. *J. Nanopart. Res.* **2016**, *18*, 127. [CrossRef]

37. Wang, C.; Chan, Q.N.; Kook, S.; Hawkes, E.R.; Lee, J.; Medwell, P.R. External irradiation effect on the growth and evolution of in-flame soot species. *Carbon* **2016**, *102*, 161–171. [CrossRef]

38. Yuen, A.C.Y.; Yeoh, G.H.; Timchenko, V.; Cheung, S.C.P.; Chan, Q.N.; Chen, T. On the influences of key modelling constants of large eddy simulations for large-scale compartment fires predictions. *Int. J. Comut. Fluid Dyn.* **2017**, *31*, 324–337. [CrossRef]

39. Yuen, A.C.Y.; Yeoh, G.H.; Timchenko, V.; Chen, T.B.Y.; Chan, Q.N.; Wang, C.; Li, D.D. Comparison of detailed soot formation models for sooty and non-sooty flames in an under-ventilated ISO room. *Int. J. Heat Mass Transf.* **2017**, *115*, 717–729. [CrossRef]

40. Tree, D.R.; Svensson, K.I. Soot processes in compression ignition engines. *Prog. Energy Combust. Sci.* **2007**, *33*, 272–309. [CrossRef]

41. Wang, H. Formation of nascent soot and other condensed-phase materials in flames. *Proc. Combust. Inst.* **2011**, *33*, 41–67. [CrossRef]

42. Jones, W.P.; Whitelaw, J.H. Calculation methods for reacting turbulent flows: A review. *Combust. Flame* **1982**, *48*, 1–26. [CrossRef]

43. Chow, W.; Han, S.S. Experimental Data on Scale Modeling Studies on Internal Fire Whirls. *Int. J. Eng. Perform. Based Fire Codes* **2011**, *10*, 63–74.

44. Chen, T.B.Y.; Yuen, A.C.Y.; Wang, C.; Yeoh, G.H.; Timchenko, V.; Cheung, S.C.P.; Chan, Q.N.; Yang, W. Predicting the fire spread rate of a sloped pine needle board utilizing pyrolysis modelling with detailed gas-phase combustion. *Int. J. Heat Mass Transf.* **2018**, *125*, 310–322. [CrossRef]

45. Yuen, A.; Chen, T.; Yang, W.; Wang, C.; Li, A.; Yeoh, G.; Chan, Q.; Chan, M. Natural Ventilated Smoke Control Simulation Case Study Using Different Settings of Smoke Vents and Curtains in a Large Atrium. *Fire* **2019**, *2*, 7. [CrossRef]

46. Yuen, A.C.Y.; Yeoh, G.H.; Timchenko, V.; Cheung, S.C.P.; Chen, T. Study of three LES subgrid-scale turbulence models for predictions of heat and mass transfer in large-scale compartment fires. *Numer. Heat Transf. Part A Appl.* **2016**, *69*, 1223–1241. [CrossRef]

47. Yuen, A.C.Y. On the Prediction of Combustion Products and Soot Particles in Compartment Fires. Ph.D. Thesis, UNSW Sydney, Kensington, Australia, 2014.

48. Wang, C. External Irradiation Effect on the Evolution of In-flame Soot Species. Ph.D. Thesis, University of New South Wales, Sydney, Australia, 2015.

49. Yuen, A.C.Y.; Yeoh, G.H.; Timchenko, V.; Barber, T. LES and multi-step chemical reaction in compartment fires. *Numer. Heat Transf. Part A Appl.* **2015**, *68*, 711–736. [CrossRef]

50. Kee, R.J.; Rupley, F.M.; Miller, J.A.; Coltrin, M.E.; Grcar, J.F.; Meeks, E.; Moffat, H.K.; Lutz, A.E.; Dixon-Lewis, G.; Smooke, M.D.; et al. *Chemkin Collection Release 3.6*; Reaction Design: San Diego, CA, USA, 2000.

51. Yuen, A.C.Y.; Yeoh, G.H.; Timchenko, V.; Cheung, S.C.P.; Barber, T.J. Importance of detailed chemical kinetics on combustion and soot modelling of ventilated and under-ventilated fires in compartment. *Int. J. Heat Mass Transf.* **2016**, *96*, 171–188. [CrossRef]

52. Chen, T.B.Y.; Yuen, A.C.Y.; Yeoh, G.H.; Timchenko, V.; Cheung, S.C.P.; Chan, Q.N.; Yang, W.; Lu, H. Numerical study of fire spread using the level-set method with large eddy simulation incorporating detailed chemical kinetics gas-phase combustion model. *J. Comput. Sci.* **2018**, *24*, 8–23. [CrossRef]
53. Wang, C.; Yuen, A.C.Y.; Chan, Q.N.; Chen, T.B.Y.; Yang, W.; Cheung, S.C.-P.; Yeoh, G.H. Sensitivity Analysis of Key Parameters for Population Balance Based Soot Model for Low-Speed Diffusion Flames. *Energies* **2019**, *12*, 910. [CrossRef]
54. Mueller, M.E.; Chan, Q.N.; Qamar, N.H.; Dally, B.B.; Pitsch, H.; Alwahabi, Z.T.; Nathan, G.J. Experimental and computational study of soot evolution in a turbulent nonpremixed bluff body ethylene flame. *Combust. Flame* **2013**, *160*, 1298–1309. [CrossRef]
55. Wang, C.; Yuen, A.C.Y.; Chan, Q.N.; Chen, T.B.Y.; Yang, W.; Cheung, S.C.P.; Yeoh, G.H. Characterisation of soot particle size distribution through population balance approach and soot diagnostic techniques for a buoyant non-premixed flame. *J. Energy Inst.* **2019**, in press. [CrossRef]
56. Zukoski, E.E.; Kubota, T.; Cetegen, B. Entrainment in fire plumes. *Fire Saf. J.* **1981**, *3*, 107–121. [CrossRef]
57. Lei, J.; Liu, N. Flame precession of fire whirls: A further experimental study. *Fire Saf. J.* **2016**, *79*, 1–9. [CrossRef]

*applied
sciences*

Article

Similarity Analysis for Time Series-Based 2D Temperature Measurement of Engine Exhaust Gas in TDLAT

Hyeonae Jang [1] and Doowon Choi [2,*]

[1] Research & Business Foundation, Dong-A University, Busan 49315, Korea; jha3562@dau.ac.kr
[2] Technical Center for High-Performance Valves, Dong-A University, Busan 49315, Korea
* Correspondence: onessam21@dau.ac.kr; Tel.: +82-51-200-1485

Received: 24 October 2019; Accepted: 29 December 2019; Published: 30 December 2019

Abstract: As regulations on the emission of pollutants from combustion systems are further tightened, it is necessary to reduce pollutant species and improve combustion efficiency to completely understand the process in the combustion field. Tunable diode laser absorption tomography (TDLAT) is a powerful tool that can analyze two-dimensional (2D) temperature and species concentration with fast-response and non-contact. In this study, stabilized spectra were implemented using the mean periodic signal technique to enable real-time 2D temperature measurement in harsh conditions. A time series statistical-based verification algorithm was introduced to select an optimal spectral cycle to track 2D reconstruction temperature. The statistical-based verification is based on the Two-sample t test, root mean square error, and time-based Mahalanobis distance, which is a technique for similarity analysis between thermocouple and reconstruction temperature of 18 candidate cycles. As a result, it was observed that the statistical-based TDLAT contribute to improving the accuracy of time series-based 2D temperature measurements.

Keywords: tunable diode laser absorption tomography; exhaust gas; temperature; similarity analysis; Mahalanobis distance; root mean square error

---

1. Introduction

With global warming prevention and environmental protection emerging as an issue all over the world, emission regulations are gradually being tightened as the effects of emissions from combustion systems on air and human bodies are highlighted. For example, EURO 6 regulations are an effort to reduce emissions such as CO, HC, NO_x, and PM by the auto industry. The recent EURO 6d emission regulations require real driving emissions (RDE) tests, which measure pollutants emitted from motor vehicles while they are driven, as an additional approval. As a result, demand for vehicles with low NO_x emissions and high engine power has increased, and several studies have been conducted to improve the performance of NO_x after treatment systems [1–3]. An important element in catalytic after treatment devices is exhaust gas temperature because catalytic performance is a function of temperature. A car engine running at high temperature tends to emit less CO_2 and more NO_x. On the other hand, a lower engine temperature will result in more CO_2 and less NO_x. However, the challenge of reducing both CO_2 and pollutant emissions simultaneously still remains. Therefore, analysis of temperature and species concentration during combustion is necessary for fundamental understanding of combustion reaction, which may reduce pollutants in the combustion system and improve combustion efficiency. However, the thermocouple (TC), a conventionally used one-point temperature measurement device, has difficulty in analyzing non-uniform flow, such as in engine exhaust. In addition, the techniques such as coherent anti-stokes Raman scattering (CARS) [4,5] and laser induced grating spectroscopy

(LIGS) [6], which are recently used in the spectroscopy field, can measure temperature at a fast response speed, but have a limit of point measurement, and filtered Rayleigh scattering (FRS) [7] and planer laser induced fluorescence (PLIF) [8,9] enable two-dimensional temperature, but their measurement systems are expensive and complicated. Tunable diode laser absorption tomography (TDLAT) is a suitable tool for understanding the combustion process by allowing 2D analysis of temperature and species concentration with non-contact fast-response. Therefore, it has been developed with various TDLAT techniques for measuring 2D temperature and species concentration in combustion fields [10–14]. Deguchi et al. demonstrated a 2D temperature distribution measurement by TDLAT on Bunsen-type flame burners and gasoline engines [10]. Ma et al. developed a hyperspectral tomography (HT) method that can simultaneously measure 2D distribution of temperature and H_2O concentration with a temporal resolution of 50 kHz at the exhaust plane of a j85 engine [11]. Busa et al. developed a measurement method of combustion efficiency that combines TDLAT and particle image velocimetry (PIV) in the scramjet engine [13]. However, these methods cannot be applied directly for real-time measurements because of excessive computational time and lack of time resolution assessment required for computed tomography (CT) analysis.

This study introduced a spectral cycle parameter improving reconstruction performance with a posteriori statistical technique in conventional absorption tomography to measure the exhaust gas temperature in milliseconds through TDLAT. Signal stabilization technique was applied in various cycle models to obtain stabilized absorption spectra in harsh environments, and statistical verification using TC measurement was performed to evaluate the results of the time series 2D reconstruction temperature distribution.

2. Theory of TDLAS

Tunable diode laser absorption spectroscopy (TDLAS) is a measurement technique commonly used in a variety of applications recently as well as temperature and density of molecules in gas [15,16]. When a laser beam permeates an absorption medium, its intensity will be attenuated by the absorbing gas species. This phenomenon is measured by TDLAS system that continuously scans laser wavelengths. The principle of TDLAS is theoretically expressed in terms of absorption by Beer-Lambert law that describes the relationship between the incident intensity I_{v0} and transmitted intensity I_v as follows:

$$\frac{I_v}{I_{v0}} = exp\{-A_v\} = exp\left\{-\int_0^L \alpha_v(s)ds\right\}, \tag{1}$$

where A_v denotes the spectral absorbance of representative wavenumber v. L denotes the optical path length, and α_v is the local mass-weighted absorption coefficient at position s along the laser beam, from 0 to L. This quantity is given by the following expression:

$$\alpha_v = \sum_i \frac{x_i p}{kT} \sum_j S_{i,j}(T) f_v\left(v; v_j, \Delta v_G, \Delta v_L\right), \tag{2}$$

where, k [J/K] denotes Boltzmann's constant, f_v is the Voigt profile, with Lorentzian and Gaussian half width at half maximums (HWHMs), Δv_L and Δv_G [17–19]. p denotes pressure, assumed to be uniform and x_i denotes mole-fraction of absorbing species i, where, x_i, p, and T are evaluated locally. $S_{i,j}(T)$ denotes temperature dependent line strength of absorption transition j and can be calculated by a function of temperature as follows:

$$S_{i,j}(T) = S_{i,j}(T_0)\frac{Q(T_0)}{Q(T)}\frac{T}{T_0}exp\left[-\frac{hcE_i''}{k}\left(\frac{1}{T} - \frac{1}{T_0}\right)\right]\left[1 - exp\left(\frac{-hcv_{0,i}}{kT}\right)\right]\left[1 - exp\left(\frac{-hcv_{0,i}}{kT_0}\right)\right]^{-1}, \tag{3}$$

where, h (J·s) and c (cm/s) denote Planck's constant and the speed of light, v_0 (cm^{-1}) denotes the line-center wavenumber, E_i'' (cm^{-1}) denotes lower-state energy of the transition, T_0 (K) is the reference temperature (296 K), and $Q(T)$ denotes the partition function of molecular absorption at a particular temperature. It is usually expressed as a 4th order polynomial function [20]. Although it is difficult to accurately describe the chemical formula of gasoline, iso octane (C_8H_{18}) was selected because it is the hydrocarbon component with similar thermal properties to the actual composition. The combustion reaction of iso octane is as follows:

$$C_8H_{18} + 12.05(O_2 + 3.76N_2) \rightarrow 8CO_2 + 9H_2O + 12.05 \cdot (3.76)N_2. \tag{4}$$

Among the combustion products, nitrogen (N_2) is an inert gas and hence, there is no chemical or absorption reaction. Therefore, water-vapor (H_2O) was chosen for exhaust gas diagnosis because of its high mole fraction and high absorption strength.

Figure 1 shows the theoretical H_2O absorption spectra calculated using the HITRAN2012 molecular spectroscopic database [21]. The patterns of the absorption spectra can be observed as the temperature changes. In this study, three peaks lines (7202.255 cm^{-1}, 7202.909 cm^{-1}, and 7203.891 cm^{-1}) with high line strengths and no interference with other molecules were chosen for calculating reconstruction temperature.

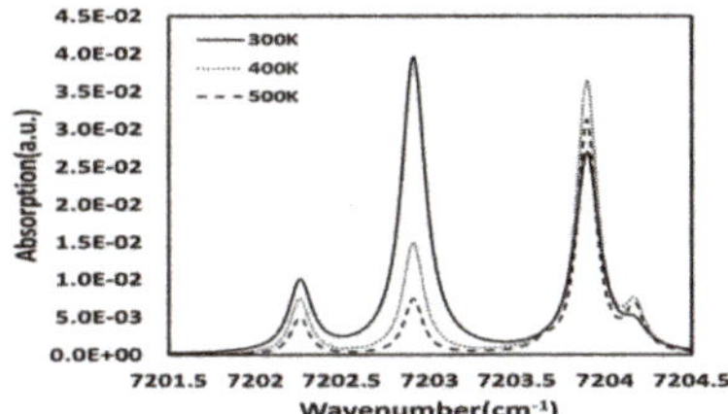

Figure 1. Theoretical H_2O absorption spectrum.

3. Experiment and Tomography Analysis

A schematic diagram of a TDLAS system measuring engine exhaust temperature is shown in Figure 2. The OHC (over head camshaft) gasoline engine (EX17, Subaru Robin, Doylestown, PA, USA) is driven in idle mode to produce emissions. The exhaust pipe installed 5 mm below the measurement cell was 20 mm in diameter and 150 mm in length. The distributed feedback (DFB) laser (NLK1E5GAAA, NTT Electronics, Japan) for H_2O gas target is tuned to the near-infrared spectral region at 1388 nm by a laser diode controller (LDC-3900, ILX Lightwave, Bozeman, MT, USA). The modulated laser beam is split into 16 channels by a 1 × 16 fiber optic coupler (SMF-28e, OPNETI, NY, USA). Among them, the 10 beams are irradiated into 10 collimators (50-1310-APC, THORLABS, Newton, NJ, USA) and the transmitted optical signals are detected by 10 photodetectors (G12180-010A, Hamamatsu Photonics, Japan). The received laser signals are amplified and filtered by low pass filters to remove signal noise. The amplified signals are transmitted into a data recorder (8861 Memory High Coda HD Analog16, HIOKI, Japan) in real time at a sampling rate of 500 kHz. Simultaneously, the real temperature for verification is measured by a B-type TC at the sampling rate of 1 kHz at the center of the target area. After absorption, signals collect by the data acquisition card are sent to the computer for reconstruction calculation.

Multiple parallel laser beams pass through the region of interest. The intersected spectral information is used to reconstruct 2D temperature by CT analysis. Figure 3 shows the laser paths and grids in the measurement region. The laser path consists of 10 real beams in a 5 × 5 layout and eight interpolated virtual beams The interpolated virtual beams imply the virtual signal data interpolated from the signal of the actual beam, and in the signal application field a widely introduced cubic spline interpolation is used in this study [22]. The cell length through which the laser passes in the measurement area is 60 mm and reconstruction area is 40 mm, and the spatial region for CT analysis consist of 81 (9 × 9) square grids with a grid size of 2.0 mm × 2.0 mm.

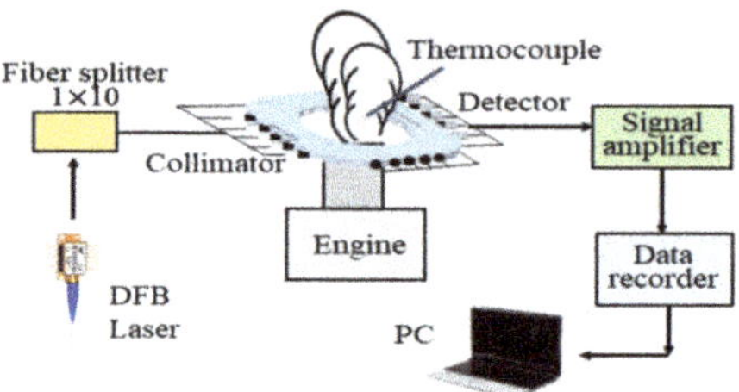

Figure 2. Schematic diagram of the experimental setup using tunable diode laser absorption spectroscopy (TDLAS).

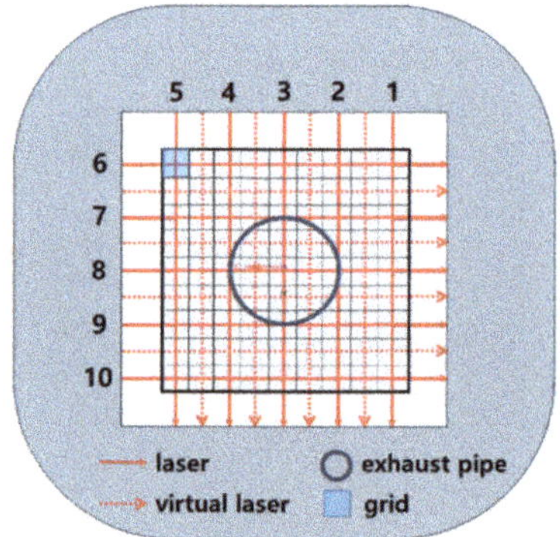

Figure 3. Laser paths and measurement cell.

Unexpected error signals due to harsh conditions were removed to reduce reconstruction calculation time for converging. The formula is as follows:

$$\delta_t = \sum_{n=1}^{N} \frac{\left|\bar{I}_{v0,i} - \bar{I}_{v,i}\right|_t}{\left|\bar{I}_{v0,i} - \bar{I}_{v,i}\right|_{t-1}}, \tag{5}$$

where, the subscript t denotes a cycle at t time and n denotes the data number in a cycle. δ denotes the error rate, which is a ratio of absolute error sum in $\bar{I}_v$ and $\bar{I}_{v0}$ of the cycle to that of the previous cycle. Here, the signal cycle was eliminated when δ_t is greater than 1.5. This method effectively reduced the error when converting the signal into the spectra. The noises generated by beam steering, window foul, and thermoelectric noises were approximated by Gaussian noise modeling, and the stabilized absorption spectra was obtained by the polynomial reduction method (PRM) averaging a number of absorption signal cycles [10,23,24] Although averaging a significant number of absorption cycles provides more stable signal, it has the drawback of degrading time-resolution and distorting raw signals. Hence, it is necessary to investigate appropriate number of cycles with good reconstruction performance and fast time-resolution simultaneously.

Figure 4 shows the result of mean transmitted signals ($\bar{I}_v$) and mean absorption spectra using a total of 201 cycles (1 reference cycle, 100 cycles in front and 100 cycle in back) measured by the TDLAS system. Here, one cycle of measured signal consists of 500 samples with a recording time of 1 ms. A total of 5 million signals were received during 10 s of engine operation. Figure 4a is the result of averaging each raw signal acquired from the photodetector at the same time when 10 lasers passed through the target area. To convert these signals into absorption spectra, $\bar{I}_v$ and the mean incident signal ($\bar{I}_{v0}$) must be known as shown in Equation (4). $\bar{I}_{v0}$, also known as the baseline, was estimated by PRM, which is a least squares fitting using 5th order polynomial curve with $\bar{I}_v$ [23,25]. To facilitate curve fitting, the saw-tooth waveform had a frequency of 1 kHz and a sampling rate of

2 µs. All measured absorption spectra generated from $\bar{I}_v$ and $\bar{I}_{v0}$ is shown in Figure 4b. These spectra consisted of 311 wavenumbers. Laser group 0 and 1 represent a group of vertical and horizontal lasers, respectively. The spectra pattern can be identified by different laser numbers. As shown in Figure 1, the measured absorption spectra are determined by the temperature of the measurement area, so the reconstructed temperature distribution in measurement zone can be traced by a tomographic reconstruction method from the information of each absorption spectrum.

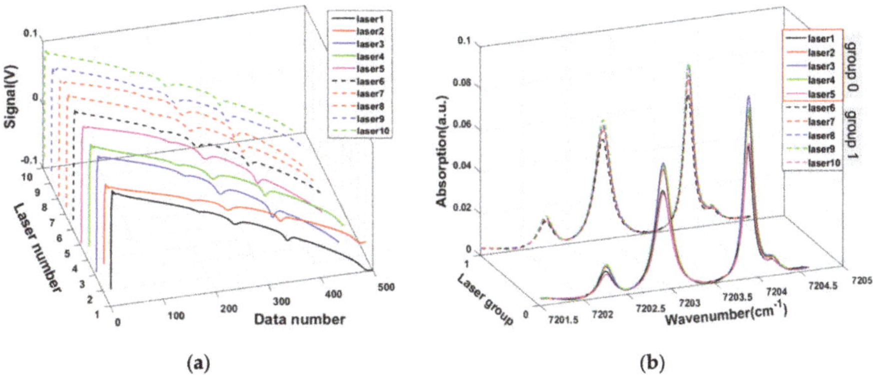

(a)   (b)

Figure 4. Results of mean transmission signals and mean absorption spectra of 10 laser beams: (**a**) 10 mean transmitted signals using 201 cycles and (**b**) 10 mean absorption spectra using 201 cycles.

The determination method of cycle and temperature is illustrated in Figure 5. First, raw absorption signals, reference time i and candidate cycle j are input. i denotes analysis time point at engine running time. A total of 990 absorption signals were used in 0.1 s from 1.0 to 9.9 s. j denotes the number of cycles of absorption signal to be averaged. Eighteen candidate cycles were chosen to evaluate the analytical performance in various cycles (31, 51, 71, 91, 111, 151, 201, 251, 301, 351, 401, 451, 501, 601, 701, 801, 901, and 1001 cycles).

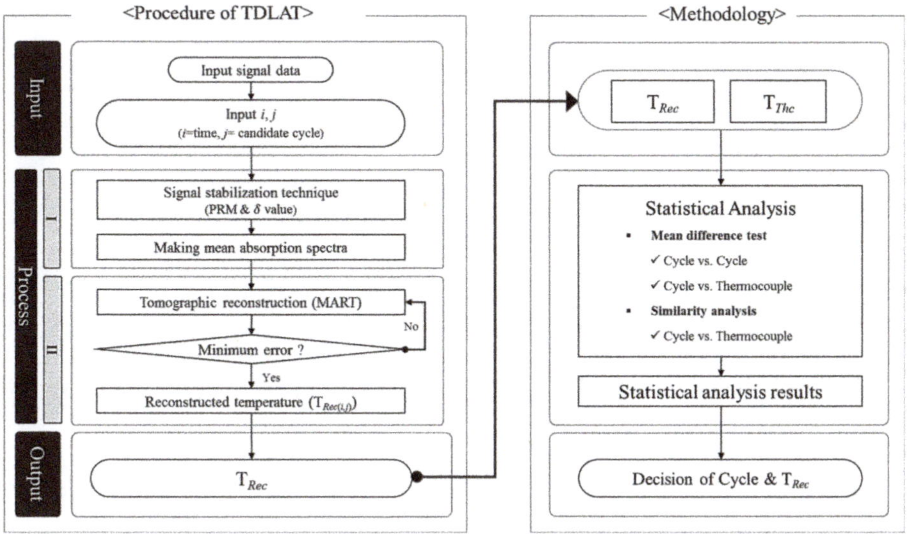

Figure 5. Flowchart for tracking cycle and reconstructed temperature using TDLAT.

In the second step, the PRM and δ value are combined to obtain mean absorption spectra, described in more detail above. This technique was able to remove noise more effectively than the conventional PRM. Reconstruction calculation consisting of 9×9 array was calculated using mean absorption spectra. The reconstructed temperature distribution was tracked by repeated calculations until errors in the experimental and theoretical absorption spectra were minimized by the multiplicative algebraic reconstruction technique (MART) algorithm [24].

$$\alpha_{v(n,m)}^{(k)} = \alpha_{v(n,m)}^{(k-1)} \times \left[\frac{A_v}{L_{(n,m)} \sum_{m=1}^{M} \sum_{n=1}^{N} \alpha_{v(n,m)}^{(k-1)}} \right]^{\beta L_{(n,m)}}, \tag{6}$$

where, the superscript k represents the number of iteration and the subscript (n, m) is the gird number. β is the relaxation parameter that affects the rate of convergence and was set to 0.1. The initial values of $\alpha_{v(n,m)}$ at each grid are an important factor for fast convergence calculation, which were approximated to target values by adopting the multiplication line of sight (MLOS) method [26]. The reconstruction calculation was terminated when difference of $\alpha_{v(n,m)}^{(k)}$ and $\alpha_{v(n,m)}^{(k-1)}$ was less than 1×10^{-6}.

Figure 6 shows a comparison of the measured absorption spectra with the calculated absorption spectra in the engine running time of 5.0 s. The average root-mean-square error (RMSE) at 311 wavenumber were 9.766×10^{-4}, 4.485×10^{-5}, 1.273×10^{-3}, and 2.144×10^{-3}, respectively, in Figure 6a–d. Measured spectra were in good agreement with spectra by the theoretical calculation, but the measured spectra for the 51 and 201 cycles were slightly different from each other, and their reconstruction calculations were also differed slightly. This implies that the number of mean absorption cycles is an important parameter for the reconstruction results.

In the third step, the reconstructed temperatures, $T_{Rec(i,j)}$, calculated for each candidate cycle and time are output. To estimate the $T_{Rec(i,j)}$ from the finally converged $\alpha_{v(n,m)}$ at all grids, the theoretical absorption coefficients at the temperature has already been calculated using the HITRAN database for the mole fraction 1.0. The correlation between the theoretical absorption coefficient, $\alpha_{v,the}$, and the reconstructed absorption coefficient, α_v, was evaluated to find the value with the most similar pattern, i.e., the nearest correlation coefficient to 1.0, and to obtain $T_{Rec(i,j)}$. The correlation coefficient, $C_{n,m}$, at each grid is described as follows.

$$C_{n,m} = \frac{\sum_{v=1}^{n} (\alpha_v - \overline{\alpha})(\alpha_{v,the} - \overline{\alpha}_{the})}{\sqrt{\sum_{v=1}^{n} (\alpha_v - \overline{\alpha})^2 \sum_{v=1}^{n} (\alpha_{v,the} - \overline{\alpha}_{the})^2}}, \tag{7}$$

where $\overline{\alpha}$ and $\overline{\alpha}_{the}$ denote the average value each of the theoretical and reconstructed absorption coefficient.

Figure 7 shows the results of 2D reconstructed temperature distribution using 51 and 201 cycles at the engine running time of 5.0 s. Comparing Figure 7a,b, the temperature difference at the center point was 12.35 K and the RMSE at all grids was 5.87 K. This is an example of the highest temperature difference compared to other engine running times. The characteristics of the temperature distribution for each time were similar, but the temperature fluctuation was shown differently. From these results, it was found that the number of mean cycles affected the results of CT analysis. Additionally, the output $\overline{T}_{Rec}$ was set to the average value of nine grids located in the center of the measurement area as the TC was installed in the center of the exhaust pipe and the temperature was measured at one point. After all, the output T_{Rec} consists of a set of elements $\overline{T}_{Rec}$ for each i and j. The number of T_{Rec} was 17820 in the i × j array (990 × 18).

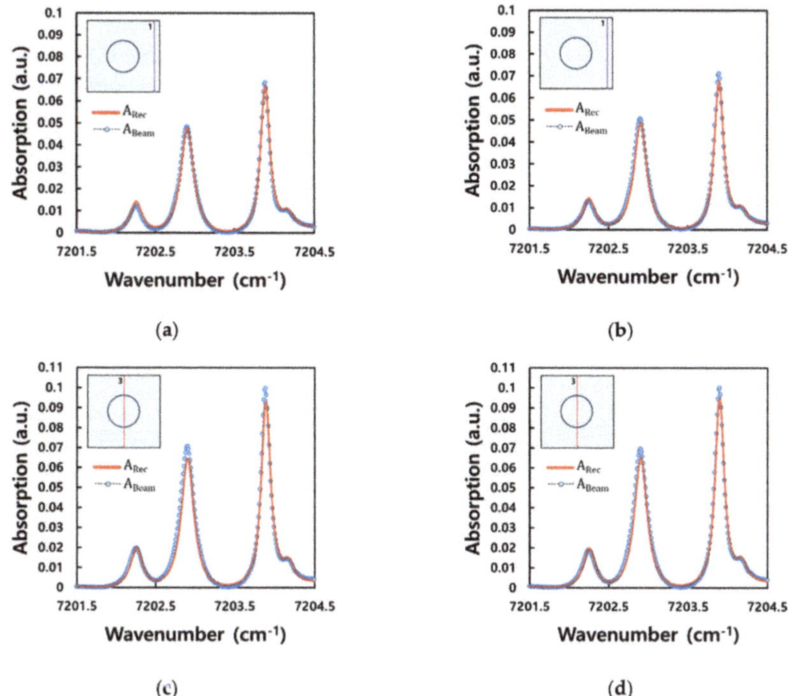

Figure 6. The spectra of reconstructed and measured absorption data at the engine running time 5.0 s:
(**a**) 51 cycles, boundary; (**b**) 201 cycles, boundary; (**c**) 51 cycles, center; and (**d**) 201 cycles, center.

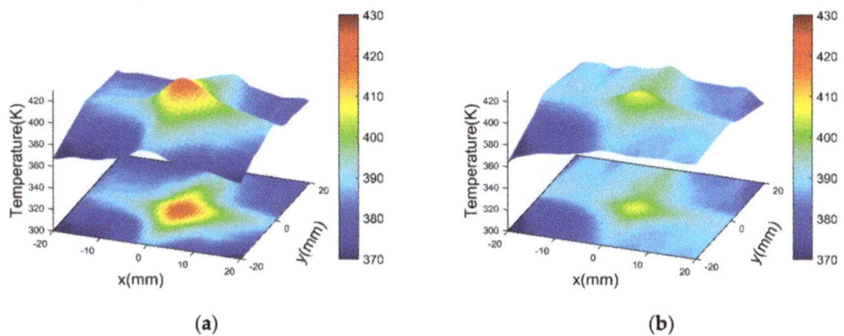

Figure 7. 2D reconstructed temperature distributions at the engine running time 5.0 s: (**a**) 51 cycles and
(**b**) 201 cycles.

In the fourth step, the statistical-based analysis was performed using T_{Rec} and T_{Thc}, which means temperature measured by TC and was used to determine the cycle and T_{Rec} most similar to T_{Thc}. Further details are described in Section 4.

The computation time of each step on an Intel Core i7 2600K CPU with 16GB RAM was as follows. The CPU time of Process I to convert signal data into mean absorption spectra was 1.38 s and 1.90 s (311 wavenumbers) respectively at 51 and 201 cycles. As the number of mean cycles increased, the CPU time increased linearly. The CPU time of Process II was scored in under 1.79 s (311 wavenumbers), mostly converging on less than 150 iterations.

4. Statistical-Based Mean Difference Test and Similarity Analysis

The statistical-based analysis is performed to identify cycles that represent T_{Rec} similar to T_{Thc}. The 18 candidate cycles were selected using suitable intervals for analysis from 31 to 1001 cycles. The procedure is shown in Figure 8. Paired t test, which is a mean difference testing method, is conducted to evaluate the similarity between the 18 candidate cycles [27].

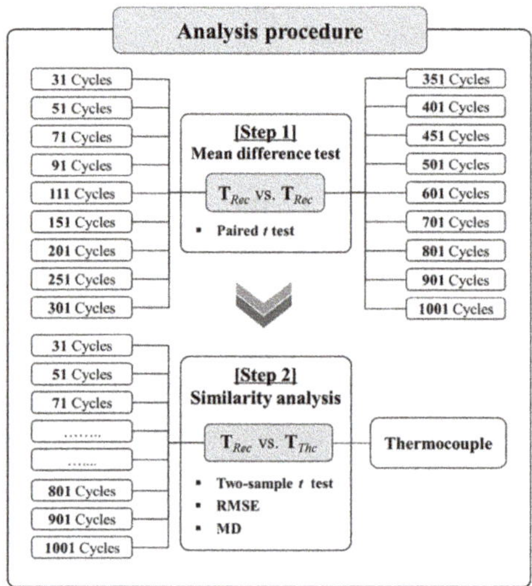

Figure 8. Procedure for mean different test and similarity analysis.

The next step is followed by performing a mean difference test and similarity analysis between T_{Thc} and T_{Rec} for all 18 candidate cycles. The mean difference test uses the representative statistical technique of Two-sample t test. RMSE and time-based Mahalanobis distance (TMD) were used for similarity analysis. RMSE is a measure of generalization of standard deviations that indicates the difference between actual and estimated values [28]. The Mahalanobis distance (MD) represents the distance between data and considers covariance [29].

Finally, the cycles that represent T_{Rec} most similar to T_{Thc} based on the analysis results were determined.

4.1. [Step 1] Mean Difference Test

T_{Rec} derived from 18 candidate cycles consists of 0.1 s intervals from 1 to 9.9 s as shown in Table 1. To test the mean temperature difference in 18 candidate cycles, the analysis was performed on nine $T_{Rec(i,j)}$, which is the reconstructed temperature of ith reference time of the j-cycles. The total number of sets comparing $T_{Rec(i,j)}$ in 18 candidate cycles to each other for the paired t test were 153. For the actual test, the analysis was performed only at the reference time (1 s, 2 s, 3 s, 4 s, 5 s, 6 s, 7 s, 8 s, and 9 s) for all 153 sets.

4.1.1. Paired t Test

The paired t-test is useful for analyzing the same set of items that were measured under two different conditions, i.e., differences in measurements made on the same subject before and after a treatment, or differences between two treatments given to the same subject.

The paired t-test is based on the test statistic as shown below:

$$T_{pt} = \frac{\overline{D} - d_0}{S_D / \sqrt{n}}, \tag{8}$$

where,

$$\overline{D} = \frac{\sum_{k=1}^{n}\left(T_{Rec(i,A)}{}^{k} - T_{Rec(i,B)}{}^{k}\right)}{n}, \tag{9}$$

$$S_D = \sqrt{\frac{\sum_{k=1}^{n}\left(T_{Rec(i,A)}{}^{k} - T_{Rec(i,B)}{}^{k} - \overline{D}\right)^2}{n-1}}, \tag{10}$$

where, $\overline{D}$ denotes the mean of the differences between the two kinds of cycles. d_0 denotes a constant that can be set depending on whether the mean of the difference between the two groups is zero or a specific non-zero value. S_D denotes the standard deviation of differences between two kinds of cycles. $T_{Rec(i,A)}{}^{k}$ denotes the kth reconstruction temperature at ith time (s) of cycle A, $i = 1.0, 1.1, 1.2, \cdots, 9.9$. $T_{Rec(i,B)}{}^{k}$ denotes the kth reconstruction temperature at ith time (s) of cycle B. n denotes the number of T_{Rec} in ith time (s).

Table 1. T_{Rec} per 18 candidate cycles used in [Step 1].

No.	Time (s)	T_{Rec}				
		31	51	$\cdots$	901	1001
1		404.833	400.000	$\cdots$	405.432	404.189
2		406.775	401.665	$\cdots$	407.081	405.842
3		404.947	399.784	$\cdots$	405.436	404.455
4		407.458	401.846	$\cdots$	407.028	405.633
5	1.0	409.400	403.506	$\cdots$	408.581	407.223
6		407.556	401.614	$\cdots$	406.820	405.720
7		404.951	399.982	$\cdots$	405.459	404.137
8		406.842	401.604	$\cdots$	406.900	405.606
9		404.999	399.722	$\cdots$	405.051	405.159
$\cdots$	$\cdots$	$\cdots$	$\cdots$	$\cdots$	$\cdots$	$\cdots$
806		398.712	399.182	$\cdots$	400.164	400.543
807		397.870	397.965	$\cdots$	402.270	402.782
808	9.9	396.147	396.544	$\cdots$	401.184	401.859
809		397.183	397.668	$\cdots$	401.351	401.646
810		396.221	396.327	$\cdots$	403.357	403.785

For paired t test, the hypothesis is as follows:

- Null hypothesis: $H_0 : \overline{D} = d_0$ (the population mean of the difference $(\overline{D})$ equals the hypothesized mean of the difference $(d_0 = 0)$.
- Alternative hypothesis: $H_1 : \overline{D} \neq d_0$ (the population mean of the difference $(\overline{D})$ does not equal the hypothesized mean of the difference $(d_0 = 0)$.

Here, the null hypothesis is rejected when the p-value is less than or equal to 0.05, i.e., when there is no difference between the mean of the two groups.

4.1.2. Results of Mean Difference Test in [Step 1]

Table 2 is a summary of the paired t-test results. For 153 sets, 1377 test results would be obtained when each test is performed from 1 to 9 s. If the p-value is less than 0.05, the null hypothesis is rejected. The p-values of 1 s, 3 s, 8 s, and 9 s were less than 0.05, i.e., the means of all cases were not equal to each other. In 2 s, the p-value was less than 0.05 except in four sets (i.e., 251 vs. 301, 451 vs. 501, 451 vs. 601, and 501 vs. 601). In 4 s, only one set (351 vs. 401) showed a p-value greater than 0.05 and the remaining 152 sets had a p-value of less than 0.05. In 5 s, 6 s, and 7 s, the p-value was greater than 0.05, while the remaining 152 sets had a p-value less than 0.05. Therefore, except for eight sets among 1377 sets, the p-value in all sets was less than 0.05 and hence, there was a mean difference between the cycles.

Table 2. Results of paired *t*-test.

Time (s)	Test Sets			*p*-Value
1.0		All sets		<0.05
	251	vs.	301	0.626
2.0	451	vs.	501	0.476
	451	vs.	601	0.402
	501	vs.	601	0.331
3.0		All sets		<0.05
4.0	351	vs.	401	0.371
5.0	451	vs.	501	0.208
6.0	401	vs.	451	0.633
7.0	601	vs.	701	0.221
8.0		All sets		<0.05
9.0		All sets		<0.05

4.2. [Step 2] Similarity Analysis

Table 3 shows the $T_{Thc(i)}$ and $\overline{T}_{Rec(i,j)}$ of all 18 candidate cycles used in the analysis where each $\overline{T}_{Rec(i,j)}$ is the mean value for nine $T_{Rec(i,j)}{}^k$ per 0.1 s interval in Table 1. For example, in Table 3, $\overline{T}_{Rec(1.0,31)}$ is 406.418, which averaged reconstruction temperatures from $T_{Rec(1.0,31)}{}^1 = 404.833$ to $T_{Rec(1.0,31)}{}^9 = 404.999$ in Table 1. Here, three methods were used for similarity analysis. First, the Two-sample *t* test was used to test the difference between the means of the two groups. Second, RMSE analysis was performed to address the difference between estimates or models and observed values in real-world settings. RMSE analysis can be used to aggregate the differences between the residuals of two groups into one measure. Finally, in TMD analysis the correlation between two groups with different means was taken into account.

Table 3. Temperatures used in [Step 2].

No.	Group	Time (s)	$T_{Thc(i)}$	$\overline{T}$			
				31	51	$\cdots$	1001
1		1.0	408.000	406.418	401.080	$\cdots$	405.204
2		1.1	408.688	402.722	401.324	$\cdots$	406.271
3		1.2	409.921	407.890	407.890	$\cdots$	406.525
4		1.3	407.798	410.285	411.842	$\cdots$	406.990
5	1	1.4	411.453	402.434	411.723	$\cdots$	406.453
6		1.5	406.759	392.685	401.679	$\cdots$	406.284
7		1.6	406.005	397.013	405.508	$\cdots$	407.419
8		1.7	405.461	418.457	416.981	$\cdots$	407.537
9		1.8	405.110	407.200	401.237	$\cdots$	406.350
10		1.9	404.842	416.607	407.599	$\cdots$	405.416
$\cdots$	$\cdots$	$\cdots$	$\cdots$	$\cdots$	$\cdots$	$\cdots$	$\cdots$
80		9.0	404.658	397.217	397.462	$\cdots$	401.795
81		9.1	405.853	394.707	393.932	$\cdots$	401.638
82		9.2	402.657	408.041	403.190	$\cdots$	400.741
83		9.3	400.688	395.484	396.351	$\cdots$	400.231
84	9	9.4	401.921	402.101	398.922	$\cdots$	401.280
85		9.5	399.798	393.647	396.263	$\cdots$	401.277
86		9.6	403.453	410.205	399.067	$\cdots$	401.559
87		9.7	400.895	390.764	398.329	$\cdots$	401.343
89		9.8	403.987	398.441	401.295	$\cdots$	400.894
90		9.9	401.752	403.739	409.931	$\cdots$	400.857

4.2.1. Two-Sample *t*-Test

Two-sample *t*-test is a widely used hypothesis test used to compare whether the averages are equal. The assumptions made while doing a Two-sample *t*-test include: (i) the data are continuous, (ii) the data follows normal probability distribution, (iii) the variances of the two populations are equal, (iv) two-sample is independent, and (v) both samples are simple random samples from their respective populations. The Two-sample *t*-test is based on the test statistic shown below:

$$T_{tw} = \frac{\overline{T}_{Thc^g} - \overline{\overline{T}}_{Rec^g}}{\sqrt{T_{S^2}}} \sqrt{\frac{n_{T^g} n_{R^g}}{n_{T^g} + n_{R^g}}}, \tag{11}$$

$$T_{S^2} = \frac{\sum_{i=1}^{n_{T^g}} \left(T_{Thc(i)} - \overline{T}_{Thc^g}\right)^2 + \sum_{i=1}^{n_{R^g}} \left(\overline{T}_{Rec(i,j)} - \overline{\overline{T}}_{Rec^g}\right)^2}{n_{T^g} + n_{R^g} - 2}, \tag{12}$$

where, T_{S^2} denotes the variance of $T_{Thc(i)}$ and $T_{Rec(i,j)}$. $T_{Thc(i)}$ denotes the TC temperature at ith time (s), $i = 1.0,\ 1.1,\ 1.2,\ \cdots,\ 9.9$. $\overline{T}_{Thc}{}^g$ denotes the mean of all $T_{Thc(i)}$ in the gth group, $g = 1, 2, \cdots, 9$. $\overline{\overline{T}}_{Rec}{}^g$ denotes the mean of all $\overline{T}_{Rec(i,j)}$ in the gth group. n_{T^g} denotes the number of all $T_{Thc(i)}$ in the gth group. $n_{Rec}{}^g$ denotes the number of all $T_{Rec(i,j)}$ in the gth group.

For the Two-sample *t* test, the hypothesis is as follows

- Null hypothesis: H_0: $= \overline{T}_{Thc}{}^g - \overline{\overline{T}}_{Rec}{}^g = \delta_0$ (the difference between the population means ($\overline{T}_{Thc}{}^g - \overline{\overline{T}}_{Rec}{}^g$) is equal to the hypothesized difference ($\delta_0 = 0$)).
- Alternative hypothesis: H_1: $= \overline{T}_{Thc}{}^g - \overline{\overline{T}}_{Rec}{}^g \neq \delta_0$ (the difference between the population means ($\overline{T}_{Thc}{}^g - \overline{\overline{T}}_{Rec}{}^g$) is not equal to the hypothesized difference ($\delta_0 \neq 0$)).

Here, the null hypothesis is rejected when the *p*-value is less than or equal to 0.05, i.e., when there is no difference between the mean of the two groups. In this study, statistical software version 18 of Minitab was used for the Two-sample *t* test.

4.2.2. Root Mean Square Error (RMSE)

RMSE is used to measure the difference between values predicted by a model or an estimator and the value observed. The RMSE is a measure of accuracy to compare forecasting errors of different models for a particular dataset and not between the dataset as it is scale-dependent. RMSE is always non-negative and a value of zero would indicate a perfect fit to the data. In general, a lower RMSE is better than a higher one. The formula is as follows:

$$T_{RMSE} = \sqrt{\frac{1}{m} \times \sum_{i=1}^{m} \left(T_{Thc(i)} - T_{Rec(i,j)}\right)^2}. \tag{13}$$

4.2.3. Time-Based Mahalanobis Distance (TMD)

MD was introduced by P. C. Mahalanobis in 1936 to measure distance between a point P and distribution D. It is a multi-dimensional generalization of the idea for measuring the amount of standard deviations P is away from the mean of D [30]. Among various methods for measuring distance (i.e., Euclidean distance, Manhattan distance, etc.), MD is used as it has the advantage of calculating the distance by considering weight according to the magnitude of covariance between variables [31]. Thus, MD is a method that is used to assess similarities between two groups. In this study, TMD analysis is performed to analyze similarities between T_{Thc} and $\overline{T}_{Rec}$ by reference time. The closer the TMD value is to zero, the more likely T_{Thc} and $\overline{T}_{Rec}$ are to be similar. The calculation of TMD procedure is as follows:

[Stage 1] Calculate the covariance matrix (S) and TMD using the following formula (14) and (15).

$$S = \begin{bmatrix} Var\left[T_{Thc(i)}\right] & Cov\left[T_{Thc(i)}, \overline{T}_{Rec(i,j)}\right] \\ Cov\left[\overline{T}_{Rec(i,j)}, T_{Thc(i)}\right] & Var\left[\overline{T}_{Rec(i,j)}\right] \end{bmatrix}, \tag{14}$$

$$MD = \sqrt{\left(X_i - \overline{X}\right)^T S^{-1}\left(X_i - \overline{X}\right)}, \tag{15}$$

$$\left(X_i - \overline{X}\right) = \begin{pmatrix} T_{Thc(i)} - \overline{T}_{Thc} \\ \overline{T}_{Rec(i,j)} - \overline{\overline{T}}_{Rec(i,j)} \end{pmatrix}, \tag{16}$$

where, $\overline{T}_{Thc}$ denotes the geometric mean of all $T_{Thc(i)}$, $i = 1.0,\ 1.1,\ 1.2,\ \cdots,\ 9.9$. $\overline{\overline{T}}_{Rec(i,j)}$ denotes the geometric mean of all $\overline{T}_{Rec(i,j)}$ in jth cycles. S^{-1} denotes inverse covariance matrix. T indicates that the vector should be transposed.

[Stage 2] The obtained TMD values are calculated using the following formula for each time (s) interval.

$$T_{MD} = (MD_1 \times MD_2 \times \cdots \times MD_n)^{\frac{1}{n}}, \ n = 1, 2, \cdots m. \tag{17}$$

4.2.4. Results of Similarity Analysis in [Step 2]

Two-sample t test, RMSE, and TMD analysis were performed to select the cycles that indicate a T_{Rec} similar to that of T_{Thc}. The results of a normality test conducted on T_{Thc} and T_{Rec} for 18 candidate cycles prior to the Two-sample t test showed that all were satisfactory for normality. However, the equivalence test between T_{Thc} and T_{Rec} showed that some results did not satisfy the equivalence. The results of the equivalence test are summarized in Appendix A. In general, Welch's t-test is used if the two groups do not meet equal variances. However, if the very small samples ($n < 10$), the Two-sample t test performs slightly better than Welch's t-test [32].

The first Two-sample t-test results are shown in Table 4 and the null hypothesis is rejected when the p-value is less than 0.05 (see Section 4.2.1). The analysis results are explained based on the cycle, all p-values from 1 to 9 s for 31, 51, 91, 111, and 251 cycles were above 0.05. For the 71 cycles, the p-value was close to 1 at 2 s, 5 s, 6 s, and 8 s. In other words, the probability of selecting null hypothesis was high. For 151 and 201 cycles, the p-value was over 0.05 for all seconds except 5 s. On the other hand, the 451 cycles showed p-value less than 0.05 at 5 s and 9 s. The p-value of the 901 cycles was less than 0.05 at 8 s and 9 s. In the 501, 601, 701, and 801 cycles, the p-values were less than 0.05 for 5 s, 8 s, and 9 s. In the case of the 1001 cycles, the p-value of 3 s, 8 s, and 9 s was found to be below 0.05.

The result of RMSE analysis is shown is Table 5. The smaller the RMSE values, the more the similarity between T_{Thc} and T_{Rec}. First, the analysis results were based on 10 temperatures per group from 1 to 9 s (see Table 3). In terms of time, the RMSE value (1.951) of 401 cycles in 1 s was the smallest. In 2 s, the RMSE value of 1001 cycles was the smallest at 2.270. RMSE values for 601 cycles were the smallest in 3 s and 4 s. The RMSE values for 251, 201, and 701 cycles were the lowest in 5 s, 6 s, and 7 s. RMSE values for 1001 cycles were the smallest in 8 s and 9 s. The analysis using 90 temperatures from 1 to 9.9 s (see Table 3) showed that the RMSE value of 601 cycles was the smallest.

Lastly, the analysis results of TMD are shown is Table 6. The smaller the TMD value, the more the similarity between T_{Thc} and T_{Rec}. The analysis results were based on 10 temperatures per group from 1 to 9 s (see Table 3). In terms of time, the TMD value (1.626) of 351 cycles in 1 s was the smallest. In 2 s, the TMD value of 111 cycles was the smallest at 0.770. The TMD value for 151, 201, 251, and 801 cycles were the lowest in 3 s, 4 s, 5 s, and 6 s. TMD values for 201 cycles were the smallest in 7 s, 8 s, and 9 s. Next, the analysis using 90 temperature readings from 1 to 9.9 s showed that the TMD value (0.933) of 201 cycles was the smallest.

Table 4. Results of Two-sample *t*-test.

Cycles	Time (s)								
	1	**2**	**3**	**4**	**5**	**6**	**7**	**8**	**9**
					p-**Value**				
31	0.647	0.463	0.839	0.273	0.692	0.496	0.924	0.976	0.169
51	0.711	0.726	0.316	0.292	0.707	0.681	0.864	0.728	0.069
71	0.774	0.985	0.147	0.292	0.982	0.896	0.727	0.997	0.164
91	0.603	0.891	0.225	0.441	0.369	0.985	0.842	0.693	0.362
111	0.531	0.983	0.390	0.667	0.092	0.901	0.828	0.352	0.482
151	0.486	0.892	0.402	0.671	<0.05	0.779	0.708	0.574	0.474
201	0.396	0.964	0.328	0.381	<0.05	0.958	0.684	0.488	0.177
251	0.510	0.858	0.271	0.289	0.074	0.743	0.791	0.467	0.078
301	0.384	0.890	0.246	0.325	<0.05	0.789	0.928	0.137	0.109
351	0.415	0.928	0.196	0.411	<0.05	0.770	0.979	0.102	0.073
401	0.406	0.790	0.177	0.253	<0.05	0.655	0.984	0.112	0.051
451	0.530	0.735	0.136	0.235	<0.05	0.537	0.946	0.081	<0.05
501	0.509	0.691	0.136	0.269	<0.05	0.622	0.865	<0.05	<0.05
601	0.478	0.620	0.097	0.284	<0.05	0.495	0.886	<0.05	<0.05
701	0.432	0.631	0.084	0.367	<0.05	0.449	0.911	<0.05	0.054
801	0.357	0.550	0.058	0.336	<0.05	0.430	0.977	<0.05	<0.05
901	0.316	0.574	0.053	0.431	0.062	0.431	0.871	<0.05	<0.05
1001	0.216	0.528	<0.05	0.446	0.057	0.413	0.773	<0.05	<0.05

Table 5. Results of root mean square error (RMSE) analysis between T_{Thc} and T_{Rec} for 18 candidate cycles.

Cycles	Time (s)									Total	
	1	**2**	**3**	**4**	**5**	**6**	**7**	**8**	**9**		
					RMSE					**RMSE**	**Rank**
31	8.487	8.039	10.325	8.441	12.622	7.869	7.320	5.887	5.887	8.578	18
51	5.514	5.592	6.370	5.994	7.759	4.776	6.912	5.526	5.526	6.049	17
71	6.269	4.497	5.404	6.334	5.316	6.007	6.028	4.420	4.420	5.538	16
91	4.879	3.775	4.331	5.559	3.894	5.921	5.073	3.620	3.620	4.822	15
111	4.702	3.860	4.282	5.046	3.580	5.265	4.156	3.496	3.496	4.393	14
151	3.755	3.129	4.216	4.833	3.564	2.957	4.314	4.065	4.065	3.918	13
201	3.177	2.978	3.349	3.376	2.900	2.529	3.810	3.736	3.736	3.199	12
251	3.024	3.063	3.496	3.400	2.338	3.166	3.243	2.755	2.755	3.120	11
301	2.603	2.884	3.159	2.961	2.749	2.844	2.541	2.267	2.267	2.809	10
351	2.056	2.733	2.799	2.686	2.955	3.333	2.529	2.099	2.099	2.703	9
401	1.951	2.583	2.641	2.451	2.849	3.149	2.546	2.456	2.456	2.584	4
451	2.047	2.909	2.662	2.403	2.785	3.439	2.512	2.156	2.156	2.641	7
501	2.433	2.580	2.703	2.278	2.894	3.543	2.433	2.140	2.140	2.643	8
601	2.096	2.453	2.561	2.265	2.947	3.368	2.182	2.039	2.039	2.491	1
701	2.201	2.591	2.899	2.426	2.925	3.544	1.905	1.987	1.987	2.555	3
801	2.209	2.318	3.055	2.629	2.771	3.492	1.907	1.870	1.870	2.549	2
901	2.438	2.499	3.174	2.568	2.735	3.693	2.054	1.715	1.715	2.624	6
1001	2.434	2.270	3.438	2.576	2.693	3.606	2.041	1.660	1.660	2.597	5

Table 6. Results of time-based Mahalanobis distance (TMD) analysis between T_{Thc} and T_{Rec} for 18 candidate cycles.

Cycles	Time (s)									Total	
	1	2	3	4	5	6	7	8	9		
					TMD					TMD	Rank
31	1.860	1.124	1.188	1.208	1.174	1.383	0.889	0.731	0.880	1.121	4
51	1.758	1.079	1.279	1.267	1.204	1.253	0.910	0.854	0.876	1.136	6
71	1.829	0.967	1.250	1.305	0.995	1.363	0.970	0.736	1.014	1.123	5
91	1.780	0.873	1.170	1.348	1.006	1.299	0.915	0.773	1.148	1.112	3
111	1.777	0.770	1.155	1.307	1.022	1.304	0.921	0.835	0.962	1.082	2
151	1.727	0.936	1.133	1.456	1.143	1.303	1.052	0.997	1.071	1.181	10
201	2.132	0.798	1.552	0.888	1.549	0.992	0.684	0.429	0.506	0.933	1
251	1.675	1.277	1.391	1.273	0.965	1.361	1.016	0.812	1.163	1.190	14
301	1.719	1.271	1.506	1.252	1.258	1.291	1.050	0.755	0.872	1.185	12
351	1.626	1.280	1.394	1.261	1.291	1.481	0.996	0.745	0.911	1.188	13
401	1.708	1.297	1.477	1.273	1.363	1.263	1.046	0.911	0.993	1.237	18
451	1.772	1.375	1.312	1.292	1.205	1.258	0.972	0.788	1.033	1.194	16
501	1.871	1.342	1.264	1.328	1.309	1.153	0.945	0.811	1.056	1.200	17
601	1.880	1.373	1.262	1.439	1.320	0.981	0.997	0.726	1.098	1.191	15
701	1.956	1.355	1.341	1.458	1.258	0.990	0.882	0.754	1.053	1.182	11
801	2.012	1.320	1.347	1.479	1.299	0.849	0.814	0.727	1.147	1.165	8
901	2.061	1.319	1.411	1.400	1.246	1.055	0.779	0.693	1.138	1.177	9
1001	2.058	1.233	1.382	1.393	1.243	1.006	0.788	0.655	1.153	1.154	7

5. Conclusions

The results of this study are summarized as follows: (*i*) the paired *t* test for 18 candidate cycles indicated that, in most cases, T_{Rec} was not the same in each cycle. This implies that T_{Rec} might vary depending on the cycle set during CT analysis. Therefore, a standard for proper cycle setting is needed. (*ii*) Three types of similarity analysis yielded different results. First, the Two-sample *t* test showed that the T_{Rec} for 71 cycles was not the much different from that of T_{Thc}. Second, the results from RMSE showed that the T_{Rec} of the 601 cycles was most similar to that of the T_{Thc}. Finally, TMD results show that the T_{Rec} for 201 cycles was most similar to that of T_{Thc}. (*iii*) The reason for different results is that all three methodologies have different characteristics. For a more visualized description, a comparison of T_{Thc} and T_{Rec} was provided based on the similarity analysis result. Figures 9–11 show a comparison of the T_{Thc} with the T_{Rec} for each 71, 601, and 201 cycles, respectively. The movement of T_{Thc} and T_{Rec} over time shows that T_{Rec} for 601 and 201 cycles moved closely with T_{Thc}. However, a closer look at the temperature change pattern shows that T_{Rec} of 201 cycles was moving similarly to T_{Thc}. On the other hand, for 71 cycle, the range of T_{Rec} changes over time was observed to be much larger than T_{Thc}. (*iv*) For the Two-sample *t*-test, it was difficult to determine similarity by sensitively reflecting changes in temperature as the difference in the mean of the overall temperature data in a given period was taken into account. Likewise, the RMSE was purely an indicator of the mean of distance between T_{Thc} and T_{Rec}, which made it difficult to analyze, including the correlation between changes in T_{Thc} and T_{Rec}. TMD was a suitable indicator for this study because it considers both the distance and correlation between T_{Thc} and T_{Rec}. Therefore, if the engine used in this study is in idle mode, setting it to 201 cycles in the CT analysis will increase the accuracy of the analysis while minimizing the analysis time and effort.

Figure 12 shows the 2D reconstruction temperature distribution in 201 cycles in time from 1.0 to 9.0 s. The temperature of the peak area of the 2D distribution in each time was equal to that of the corresponding time at the 201 cycle reconstruction temperature in Figure 11. As illustrated in Figure 12, it can be seen that the maximum temperature was continuously distributed in the central area. For example, Figure 12b,h show a reconstruction temperature of approximately 410 K and 408 K in the central area, and Figure 10 shows also the same temperature at that time. This was the same in other times.

This study presented the statistical analysis procedure and the appropriate cycle to improve TDLAT performance when measuring time series-based 2D temperature in idle mode of engine. To the best of our knowledge, this is the first attempt to present the criteria for cycle selection through the similarity analysis when based on signal stabilization technique. This signal stabilization technique was more effective in error reduction than the conventional polynomial reduction method when converting signals into absorption spectra.

However, limitations of this study include (*i*) it was only tested in idle mode of a particular engine, so additional experiments are required in operation mode with relatively large temperature variations; and (*ii*) in idle mode, the temperature variation in the peak area was uniformly measured only for that area, but it is also necessary to check the need for measurements in other areas.

To overcome these limitations, the following studies will be carried out in the future: (*i*) conduct a time series-based 2D temperature measurement study by applying the latest statistical techniques (i.e., linear Bayesian absorbance inference, Tikhonov regularization, and so on) to provide a fast and accurate case of reconstruction temperature calculation and (*ii*) typically, the reconstruction temperature is verified by TC measurements. The reliability of verification will depend on the TC measurement method, therefore studies on the TC verification methodology will also be conducted; and (*iii*) the size of the laser grid will affect the bias of reconstruction temperature. We would like to conduct a study on the selection of optimal grid sizes to minimize bias.

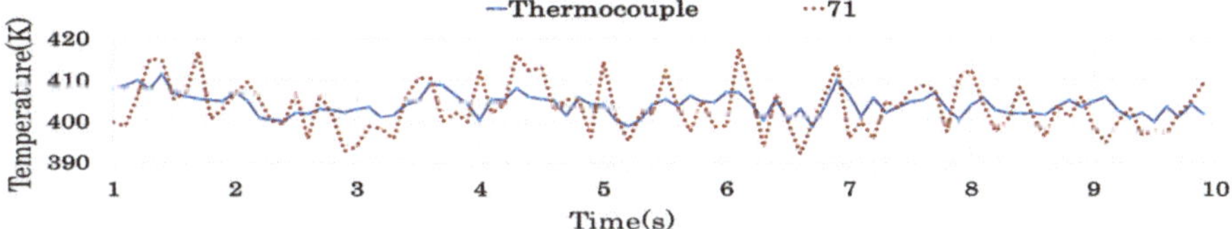

Figure 9. Comparison of T_{Thc} and T_{Rec} for 71 cycles.

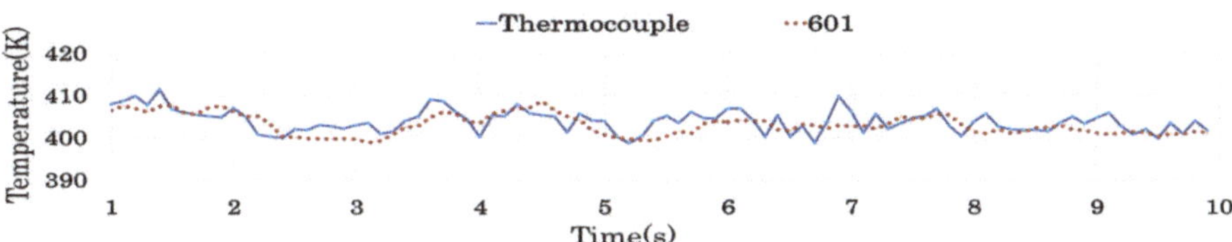

Figure 10. Comparison of T_{Thc} and T_{Rec} for 601 cycles.

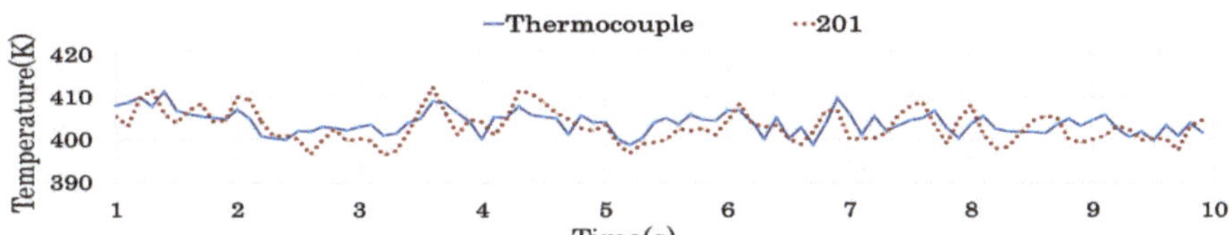

Figure 11. Comparison of T_{Thc} and T_{Rec} for 201 cycles.

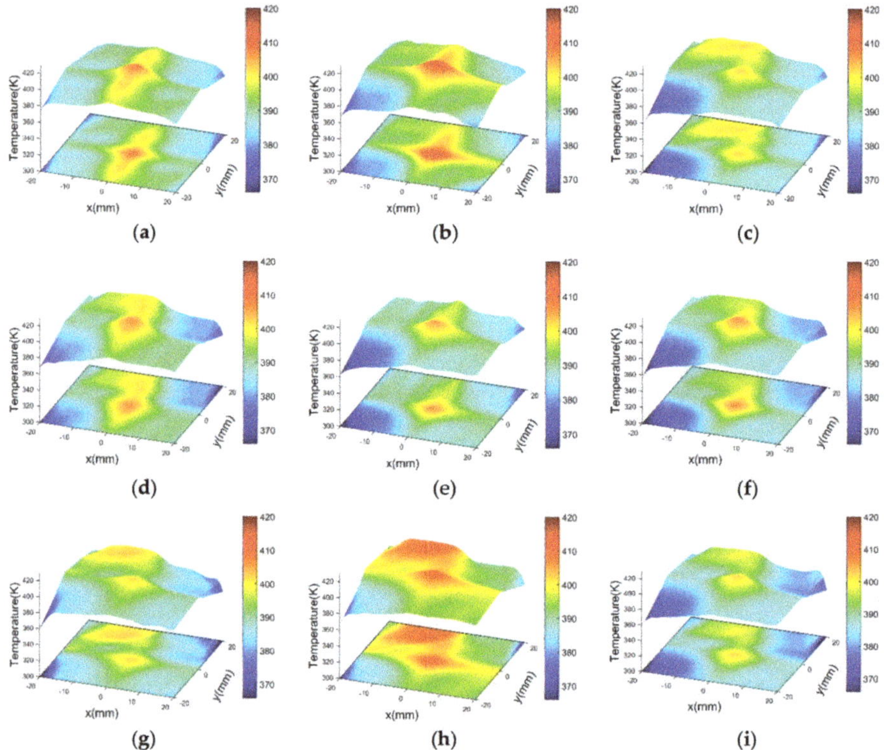

Figure 12. 2D reconstruction temperature distributions at 201 cycles: (**a**) 1.0 s; (**b**) 2.0 s; (**c**) 3.0 s; (**d**) 4.0 s; (**e**) 5.0 s; (**f**) 6.0 s; (**g**) 7.0 s; (**h**) 8.0 s; and (**i**) 9.0 s.

Author Contributions: Conceptualization, H.J., D.C.; Methodology, H.J., D.C.; Modeling, H.J., D.C.; Experiment, D.C.; Analysis, H.J.; validation, H.J., D.C.; Writing—original draft preparation, H.J., D.C.; Writing—Review and Edition, H.J., D.C.; Funding Acquisition, H.J., D.C. All authors have read and agreed to the published version of the manuscript.

Funding: This work was supported by the National Research Foundation of Korea (NRF) grant funded by the Korea government (MSIT) (No. 2018R1C1B5085281, 2019R1G1A1010335).

Conflicts of Interest: The authors declare no conflicts of interest.

Appendix A

The Two-sample t test assumes a normality and equivalence. Since normality was satisfied, only the results of the equivalence test were summarized in Table A1. The equivalence is satisfied when p-value is more than 0.05.

Table A1. Result of the equal variance test between T_{Thc} and T_{Rec} for 18 candidate cycles.

Cycles	Time (s)								
	1	**2**	**3**	**4**	**5**	**6**	**7**	**8**	**9**
	P-Value								
31	<0.05	0.055	<0.05	<0.05	0.094	<0.05	<0.05	<0.05	<0.05
51	<0.05	<0.05	0.175	<0.05	0.062	0.105	<0.05	<0.05	0.205
71	<0.05	<0.05	0.127	<0.05	0.117	<0.05	<0.05	<0.05	0.071
91	0.071	<0.05	0.109	<0.05	0.265	0.073	0.071	<0.05	0.099
111	0.160	0.053	0.142	<0.05	0.426	0.162	0.392	0.060	0.207
151	0.099	0.213	0.218	<0.05	0.401	0.706	0.106	<0.05	0.143
201	0.581	0.215	0.152	0.168	0.997	0.617	0.104	<0.05	0.832
251	0.899	0.148	0.115	0.280	0.645	0.688	0.388	0.302	0.678
301	0.349	0.276	0.066	0.355	0.466	0.173	0.801	0.635	0.537
351	0.398	0.285	0.220	0.181	0.614	0.271	0.660	0.758	0.391
401	<0.05	0.248	0.153	0.358	0.499	<0.05	0.529	0.565	0.062
451	0.054	0.243	0.284	0.227	0.348	0.064	0.358	0.867	0.125
501	<0.05	0.427	0.379	0.562	0.332	<0.05	0.114	0.127	<0.05
601	<0.05	0.548	0.794	0.854	0.387	<0.05	<0.05	<0.05	<0.05
701	<0.05	0.247	0.832	0.833	0.281	<0.05	<0.05	<0.05	<0.05
801	<0.05	0.413	0.607	0.565	0.231	<0.05	<0.05	<0.05	<0.05
901	<0.05	0.352	0.729	0.441	0.210	<0.05	<0.05	<0.05	<0.05
1001	<0.05	0.687	0.343	0.383	0.103	<0.05	<0.05	<0.05	<0.05

References

1. Amin, R.; Galen, B.F.; John, W.H.; Joseph, R.T.; James, D.P.; Christine, K.L. Rapidly pulsed reductants for diesel NO_x reduction with lean NO_x traps: Effects of pulsing parameters on performance. *Appl. Culul. B Environ.* **2018**, *223*, 177–191.
2. Perng, C.C.; Easterling, V.G.; Harold, M.P. Fast lean-rich cycling for enhanced NO_x conversion on storage and reduction catalysts. *Catal. Today* **2014**, *231*, 125–134. [CrossRef]
3. Lietti, L.; Artioli, N.; Righini, L.; Castoldi, L.; Forzatti, P. Pathways for N_2 and N_2O formation during the reduction of NO_x over Pt–Ba/Al_2O_3 LNT catalysts investigated by labeling isotopic experiments. *Ind. Eng. Chem. Res.* **2012**, *51*, 7597–7605. [CrossRef]
4. Roy, S.; Gord, J.R.; Patnaik, A.K. Recent advances incoherent anti-Stokes Raman scattering spectroscopy: Fundamental developments and applications in reacting flows. *Prog. Energy Combust. Sci.* **2010**, *36*, 280–306. [CrossRef]
5. Steuwe, C.; Kaminski, C.F.; Baumberg, J.J.; Mahajan, S. Surface enhanced coherent anti-Stokes Raman scattering on nanostructured gold surfaces. *Nano Lett.* **2011**, *11*, 5339–5343. [CrossRef] [PubMed]
6. Williams, B.; Edwards, M.; Stone, R.; Williams, J.; Ewart, P. High precision in-cylinder gas thermometry using laser induced gratings: Quantitative measurement of evaporative cooling with gasoline/alcohol blends in a GDI optical engine. *Combust. Flame* **2014**, *161*, 270–279. [CrossRef]
7. Hoffman, D.; Münch, K.U.; Leipertz, A. Two-dimensional temperature determination in sooting flames by filtered Rayleigh scattering. *Opt. Lett.* **1996**, *21*, 525–527. [CrossRef]
8. Itani, L.; Bruneaux, G.; Lella, A.; Schulz, C. Two-tracer LIF imaging of preferential evaporation of multi-component gasoline fuel sprays under engine conditions. *Proc. Combust. Inst.* **2015**, *35*, 2915–2922. [CrossRef]
9. Cho, K.Y.; Satija, A.; Pourpoint, T.L.; Son, S.F.; Lucht, R.P. High-repetition-rate three-dimensional OH imaging using scanned planar laser-induced fluorescence system for multiphase combustion. *Appl. Opt.* **2014**, *53*, 316–326. [CrossRef]
10. Deguchi, Y.; Yasui, D.; Adachi, A. Development of 2D temperature and concentration measurement method using tunable diode laser absorption spectroscopy. *J. Mech. Eng. Autom.* **2012**, *2*, 543–549.
11. Ma, L.; Li, X.; Sanders, S.T.; Caswell, A.W.; Roy, S.; Plemmons, D.H.; Gord, J.R. 50-kHz-rate 2D imaging of temperature and H_2O concentration at the exhaust plane of a J85 engine using hyperspectral tomography. *Opt. Express* **2013**, *21*, 1152–1162. [CrossRef] [PubMed]

12. Busa, K.M.; Ellison, E.N.; McGovern, B.J.; McDaniel, J.C.; Diskin, G.S.; DePiro, M.J.; Capriotti, D.P.; Gaffney, R.L. Measurements on NASA Langley Durable Combustor Rig by TDLAT: Preliminary Results. In Proceedings of the 51st AIAA Aerospace Sciences Meeting Including New Horizons Forum and Aerospace Exposition, Grapevine, TX, USA, 7–10 January 2013.

13. Busa, K.M.; Rice, B.E.; McDaniel, J.C.; Goyne, C.P.; Rockwell, R.D.; Fulton, J.A.; Edwards, J.R.; Diskin, G.S. Scramjet combustion efficiency measurement via tomographic absorption spectroscopy and particle image velocimetry. *AIAA J.* **2016**, *54*, 2463–2471. [CrossRef]

14. Sun, P.; Zhang, Z.; Li, Z.; Gou, Q.; Dong, F. Study of two dimensional tomography reconstruction of temperature and gas concentration in combustion field using TDLAS. *Appl. Sci.* **2017**, *7*, 990. [CrossRef]

15. Zhang, Z.R.; Pang, T.; Yang, Y.; Xia, H.; Cui, X.; Sun, P.; Wu, B.; Wang, Y.; Sigrist, M.W.; Dong, F. Development of a tunable diode laser absorption sensor for online monitoring of industrial gas total emissions based on optical scintillation cross-correlation technique. *Opt. Express* **2016**, *24*, 943–955. [CrossRef] [PubMed]

16. Dickheuer, S.; Marchuk, O.; Tsankov, T.V.; Luggenhölscher, D.; Czarnetzki, W.; Ertmer, S.; Kreter, A. Measurement of the magnetic field in a linear magnetized plasma by tunable diode laser absorption spectroscopy. *Atoms* **2019**, *7*, 48. [CrossRef]

17. Stritzke, F.; Kely, S.V.D.; Feiling, A.; Dreizler, A.; Wagner, S. TDLAS-based NH_3 mole fraction measurement for exhaust diagnostics during selective catalytic reduction using a fiber-coupled 2.2-µm DFB diode laser. *Opt. Express* **2015**, *119*, 143–152. [CrossRef]

18. Bolshov, M.A.; Kuritsyn, Y.A.; Romanovskii, Y.V. Tunable diode laser spectroscopy as a technique for combustion diagnostics. *Spectrochim. Acta Part B* **2015**, *106*, 45–66. [CrossRef]

19. Xu, L.; Liu, C.; Jing, W.; Cao, Z.; Xue, X.; Lin, Y. Tunable diode laser absorption spectroscopy-based tomography system for on-line monitoring of two-dimensional distributions of temperature and H_2O mole fraction. *Rev. Sci. Instrum.* **2016**, *87*, 013101. [CrossRef]

20. Gamache, R.R.; Kennedy, S.; Hawkins, R.; Rothman, L.S. Total internal partition sums for molecules in the terrestrial atmosphere. *J. Mol. Struct.* **2000**, *517–518*, 407–425. [CrossRef]

21. Rothman, L.S.; Gordon, I.E.; Barbe, A.; Benner, D.C.; Bernath, P.F.; Birk, M.; Bizzocchi, L.; Boudon, V.; Brown, L.R.; Campargue, A.; et al. The HITRAN2012 molecular spectroscopic database. *J. Quant. Spectrosc. Radiat. Transf.* **2013**, *130*, 4–50. [CrossRef]

22. Hong, D.H.; Wang, L.; Truong, T.K. Low-complexity direct computation algorithm for cubic-spline interpolation scheme. *J. Vis. Commun. Image Represent.* **2018**, *50*, 159–166. [CrossRef]

23. Deguchi, Y.; Noda, M.; Abe, M.; Abe, M. Improvement of combustion control through real-time measurement of O_2 and CO concentrations in incinerators using diode laser absorption spectroscopy. *Proc. Combust. Inst.* **2002**, *29*, 147–153. [CrossRef]

24. Choi, D.W.; Jeon, M.G.; Cho, G.R.; Kamimoto, T.; Deguchi, Y.; Doh, D.H. Performance improvements in temperature reconstructions of 2-D tunable diode laser absorption spectroscopy (TDLAS). *J. Therm. Sci.* **2016**, *25*, 84–89. [CrossRef]

25. Zaatar, Y.; Bechara, J.; Khoury, A.; Zaouk, D.; Charles, J.-P. Diode laser sensor for process control and environmental monitoring. *Appl. Energy* **2000**, *65*, 107–113. [CrossRef]

26. Doh, D.H.; Lee, C.J.; Cho, G.R.; Moon, K.R. Performances of Volume-PTV and Tomo-PIV. *Open J. Fluid Dyn.* **2012**, *2*, 368–374. [CrossRef]

27. Choi, K.M.; Lee, J.T.; Jeon, S.G.; Hong, T.G.; Paek, J.K.; Han, S.H.; Yim, D.S. A review of fundamentals of statistical concepts in clinical trials. *J. Korean Soc. Clin. Pharmacol. Ther.* **2012**, *20*, 109–124. [CrossRef]

28. Cort, J.W.; Kenhi, M. Advantages of the mean absolute error (MAE) over the root mean square error (RMSE) in assessing average model performance. *Clim. Res.* **2015**, *30*, 79–82.

29. Seo, M.K.; Yun, W.Y. Clustering based hot strip roughing mill diagnosis using Mahalanobis distance. *J. Korean Inst. Ind. Eng.* **2017**, *43*, 298–307.

30. Wikipedia. Available online: https://en.wikipedia.org/wiki/Mahalanobis_distance (accessed on 23 September 2019).

31. Yu, J.W.; Jang, J.Y.; Yoo, J.Y.; Kim, S.S. Fault detection method for steam boiler tube using Mahalanobis distance. *J. Korean Inst. Intell. Syst.* **2016**, *26*, 246–252. [CrossRef]
32. Minitab 19 Support. Available online: https://support.minitab.com/en-us/minitab/18/Assistant_Two_Sample_t.pdf (accessed on 20 November 2019).

Article

Combustion Process of Canola Oil and n-Hexane Mixtures in Dynamic Diesel Engine Operating Conditions

Rafał Longwic [1,*], Przemysław Sander [1], Anna Zdziennicka [2], Katarzyna Szymczyk [2] and Bronisław Jańczuk [2]

1   Department of Vehicles, Faculty of Mechanical Engineering, Lublin University of Technology, 20-618 Lublin, Poland; p.sander@pollub.pl
2   Department of Interfacial Phenomena, Institute of Chemical Sciences, Faculty of Chemistry, Maria Curie-Sklodowska University in Lublin, Maria Curie-Sklodowska Sq. 3, 20-031 Lublin, Poland; aniaz@hektor.umcs.lublin.pl (A.Z.); katarzyna.szymczyk@poczta.umcs.lublin.pl (K.S.); bronislaw.janczuk@poczta.umcs.lublin.pl (B.J.)
*   Correspondence: r.longwic@pollub.pl; Tel.: +48-606-785-513

Received: 6 November 2019; Accepted: 15 December 2019; Published: 20 December 2019

Abstract: The article discusses the problem of using canola oil and n-hexane mixtures in diesel engines with storage tank fuel injection systems (common rail). The tests results of the combustion process in the dynamic operating conditions of an engine powered by these mixtures are presented. On the basis of the conducted considerations, it was found that the addition of n-hexane to canola oil does not change its energy properties and significantly improves physicochemical properties such as the surface tension and viscosity. It contributes to the improvement of the flammable mixture preparation process and influences the course of the combustion process.

Keywords: combustion; alternative fuel; canola oil (Co); diesel engine; common rail; diesel fuel (Df); n-hexane; injection

---

1. Introduction

In general use, the diesel engine will be used for many years to come (road, sea, rail, stationary engines of working machines) [1,2]. However, the sharpening of the standards concerning the emission of toxic compounds and the unstable situation on the market of petroleum-derivative fuels make it necessary to conduct research on the improvement of fuel supply systems and the construction of the internal combustion engine itself. This is done for obvious reasons, as it may have a positive impact on the improvement of the injection and combustion processes, which may have an influence on the reduction of fuel consumption and indirectly on the reduction of exhaust gas emissions. Another direction of research, leading to compliance with rigorous exhaust emission standards, is to improve the properties of fuels used in internal combustion engines. The physicochemical properties of fuels have a direct impact on the course of the injection and combustion processes [3]. Hydrocarbon alternative fuels not derived from crude oil processing, i.e., diesel, gasoline, etc., have an increasing share in the market for diesel fuels: CNG (compressed natural gas); alcohols (methanol, ethanol, and butanol); vegetable oil (canola oil, soya oil, sunflower oil, palm oil, and peanut oil); and methyl esters of vegetable oils such as canola oil and palm oil.

Additives to diesel or vegetable oils, which are intended to improve the combustion process and contribute to reducing the amount of toxic components in the exhaust gases, are also used. The main additives that have been tested are ethanol, methanol, ethers (methyl tertiary butyl, EMTB; ethyl tertiary butyl, EETB; dimethyl, DME; diethyl, DEE), n-butanol, and FAME additives (fatty acid methyl esters).

The power supply of motor vehicles with alcohol fuels (methanol, ethanol, butanol), ethers (ETTB, DME, DEE), or synthetic fuels is not common and is usually limited to vehicles or research engines. The use of alcohol as a fuel is made more difficult by its low auto-ignition capacity, low calorific value, poor blending with diesel, and high hygroscopicity, so a suitable emulsifier would have to be used to obtain stable diesel–alcohol mixtures. Among alcohols, ethanol is the most often used because it has an advantage over methanol in the range of calorific value and octane number, and in a mixture with petrol, it becomes more difficult to stratify (also at lower temperatures). In order to solve this problem, studies are being conducted on the development of microemulsions consisting of insoluble liquids, i.e.,vegetable oil and methanol, ethanol, or ionic/non-ionic amphiphilic compounds [2–17].

Particularly worth mentioning are fuels derived from oil plants (palm oil, coconut oil, canola oil, soybean oil, linseed oil, peanut oil). Plant oils belong to the group of unconventional liquid fuels. Plant fuels and their esters are obtained from plant seeds. In Polish conditions, it is canola oil, which can be used to power a diesel engine. Research works on the use of canola oil as fuel were conducted in one of three ways, i.e., the use of refined canola oil; use of Co mixtures with Df and alcohols; and the use of canola oil methyl ester (FAME, biocomponent for Df).

The concept of using canola oil as a fuel presented in the article assumes supplying vegetable fuel to diesel engines without their structural changes. Due to known problems in the use of canola oil as fuel and the different physicochemical properties of canola oil, a minimum amount of additive (n-hexane) was applied to canola oil, which brings the physicochemical properties of canola oil closer to the physicochemical properties of diesel fuel [18,19]. The chemical additive used in a small amount in the mixture with canola oil enables the use of canola oil as fuel for diesel engines, taking into account the following functions of the purpose: the physicochemical parameters of canola oil mixtures with additions are similar to the physicochemical parameters of diesel oil in a wide range of ambient temperatures, including the maximum energy parameters of the engine operation and the minimum emission value of the toxic components of the exhaust gas.

The use of an n-hexane additive enables canola oil to be used as fuel in the common use for means of transport. In the first phase of the research, physicochemical tests of canola oil mixtures with n-hexane addition were carried out, and the possibility of using the above solution in a group of engines with conventional injection systems, i.e., engines of agricultural tractors, passenger cars, trucks, and stationary engines equipped with various types of in-line and rotary injection pumps [20,21].

In the presented article, the possibility of using canola oil and n-hexane mixtures in compression ignition engines with storage injection systems was examined, and the course of combustion processes in dynamic engine operating conditions was determined.

2. Materials and Methods

2.1. Fuels Tested

Diesel fuel (Df) complying with EN590 [22], commercial canola oil (Co), and non-reactive solvent n-hexane (Hex), whose main physicochemical properties are shown in Table 1, were used in the tests. N-hexane (C_6H_{14}) is an organic chemical compound from the alkane group. N-hexane isomers are very unreactive and are often used as solvents in organic reactions as they are highly non-polar. On the basis of canola oil (Co), two mixtures with n-hexane were prepared in the following proportions (*v/v*): 10% (10%Hex90%Co) and 15% (15%Hex85%Co). For the fuels tested, their basic physicochemical properties are specified in Table 2.

The equilibrium surface tension (γ_{LV}) of canola oil+n-hexane mixtures (5-20%Hex) was measured by the Krüss K9 tensiometer according to the platinum ring detachment method (duNouy's method) at 293K. Before the surface tension measurements, the tensiometer was calibrated using water (γ_{LV} = 72.8 mN/m at 293 K) and methanol (γ_{LV} = 22.5 mN/m at 293 K). The ring was cleaned with distilled water and heated to a red color with a Bunsen burner before each measurement. In all cases, more than 10 successive measurements were performed. The standard deviation was ±0.1 mN/m.

The density of the aqueous solution of studied canola oil+n-hexane mixtures was measured with a U-tube densitometer, DMA 5000 Anton Paar, with the precision of the density measurements equal to ± 0.000005g cm^{-3}. The uncertainty was calculated to be equal to 0.01%. The viscosity measurements of the canola oil+n-hexane mixtures were performed with the Anton Paar viscometer, AMVn, with the precision of 0.0001mPas and an uncertainty of 0.3%. All density and viscosity measurements were made at 293 K.

Table 1. Physicochemical properties of n-hexane [23].

Parameter	Unit	Value
Kinematic viscosity index in 20 °C	Mm^2/s	0.50
Vapour pressure in 20 °C	Mbar	160
Dynamic viscosity index in 20 °C	mPa·s	0.326
Density in 20 °C	g/mL	0.66
Solubility in water in 20 °C	g/dm^3	0.00095
Ignition temperature	°C	−22
Boiling point temperature	°C	68
Self-ignition temperature	°C	240
Melting temperature	°C	−94
Explosiveness limits	%	low: 1.0 obj.; high: 8.1obj.

Table 2. Selected physicochemical properties of the fuels studied [20,22].

Parameter	Unit	Value		
		Df	10%Hex	15%Hex
Density in 15 °C	(kg/m^3)	835	895	887
Kinematic viscosity in 40 °C	(mm^2/s)	2.7	19.6	15.2
Cold filter block age temperature	(^C)	12	−3	−7
Ignition temperature	(°C)	72	<40	<40
Surface tension	(mN/m)	29.2	28.4	27.0

2.2. Research Methodology and Research Station

The test object was a diesel engine with a common rail storage injection system installed in a Fiat Qubo vehicle, which met the Euro 5 emission standards. The vehicle was equipped with a five-stage gearbox. The test vehicle was equipped with an additional (external) independent fuel tank with an additional fuel pump, allowing for the quick replacement of the tested fuels. The modification of the fuel system of the vehicle concerned a low pressure system (about 3 bar). After switching to the auxiliary fuel system (low pressure), the main fuel system was automatically disconnected. The high-pressure fuel system was not modified, so the fuel pressure in the high-pressure system was the same for all the tests performed. The technical data of the engine and the view of the test vehicle are shown in Table 3 and Figure 1.

Table 3. Technical data 1.3 multijet test vehicle Fiat Qubo [24].

The Number of Cylinders	4
Cylinder diameter (mm)	69.6
Piston stroke (mm)	82
Total capacity (cm^3)	1248
Maximum power (kW CEE)	55
Maximum power (HP CEE)	75
Operating at maximum power (rpm)	4000
Maximal moment (Nm CEE)	190
Maximal moment (kgm CEE)	19.4
Speed at maximum torque (rpm/1 min)	1500
Idle rotation speed (rpm)	850 ± 20
Compression degree	16.8:1

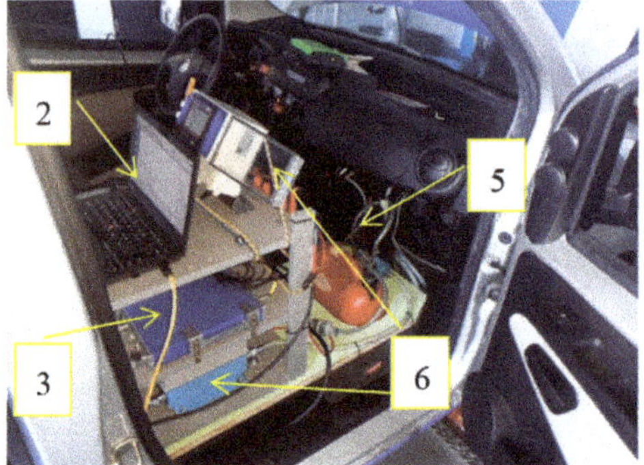

Figure 1. Research station simulating vehicle motion under traction conditions: Fiat Qubo test car with 1.3 Multijet engine. 2. Computer with installed AVL Indicom V2.7 software. 3. Indimicro 602 engine indicator system. 4. DF4FS-HLS chassis dynamometer. 5. Additional fuel system tank.6. Rotameter with Multicon CMC-99 module by Simex.

The Indimicro602 recording system of the AVL company with a built-in signal amplifier, cooperating with four analog input channels and two digital inputs, allowing for the recording of quick-change parameters in real time, was used to indicate the engine of the test vehicle. The connection diagram of the system is shown in Figure 2. The signals recorded by the AVL Indimicro system include the following. First, there is the pressure course inside the cylinder, which was recorded by the piezoelectric sensor AVL GH13P installed in the glow plug socket of the first cylinder by means of an adapter (Figure 2, position 3) and whose signal was processed in the amplifier—the AVL measuring module (Figure 2, position 6). Then, the engine crankshaft position signal informing about the crankshaft position was obtained from the induction sensor cooperating with the toothed flywheel by means of the AVL universal pulse conditioner 389Z01 analog–digital converter (Figure 2 position 5).In addition, the injection parameters were analyzed on the basis of the analog control signal of the electromagnetic injector after conversion into a digital signal (Figure 2 item 7).

An indication of a compression ignition engine in conditions simulating vehicle motion in traction conditions was performed by simulating driving on a DF4FS-HLS chassis dynamometer. The diagram of the test stand is shown in Figure 1. The DF4FS-HLS chassis dynamometer was part of the system, which included a stationary (first) roller set with an electro-vacuum brake and hydraulic pump; a mobile (second) roller set with an electro-vacuum brake, hydraulic pump, and gear motor for drive; a control panel (dashboard);control of the hydraulic system; an axial fan for cooling the vehicle; and a PC with dynamometer software.

The combustion engine of the traction unit is exploited mainly under dynamic conditions [25]. Vehicle driving tests are currently used to evaluate the energy and ecological parameters of engines. The most commonly used driving test is the Worldwide Harmonized Light Vehicles Test Procedure (WLTP). The characteristic points of the WLTP driving test corresponding to the dynamic conditions of engine operation have been selected.

Therefore, the tests consisted of a series of measurements for each of the fuels, which allowed the recording of fast-changing parameters during the acceleration of the engine with diesel. First of all, the torque (Mo) and power (Ne) values of the engine that were supplied with the tested fuels were determined. In the next stage of the research, a series of accelerations was performed, during which the following test methodology was applied (the measurement conditions result from the selection of characteristic points of the WLTP test running in dynamic conditions): gearbox ratio—fourth gear, initial vehicle speed 60 km/h, final speed—80 km/h, acceleration lever pitch during testing—40% constant for all fuels. The vehicle was loaded with rolling resistance forces.

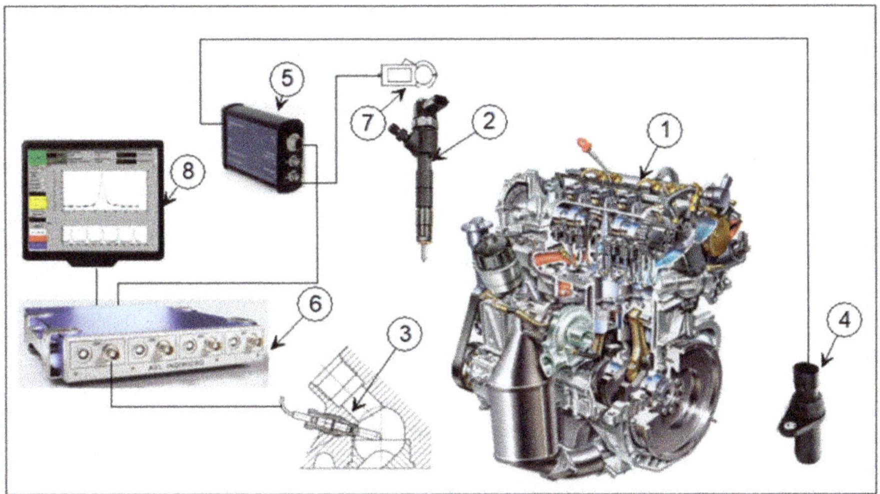

Figure 2. AVL connection diagram: 1. Diesel engine of the test vehicle, 2. Electromagnetic injector and cylinder, 3. Piezoelectric sensor AVLGH13P integrated with an adapter for glow plug and cylinder socket, 4. Crankshaft inductive sensor, 5. AVL Universal Pulse Conditioner 389Z01 transmitter, 6. AVL Indimicro 602 measuring module, 7. Measuring rod, 8. Mobile computer with AVL Indicom software.

Selected parameters of the combustion process, including the mean indexed pressure (p_i), maximum combustion pressure (P_{cmax}), maximum rate of increase of combustion pressure ($dp/d\alpha_{max}$), and the amount of heat produced, were calculated with the use of commonly known laws of thermodynamics, which were implemented in the AVL Indicom software. The calculations were performed at a frequency of 1^0CA.

In the presented paper, the authors of each fuel performed 10 dynamic acceleration tests of the vehicle, during which the engine operating parameters were recorded. Table 4 presents the values of engine parameters averaged over 10 acceleration tests (at selected engine speed points). Figures 3–8 show representative graphs from the conducted tests.

Table 4. Average results of 10 tests of the engine for operating parameters at selected vehicle speeds under dynamic conditions, simulation of extra-urban traffic "High" according to the Worldwide Harmonized Light Vehicles Test Procedure (WLTP) driving test standard.

Fuel	"I" 60 km/h				"II" 70 km/h			
	N [rpm]	p_i [Mpa]	P_{cmax} [Mpa]	$dp/d\alpha_{max}$ MPa/°CA	n [rpm]	p_i [Mpa]	P_{cmax} [Mpa]	$dp/d\alpha_{max}$ MPa/°CA
Df	1928.6	0.95	8.71	0.458	2242.2	1.04	8.52	0.406
10%Hex90%Co	1928.6	0.953	9.05	0.314	2242.2	0.935	7.9	0.32
15%Hex85%Co	1928.6	0.915	8.06	0.488	2242.2	0.943	7.78	0.336
	"III" 74 km/h				"IV" 80 km/h			
	N [rpm]	p_i [Mpa]	P_{cmax} [Mpa]	$dp/d\alpha_{max}$ MPa/°CA	n [rpm]	p_i [Mpa]	P_{cmax} [Mpa]	$dp/d\alpha_{max}$ MPa/°CA
Df	2389.5	1.07	8.99	0.365	2509.4	1.06	9.61	0.399
10%Hex90%Co	2389.5	0.936	8.49	0.359	2509.4	0.929	9.12	0.468
15%Hex85%Co	2389.5	0.924	8.14	0.389	2509.4	0.931	8.8	0.516

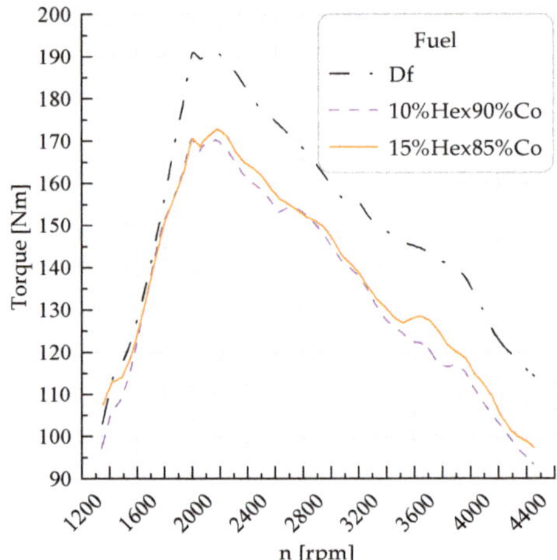

Figure 3. Torque values depending on the rotational.

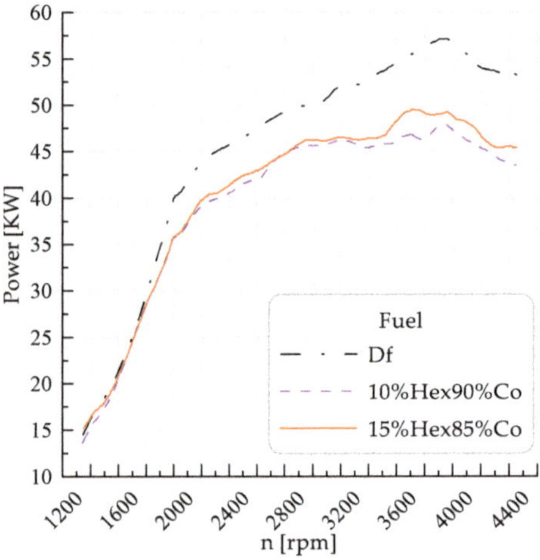

Figure 4. Power values depending on the rotationalspeed, engine with diesel supplied with the testedspeed supplied with the tested fuels, i.e., Df, fuels, i.e., Df, 10%Hex90%Co, 15%Hex85%Co, 10%Hex90%Co, and 15%Hex85%Co.

3. Results

Maximum torque (Nm) and maximum power (KW) were determined during engine tests. For diesel fuel (Df), the maximum torque and power values were 191.3 Nm at 2123 rpm and 57 kW at 3916 rpm, respectively. For vegetable fuels, lower torque values were obtained of about 11% (10%Hex90%Co) and 10% (15%Hex85%Co) and power of about 17% (10%Hex90%Co) and 14% (15%Hex85%Co). The accelerations obtained were the lowest for the canola oil with a 10% addition of n-hexane. The effect of the amount of n-hexane addition in Co on the obtained values of torque and engine power as well as acceleration, which may be related to the influence of physicochemical properties of fuels, was observed. Detailed values obtained during tests of the test vehicle powered by the tested fuels are presented in Table 5.

During the tests conducted on the diesel engine fueled with the tested fuels, the average acceleration of the test vehicle was obtained, the course of which is shown in Figure 5:Df—0.55 m/s^2, 10%Hex90%Co—0.39 m/s^2, and 15%Hex85%Co—0.50 m/s^2.

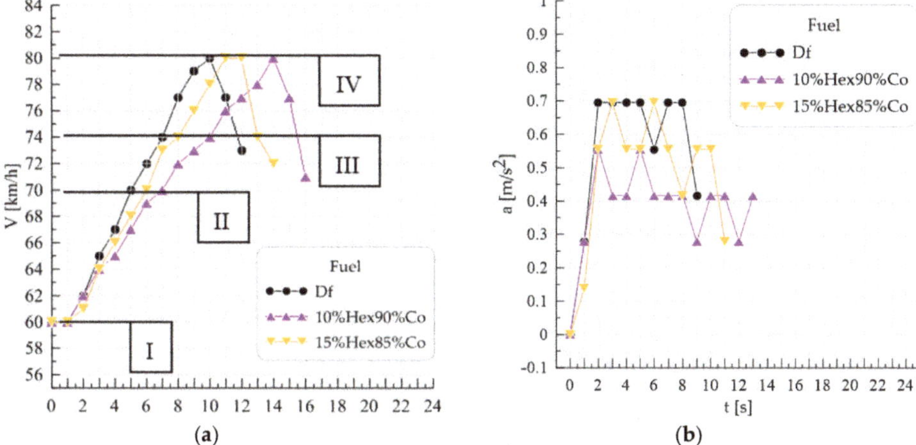

Figure 5. The speed of the test vehicle (**a**) and acceleration of the vehicle (**b**) loaded with resistance to movement of the vehicle, powered by the tested fuels, i.e., Df, 10%Hex90%Co, and 15%Hex85%Co. Initial conditions: gearbox ratio—IV gear, initial vehicle speed—60 km/h, final speed—80 km/h, acceleration lever pitch—constant for all fuels—40%. Sampling frequency in 1s.

Table 5. The maximum torque and maximum engine power obtained when performing a test on a chassis dynamometer. Df: diesel fuel.

Parameter	Df	10%Hex90%Co	15%Hex85%Co
Power of the vehicle	57 KW/3916 rpm	47.8 KW/3995 rpm	49.5 KW/3714 rpm
Torque	191.3Nm/2123 rpm	170.5 Nm/1998 rpm	172.8 Nm/2186 rpm
Power on the wheels	45.5 KW/3893 rpm	39.2 KW/2891 rpm	40.3 KW/3703 rpm
Maximumspeed	143km/h/4520 rpm	143 km/h/4509 rpm	143 km/h/4516 rpm
Acceleration time	35.30s	42.88s	37.98s

During the performed accelerations of the vehicle, points were selected (shown in Figure 5a) at which combustion process parameters were determined, i.e., the mean index pressure (p_i), maximum combustion pressure (P_{cmax}), and maximum rate of increase of combustion pressure ($dp/d\alpha_{max}$). Detailed values are shown in Table 4. It was found that the average indexed pressure at each of the measurement points (I, II, III, and IV) was the highest for diesel oil. The biggest difference was observed in relation to 15%Hex85%Co and reacheda maximum of 13%. It was observed that increasing the content of n-hexane in the mixture with canola oil led in most of the cases to a decrease in the mean index pressure and maximum combustion pressure and to an increase in the maximum rate of increase of combustion pressure. Similar trends in the observed parameters were observed for other engine speeds analyzed.

Figure 6 shows the course of the main parameters of the combustion process of an engine fueled with diesel oil (position a), canola oil with 10% n-hexane (position b), and canola oil with 15% n-hexane (position c). It was found that under dynamic conditions, the self-ignition delay angle (αID) was the lowest for Df in the whole engine speed range. For 15% of the n-hexane content in Co, the angle of auto-ignition delay was higher by approximately 13% in relation to Df. Rapeseed fuels were characterized by comparable self-ignition delay values.

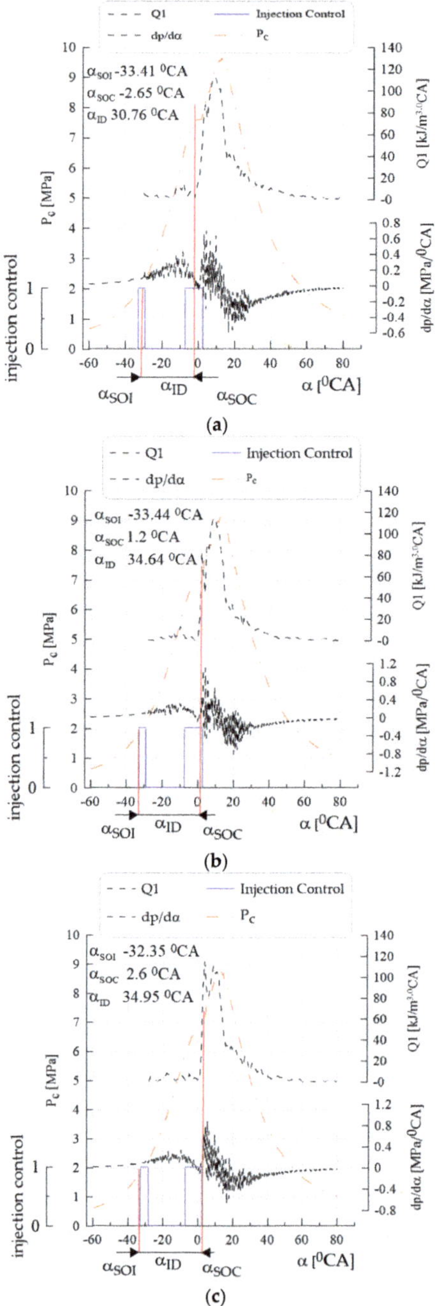

Figure 6. The course of the main injection and combustion parameters, i.e., the rate of heat transfer (Q1), pressure inside the combustion chamber (Pc), and the injector control signal—in dynamic engine operating conditions, for a selected engine speed of approximately 2509 rpm (for a speed of 80 km/h). The following fuels were tested: (**a**) Df, (**b**) 10%Hex90%Co, and (**c**) 15%Hex85%Co.Additionally, the following angles were given: αSOI(Start of Injection), αSOC(Start of Combustion), and αID(Ignition Delay). Sampling frequency in 1°CA.

4. Discussion

The different physicochemical properties of mixtures of canola oil and n-hexane in comparison with diesel oil, as shown in Table 2, cause different coursesin the injection and combustion processesof these fuels. In diesel engines, one of the most important parameters affecting the combustion process is the self-ignition delay [26]. Differences in the calorific value and the heat of combustion of mixtures cause changes in the operational indicators of engine operation. In the dynamic conditions of engine operation supplied with the tested fuels, the injection time was similar (the differences reached a maximum of 5%). This resulted in the delivery of a volume-comparable amount of fuel to the combustion chamber.

The reason for the lower values of the average indicated pressure (Table 4) that were obtained when supplying canola oil with an addition of n-hexane to Df was the generation of less heat during combustion, which is shown, among others, in Figure 7. During combustion processes of diesel oil on the heat-excitation rate curves, the classical kinetic and diffusion phase could be distinguished, whereas during the combustion of canola oil with the addition of n-hexane, the kinetic phase disappeared with increased n-hexane content, which resulted in higher pressure rise velocities—see Figure 8. This also significantly affected the beginning of combustion, whose angle for diesel oil appeared the earliest (before the kinetic phase of combustion). For canola oil with added n-hexane, the start of combustion appeared later by approximately 4–12°CA in terms of Df(depending on engine speed), while an increase in the amount of n-hexane caused further delay in the start of combustion. A similar tendency was observed forthe angle of self-ignition delay. Figure 6 illustrates the differences in the occurrence of the angle of the beginning of combustion and other parameters of the combustion process when feeding the fuels under study. Another reason for the observed changes could have been a different course of the flammable mixture formation process, due to the different viscosity of the fuels tested. The occurrence of the kinetic phase of combustion resulted in a decrease in the maximum rate of increase of combustion pressure (dp/dα) in the engine fueled with Df. For canola oil, with the addition of n-hexane, the maximum rate of pressure rise was higher than for Df, and brewing increased with increase of n-hexane content in the mixture.

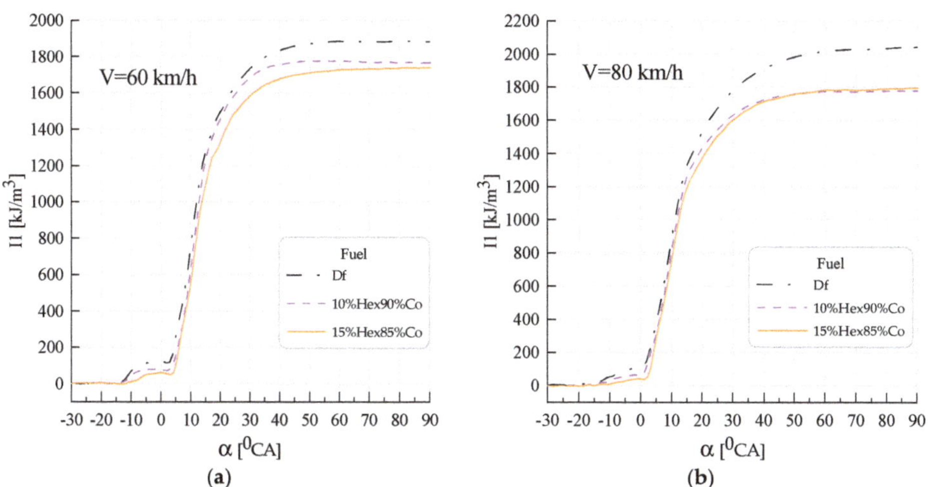

Figure 7. Heat discharge (I1) under dynamic engine operating conditions, (**a**) vehicle speed 60 km/h and engine speed approximately 1928 rpm, (**b**) vehicle speed 80 km/h and engine speed approximately 2509 rpm, tested fuels: Df, 10%Hex90%Co, and 15%Hex85%Co. Sampling frequency in 1°CA.

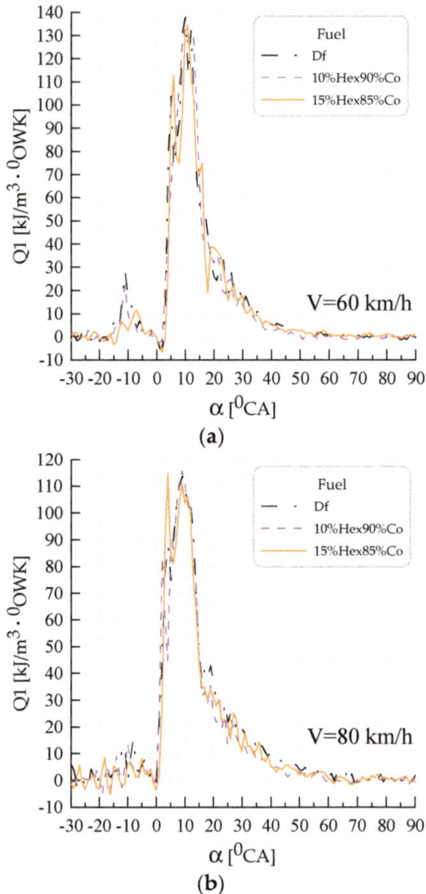

Figure 8. Combustion heat transfer rate (Q1) under dynamic engine operating conditions (**a**) vehicle speed 60 km/h and engine speed approximately 1928 rpm, (**b**) vehicle speed 80 km/h and engine speed approximately 2509 rpm, tested fuels: Df, 10%Hex90%Co, and 15%Hex85%Co. Sampling frequency in 1°CA.

The addition of n-hexane to canola oil results in a reduction of surface tension, slight changes in density, and significant changes in viscosity (Figures 9 and 10). The surface tension value of the oil mixture and 5% n-hexane is only 0.8 mN/m higher than the surface tension of the diesel oil. In turn, for the mixture of oil and 10% n-hexane, it is 0.8 mN/m lower. Changes in surface tension as a function of n-hexane concentration in the mixture with canola oil indicate that the mixture does not behave as an ideal mixture (Figure 9) [27], i.e., changes in this tension are not directly proportional to the composition of the mixture. According to the suggestion of Fowkes and van Oss, the surface tension results from Lifshitz-van der Waals interactions, hydrogen bonds, and electrostatic interactions [28,29]. Therefore, it depends on the type of functional groups occurring in the liquid molecules at the phase boundaries. In the case of a mixture of canola oil and n-hexane, the dominant functional groups that may occur at the liquid–air interface are $-CH_3$, $=CH_2$, $\equiv CH$, $=CO$, and $-COOH$. The surface tension of the canola oil/n-hexane mixture depends on the density of the mentioned functional groups. Since the main components of canola oil are unsaturated higher fatty acids (>90%), the contribution of the $=CO$ and $-COOH$ groups to the surface tension of the oil is significant. Therefore, the orientation of oleic, linoleic, and linoleic acid molecules at the phase boundary has a major influence on the surface tension of canola oil. The n-hexane molecules present in the mixture with canola oil due to the strong hydrophobic interactions between the n-hexane molecules and the apolar part of unsaturated fatty acids increase the likelihood of orientation of the formed acid–n-hexane complexes with the hydrophobic part directed toward the gas phase, which in turn lowers the surface tension of the mixture. This may explain the nonlinear relationship between the surface tension of the mixture and its composition (Figure 9).

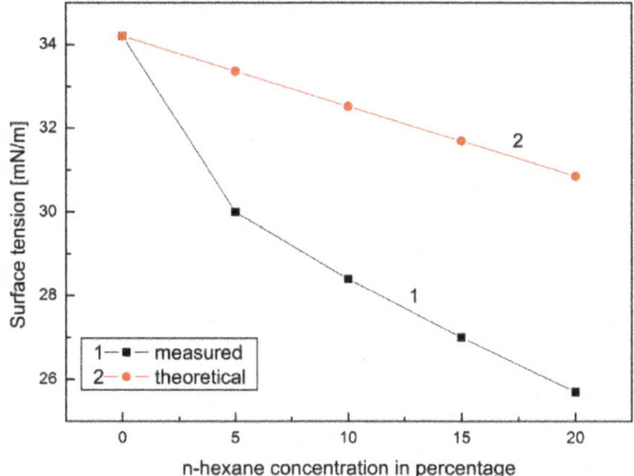

Figure 9. A plot of the surface tension of canola oil+n-hexane solution vs. n-hexane concentration. Curve 1 corresponds to the measured values, while curve 2 corresponds to the theoretical values.

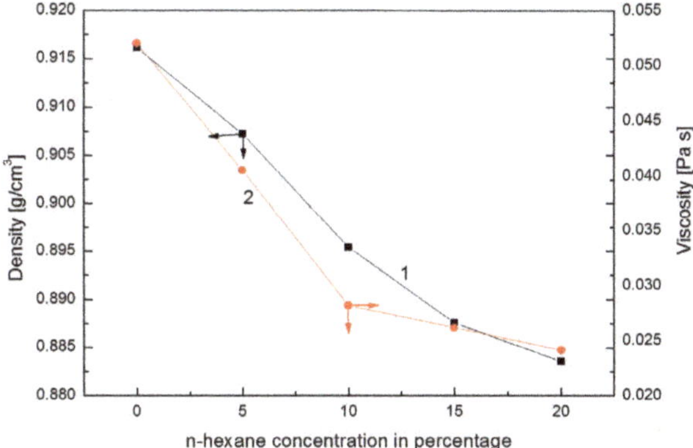

Figure 10. A plot of density (curve 1) and viscosity (curve 2) of canola oil+n-hexane solution. The x-axis indicates n-hexane concentration.

The value of surface tension, which is closely related to the interactions between functional groups in the surface layer, can be a decisive factor for the vapor pressure of individual components of a mixture. Since the changes in the surface tension of the tested mixtures as a function of n-hexane concentration indicate non-ideal behaviour of the mixture, it can be expected that the vapor pressure of n-hexane and canola oil components does not comply with the Raoult law [27]. This, in turn, may have a considerable influence on the obtained flash temperature values of the tested blends (<40°C). Moreover, the decrease in surface tension of the canola oil/n-hexane mixture as a function of the n-hexane volume concentration results in a decrease in the volume of the mixture droplets flowing out of the injector tip, which may have a significant impact on the injection process.

The density of the mixture canola oil and 10% n-hexane is slightly lower than the density of diesel oil (Figure 10). Although the addition of n-hexane significantly reduces the viscosity of canola oil, even with its content in the mixture of 20%, the viscosity of the blend is slightly higher than that of diesel oil. On the other hand, the decrease in viscosity affects the change in the angle at the beginning of combustion.

From the point of view of using a mixture of canola oil and n-hexane for diesel engines, it is important that the combustion heat of canola oil under the influence of n-hexane and the oxygen consumption in this process are changed. It is known that canola oil is composed of many chemical compounds; however, the content of oleic, linoleic, and linolenic acid is over 90% [30]. The process of combustion of these acids can be represented by a reaction (1)–(9):

$$C_{17}H_{33}COOH + 25.5O_2 = 18CO_2 + 17H_2O \tag{1}$$

$$C_{17}H_{33}COOH + 16.5O_2 = 18CO + 17H_2O \tag{2}$$

$$C_{17}H_{33}COOH + 7.5O_2 = 18C + 17H_2O \tag{3}$$

$$C_{17}H_{31}COOH + 25O_2 = 18CO_2 + 16H_2O \tag{4}$$

$$C_{17}H_{31}COOH + 16O_2 = 18CO + 16H_2O \tag{5}$$

$$C_{17}H_{31}COOH + 7O_2 = 18C + 16H_2O \tag{6}$$

$$C_{17}H_{29}COOH + 24.5O_2 = 18CO_2 + 15H_2O \tag{7}$$

$$C_{17}H_{29}COOH + 15.5O_2 = 18CO + 15H_2O \tag{8}$$

$$C_{17}H_{29}COOH + 6.5O_2 = 18C + 15H_2O \tag{9}$$

The following chemical reactions (10)–(12) were used to compare the amount of oxygen needed to burn n-hexane:

$$C_6H_{14} + 9.5O_2 = 6CO_2 + 7H_2O \tag{10}$$

$$C_6H_{14} + 6.5O_2 = 6CO + 7H_2O \tag{11}$$

$$C_6H_{14} + 3.5O_2 = 6C + 7H_2O \tag{12}$$

Based on the combustion reaction of 1 mole of n-hexane and the three acids mentioned above, it can be concluded that the amount of oxygen needed to burn n-hexane is much lower than in the case of acids, but such a comparison does not reflect the amount of oxygen needed to burn equal volumes of these substances. Therefore, the number of oxygen moles needed to burn each of these compounds has been calculated by taking into account the density values. If the product of combustion of all compounds is carbon dioxide then the number of oxygen moles needed to burn 1 dm^3 of the compound is 80.34, 80.25, 80.43, 80.43, and 72.75, respectively, for oleic acid, linoleic acid, linoleic acid, linolenic acid, and n-hexane. This comparison shows that the number of oxygen moles necessary to burn the same amount of n-hexane is significantly lower than for the three acids mentioned above. These differences should be reflected in the values of heat of combustion of n-hexane and tested acids, and thus canola oil. The heat value of canola oil combustion per 1 dm^3 is 35284.32 kJ. It turned out that this value did not differ much from the value for oleic acid (35147.60 kJ). However, the combustion heat of n-hexane, as expected from the amount of oxygen needed for its combustion, is lower and amounts to 31911 kJ. Taking these values into consideration, the heat of combustion of the canola oil/n-hexane mixture as a function of its composition was calculated. The calculations show that in the range of concentrations (from 0 to 20% of n-hexane in the mixture) the heat of combustion slightly decreases from 35284.32 to 34609.66 kJ/dm^3 (Figure 11). Changes in the combustion heat of a canola oil and n-hexane mixture are related to the amount of oxygen needed to burn the mixture. Considering that more than 90% of the oil composition is made up of the acids tested, the average value of oxygen mole needed to burn 1 dm^3 of oil was used to calculate the number of oxygen needed to burn 1 dm^3 of oil. The calculated amount of oxygen mole needed to burn 1 dm^3 of the mixture, similarly to the heat of combustion, changes slightly (from 80.34 to 78.82). On the basis of the conducted considerations, it can be stated that the addition of n-hexane to canola oil slightly changed the heat of combustion and significantly improved the physicochemical properties such as the surface tension and viscosity.

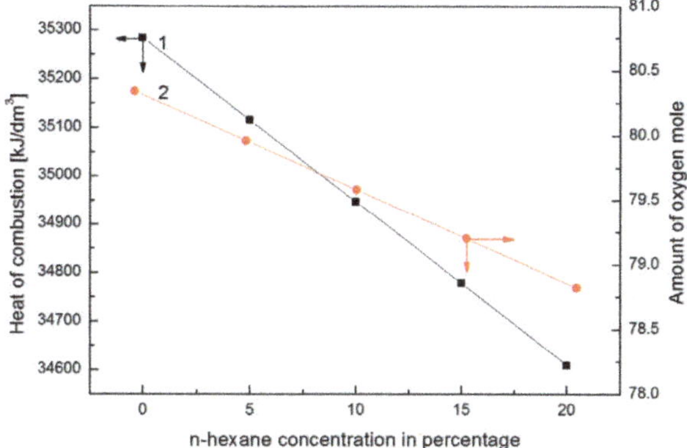

Figure 11. A plot of heat of combustion (curve 1) and amount of oxygen mole needed to burn off 1 dm^3 of canola oil–n-hexane mixture (curve 2) vs. n-hexane concentration.

The addition of n-hexane to canola oil caused a significant decrease in its viscosity and surface tension. Changes in the surface tension of canola oil with the addition of n-hexane as a function of composition are not linear, and synergy in the reduction of surface tension is observed. It was found that the mixture of canola oil and n-hexane is not ideal. A significant decrease in the viscosity and surface tension of oleic acid, the main component of canola oil, by adding n-hexane, indicates that such a mixture can be used in diesel engines.

5. Conclusions

The conducted research and its analysis allow us to formulate a few final conclusions:

- The vapor pressure of n-hexane and canola oil components does not comply with Raoult law, and this in turn may have a significant influence on the obtained values of ignition temperature of the tested blends (<40°C); the reduction of surface tension of the canola oil mixture with n-hexane as a function of n-hexane volume concentration causes a decrease in the volume of mixture drops flowing out of the injector tip, which may have a significant influence on the injection process;
- Based on the calculations carried out, it can be concluded that the addition of n-hexane to canola oil slightly changed the heat of combustion of the same volume of the prepared mixture and significantly improved physicochemical properties such as the surface tension and viscosity;
- Under dynamic conditions of engine operation supplied with the tested fuels, the injection time was similar (the differences reached a maximum of 5%); this resulted in supplying the combustion chamber with a volume comparable amount of fuel;
- Obtaining lower values of the average pressure indexed when supplying canola oil with n-hexane in relation to Df was caused by lower heat generation during combustion; during diesel combustion, the classical kinetic and diffusion phase could be distinguished on the heat transfer rate curves, whereas during the combustion of canola oil with n-hexane, the kinetic phase disappeared with increased n-hexane content, which resulted in higher pressure increase rates; for canola oil with n-hexane addition, the beginning of combustion occurred later by approximately 4–12°CA with respect to Df (depending on engine speed), while an increase in the share of n-hexane caused

further delay in the beginning of combustion—a similar tendency was observed for the angle of self-ignition delay.

Author Contributions: R.L., P.S., A.Z., K.S. and B.J. designed the experiments, analyzed the experimental data, made figures, participated in the preparation of the manuscript, and wrote the part of the manuscript. R.L., P.S. and B.J. conceived the concept of the studies, wrote the main part of the manuscript, supervised the studies, participated in the manuscript preparation, signed the experiments, analyzed the experimental data, and made figures. All authors have read and agreed to the published version of the manuscript.

Funding: This project/research was financed in the framework of the project Lublin University of Technology-Regional Excellence Initiative, funded by the Polish Ministry of Science and Higher Education (contract No. 030/RID/2018/19).

Conflicts of Interest: The authors declare no conflicts of interest.

References

1. Kalghatgi, G. Is it really the end of internal combustion engines petroleum in transport? *Appl. Energy* **2018**, *225*, 965–974. [CrossRef]
2. Ho, R.J.; Yusoff, M.Z.; Palanisamy, K. Trend future of diesel engine: Development of high efficiency low emission low temperature combustion diesel engine. *IOP Conf. Ser. Earth Env. Sci.* **2013**, *16*. [CrossRef]
3. Bergthorson, J.M.; Thomson, M.J. A review of the combustion and emissions properties of advanced Transportation biofuels and their impact on existing and future engines. *Renew. Sustain. Energy Rev.* **2015**, *42*, 1393–1417. [CrossRef]
4. Alptekin, E.; Canakci, M. Characterization of the key fuel properties of methyl ester–diesel fuel blends. *Fuel* **2009**, *88*, 75–80. [CrossRef]
5. Chen, Z.; Wu, Z.; Liu, J.; Lee, C. Combustion and emissions characteristics of high n-butanol/diesel ratio blend in a heavy-duty diesel engine and EGR impact. *Energy Convers Manag.* **2014**, *78*, 787–795. [CrossRef]
6. Gray, C.; Webster, G. *A study of Dimethyl Ether (DME) as an alternative fuel for Diesel Engine, Aplications, Advanced Engine Technology LTD*; Project No. 9493-94 Canada – Final Raport; Advanced Engine Technology Limited: Nepean, ON, Canada, 2001.
7. Lapuerta, M.; Garcıa-Contreras, R.; Campos-Fernandez, J.; Dorado, M.P. Stability, Lubricity, Viscosity, and Cold-Flow Properties of Alcohol-Diesel Blends. *Energy Fuels* **2010**, *24*, 4497–4502. [CrossRef]
8. Rakopoulos, D.C.; Rakopoulos, C.D.; Giakoumis, E.G.; Dimaratos, A.M.; Kyritsis, D.C. Effects of butanol–diesel fuel blends on the performance and emissions of a high-speed DI diesel engine. *Energy Convers. Manag.* **2010**, *51*, 1989–1997. [CrossRef]
9. Bailey, B.; Eberhardt, J.; Goguen, S.; Erwin, J. Diethyl Ether (DEE) as a Renewable Diesel Fuel. *SAE Pap.* **1997**, *106*, 1578–1584.
10. Sinha, S.; Agarwal, A.K. *Performance Evaluation of a Biodiesel (Rice Bran Oil Methyl Ester) Fueled Transport Diesel Engines*; SAE: Warrendale, PA, USA, 2005; pp. 1–9.
11. Woźniak, G.; Longwic, R. Initial assessment of the course of combustion process in a diesel engine powered by diesel oil and Brown's gas. *IOP Conf. Ser. Mater. Sci. Eng.* **2018**, *421*, 042085. [CrossRef]
12. Atmanli, A.; Ileri, E.; Yuksel, B.; Yilmaz, N. Extensive analyses of diesel–vegetable oil–n-butanol ternary blends in a diesel engine. *Appl. Energy* **2015**, *145*, 155–162. [CrossRef]
13. Yerrennagoudaru, H.; Manjunatha, K.; Nobin, K. Rapeseed oil blended with Methanol on Performances Emissions in Turbocharged Diesel Engine. *IOSR J. Mech. Civ. Eng. (IOSR-JMCE)* **2016**, *13*, 91–96.
14. Ileri, E.; Atmanli, A.; Yilmaz, N. Comparative analyses of n-butanol-rapeseed oil-diesel engine blend with biodiesel diesel biodiesel-diesel fuels in a turbocharged direct injection diesel engine. *J. Energy Inst.* **2016**, *89*, 586–593. [CrossRef]
15. Beroun, S.; Martins, J. The Development of Gas, (.C.N.G.; LPG; H2) Engines for Buses Trucks their Emission Cycle Variability Characteristics. *SAE Pap.* **2001**. [CrossRef]
16. Selim, M.Y. Pressure-time characteristics in diesel engine fueled with natural gas. *Renew. Energy* **2001**, *22*, 473–489. [CrossRef]
17. Verhelst, S.; Turner, J.W.; Sileghem, L.; Vancoillie, J. Methanol as a fuel for internal combustion engines. *Prog. Energy Combust. Sci.* **2019**, *70*, 43–88. [CrossRef]

18. Longwic, R.; Sander, P. The course of combustion process under real conditions of work of a traction diesel engine supplied by mixtures of canola oil containing n-hexane. *IOP Conf. Ser. Mater. Sci. Eng.* **2018**, *421*, 042050. [CrossRef]

19. Górski, K.; Sander, P.; Longwic, R. The assessment of ecological parameters of diesel engine supplied with mixtures of canola oil with n-hexane. *IOP Conf. Ser. Mater. Sci. Eng.* **2018**, *421*, 042025. [CrossRef]

20. Zdziennicka, A.; Szymczyk, K.; Jańczuk, B.; Longwic, R.; Sander, P. Adhesion of canola and diesel oils to some parts of diesel engine in the light of surface tension components and parameters of these substrates. *Int. J. Adhes. Adhes.* **2015**, *60*, 23–30. [CrossRef]

21. Zdziennicka, A.; Szymczyk, K.; Jańczuk, B.; Longwic, R.; Sander, P. Surface, Volumetric, and Wetting Properties of Oleic, Linoleic, and Linolenic Acids with Regards to Application of Canola Oil in Diesel Engines. *Appl. Sci.* **2019**, *9*, 3445. [CrossRef]

22. *Polish Standard PN-EN 590*; Fuels for motor vehicles - Gas oils - Requirements and test methods; European Committee: Poland, 2011.

23. *Safety Data Sheet for Preparation*; update date 01.06; POCH LTD: Poland, 2008.

24. Longwic, R.; Sander, P. The characteristics of the combustion process occurring under real operating conditions of traction. *IOP Conf. Ser. Mater. Sci. Eng.* **2016**, *148*, 012071. [CrossRef]

25. Longwic, R.; Sen, A.K.; Górski, K.; Lotko, W.; Litak, G. Cycle-to-Cycle Variation of the Combustion Process in a Diesel Engine Powered by Different Fuels. *J. Vibro Eng.* **2011**, *13*, 120–127.

26. Longwic, R.; Sander, P.; Nieoczym, A.; Lotko, W.; Krzysiak, Z.; Samociuk, W.; Bąkowski, H. Effect of some properties of hydrocarbon fuels on self-ignition delay. *Przemysł Chem.* **2017**, *96*, 1123–1127.

27. Atkins, P.; de Paula, J.; Keeler, J. *Physical Chemistry*; Oxford University Press: Oxford, UK, 2017.

28. Fowkes, F.M. Attractive forces at interfaces. *Ind. Eng. Chem.* **1964**, *56*, 40–52. [CrossRef]

29. Van Oss, C.J.; Chaudhury, M.K.; Good, R.J. Monopolar surfaces. *Adv. Colloid Interface Sci.* **1987**, *28*, 35–64. [CrossRef]

30. Liu, H.; Przybylski, R.; Dawson, K.; Eskin, N.A.M.C.; Biliaderis, G. Comparison of the composition and properties of canola and sunflower oil sediments with canola seed hull lipids. *J. Am. Chem. Soc.* **1996**, *73*, 493–498. [CrossRef]

Article

Model Issues Regarding Modification of Fuel Injector Components to Improve the Injection Parameters of a Modern Compression Ignition Engine Powered by Biofuel

Jacek Eliasz, Tomasz Osipowicz *, Karol Franciszek Abramek and Łukasz Mozga

Department of Automotive Engineering, West Pomeranian University of technology, 70–310 Szczecin, Poland; Jacek.Eliasz@zut.edu.pl (J.E.); Karol.Abramek@zut.edu.pl (K.F.A.); ml23722@zut.edu.pl (Ł.M.)
* Correspondence: tosipowicz@zut.edu.pl; Tel.: +48-503420350

Received: 15 October 2019; Accepted: 11 December 2019; Published: 13 December 2019

Abstract: This article presents a theoretical analysis of the use of spiral-elliptical ducts in the atomizer of a modern fuel injector. The parameters of the injected fuel stream can be divided into quantitative and qualitative. The quantitative parameter is the injection dose amount, and the qualitative parameter is characterized by the stream of injected fuel (width, atomization, opening angle, and range). The purpose of atomizer modification is to cause additional flow turbulence, which may affect the stream parameters and improve the combustion process of the combustible mixture in a diesel engine. The spiral-elliptical ducts discussed here could be used in engines powered by vegetable fuels. The stream of such fuels has worse quality parameters than conventional fuels, due to their higher viscosity and density. The proposal to use spiral-elliptical ducts is an innovative idea for diesel engines.

Keywords: fuel injector; CI engine; biofuel; fuel combustion

1. Introduction

The research problems of modern compression ignition engines concern issues related to ecology. The most important aspect is the emission of pollutants in exhaust gases. Experiments on reducing the toxicity of exhaust gases have been carried out in various directions, such as the use of electronic engine management systems, the use of exhaust after-treatment systems, various modifications of engine components, and the use of alternative fuels. In order to meet stringent standards regarding the emission of pollutants in the exhaust gases of diesel engines, a common rail fuel supply system has been introduced. The advantage of this system is the ability to adjust the injection torque and pressure of the engine operating conditions by separating the fuel accumulating element (injection pump) with fuel injectors with a pressure accumulator (rail). Various design solutions for the Common Rail system have been developed, but their operating principle is the same. The essence of this solution is to divide the injection doses into one engine cycle, and to control the desired pressure and fuel injection moment. The purpose of this solution to appropriately organize the combustion process of the combustible mixture in the engine compartment so that the emission of toxic substances to the atmosphere, such as nitrogen oxides, carbon monoxide, unburned hydrocarbons and solid particles, is reduced. The element of the injection system responsible for organizing the combustion process of the combustible mixture in the cylinder workspace is the injectors. Their task is to distribute and atomize fuel in the combustion chamber. Two aspects of the injection process are considered: Quantitative and qualitative. The quality of the injected fuel stream plays an important role in the process of creating a combustible mixture. Its physical properties, such as viscosity and density, are responsible for this parameter. Fuels of plant origin have a higher viscosity and density, which affects the parameters of the stream, such as atomization, width, length, or angle of opening. However, it is possible to change these parameters by using spiral-elliptical ducts on the non-working part of the sprayer. This work analyzes the possibility of using this solution in the injection systems of compression ignition engines.

Improving the quality parameters of the fuel stream can be achieved by increasing the pressure in the injection system. The issues related to the influence of pressure were analyzed in Wang et al. work [1]. They showed that the range of the stream depends on the pressure in the system. The most favorable penetration result was obtained from a pressure of 150 MPa to 210 MPa. The spray angle is the largest at low pressures. With increasing pressure, it decreases, and remains constant. The speed of fuel outflow from the atomizer can reach 320 m/s at a pressure of 250 MPa. Authors Han et al. [2] characterized how fuel properties affect the droplet size during spraying. Conventional diesel oil, rapeseed oil methyl ester, rapeseed oil ethyl ester, and biofuel made from coconut oil were used for their research. Studies have shown that fuel with higher surface tension and viscosity, such as ethyl and methyl ester of rapeseed oil, is characterized by a greater ability to penetrate the stream, but the width of the opening angle and the flow area of the fuel spray is lower. Coconut oil methyl ester and conventional diesel oil were characterized by better atomization and spray surface. Microscopic studies have also shown that fuels with lower viscosity and surface tension (diesel oil coconut methyl ester) have smaller Sauter diameters (SMD) in the sprayed stream. Research conducted by Anis, et al. [3] showed that preheating the fuel improves spraying parameters. The analysis was carried out for pure biofuels and mixtures of biofuels with conventional fuel. The test results showed that heating up the fuel in fuel injectors to up to 343 K improves the quality characteristics of the injected fuel stream, and heating up to 323 K in the injection pump increases its efficiency. Other research Han et al. [2] compared how the internal velocity of the flow influences the division and fragmentation of fuel droplets during atomization. During fuel injection into the engine compartment, turbulent flows occur, which affect the quality parameters of the stream. The purpose of the simulations was to help understand the process of fragmentation of fuel droplets in the combustion chambers of a compression ignition engine. The analysis showed that intensity of turbulent flow affects the process of fragmentation of the fuel stream. Increasing the pressure in the atomizer increases the fuel speed at the atomizer outlet, which increases turbulence. In another article, Anantha et al. [4], the process of injection of two diesel fuels and soybean oil methyl ester was compared. Computer analysis of fuel flow through the atomizer and simulation of the atomization process was carried out for a Bosch 1.0 fuel injector at 135 MPa injection pressure, 1 MPa overflow pressure, 300 K ambient temperature, and 0.5 ms injection time. Two types of fuel atomizers, with cylindrical and conical holes, were used for the tests. The analysis showed that a much higher occurrence of cavitation appears when using cylindrical holes in fuel atomizers, regardless of the type of fuel. The fuel flow through the atomizer is higher for tapered holes, though minimally so for diesel, due to the lower density compared to soybean oil methyl ester. The average fuel flow rate is higher for diesel and comparable for the type of holes in the fuel atomizer. This is due to the lower viscosity of conventional fuel. Studies have shown that plant biofuels have lower tendencies towards cavitation, due to their higher density and viscosity Bishop et al. [5] and Yu et al. [6]. Observation of the stream of injected fuel showed that the angle of opening with the use of cylindrical holes in the atomizer is similar for diesel and vegetable fuel, while with the cone atomizer it is larger for biodiesel, because it produces a larger cloud during scattering in the combustion chamber. The largest range of the fuel stream was recorded for diesel with tapered holes, while for biofuel the design of the hole was irrelevant. The Sauter diameter was definitely lower for conventional fuel regardless of the type of hole used in the fuel atomizer. The difference during fuel flow through the atomizer openings and the atomization process is mainly influenced by the density and viscosity of the tested fuels Bishop et al. [5]. Research carried out in Tinprabath et al. [7] presents a theoretical and experimental analysis of flow in the atomizer, and the injection characteristics of conventional fuel without bioadditives (B0) and biofuel (B100). The simulation results showed that the phenomenon of cavitation and turbulence during flow is more common with conventional fuels. As the fuel pressure in the atomizer increases, the kinetic energy of the liquid increases, which increases turbulence and cavitation.

Authors Tinprabath et al. [7] present an experimental analysis of the impact of diesel oil, vegetable fuel, and their common mixtures on the injection flow rate. Bosch piezoelectric injectors and five types of fuel were used for the tests: Diesel oil, winter diesel oil, mixtures of diesel oil and B20 biofuels (80% diesel fuel, 20% biofuels), B50 (50% diesel fuel, 50% biofuels), and clean B100 biofuel. The fuel injector pressure was 30–60 MPa at room temperature, at which no evaporation of the mentioned fuels takes place, and below zero (−5 °C and −8 °C). The experiment showed that low ambient temperatures do not affect the auto-ignition delay period. Fuel viscosity affects the duration of the injection, therefore, there are differences in the injection dose for individual fuels. At positive temperatures for B100 vegetable fuels, the flow rate decreases linearly until the viscosity increases. At negative temperatures for fuel with additives (diesel fuel and diesel fuel winter mixed with biofuel), viscosity, measured at atmospheric pressure, is not the only property that affects the flow rate. Pure diesel, B20 and B50 had similar viscosities. Only winter diesel fuel had better parameters.

The injection parameters are affected by the physico-chemical properties of the fuel, such as density, viscosity, surface tension, the heat of evaporation, boiling point, liquidity point, and cetane number Pandey et al. [8]. Density is a parameter directly proportional to the volume elasticity coefficient. Factors, such as density, speed of the propagating wave sound, and the volumetric elasticity coefficient affect the moment of fuel injection into the combustion chamber (injection advance angle). The higher the fuel density and volumetric coefficient of elasticity, the earlier the fuel begins to oxidize in the cylinder space, which causes the cylinder to have a higher temperature and higher nitrogen oxide emissions in exhaust gases Pandey et al. [8]. Research Knothe [9] has shown that biodiesel made of saturated fatty acids (low iodine number) has a higher cetane number, but worse low temperature properties, and biodiesel made of unsaturated fatty acids, inversely has a low cetane number and better properties at low temperatures. In addition, the volume modulus has a higher value for unsaturated fatty acid methyl esters than for saturated fatty acid methyl esters. Saturated fatty acid esters have a lower density, due to the smaller number of carbon bonds in the chain Pandey et al. [8]. Kinematic viscosity determines the liquid's fluidity. The fluidity decreases as the kinematic viscosity increases. This is another very important parameter affecting the quality parameters of the fuel injection process into the combustion chamber. This property affects the atomization of the fuel stream. The higher the kinematic viscosity, the lower the spray angle and the droplet diameter during the injection process. This results in the deposition of soot in the combustion chamber of the engine and its components parts, contamination of the engine oil, and startup problems at lower temperatures. In work Knothe [9], the influence of the number of fatty acid bonds on the magnitude of kinematic viscosity was examined. Studies have shown that the chain length, position and number of double bonds affect kinematic viscosity and ease of oxidation. The longer the chain, the more the viscosity increases, the greater the number of unsaturated fatty acids in methyl esters of vegetable oils, and the smaller the decrease. Too high a viscosity of the fuel causes earlier ignition, which adversely affects the combustion process and causes increased emissions of nitrogen oxides into the atmosphere Pandey et al. [8]. The boiling point affects the evaporation properties of the fuel. The smaller it is, the easier it is to create a combustible mixture. The fuel distillation temperature increases with the content of carbon molecules in the fuel, and decreases with the increase of unsaturated fatty acids Kegl et al. [10], Knothe [9]. The liquidity point determines how low the fuel can flow. The disadvantage of vegetable fuels is insufficient liquid properties at low temperatures. This is affected by the chain length of higher fatty acid methyl esters. If the ester contains more saturated fatty acids, the liquidity temperature increases, and if it contains more unsaturated fatty acids, it decreases Pandey et al. [8]. The parameters determine the injection properties of a modern fuel injector, such as injection time, injection advance angle, injection dose, fuel injection delay, and atomizer opening pressure, and play an important role in the combustion process. The fuel injection time depends on the chemical properties of the fuel, especially its calorific value. Lower calorific fuels require longer fuel injection times, while higher ones require the opposite. Longer injection times cause an incomplete combustion process, which increases the emission of carbon oxides, soot, and hydrocarbons in the exhaust gas. Another parameter determining the fuel injection properties is the injection advance angle. It is affected by the physical fuel properties, such as the density, kinematic viscosity, and volumetric coefficient of elasticity of the fuel. If the mentioned parameters increase, then the attenuation of fuel in the injector decreases, and it causes faster needle rising in the atomizer, which extends the injection advance angle Salvador et al. [11]. This results in increased pressure in the combustion chamber and a higher emission of nitrogen oxides. Density, viscosity and sound velocity value have an impact on the compressibility factor of the fuel during flow through the fuel atomizer holes, but they do not affect the Reynolds number Salvador et al. [12]. The fuel injection CFD simulation was carried out in two phases: Incompressible flow at constant liquid properties and compressible flow at liquid properties, calculated locally as a function of flow pressure conditions. Analyzing the research carried out by the authors, it can be stated that the issues related to fuel flow through the injector atomizer should be considered in terms of compressibility of the liquid, depending on the amount of pressure and flow temperature. At low system pressures, the flow is moderate. The liquid is in a state of transition between the laminar and turbulent flow Payri at el. [13]. The injection dose factor is then dependent on the Reynolds number. However, when the pressure in the system begins to increase, the flow passes into the turbulent phase, then the injection dose factor is independent of the Reynolds number Salvador et al. [14]. In work Desantes et al. [15], the process of injection of methyl esters of rapeseed oil mixed with diesel oil was compared in the proportions—5% bio-additive (B5), 30% bio-additive (B30) and pure methyl ester (B100). Studies have shown that the parameters of the fuel stream for B5 and B30 fuels are similar. The pure rapeseed oil methyl ester stream has a higher flow rate, due to the higher density, but similar momentum. The analysis of the fuel injector operation showed that its speed depends on the fuel used. The increased viscosity of the biofuel causes the needle to rise more slowly. This is very important when generating small doses of fuel, which depend mainly on the dynamics of the fuel injector. It was noticed that clean biofuel has a slightly larger penetration range, but a narrower angle of opening of the stream, which affects the dose of injected fuel Saltas et al. [16]. Due to the increased density, the speed of the injected fuel is lower, and the surface tension causes poorer atomization of the droplets. These factors cause worse mixing conditions with air and directly influence the combustion process. Biofuels have an inclination towards dampness absorption Sundus et al. [17] and Othmana et al. [18]. Research conducted by Antonov et al. [19] discussed the spray effects two fluids: Diesel and water, and rape oil and water. The results showed that the micro-explosion of drops in the biofuel stream needs more energy because of higher viscosity and surface tension compared to diesel fuel. The structure of primary and secondary breakup droplets depends on the Weber number Minakov et al. [20].

After analyzing the literature on the subject of this work, it can be stated that phenomena occurring during the course of fuel atomization, such as the formation of waves, their amplitude increase, and loss of stability, are affected by fluid vibrations during motion. The frequency of vibrations varies, and their causes can be external and internal. Internal factors are disturbances arising in the atomizer itself, such as liquid turbulences, atomizer vibrations (because the work of the needle), expansion of the liquid, due to pressure changes, and disturbances in the movement of liquids on the unevenness and edges of holes. External factors depend on the liquid speed, gas density, liquid surface and pressure. On the basis of the authors' preliminary research Osipowicz et al. [21], it can be concluded that the implementation of changes to the non-working part of the needle affects the course of the process of atomizing fuel in the combustion chamber. Fuels of plant origin, due to their physical properties, have poorer spray ability in comparison with conventional diesel oil. Spiral-elliptical ducts may cause additional turbulence during the liquid flow through the atomizer immediately before the injection process. Due to the additional turbulence caused, particles may break down, and the quality of the fuel stream may improve, which will affect the combustion of the combustible mixture. Increased flow turbulence can locally increase the temperature of the liquid in the atomizer. By analyzing the literature, we can conclude that temperature has a significant impact on the density and viscosity of biofuels. The proposed changes are innovative and have not been used anywhere before.

None of the above works analyzed how the additional turbulence caused by the fuel flow before flowing into the atomizer holes will affect the atomization, range, angle, and speed of the injected fuel stream. Additionally, it is possible to change parameters directly in the fuel atomizer, due to turbulence and increased temperature. In previous studies in this field, no attempts have been made to carry out modifications of fuel atomizers. Increasing the pressure in the system, heating the fuel, and changing the design of the combustion chambers are known strategies that have been extensively explored. Therefore, the innovative element of this project is the analysis of the fuel flow in a modified atomizer.

The phenomena of fuel flow in the atomizer are variable. During the injector operation, the atomizer closes and opens continuously. The type of flow depends on the pressure in the whole system. In modern Common Rail systems, the pressure in the fuel tank oscillates between 25–200 MPa, and the value depends on the engine load.

This paper analyzes an electromagnetic fuel injector of the Bosch 1.0 generation, working on pressures in the range of 30–135 MPa. The fuel flow model was created based on a test of the injector on a Zapp Carbontech CRU2i test bench.

2. Impact of Fuel Injection Parameters on the Combustion Process in a Compression Ignition Engine

The parameters of the injected fuel stream have a significant impact on the combustion process in a compression ignition engine. The process of creating a combustible mixture consists of atomizing and spreading the fuel stream, heating, evaporation, and mixing of the fuel with air. The preparation of the combustible mixture lasts from the moment of starting the fuel injection to the end of its combustion. Combustion of a combustible mixture in a compression ignition engine depends on the quantitative and qualitative characteristics of the injection, design of the combustion chamber, fuel properties, speed and direction of the load propagation, location of fuel injectors, and mutual orientation of the fuel streams. Ignition of atomized fuel is a chain; a multi-stage process. The first ignition foci are formed according to the volumetric combustion process. The flame spreads, and the combustion of the fuel-air mixture prepared during the auto-ignition delay period begins. The combustible mixture goes into the diffusion combustion stage and afterburning takes place. The process of fuel combustion in a diesel engine takes place in several stages.

The first stage of the combustion process is the auto-ignition delay period (τ_s). It is the time from the beginning of fuel injection until the appearance of the first auto-ignition foci (on the indicator graph, the point at which the pressure in the cylinder, due to heat release is greater than the pressure in the cylinder when compressing air without injection). The auto-ignition delay is the time when the stream breaks down into droplets, partial evaporation and mixing of fuel vapors with air, and a period of acceleration of chemical reactions. This stage is intended to be shortened, because a shorter delay of self-ignition causes less fuel flow to the chamber, which causes a slower pressure increase in the combustion chamber. A pressure increase during the combustion process that is too rapid increases the combustion temperature, which promotes increased emission of nitrogen oxides in exhaust gases, increases the impact loads of engine components, and increases the noise associated with the combustion process. The average rate of pressure rise should not exceed 0.3–0.8 MPa/°OWK. The following factors affect the auto-ignition delay period.

1. Cetane number of the fuel, which determines the ability of a given fuel to ignite. The higher its value, the better the fuel's flammable properties. The cetane number of fuels can be increased by using various additives.
2. The type of combustion chamber affects the self-ignition delay because there are differences in the distribution of fuel in the space and layer at the wall, and in the temperature distribution of the walls of the combustion chamber.
3. The fuel pressure and temperature at the beginning of the fuel injection process affect the first stage of the combustion process. Increasing the pressure, and as a result, the temperature of the lower injection timing shortens the auto-ignition delay.

4. Quantitative and qualitative characteristics of the fuel injection. The dose and shape of the stream affect the auto-ignition delay period. Correct injection rates (pilot, basic) generated by fuel injectors and appropriate atomization of the fuel jets shorten the auto-ignition delay.

5. Increasing the engine speed improves atomization and increases the pressure and temperature in the combustion chamber, which reduces the ignition delay.

The targeted intensity of the charge movement in the combustion chamber shortens the auto-ignition delay period.

The auto-ignition delay period according to Heywood is determined by the following relationship [9],

$$\tau_s = (0.36 + 0.22 \cdot c_m) \cdot \exp\left[E_a \left(\frac{1}{RT_2} - \frac{1}{17.19}\right) \cdot \left(\frac{21.2}{P_2 - 12.4}\right)^{0.63}\right], \tag{1}$$

where P_2 is the temperature in the combustion chamber, T_2 is the pressure in the combustion chamber, E_a is the amount of activation energy, and c_m is the average piston speed.

The activation energy value is calculated from the formula

$$E_a = \frac{618.840}{LC + 25} \ [\text{J}/\text{mol}], \tag{2}$$

where LC is the cetane number.

Based on relationship (1), it can be stated that increasing the cetane number of fuel extends the delay of auto-ignition. According to theoretical considerations, the quality of fuel atomization in the combustion chamber does not affect the auto-ignition delay period. However, the considerations apply to older design engines operating at low pressures in the injection system, and the old generation fuel injectors. The Common Rail system works at working pressures of 25–200 MPa. Phenomena inside the system are very variable. The occurring turbulent flow of liquid in the fuel atomizer means that the atomization, width, range, and angle of the stream of flowing fuel have different values than in classic systems. Additionally, the changes made to the fuel atomizer can affect the additional turbulence of the liquid and improve its mixing with air. The second stage of the combustion process, called kinetic combustion (rapid combustion), lasts from the moment of the ignition of the combustible mixture until reaching the maximum pressure in the cylinder. During this phase, oxidation of part of the fuel–air mixture prepared during auto-ignition delay occurs, resulting in rapid heat release and pressure build-up. At the end of this stage, the combustion process is limited by the speed of mixing of the fuel with air. The following factors influence the kinetic combustion period.

1. The injection dose introduced during the auto-ignition delay period and during kinetic combustion, and the characteristics of the injected fuel stream. A smaller amount of fuel supplied during the auto-ignition delay causes a smaller value of pressure increase in the combustion chamber in relation to the crankshaft rotation. If the atomization is finer, the faster the first doses of fuel mix with the air, and the more gently the pressure in the cylinder in the second phase increases.

2. The design of the combustion chamber significantly affects the development of the second stage, and in particular, its impact on the auto-ignition delay period and the amount of air-fuel mixture prepared for combustion after it has started.

3. Increasing the charge movement speed to a certain value in the combustion chamber intensifies heat release in the kinetic combustion phase.

4. When the engine speed increases, the atomization improves, the auto-ignition delay decreases, the charge movement speed in the combustion chamber increases, and pressure and temperature increase. These factors accelerate chemical reactions.

5. By reducing the engine load, the duration of the second stage is shortened, resulting in a reduction of the fuel dose and a shorter time for its supply to the cylinder.

The third stage of combustion begins when the reaction speed is much higher than the mixing speed of the reactants, then the duration of the combustion process depends on the mixing speed of the reactants. This period is called diffusion combustion. It begins at the moment of the greatest pressure in the cylinder, and continues until the maximum temperature is reached. In modern diesel engines, the temperature reaches a maximum of 20–40 °C according to GMP. This is due to the fact that intense heat production is caused by the second period of the combustion process. The third stage of combustion of the combustible mixture in a compression ignition engine depends on the following factors.

1. The use of engine boost increases the amount of heat generated. By increasing the engine boost level, the duration of the third phase is extended. This is related to the amount and speed of heat released.
2. Injection dose size and quality of the fuel atomization process after starting the combustion process. At low engine loads, the fuel injection ends before the beginning of the third combustion phase, then the amount of heat released is small.
3. By increasing the speed of movement of the load in the cylinder to a certain value, heat release in the third stage increases. However, too much air turbulence reduces heat generation, due to deterioration of fuel decomposition in the combustion chamber, and movement of combustion products from the zone of one fuel stream to another. This may result in incomplete combustion and an increase in smoke opacity.
4. Increasing the engine speed increases the supply, fuel atomization, and load speed in the combustion chamber, which shortens the third stage of combustion.

The fourth stage of combustion in a compression ignition engine called afterburning begins when the maximum temperature of the cycle is reached, and lasts until the end of heat release. A soot burning reaction occurs during this stage. The following factors affect the course of the fourth stage.

1. The phenomenon of fuel falling on cool surfaces of the cylinder space, which causes the extension of the duration of this stage.
2. Supercharging the engine leads to a prolongation of the period of combustion of fuel, due to the longer fuel injection time and the deterioration of the spread of the fuel in the combustion chamber.
3. The spraying process in the final stage of combustion. The large maximum diameter of the fuel droplets results in extended fuel injection, which means that the afterburning time of the combustible mixture is longer. This deteriorates the use of thermal energy resulting from combustion, which affects the operation and reliability of the engine by depositing carbon combustion deposits and coking the fuel atomizers.
4. Turbulent charge fluctuation contributes to improving the mixing of fuel and air.

Analyzing the combustion process of the fuel–air mixture in a diesel engine, it can be concluded that the qualitative characteristics of fuel atomization in the cylinder space affect each stage of combustion.

3. Presentation of the Modifications Made to the Fuel Injector Sprays

The purpose of spray modification is to improve the fuel injection process in a modern diesel engine with a Common Rail system powered by vegetable fuel. The fuel injection process can be considered in terms of quantity and quality. The quantity, describing the quantitative aspect, is the injection dose, while the qualitative aspect includes the shape, range, width, opening angle, speed, and atomization of the stream of fuel flowing from the atomizer. Figure 1 shows a modified fuel atomizer with spiral-elliptical ducts Osipowicz [22].

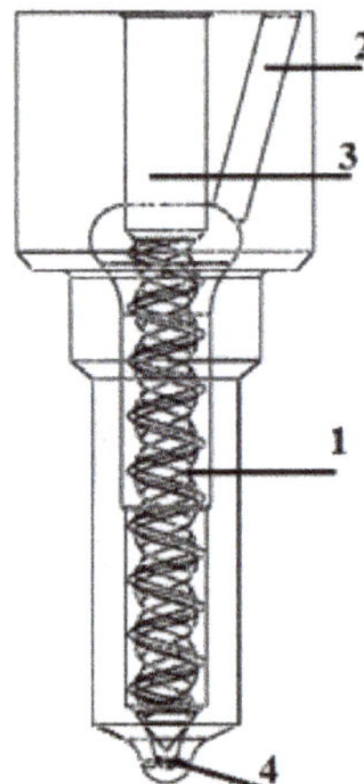

Figure 1. The modified fuel atomizer. (**1**) Channels in the non-working part of the needle, (**2**) the fuel supply channel, (**3**) the precision atomizer pair, (**4**) the injection holes [22].

The element of the fuel injector responsible for the correct process of fuel injection into the combustion chamber is the atomizer (Figure 1). It consists of a body and a needle. The fuel flows into the needle chamber through channel 2, then flows around the non-working part of needle 1. Annular channels made on the non-working part of the needle, as a result of the reciprocating movement of the needle, cause additional fuel turbulence. When the cone valve 4 opens the holes in the fuel atomizer, injection into the combustion chamber takes place. The purpose of the atomizer modification is to improve the atomization process of vegetable fuels in the combustion chamber of the engine, due to the turbulence caused by the atomizer and the accompanying phenomena. The parameters affecting the quality of the injected fuel stream are viscosity and density. Table 1 presents selected physical parameters of conventional and vegetable fuels. As can be seen, the viscosity and density of vegetable fuels have higher values than conventional fuel, therefore, according to the authors, the use of spiral-elliptical ducts on the non-working part of the needle will affect the physical parameters of the fuel and improve the atomization process in the combustion chamber of the compression ignition engine.

Table 1. Physical parameters of diesel fuel B100 and raps oil

Parameter	Diesel Fuel	Rapeseed Oil Methyl Ester	Rapeseed Oil
Density [kg/dm^3] 20 °C	0.817–0.856	0.86–0.9	0.91–0.92
Kinetic viscosity [mm^2/s] 15 °C	2.90–5.50	6–9	68–97.7
Ignition temperature [°C]	20–84	111–175	317–324

One of the features of turbulent flows is the fluctuation of momentum and kinetic energy of turbulence Hinze [23], and the phenomenon of heat exchange during flow Elsner [24], Jiao [25]. The viscosity and density of liquids depending on the temperature. If the temperature increases locally under the influence of additional vortices, then the physical parameters of the fuel will also change, which will affect the spraying process Suh et al. [26]. Previous work related to the analysis of fuel flows in atomizers, and its atomization have not taken into account the induction of additional turbulence in the area of injectors. The concept of creating additional turbulence is innovative, and has not been used anywhere until now.

4. Fuel Flow Analysis in the Fuel Atomizer

In order to illustrate the phenomena occurring before the injection process, a fuel flow model was made for a standard and modified atomizer according to the received patent Osipowicz [19]. The analysis was performed using Solidworks Flow Simulation. This program enables the study of a wide range of fluid flow and heat transfer phenomena. The model has a demonstrative function, and its task is to examine whether the changes made have affected the fuel turbulence in the atomizer. The flow was tested when the atomizer was opened at the moment when the needle was raised. This experiment is the first stage of the design and implementation of spiral-elliptical ducts in modern fuel injectors.

Based on the tests carried out, it can be concluded that the tested fuel injectors are in working order. The simulation of the fuel flow through the atomizer was carried out for three measurements with the same actuation times, with 500 µs fuel injectors at 30, 60, and 120 MPa pressure (Tables 2 and 3).

Table 2. Measurements of the injection doses according to the standard test of the standard fuel injector.

No.	Pressure [MPa]	Injection Time [µs]	Range [mm^3/H]	Dosage [mm^3/H]
1	135	780	34.71–49.69	35.3
2	30	420	0.31–3.89	0.7
3	80	260	0.31–4.09	1.2

Table 3. Measurements of the injection doses according to the standard test of the modified fuel injector.

No.	Pressure [MPa]	Injection Time [µs]	Range [mm^3/H]	Dosage [mm^3/H]
1	135	780	34.71–49.69	34.8
2	30	420	0.31–3.89	0.9
3	80	260	0.31–4.09	1.1

During the tests the phenomena prevailing in the atomizer were described. The following figure shows the tip of the fuel injector divided into zones (Figure 2).

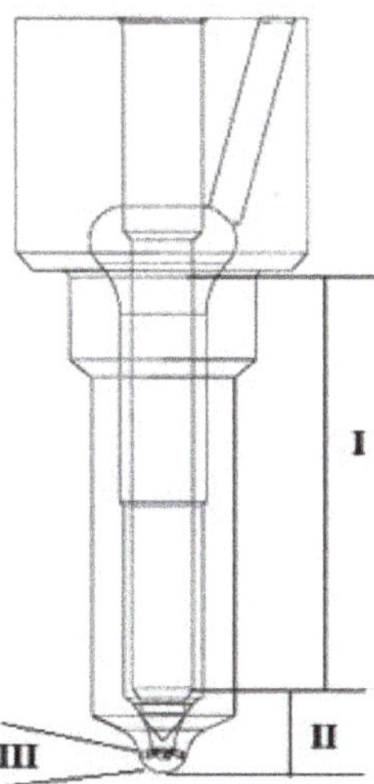

Figure 2. Fuel atomizer divided into zones: (**I**) The non-working part of the needle, (**II**) the needle well, (**III**) the outlet from the atomizer.

For the analysis of phenomena occurring in the atomizer of a modern fuel injector, general equations of the motion of viscous fluids, resulting from the momentum conservation principle, can be used. These phenomena are described by the Navier-Stokes equations in scalar form, where $V_{x, y, z}$ are fluid velocities in a Cartesian system, $\vec{V}$ is the fluid flow velocity, p is the pressure, X, Y, Z are the individual mass forces in the Cartesian system components, $\vec{F}$ is the individual mass force, and ρ is the fluid density:

$$\frac{dV_x}{dt} = X - \frac{1}{\rho}\cdot\frac{\partial p}{\partial x} + v\cdot\Delta V_x + \frac{1}{3}\cdot v\cdot\frac{\partial}{\partial x}(div\vec{V}), \tag{3}$$

$$\frac{dV_y}{dt} = Y - \frac{1}{\rho}\cdot\frac{\partial p}{\partial y} + v\cdot\Delta V_y + \frac{1}{3}\cdot v\cdot\frac{\partial}{\partial y}(div\vec{V}), \tag{4}$$

$$\frac{dV_z}{dt} = Z - \frac{1}{\rho}\cdot\frac{\partial p}{\partial z} + v\cdot\Delta V_z + \frac{1}{3}\cdot v\cdot\frac{\partial}{\partial z}(div\vec{V}), \tag{5}$$

and in vector form:

$$\frac{d\vec{V}}{dt} = \vec{F} - \frac{1}{\rho}\cdot grad(p) + v\cdot\Delta\vec{V} + \frac{1}{3}\cdot\gamma\cdot grad(div\vec{V}). \tag{6}$$

The following data was used to make the model: Channel (atomizer) parameters, flow rate (fuel injection dose), and fuel parameters (density and viscosity). The Bernoulli equation was used to calculate the speed of fuel flow through the atomizer and stream, where V, V_1,V_2 are fluid flows, p_1 and p_2 are the pressure, ρ is the fluid density, μ is the fluid dynamic viscosity coefficient, and d_0 is the flow dimension.

$$\frac{\rho V_1^2}{2} + p_1 = \frac{\rho V_2^2}{2} + p_2, \tag{7}$$

$$Re = \frac{\rho\cdot V\cdot d_0}{\mu}. \tag{8}$$

Analyzing the figure, the pressure in the atomizer in zones I and II is at the same level. Based on relationships (7) and (8), the Reynolds number and fuel flow rate will also be the same. The flow parameters of the tested fuel injectors during the analysis are presented in the following table (Table 4).

Table 4. Fuel flow parameters through the atomizer during analysis.

P [MPa]	Q [m^3/s]	Re$_1$	Re$_2$	V$_1$ [m/s]	V$_2$ [m/s]
30	0.0000186	218	4,298,326	0.0253	258
60	0.0000408	480	6,229,029	0.0554	434
120	0.0000602	708	8,913,541	0.0818	535

Based on the calculations, simulations of fuel flow through a standard and modified atomizer in a Solid Works Flow Simulation environment were undertaken (Figures 3–7).

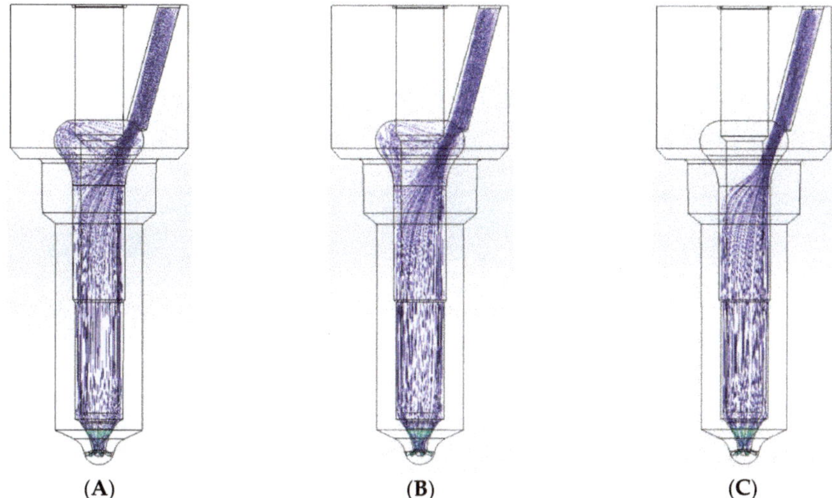

(A)   (B)   (C)

Figure 3. Simulation of fuel flow through a standard atomizer with an injection time of 500 µs and fuel temperature of 313 K. (**A**) System pressure of 120 MPa, injected dosage 20.9 mm^3/H; (**B**) system pressure of 60 MPa, injected dosage 10.2 mm^3/H; (**C**) system pressure of 30 MPa, injected dosage 2.8 mm^3/H.

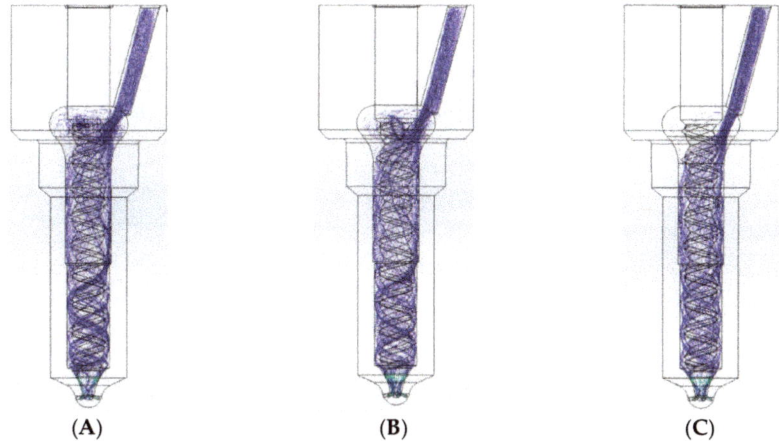

(A)   (B)   (C)

Figure 4. Simulation of fuel flow through a modified atomizer at 500 µs injection time and fuel temperature of 313 K. (**A**) System pressure of 120 MPa, injected dosage 20.9 mm^3/H; (**B**) system pressure of 60 MPa, injected dosage 10.2 mm^3/H; (**C**) system pressure of 30 MPa, injected dosage 2.8 mm^3/H.

(A)   (B)

Figure 5. Simulation of fuel flow as a function of speed in the zone of valve closing and opening of the atomizer at 120 MPa system pressure. (**A**) Classical atomizer; (**B**) atomizer with spiral-elliptical ducts.

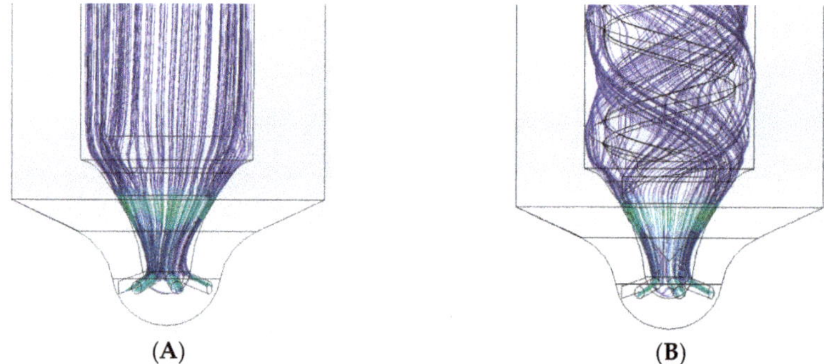

(A)   (B)

Figure 6. Simulation of fuel flow as a function of speed in the zone of valve closing and opening of the atomizer at 60 MPa system pressure. (**A**) Classical atomizer; (**B**) atomizer with spiral-elliptical ducts.

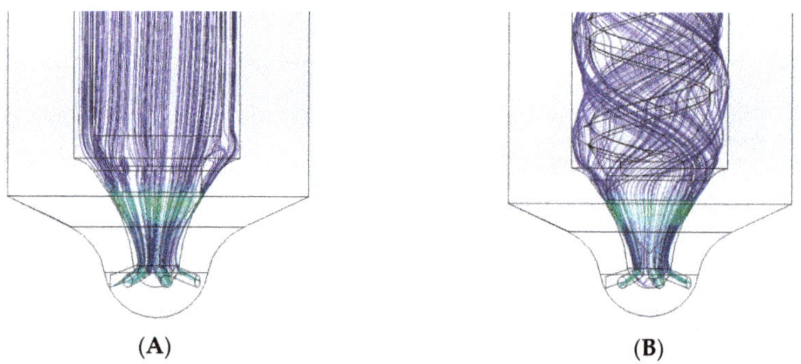

(A)   (B)

Figure 7. Simulation of fuel flow as a function of speed in the zone of valve closing and opening of the atomizer at 30 MPa system pressure. (**A**) Classical atomizer; (**B**) atomizer with spiral-elliptical ducts.

The aim of spiral-elliptical ducts is to trigger eddies along the non-working part of the injector needle. If the fuel moves in the nozzle, it will increase the temperature, due to additional turbulence. A higher temperature will lower fluid kinematic viscosity. Spiral-elliptical ducts increase the flow dimension. The Reynolds number (R_e) increases according to relationship (8), which has an influence on the fuel stream turbulence. The density of compressible liquids depends on fluid temperature and pressure, and can be considered within approximation (9), where ρ_0 is the liquid density under the reference pressure P_0, C and B are coefficients, ρ_0, B and C depend on the temperature, and P is the calculated pressure [27].

$$\rho = \frac{\rho_0}{1 - c \cdot \ln \frac{B+P}{B+P_0}}.$$ (9)

Liquid dynamic viscosity depends on fluid pressure, and can be considered within approximation (10), where μ_0 is the liquid density under the reference pressure P_0, a is a coefficient that depends on the temperature, and $P^l = 10^5$ Pa is a constant value.

$$\mu = \mu_0 e^{a(p-p^l)}.$$ (10)

Flow simulation allows us to simulate heat transfer between solids and fluids. Heat transfer conversion has been described by energy conservation Equations (11) and (12). Heat flux is defined by Equation (13). Anisotropic solid media heat conductivity is described by Equation (14), where e is the specific internal energy, Q_H is the specific heat release per unit volume, λ_i is the eigenvalues of the thermal conductivity tensor, Pr is the Prandtl number, h is the thermal enthalpy, $\sigma_c = 0.9$ constant, u is the fluid velocity, Ω is the angular velocity, and S_i is the mass distributed external force per unit mass, due to porous media resistance [27].

$$\frac{\partial \rho H}{\partial t} + \frac{\partial \rho u_i H}{\partial x_i} = \frac{\partial}{\partial x_i}(u_j(t_{ij} + t_{ij}^R) + q_i) + \frac{\partial p}{\partial t} - \tau_{ij}^R \frac{\partial u_i}{\partial x_j} + \rho \varepsilon + S_i u_i + Q_H,$$ (11)

$$H = h + \frac{u^2}{2} + \frac{5}{3}k - \frac{\Omega^2 r^2}{2} - \Sigma_m h^0 y_m,$$ (12)

$$q_i - (\frac{\mu}{Pr} + \frac{\mu_t}{\sigma_c})\frac{\partial h}{\partial x_i}, i = 1, 2, 3,$$ (13)

$$\frac{\partial \rho e}{\partial t} = \frac{\partial}{\partial x_i}(\lambda_i \frac{\partial T}{\partial x_i}) + Q_H.$$ (14)

Figure 8 presents fuel temperature during flow through classical and spiral-elliptical duct atomizers. There is a noticeable difference in heat distribution between nozzles. The atomizer with spiral-elliptical ducts has higher fluid around the needle temperature because of additional eddies.

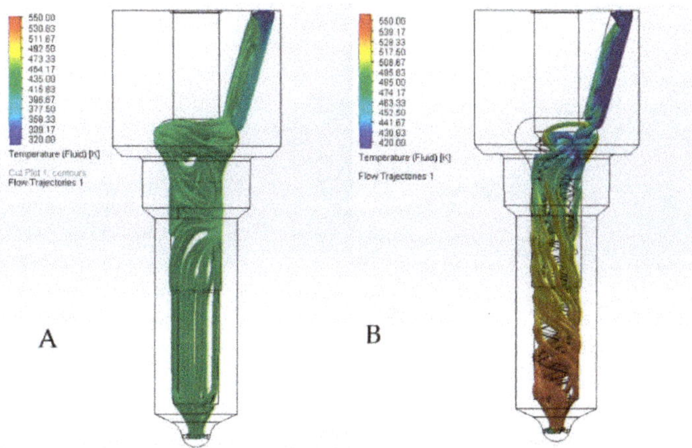

Figure 8. Fuel flow temperature distribution, pressure 120 MPa, fluid temperature 313 K, initial solid temperature 573 K, solid material tool steel X40Cr14. (**A**) Classical atomizer, (**B**) atomizer with spiral-elliptical ducts.

A simulation presented the fuel injector nozzle temperature distribution (Figure 9).

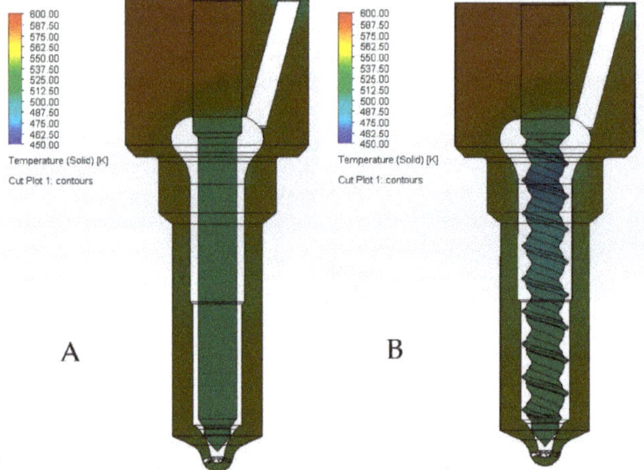

Figure 9. Fuel injector nozzle temperature distribution, pressure 120 MPa, fluid temperature 313 K, initial solid temperature 573 K, and solid material tool steel X40Cr14. (**A**) Classical atomizer, (**B**) atomizer with spiral-elliptical ducts.

Both nozzles have a similar temperature. The difference in the fluid temperature (Figure 8) triggered turbulence.

The research was undertaken using an infrared camera. Both injectors were working on the test bench. It was possible to compare the temperature of the chassis of standard and modified fuel injectors (Figure 10).

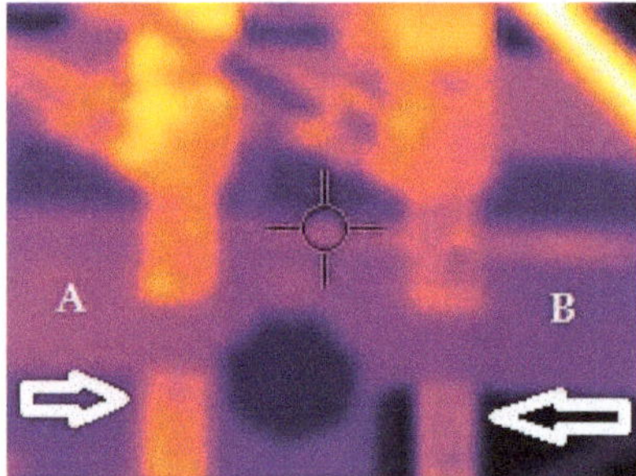

Figure 10. Fuel injector during test temperature distribution. (**A**) Atomizer with spiral-elliptical ducts; (**B**) classical atomizer.

The initial fuel spray visualization is shown in Figure 11. There was 30 MPa system pressure and a 300 µs injection time test. These are pilot dosage parameters for engine idle speed. This is also initial research, and the aim was to analyze if injector nozzle modification influences the fuel injected stream.

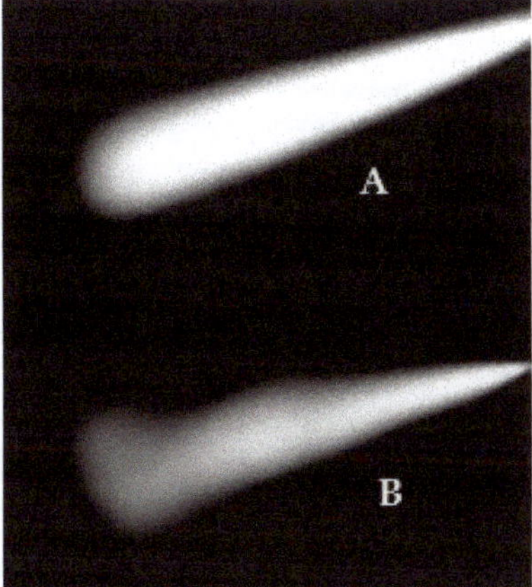

Figure 11. Initial fuel spray research, pressure 30 MPa, injection time 300 μs. (**A**) Fuel injector with a standard nozzle; (**B**) fuel injector with spiral-elliptical ducts.

5. Conclusions

Analyzing the results of simulation tests, it can be seen that modification of the non-working part of the atomizer needle changes the characteristics of the fuel flow. Spiral-elliptical ducts created additional turbulence around the spire (Figure 4). The qualitative parameters of the fuel stream depend on the velocity of the liquid at the outlet and the degree of its turbulence. The flow velocity depends on the pressure, while the degree of stream penetration is influenced by the physical parameters of the fuel, such as density and viscosity. The purpose of the modification is to intensify turbulence during fuel flow through the atomizer. The most important property of turbulent motion is the presence of many vortices of varying sizes that transfer momentum, mass and heat from one flow point to another. Vorticity Ω is the rotation of the velocity vector V. It is possible to use the kinetic energy of the rotational motion (imparted by modification of the non-working part of the needle) in order to increase the kinetic energy of the translational motion of the stream of injected fuel into the working space of the cylinder. Additionally, rotating flow around the needle has an influence on the temperature, leading to a higher Reynolds number. This phenomenon causes local changes in the physical fuel parameters, like viscosity and density. Viscosity is a parameter that affects the characteristics of the stream of fuel injected into the combustion chamber. By reducing its value as a result of the temperature rising, it is possible to improve the qualitative characteristics of the fuel injection. This process is of particular importance for a plant fueled engine, whose physical parameters are different from those of standard diesel fuel.

Qualitative parameters of the injected fuel stream play an important role during the combustion process of fuel in compression ignition engines. The results of laboratory tests have shown that the changes made to the non-working part of the fuel atomizer needle do not affect the amount of injection and overflow doses. Based on the results of the measurements, the fuel flow parameters (Reynolds number and flow speed) in the atomizer and stream were calculated. The nature of the liquid flow is determined by the Reynolds number, which depends on the speed. According to the Bernoulli equation, the results of calculations showed that in the fuel atomizer, due to high pressures, laminar fuel flow prevails (Figure 3), while in the stream the flow is hyper turbulent. The modification of the inoperative part of the needle caused additional fuel turbulence in the atomizer (Figure 4). Increased turbulence causes the temperature of the entire system to rise (Figure 8). As a result of this temperature increase, the viscosity of the fuel decreases, which affects the qualitative course of the fuel injection process. In addition, additional vortices affect the kinetic energy of the forward movement of the stream. The results of research using an infrared camera show that the temperature of the modified fuel injector is higher than that in a standard fuel injector during tests (Figure 10). Initial fuel stream visualization (Figure 11) indicates the changes in the injected fuel quality. The fuel stream was dispersed during the injection process. The analytical simulation shows that fuel in spiral-elliptical ducts has an increased temperature, due to additional turbulence. This causes a lowering of the fuel density and dynamic viscosity. These parameters have an influence on the fuel stream shape.

Based on the theoretical analysis, it can be concluded that it is possible to change the quality parameters of the stream of injected fuel into the cylinder working space. The spiral-elliptical ducts created caused additional turbulence in the fuel flow. The proposed solution is innovative, and should find particular application in compression ignition engines fueled with vegetable fuels, due to their physical properties.

In order to further, and more comprehensively, examine the effect of spiral-elliptical ducts on the operating parameters of a compression ignition engine fueled with either conventional fuel or fuel of plant origin, the following stages of testing are planned:

1. Tests of the stream of injected fuel using an injector with a standard and modified atomizer using conventional and vegetable fuels. During laboratory tests, the quality parameters of the injected fuel stream will be analyzed.
2. Tests on the engine test bench using standard and modified fuel injectors using conventional and vegetable fuels. During the tests, the operating and ecological parameters of the compression ignition engine will be analyzed.

Author Contributions: Supervision, writing—review, J.E. Conceptualization, data acquisition and interpretation, investigation and writing—original draft preparation and editing, T.O.; conceptualization, data interpretation and writing—review and editing, K.F.A.; software, investigation, writing—review and editing, L.M.

Funding: This research received no external funding.

Conflicts of Interest: We do not have any personal conflicts of interest in communicating the findings of this study, and we had no sponsor who would make claims to the findings being presented.

References

1. Wang, L.; Lowrie, J.; Ngaile, G.; Fang, T. High injection pressure diesel sprays from a piezoelectric fuel injector. *Appl. Therm. Eng.* **2019**, *152*, 807–824. [CrossRef]
2. Han, D.; Zhai, J.; Duan, Y.; Ju, D.; Lin, H.; Huang, Z. Macroscopic and microscopic spray characteristic of fatty acid esters on a common rail injection system. *Fuel* **2017**, *203*, 370–379. [CrossRef]
3. Anis, S.; Budiandono, G.N. Investigation of the effects of preheating temperature of biodiesel—Diesel fuel blends on spray characteristics and injection pump performances. *Renew. Energy* **2019**, *140*, 274–280. [CrossRef]
4. Anantha, R.L.; Deepanraj, D.; Rajakumar, S.; Sivasubramanian, V. Experimental investigation on performance, combustion and emission analysis of a direct injection diesel engine fueled with rapeseed oil biodiesel. *Fuel* **2019**, *246*, 69–74.
5. Bishop, D.; Situ, R.; Brown, R.; Surawski, N. Numerical modelling of biodiesel blends in a diesel engine. *Energy Procedia* **2016**, *110*, 402–407. [CrossRef]
6. Yu, S.; Yin, B.; Jia, H.; Wen, S.; Li, X.; Yu, J. Theoretical and experimental comparison of internal flow and spray characteristics between diesel and biodiesel. *Fuel* **2017**, *208*, 20–29. [CrossRef]
7. Tinprabath, P.; Hespel, C.; Chanchaona, S.; Foucher, F. Influence of biodiesel and diesel fuel blends on the injection rate under cold condition. *Fuel* **2015**, *144*, 80–89. [CrossRef]
8. Pandey, R.K.; Rehman, A.; Sarviya, R.M. Impact of alternative fuel properties on fuel spray behavior and atomization. *Renew. Sustain. Energy Rev.* **2012**, *16*, 1762–1778. [CrossRef]
9. Knothe, G. A comprehensive evaluation of the cetane numbers of fatty acid methyl esters. *Fuel* **2014**, *119*, 6–13. [CrossRef]
10. Kegl, B.; Lešnik, L. Modelling of macroscopic mineral diesel and biodiesel spray characteristics. *Fuel* **2018**, *222*, 810–820. [CrossRef]

11. Salvador, F.J.; Gimeno, J.; Carreres, M.; Crialesi-Esposito, M. Fuel temperature influence on the performance of last generation Common—Rail Diesel ballistic injector. Part I: 1D model development, validation and analysis. *Energy Convers. Manag.* **2016**, *114*, 376–391. [CrossRef]

12. Salvador, F.J.; De la Morena, J.; Martinez-Lopez, J.; Jaramillo, D. Assessment of compressibility effects on internal nozzle flow in diesel injectors at very high injection pressures. *Energy Convers. Manag.* **2017**, *132*, 221–230. [CrossRef]

13. Payri, F.; Bermudez, V.; Payri, R.; Salvadore, F.J. The influence of cavitation on the internal flow and the spray characteristics in diesel injections nozzles. *Fuel* **2004**, *83*, 419–431. [CrossRef]

14. Salvador, F.J.; Gimeno, J.; Carreres, M.; Crialesi-Esposito, M. Fuel temperature influence on the performance of last generation Common—Rail Diesel ballistic injector. Part I: Experimental mass flow rate measurements and discussion. *Energy Convers. Manag.* **2016**, *114*, 364–375. [CrossRef]

15. Desantes, J.M.; Payri, R.; Garcia, A.; Manin, J. Experimental study biodiesel blends effects on diesel injection process. *Energy Fuels* **2009**, *23*, 3227–3235. [CrossRef]

16. Saltas, E.; Bouilly, J.; Geivanidis, S.; Samaras, Z.; Mohammadi, A.; Yukata, L. Investigation of the effects of biodiesel aging on the degradation of Common Rail fuel injection system. *Fuel* **2017**, *200*, 357–370. [CrossRef]

17. Sundus, F.; Fazal, M.A.; Masjuki, H.H. Tribology with biodiesel: A study on enhancing biodiesel stability and its fuel properties. *Renew. Sustain. Energy Rev.* **2017**, *70*, 399–412. [CrossRef]

18. Othmana, M.F.; Adama, A.; Najafic, G.; Mamata, R. Green fuel as alternative fuel for diesel engine: A review. *Renew. Sustain. Energy Rev.* **2017**, *80*, 694–709. [CrossRef]

19. Antonov, D.V.; Kuznetsov, G.V.; Strizhak, P.A. Comparison of the characteristics of micro-explosion and ignition of two fluid water-based droplets, emulsions and suspensions, moving in the high-temperature oxidizer medium. *Acta Astronaut.* **2019**, *160*, 258–269. [CrossRef]

20. Minakov, A.V.; Shebeleva, A.A.; Strizhak, P.A.; Chernetskiy, M.Y.; Volkov, R.S. Study of the Weber number impact on secondary breakup of droplets of coal water slurries containing petrochemicals. *Fuel* **2019**, *254*, 115606. [CrossRef]

21. Osipowicz, T.; Abramek, K. Catalytic treatment in Diesel engine injectors. *Eksploat. Niezawodn. Maint. Realiability* **2014**, *16*, 22–28.

22. Osipowicz, T. Fuel Injector Atomizer. Patent No. 222791, 30 September 2016.

23. Hinze, J.O. *Turbulence*, 2nd ed.; Mc Graw-Hill: New York, NY, USA, 1975.

24. Elsner, J. An Analysis of Hot–Wire Sensitivity in Nonisothermal Flow. In *Fluid Dynamic Measurements in the Industrial and Medical Environments; Proceedings of the Disa Conference, Held at the University of Leicester, April 1972*; Humanities Press: New York, NY, USA, 1973; pp. 156–159.

25. Jiao, D.; Jiao, K.; Zhang, F.; Du, Q. Direct numerical simulation of droplet deformation in turbulent flows with different velocity profiles. *Fuel* **2019**, *247*, 302–314. [CrossRef]

26. Suh, H.K.; Lee, C.S. A review on atomization and exhaust emissions of a biodiesel—Fueled compression ignition engine. *Renew. Sustain. Energy Rev.* **2016**, *58*, 1601–1620. [CrossRef]

27. Technical Reference. *Solidworks Flow Simulation*; Dassault Systemes: Boston, MA, USA, 2018.

applied sciences

Article

Numerical Simulation of Hot Jet Detonation with Different Ignition Positions

Hongtao Zheng, Shizheng Liu, Ningbo Zhao *, Xiang Chen, Xiongbin Jia and Zhiming Li

College of Power and Energy Engineering, Harbin Engineering University, Harbin 150001, China; zhenghongtao9000@163.com (H.Z.); liushizheng1990@163.com (S.L.); chenxiang344@163.com (X.C.); xiongbinjia@hrbeu.edu.cn (X.J.); lizhimingheu@126.com (Z.L.)
* Correspondence: zhaoningbo314@hrbeu.edu.cn; Tel.: +86-0451-8251-9647

Received: 23 September 2019; Accepted: 27 October 2019; Published: 29 October 2019

Abstract: Ignition position is an important factor affecting flame propagation and deflagration-to-detonation transition (DDT). In this study, 2D reactive Navier–Stokes numerical studies have been performed to investigate the effects of ignition position on hot jet detonation initiation. Through the stages of hot jet formation, vortex-flame interaction and detonation wave formation, the mechanism of the hot jet detonation initiation is analyzed in detail. The results indicate that the vortexes formed by hot jet entrain flame to increase the flame area rapidly, thus accelerating energy release and the formation of the detonation wave. With changing the ignition position from top to wall inside the hot jet tube, the faster velocity of hot jet will promote the vortex to entrain jet flame earlier, and the DDT time and distance will decrease. In addition, the effect of different wall ignition positions (from 0 mm to 150 mm away from top of hot jet tube) on DDT is also studied. When the ignition source is 30 mm away from the top of hot jet tube, the distance to initiate detonation wave is the shortest due to the highest jet intensity, the DDT time and distance are about 41.45% and 30.77% less than the top ignition.

Keywords: hot jet detonation initiation technique; flame acceleration; detonation combustion; vortex; ignition position

---

1. Introduction

Detonation combustion has attracted plenty of attention from researchers because of its high thermal efficiency, low entropy generation and self-pressurization characteristic [1,2]. According to the formation process of detonation waves and operating characteristics in the engines, detonation engines can be divided into rotating detonation engine (RDE) [3,4], pulse detonation engine (PDE) [5,6] and standing detonation wave engine (SDWE) [7]. The detonation initiation technology is one of the bottlenecks and key technologies that restrict the engineering application of any detonation engines.

The common detonation initiation techniques are mainly divided into two categories: one is direct detonation, and the other is indirect detonation initiation. Compared with direct detonation, indirect detonation initiation requires less ignition energy, thus becoming the main direction of the detonation domain. A weak energy ignition source triggers combustion and then leads to a transition to detonation through the accumulation of energy, which is a commonly used indirect detonation initiation method [8]. Deflagration-to-detonation transition (DDT) usually requires a transition distance. However, too long transition distance may cause oversize engines and performance loss. Therefore, it is necessary to explore suitable short-range detonation initiation technology [9]. In published literature, studies have been done on detonation initiation mechanism and enhancement approach, such as hot jet [10,11], solid obstacle [12,13], fluidic obstacle [14,15], plasma [16] and shock focusing detonation initiation technology [17].

Especially, the hot jet detonation initiation firstly forms a high-energy flame in the jet tube, and then the jet flame rapidly forms a high-intensity turbulent flame in the detonation chamber, eventually forming a detonation wave in a short distance. Since the hot jet detonation initiation is an effective detonation technique with short distance and low flow loss, it is one focus of current research. Shimada et al. [18] firstly applied the hot jet tube on the detonation chamber to achieve a reliable detonation. Through visual experiments, it was found that a hot jet could quickly form a turbulent flame at the head of the detonation chamber and promote the formation of a detonation wave. Zhao et al. [19] used the numerical simulation method to study the hot jet detonation initiation, their results showed that the energy provided by the hot jet was 20 times that of the spark, and the hot jet technology could effectively reduce the initiation distance of DDT. Other literature got the same conclusion by experiments [20,21]. Subsequently, plenty attention had been poured on effect of jet intensity on DDT. Lots of research had been investigated jet intensity by changing the structure of the hot jet tube, and the same conclusion was obtained that the initiation time and distance of DDT were short when the jet intensity is sufficient [22–25]. Using ethylene/oxygen with nitrogen diluted, He et al. [26] explored the effects of different jet velocity on DDT distance. From their experimental results, it was clearly seen that jet velocity played an important role in accelerating DDT, the faster jet velocity the shorter DDT distance. Wang et al [27] numerically investigated propane/air hot jet detonation initiation process by changing the length of jet tube, it was found that the increasing length of jet tube resulted in faster jet velocity, thus decreasing the initiation time and distance of DDT.

According to the review of previous work, it is concluded that the faster jet flame velocity was, the shorter DDT distance. Up to now, the common method to obtain fast jet flame was increasing the length of jet tube. However, too long length of the hot jet tube would increase the formation time of hot jet and increase the DDT time, so that the performance of the engines was seriously affected. In addition, oversize was inconvenient for practical applications. So, alternatives to increase the length of hot jet tubes needed to be explored. Previous studies had shown the ignition position inside the detonation chamber was an important factor in flame propagation and formation of DDT [28–30]. Peng and Weng [31] numerically investigated the effects of different ignition position on DDT. They indicated that wall ignition had advantages over closed-end ignition in the initiation time and distance of DDT. Blanchard et al. [32] studied the effects of different wall ignition positions on flame propagation and DDT by hydrogen/air experiments. The results showed that the expansion of burnt fuel against the closed section of the tube behind flame front increased flame speed and turbulence when the ignition position was a certain distance from the closed section. From these studies of the ignition position, it was clearly seen that ignition position inside the detonation chamber had a great influence on the detonation initiation time and distance, and wall ignition was more conducive to flame acceleration. Using this principle, optimization of the ignition position inside the jet tube may also obtain a fast jet velocity to reduce the detonation distance without changing the length of the jet tube. However, the influence of the different ignition positions inside the jet tube on hot jet detonation initiation had not been fully studied and needed more detailed analysis and discussion.

Motivated by the above considerations, the present study performs the 2D numerical simulations to investigate the effect of different ignition positions inside the hot jet tube on DDT. Firstly, the mechanism of hot jet detonation initiation and flow characteristics are analyzed in detail through the study of the vortex-flame interaction, temperature, pressure, and velocity. Secondly, the jet parameters and flame acceleration performance of top and wall ignition are compared to investigate the reasons why wall ignition is more favorable for DDT. Finally, the wall ignition position is further optimized, it expects to lay the foundation for the design and application of the efficient and compact hot jet detonation initiation device.

2. Numerical Model and Methods

2.1. Physical Model

Figure 1 presents the physical model of the detonation chamber analyzed in this paper. As shown in the Figure, the chamber is 100 mm in diameter and 1800 mm in length, and the left wall of the detonation chamber is closed. Six obstacles are arranged inside the detonation chamber which the blocking ratio is 0.35 [33]. The distance between the first obstacle and the left wall of the detonation chamber is 380 mm. The previous three obstacles are equidistant, the distance between adjacent obstacles is 175 mm. The distance between the last three obstacles is 250 mm. Four monitoring points (P1–P4) are set to monitor the changing of parameter. The monitoring point is 100 mm behind the obstacle. The hot jet tube with the 32 mm inner diameter and the 200 mm length is aligned perpendicular to the centerline of the detonation chamber. The distance between the hot jet tube and the left wall of the detonation chamber is 114 mm, the ignition is located in the hot jet tube. As shown in Figure 1b, top ignition or wall ignition will be investigated. Flame accelerates in hot jet tubes first, and then jet flame propagates into detonation chamber and produces a detonation wave in it.

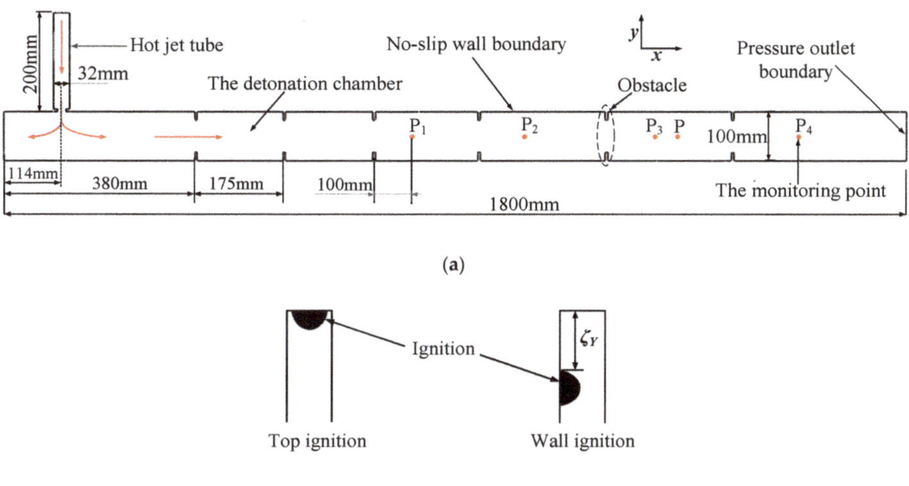

Figure 1. Schematic of geometry, (**a**) computational domain and (**b**) ignition position.

2.2. Numerical Method

In this study, the numerical simulations are performed using ANSYS Fluent software. The calculation is solved on the basis of two-dimensional Navier-Stokes equations for a viscous compressible gas coupled with chemical kinetics and the equation of state of ideal gas. The vertical jet tube causes obvious shear flows in the detonation chamber, so SST k-ω turbulence model is employed which defines the transport of the turbulence shear stress in the turbulent viscosity to resolve the unsteady turbulent flow equations [34,35]. The combustion model uses the eddy-dissipation concept model (EDC) [19,36]. The pressure correction equation is solved by a PISO algorithm coupling with second-order upwind, which has an advantage in shock capture and accurately simulate detonation [19,37,38]. The 26-species 34-steps skeletal reaction mechanism of propane/air is selected [39]. This mechanism is believed to better reflect flow field characteristics, detonation wave structure and chemical kinetics of detonation.

2.3. Initial Parameters and Boundary Conditions

The flow field is initially filled with a mixture of propane oxygen and nitrogen at a temperature of 300K and a pressure of 0.128MPa. Among them, the mass fraction of propane is 9.4% and the oxygen was 34%, the rest is nitrogen. The ignition zone is simplified as a semicircle with a diameter of 15 mm, whose temperature is 2500 K. The pressure-outlet is chosen as the exit boundary condition of calculation domain. The walls are adiabatic and no-slip boundary conditions [40].

2.4. Grid-Independent and Model Validation

According to the geometric characteristics of the physical model, the quadrangular structured meshes are chosen in this paper. Selecting different mesh sizes (ranging from 0.5 mm to 2 mm). Figure 2 shows mesh size independence analysis. By comparing the variations of pressure with time at the same monitor point (P shows in Figure 1a), the monitoring pressure no longer changes significantly with decreasing mesh size when mesh size is reduced to 1 mm. Therefore, the parameters of 1 mm for mesh size is chosen in the following numerical simulation, which meets the requirements of independence.

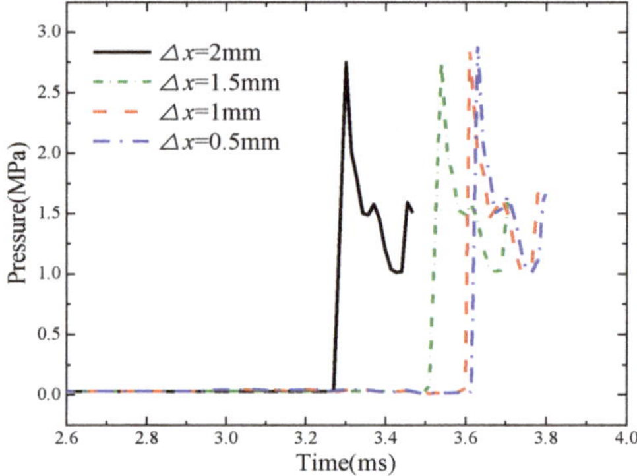

Figure 2. Mesh size independence analysis.

Experiments have been done to verify the validity of the numerical simulation. The experimental equipment schematic is shown in Figure 3. The PCB 113B24 pressure sensor is selected. Figure 4 compares the pressure curve of the experiment and numerical simulation at the same positions (P_2 and P_3 are located at 1035 mm and 1285 mm away from the left wall of detonation chamber), the pressure variation trends and peaks of them are very similar. According to Figure 4, we can confirm the experimental and simulated results are both the deflagration wave by the shape of the wave before the P_2, and there are the detonation waves at P_3. So DDT is completed between P_2 and P_3. In addition, the high detonation waves prove that the detonation initiation point is after P_2 [8,32]. The DDT time and distance of numerical simulation have good similarity with the experiment. Therefore, the validity of the numerical simulation used in this paper is proved.

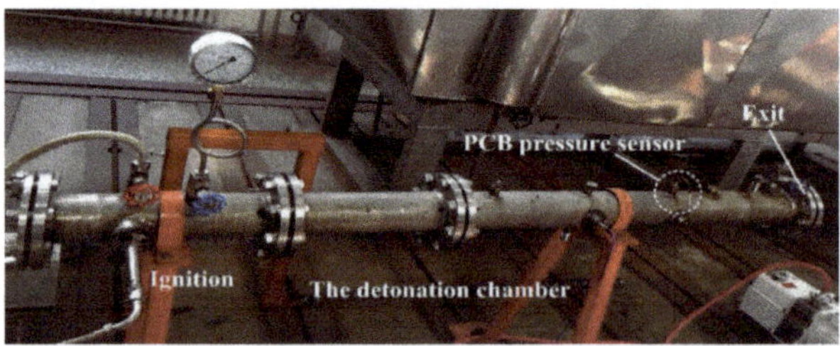

Figure 3. Schematic of the experience model.

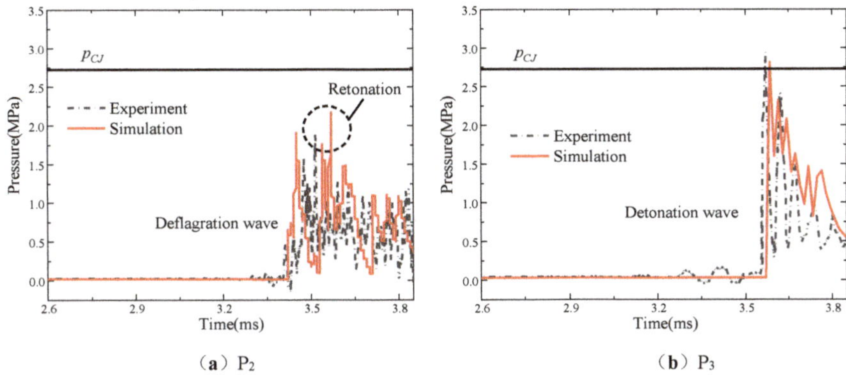

Figure 4. Pressure curves of experiment and numerical simulation.

Table 1 shows the comparison of theoretical and numerical detonation wave parameters. The theoretical parameters of detonation waves are obtained by chemical equilibrium with applications (NASA CEA) [41]. The value of the detonation parameters of this simulation method used in this paper is slightly larger than CEA, and the maximum deviation is only 3.4%. The comparison further verifies the validity of the numerical simulation method used in this paper.

Table 1. Comparisons of theoretical and numerical detonation wave parameters.

Parameters	Theoretical	Numerical	Deviation
p/p_1	21.15	21.87	3.40%
T/T_1	9.94	10.08	1.41%
V_{CJ} (m/s)	1904.4	1909.5	0.27%

Note: p is detonation front pressure. T is the detonation front temperature. p_1 is unburned pressure. T_1 is the unburned temperature. V_{CJ} is the detonation wave velocity.

3. Results Analysis and Discussion

The distance from the left wall of the detonation chamber to the position of detonation formation is defined as the initiation distance of DDT, represents by x_{DDT}. The time between ignition and hot jet propagating into the detonation chamber is defined as the time of hot jet formation, represents by t_{HJ}. The t_{TDC} is the time between the hot jet entering into the detonation chamber and the detonation initiation formation. So the time from ignition to detonation (t_{DDT}) is expressed as $t_{DDT} = t_{HJ} + t_{TDC}$. This paper mainly studies the influence of the ignition position inside the hot jet tube on the initiation distance and time of DDT. Before that, the mechanism of hot jet detonation initiation by the top ignition is necessary to be discussed firstly in detail.

3.1. Mechanism of Hot Jet Detonation Initiation

The hot jet detonation process is divided into three stages to study according to the combustion characteristic and temperature distribution. They are the stage of hot jet formation, vortex-flame interaction and detonation wave formation respectively. The flame propagation in hot jet tubes directly impacts the jet velocity and pressure, which might further affect the following formation of the final detonation wave. Therefore, the hot jet formation stage is discussed firstly. Figure 5 shows the hot jet formation in hot jet tube. Since the hot jet tube is a smooth tube, the flame-wall boundary layer interaction is an important factor affecting the flame propagation [42,43].

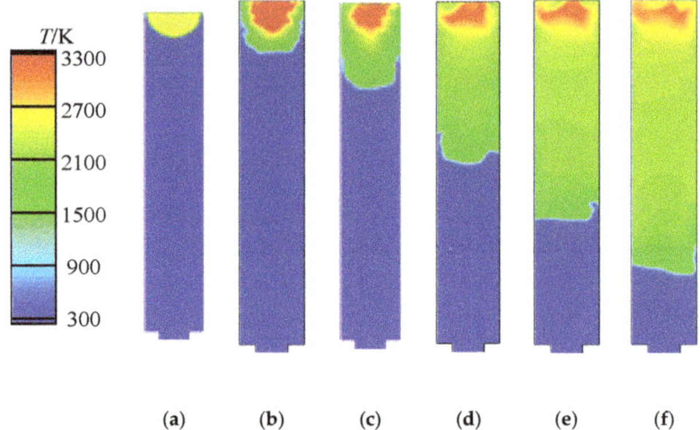

Figure 5. Hot jet formation process (**a**) t_{HJ} = 0 ms, (**b**) t_{HJ} = 0.15 ms, (**c**) t_{HJ} = 0.3 ms, (**d**) t_{HJ} = 0.6 ms, (**e**) t_{HJ} = 0.9 ms, (**f**) t_{HJ} = 1.2 ms.

From Figure 5a–c, high-temperature ignition source triggers laminar flame, and an expansion wave is generated during thermal mixture expansion [44]. Due to the obstructive effect of the wall boundary, the expansion waves are continuously superimposed and reflect on the flame. The flame wrinkles and the convex flame front propagates downstream at a laminar flow velocity of about 90m/s. In Figure 5d, since the pressure wave generated by the convex flame pushes unburned mixture to move towards near the wall, the high-density unburned mixture is formed near the wall and accelerates the flame propagation in this area. Then flame propagation speed near the wall increases rapidly, the contact area between flame and unburned mixture begins to decrease. Mass and thermal diffusion will also decrease. Flame front develops into an approximate plane, and flame propagation speed slows down.

Figure 6 shows the field of temperature and turbulent kinetic energy at 1.575 ms. The jet flame enters the detonation chamber, so t_{TDC} is assumed to be 0ms at this time. To further increase the jet velocity entering chamber, inspired by Ref. [18], the outlet of the hot jet tube has been designed as a sudden shrunken form as shown in Figure 6. Due to the narrow structure, convex flame propagates into the detonation chamber at a velocity of 333m/s. According to turbulent kinetic energy field, it is found that not only the flame front forms turbulence but also there are some high turbulence areas in the detonation chamber. The reason is that the pressure wave enters the detonation chamber ahead of flame and inevitably produces some disturbances in the detonation chamber. The Q criterion field is used to analyze these disturbances as shown in Figure 7. The Q criterion is defined by Hunt [45]:

$$Q = (\Omega_{ij}\Omega_{ji} - S_{ij}S_{ji})/2 \tag{1}$$

$$\Omega_{ij} = (\mu_{ij} - \mu_{ji})/2 \tag{2}$$

$$S_{ij} = (\mu_{ij} + \mu_{ji})/2 \tag{3}$$

where Ω_{ij} and S_{ij} are the rotate-rate and the strain-rate tensor of the velocity components, μ_{ij} and μ_{ji} are the partial derivatives of the velocity in the x and y-direction. The Q criterion can describe the structural characteristics of the vortex in the flow field.

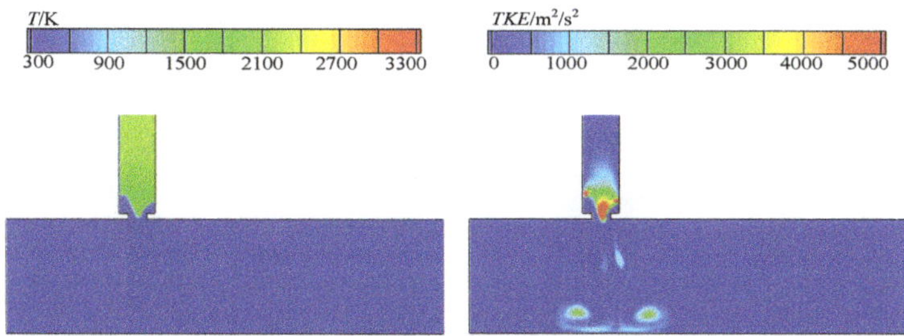

Figure 6. Temperature field and turbulent kinetic energy field at 1.575 ms.

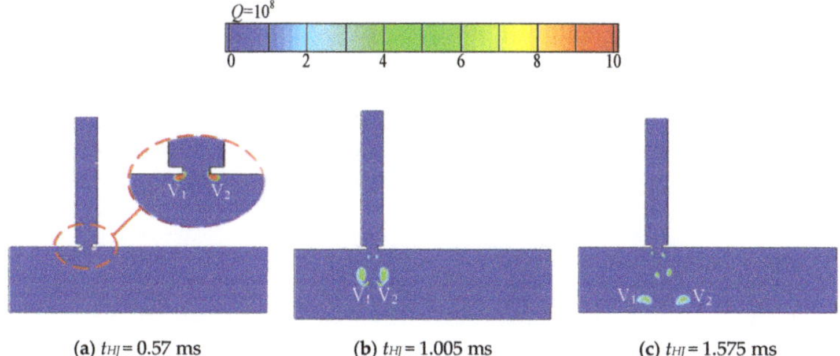

(a) t_{HJ} = 0.57 ms   (b) t_{HJ} = 1.005 ms   (c) t_{HJ} = 1.575 ms

Figure 7. Variations of Q criterion field.

Two vortexes (V_1 and V_2) can be found at the corners of the hot jet tube in Figure 7a. The pressure wave compresses the unburned mixture inside the hot jet tube into the detonation chamber. The velocity of the moving unburned mixture is significantly faster than the surrounding fluid. The intermittent velocity causes fluctuations, and the vortex occurs after the interface of the gas layer is destabilized [46]. The V_1 and V_2 continue to expand and move downstream in the detonation chamber with time. Comparing the Q criterion and turbulent kinetic energy at t_{HJ} = 1.575 ms (Figure 7c) and the *TKE* filed as shown in Figure 6, it is found that the vortex will form local turbulence.

The vortexes have formed before flame enters into the detonation chamber, so they inevitably affect the propagation of hot jet when flame propagates into the detonation chamber. Figure 8 shows the variations of temperature and Q criterion field, which reflects the interaction of vortex and jet flame. The O indicates the obstacle in the figure, and the O_i represents the *ith* obstacle. The gray area represents the vortex. At t_{TDC} = 0.03 ms, the jet flame propagates into the detonation chamber, the vertical distance between the largest scale vortexes (V_1 and V_2) and the flame front is 70 mm. From 0.03 ms to 0.24 ms, the flame surface area increases accordingly because of the sudden expansion of flow field structure and the small-scale vortexes. As the influence of the left wall, the jet flame propagates to the right in the detonation chamber. Flame reaches the down wall of the detonation chamber as shown in Figure 8d, and the V_1 and V_2 move to the sides by the hot reaction products.

Subsequently, the V_2 enhances the local turbulent fluctuation and entrains flame when t_{TDC} = 0.66 ms. Then V_2 accelerates the mixing between hot reaction products and cold unburned mixture [47], which is beneficial to promote the chemical reaction rate and diffusion rate of mass and heat. As a result, the vortex-flame interaction increases flame wrinkle surface area and accelerates the flame propagation. Since V_2 entrains the flame surface and V_1 does not touch the flame yet, the right flame surface increases more. Such actions lead to the formation of a "hook-type" flame as shown in Figure 8e. In addition, the vortex enhances the superposition between the pressure waves. The increasing pressure wave results in the stronger internal energy of unburned mixture, thus improving the flame propagation. According to Figure 8f, the right flame front propagates to the first obstacle, and the V_1 begins to entrain the left flame. The unburned mixture propagates with the promotion of the high-temperature products and creates vortexes in the boundary layer of obstacle as the Kelvin-Helmholtz (K-H) instability [48].

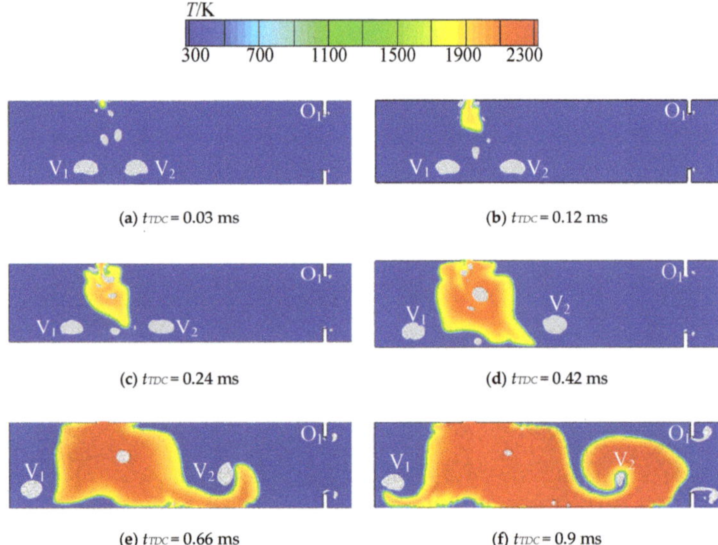

(a) t_{TDC} = 0.03 ms

(b) t_{TDC} = 0.12 ms

(c) t_{TDC} = 0.24 ms

(d) t_{TDC} = 0.42 ms

(e) t_{TDC} = 0.66 ms

(f) t_{TDC} = 0.9 ms

Figure 8. Interaction of jet flame and vortex in the detonation chamber.

As flame passes through the obstacles, the flame surface area, and flame propagation velocity both increase due to the Rayleigh-Taylor (RT) and K-H instabilities [49,50]. The flame becomes fast flame as shown in Figure 9 (a) [51], just arrives at the O_4. Figure 9 describes the formation of detonation wave by the variations of pressure and temperature field in detail. Obviously, a high-pressure zone occurs near the fourth obstacle at this moment. Pressure waves superimpose to form a shock in the leading edge of flame front. Since the leading shock compresses unburned mixture at front of flame to raises the temperature surrounding it, the chemical reaction and combustion are both promoted. Then plenty energies released from rapid combustion will strengthen the leading shock. This creates a positive feedback effect.

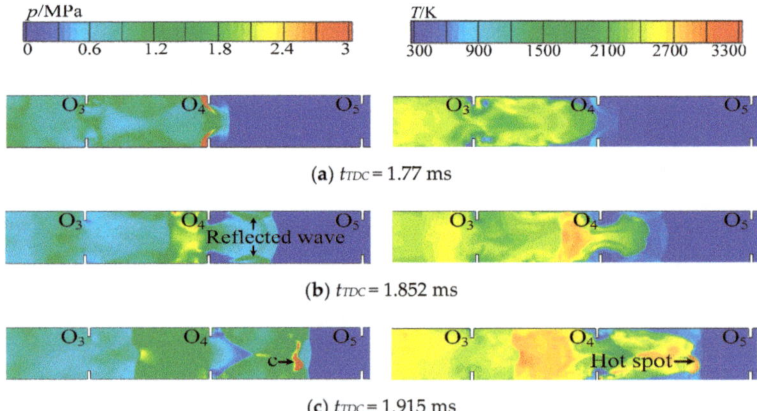

(a) t_{TDC} = 1.77 ms

(b) t_{TDC} = 1.852 ms

(c) t_{TDC} = 1.915 ms

Figure 9. *Cont.*

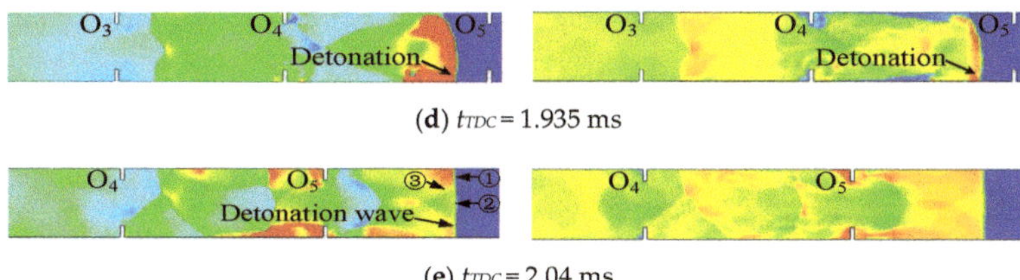

(**d**) t_{TDC} = 1.935 ms

(**e**) t_{TDC} = 2.04 ms

Figure 9. Variations of pressure and temperature field at detonation transition (DDT).

According to Figure 9b, the leading shock and flame pass the O_4, and the leading shock reflects from the wall. Then the leading and reflected waves collide at point "c" as shown in Figure 9c, the maximum pressure of "c" is over 6MPa. From the temperature field at this time, a hot spot is formed at the point "c". At t_{TDC} = 1.935 ms, the flame surface is coupled to the front shock, the hot spot develops to detonation wave at 1170 mm. The instabilities caused by internal combustion and walls lead to the generation of hot spots, which is the mechanism of the detonation wave formation [52]. The detonation wave coupled with shock and flame propagates downstream at velocity of 1909 m/s. Based on Figure 9e, ①Mach stem, ②incident shock, and ③transverse shock intersect to form the detonation wave system, the focus of the three shocks is called the "triple point" [53]. The key to the existence of stable and persistent detonation is that the "triple point" provides continuous energy to ensure the detonation wave velocity and propagate downstream successively.

3.2. The Influence of Ignition Position on the Hot Jet Detonation Initiation

Through the above study, it is found that the hot jet can form vortexes to accelerate flame propagation and promote DDT in the detonation chamber. However, the formation time of the hot jet is 44.87% of DDT time, which will greatly affect the performance of the detonation engine. In the previous paper, the ignition position inside the detonation chamber affects flame propagation [8]. So, the differences between top ignition and wall ignition are the following focus and research (ζ_Y = 0 mm as shown in Figure 1b. Figure 10 shows the time and distances to detonation initiation of two ignition positions. Simulation results clearly display that wall ignition can not only effectively reduce the t_{HJ}, but also shorten DDT time and distance. The top ignition t_{DDT} = 1.575 ms (t_{HJ}) + 1.935ms (t_{TDC}) = 3.51 ms, and x_{DDT} = 1170 mm. The wall ignition t_{DDT} = 1.005 ms (t_{HJ}) + 1.32 ms (t_{TDC}) = 2.325 ms, and x_{DDT} = 907 mm, the t_{DDT} and x_{DDT} are reduced by 33.76% and 22.48% comparing with top ignition.

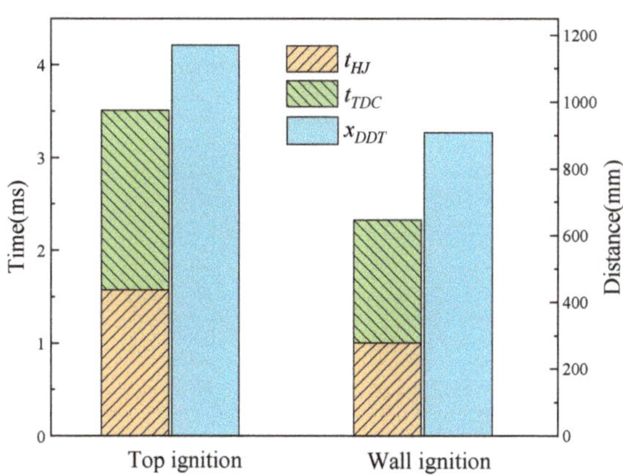

Figure 10. Time and distance to detonation initiation of top and wall ignition.

Figure 11 compares the flame propagation speed inside the jet tube of the two ignition positions, which can account for the reason of t_{HJ} reduction. In this figure, the distance represents the flame propagation distance in the hot jet tube, the left side is the ignition position, and the right side is the exit of the hot jet. The initial flame development of the two ignition positions is similar, then flame propagation speed of wall ignition is obviously higher than that of top ignition. Finally, the jet flame of wall ignition propagates into the detonation chamber at a velocity of 550m/s, which is about 1.65 times faster than that of top ignition.

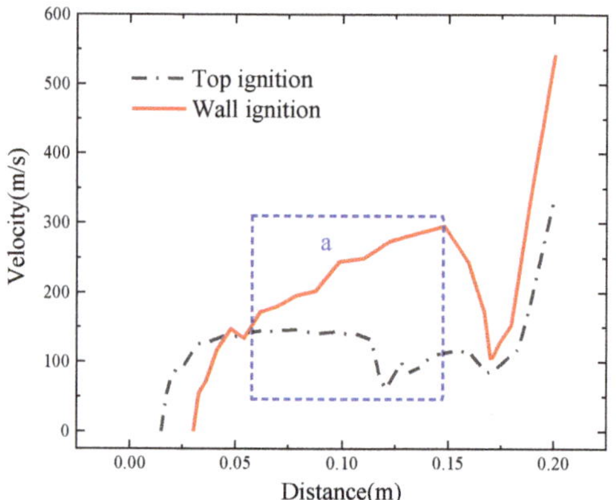

Figure 11. Flame propagation speed in the hot jet tube of two ignition positions.

Base on the "a" region of Figure 11, it is found that the flame propagation velocity of wall ignition is more than 2 times of the wall ignition. To investigate this issue, the variations of temperature field of wall ignition are studied as shown in Figure 12. At t_{HJ} = 0.3 ms, the pressure wave reaches right wall and then reflects. The reflected wave propagates to left, and it prevents the flame from contacting the right wall, thus ensuring the contact area between flame and unburned mixture. Comparing with Figure 5, there is more contact area between flame and unburned mixture in the flow field of wall ignition at this moment. More contact area releases more energy to increase the flame propagation speed. Therefore, the flame propagates faster and flame front reaches the exit of the jet tube in only 0.9 ms as shown in Figure 12c, the propagation distance is significantly farther than the flame of Figure 5e. Then flame front transforms from "fingertip" to "planar" shape. Flame contacting with wall results in reducing the contact area between flame and unburned mixture, thus slowing down the flame propagation speed. However, the flame propagation speed of wall ignition is faster than that of top wall, and wall ignition takes only 1.005 ms to form a hot jet into the detonation chamber.

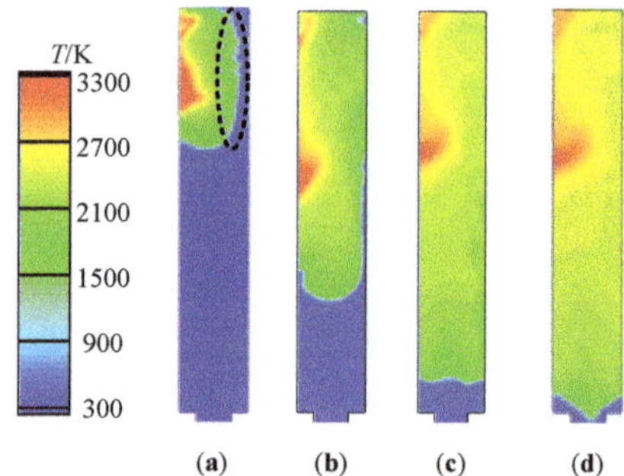

Figure 12. Variations of temperature field of the wall ignition (**a**) t_{HJ} = 0.3 ms, (**b**) t_{HJ} = 0.6 ms, (**c**) t_{HJ} = 0.9 ms, (**d**) t_{HJ} = 1.005 ms.

The jet pressure also affects the jet flame propagation into the detonation chamber, so a detailed comparison of the jet pressure changes with time at the center point of the hot jet tube is shown in Figure 13. The first pressure peak appears at 0.5 ms. The pressure peak of wall ignition is 1.35 times that of top ignition. Subsequently, the pressure of wall ignition is always higher than the top ignition. The second pressure peak appears at 0.75 ms and the peak value can reach 0.2 MPa. The jet flame of wall ignition enters detonation chamber with 0.1 MPa at 1.005 ms, which is obviously higher than 0.06 MPa of top ignition. The average pressure of the wall ignition is twice than top ignition due to more intense burning and wave superimposition. On the basis of above discussions of jet velocity, it is found wall ignition brings a faster hot jet with higher pressure.

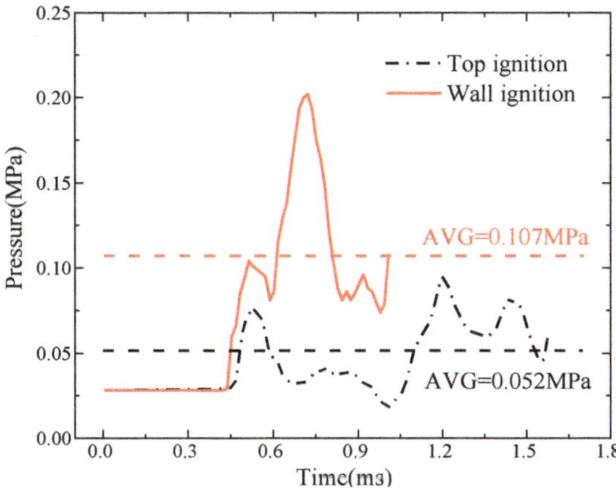

Figure 13. Pressure changes of two ignition positions at the exit of the hot jet tube.

In order to understand flame propagation and detonation wave formation of the two ignition positions. The comparison of flame propagation speed is shown in Figure 14. The trend of these two flame propagations is generally the same due to the same structure, but the flame propagation of wall ignition is substantially faster than that of top ignition. The wall ignition forms a detonation wave first. According to the characteristics of the flame speed, this figure is divided into two regions to study. **(I)** In this region, the vortex entrains the jet flame to increase the flame speed before the first obstacle. **(II)** The flame constantly accelerates because of the obstacles, and eventually forms a detonation wave.

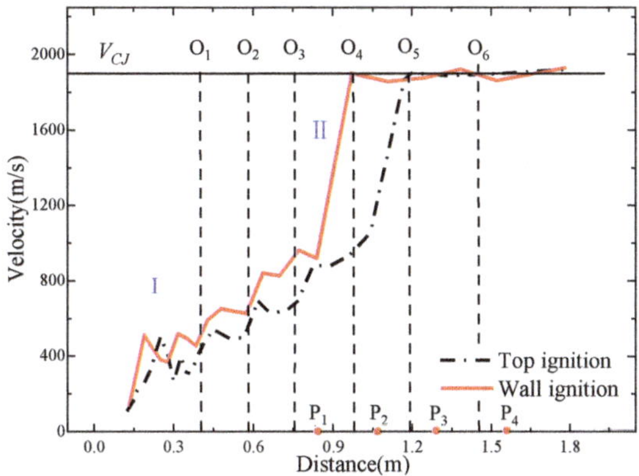

Figure 14. Flame propagation speed in the detonation chamber.

In the first region, the flame propagation speed of wall ignition increases significantly fast and remains at big speed. Except for the area around 0.23 m (the V_1 entrains the flame of top ignition as shown in Figure 8e), the flame speed of wall ignition is always faster than that of top ignition. Figure 15 displays the variations of temperature and Q criterion field of wall ignition before the first obstacle to investigate why the flame speed of wall ignition is faster. Comparing with Figure 8a, it can be seen that the scale of the vortexes (V_1 and V_2) is larger at t_{TDC} = 0.03 ms due to the higher jet pressure and the faster jet velocity. Additionally, the reducing formation time of hot jet results in short moving time of V_1 and V_2, so the vertical distance between the vortexes (V_1 and V_2) and the flame front is only 14 mm, which is obviously closer than that of top ignition. Base on Figure 15b,c, flame quickly contacts the vortexes (V_1 and V_2). The two same scale vortexes (V_1 and V_2) simultaneously entrain the flame. Then flame temperature and wrinkling surface both increase rapidly and a symmetrical "mushroom-shaped" flame is produced at t_{TDC} = 0.24 ms. Subsequently, flame reaches the first obstacle by only 0.66 ms, which is 26.67% less than top ignition. Moreover, a large amount of mixture is burned to release energy, which is more conducive to the next acceleration of the flame. Comparing with Figure 8, vortex entrains flame earlier and the scale of vortex is significantly larger, so the flame propagation speed of wall ignition is faster than that of top ignition.

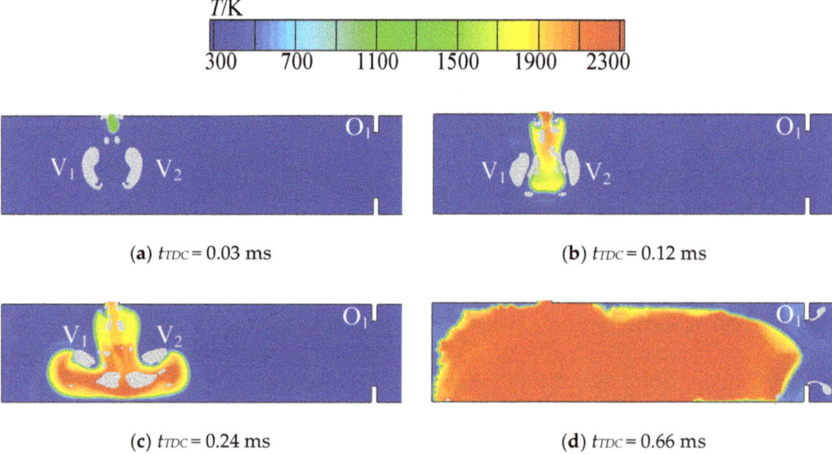

(a) t_{TDC} = 0.03 ms

(b) t_{TDC} = 0.12 ms

(c) t_{TDC} = 0.24 ms

(d) t_{TDC} = 0.66 ms

Figure 15. Variations of temperature and Q criterion field of wall ignition.

In the second region, as the flame propagation speed of wall ignition is faster than that of top ignition when the flame passes the first obstacle, its flame propagation speed is always faster at the second stage as shown in Figure 14. Subsequently, the wall ignition forms a detonation wave in advance as shown in Figure 16. Wall ignition burns more intensely at the early stage in the detonation chamber, so superposition of pressure waves is more intense. According to Figure 16a, two high-intensity reflected waves have been formed before the third obstacle. Then the reflected waves and leading shock collide at point "c", and a hot spot is formed which is significantly ahead of the top ignition. The high temperature and pressure point accelerate energy release to accelerate DDT which precedes than top ignition. The triple point(① Mach stem, ② incident shock, and ③ transverse shock) is promoted for the case of ignition at the wall because of the higher pressure waves and flame strength than that of top ignition. On the basis of above discussions, it is found that wall ignition forms a fast hot jet with high pressure, shortens the formation time of hot jet, and forms large-scale vortexes. These all contribute to accelerating flame and superimposing pressure waves. So the DDT distance and time of wall ignition are significantly shorter than top ignition.

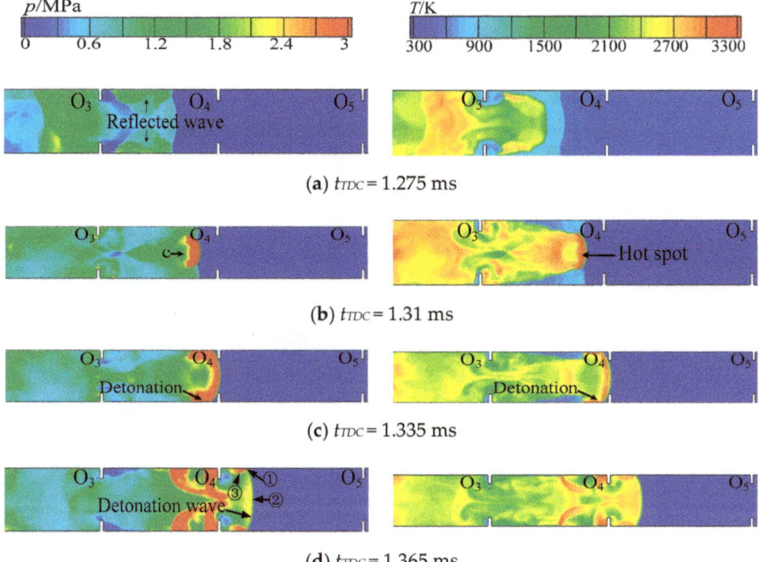

(**a**) t_{TDC} = 1.275 ms

(**b**) t_{TDC} = 1.31 ms

(**c**) t_{TDC} = 1.335 ms

(**d**) t_{TDC} = 1.365 ms

Figure 16. Variations of pressure and temperature field at deflagration-to-detonation transition (DDT) of wall ignition.

3.3. Optimization of Ignition Position

Through the above study, it is found that wall ignition is more conducive to the acceleration of DDT. To further study the performance advantage of wall ignition, the effect of different wall ignition positions on DDT is studied in this section. The ignition positions of ζ_Y = 30 mm, 60 mm, 90 mm, 120 mm, and 150 mm are simulated as well. Figure 17 displays the jet velocity and pressure of different ignition positions when jet flame front enters into the detonation chamber. Since too big ζ_Y results in less mixture combustion in the jet tube, thus weakening the jet velocity and pressure. When ζ_Y is large, the jet velocity and pressure are both relatively small. As ζ_Y increases, the jet velocity maintains at about 550 m/s and then an obvious reduction occurs. The jet pressure increases firstly and then weakens with increase of ζ_Y, and the highest peak is observed ζ_Y = 30 mm. The pressure wave is reflected not only on the left and right walls but also on the top of jet tube [46]. Therefore, it is more advantageous to obtain larger jet pressure when ζ_Y = 30 mm.

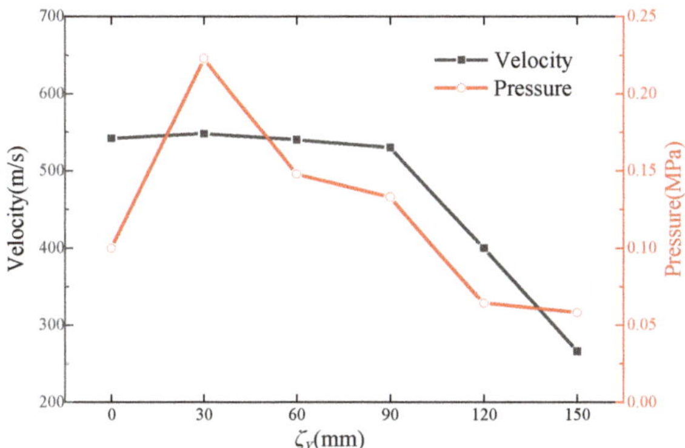

Figure 17. Jet velocity and pressure of different ignition positions.

Then jet flame propagates into the detonation chamber, flame wrinkling surface area and propagation speed both increase by vortexes and obstacles. In order to compare the flame propagation in the detonation chamber for different ignition positions, distributions of pressure and temperature on the axis at different ignition positions at t_{TDC} = 1.44 ms are shown in Figure 18 ($\zeta_Y \leq$ 60 mm) and Figure 19 ($\zeta_Y \geq$ 90 mm). Since the fast hot jet, flame and pressure propagate farther in Figure 18 than those of Figure 19. The propagation distances of flame and pressure are farthest when ζ_Y = 30 mm. The values of leading shock already meet the detonation parameter in Figure 18, and flame front couple with the leading shock, so the detonation wave has been formed in the detonation chamber.

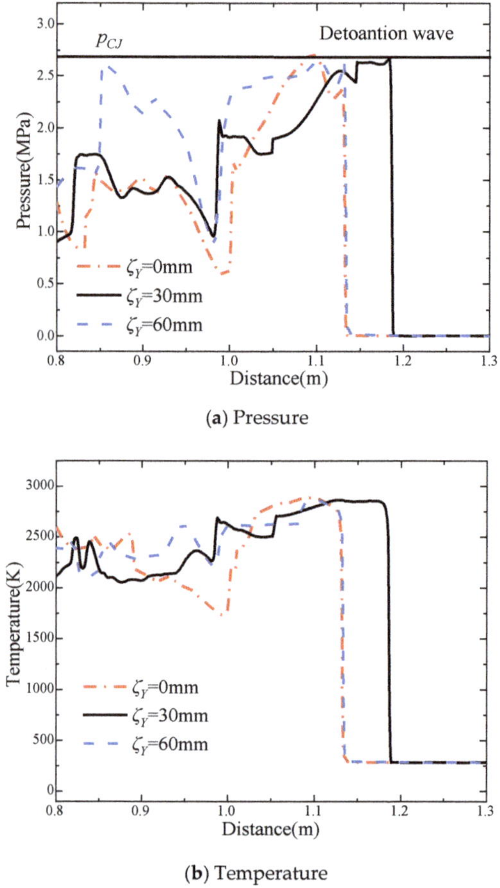

(**a**) Pressure

(**b**) Temperature

Figure 18. Distributions of (**a**) pressure and (**b**) temperature of $\zeta_Y \leq$ 60 mm at t_{TDC} = 1.44 ms.

Base on Figure 19, the propagation distance of pressure wave and flame decrease as ζ_Y increasing, and the pressure wave propagate slightly further than flame front. It is clearly seen that the pressure of deflagration wave is weak at this time, and value of pressure wave decreases as ζ_Y increasing. The pressure of ζ_Y = 150 mm is obviously lower than other ignition positions, which does not form the leading shock.

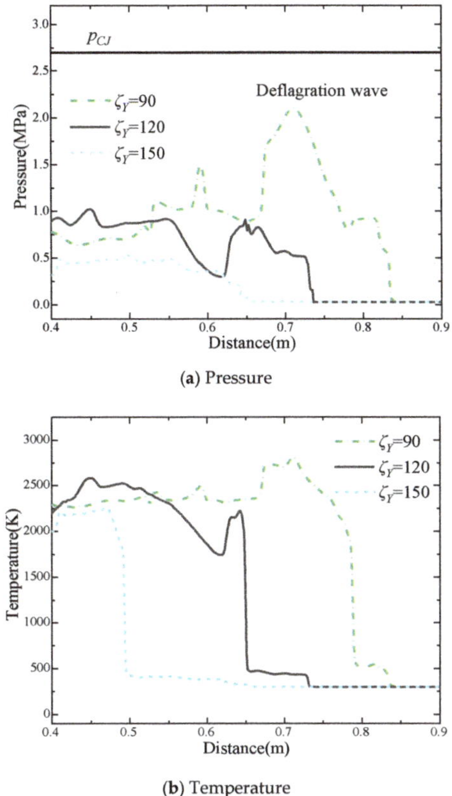

(a) Pressure

(b) Temperature

Figure 19. Distributions of (**a**) pressure and (**b**) temperature of $\zeta_Y \geq 90$ mm at $t_{TDC} = 1.44$ ms.

Figure 20 shows the DDT distance and time of different ignition positions. The t_{DDT} ($t_{HJ} + t_{TDC}$) and x_{DDT} both decrease firstly and then increase with ζ_Y increasing. The ignition position is near the exit of the hot jet tube when the ζ_Y is big, so the t_{HJ} is small. However, the reduction of jet intensity leads to a long time and distance to complete DDT. As the jet velocity and pressure are the largest at $\zeta_Y = 30$ mm, the shortest x_{DDT} is 810 mm and the fastest t_{DDT} is 2.055 ms at this time.

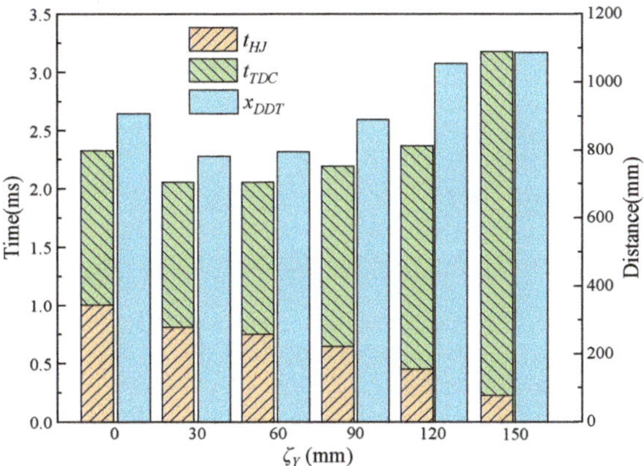

Figure 20. DDT distance and time of different ignition positions.

4. Conclusions

This paper uses a two-dimensional numerical simulation method to study the effect of different ignition positions inside the hot jet tube on the distance and time of detonation initiation. The hot jet formation, the vortex-flame interaction and character of the initiation time and distance to detonation initiation are analyzed in detailed. The primary conclusions of this study include as following:

(1) The mechanism of the hot jet detonation initiation is fast jet flame and the vortex-flame interaction. The vortexes increase turbulence intensity to accelerate blending between the unburned mixture and high-temperature products and increase superposition of the waves. Therefore, the increasing flame wrinkling surface area results in faster mass and energy release, thus increasing the flame propagation speed and accelerating DDT.

(2) Wall ignition is significantly better than top ignition on DDT. When the ignition position changes from the top to the wall inside the hot jet tube, the DDT distance and time are both showing an obvious reduction. Two differences should be paid special attention: one is the vortexes entrain flame early. Another is the scale of vortex is big and two large-scale vortexes simultaneously entrain flame. The reason mainly lies in the formation of hot jet with stronger intensity in cause of wall ignition.

(3) The different wall ignition positions also affect the detonation wave formation. The t_{DDT} and x_{DDT} all reduce firstly and then increase with ζ_Y increasing. The shortest x_{DDT} is 810 mm and the fastest t_{DDT} is 2.055 ms at $\zeta_Y = 30$ mm. This is because not only the jet velocity is ensured to be about 550 m/s, but also the jet pressure is the largest due to the pressure wave is reflected on the wall and top of the hot jet tube.

Author Contributions: Conceptualization, H.Z.; formal analysis, S.L.; methodology, N.Z.; investigation, X.C.; writing—original draft preparation, X.J.; writing—review and editing, Z.L.

Funding: The authors would like to acknowledge the Fundamental Research Funds for the Central Universities (Grant No. HEUCFP201719) for supporting this work.

Conflicts of Interest: The authors declared that there was no conflict of interest.

References

1. Kailasanath, K. Review of propulsion applications of detonation waves. *AIAA J.* **2000**, *38*, 1698–1708. [CrossRef]
2. Zheng, H.T.; Qi, L.; Zhao, N.B.; Li, Z.M.; Liu, X. A Thermodynamic Analysis of the Pressure Gain of Continuously Rotating Detonation Combustor for Gas Turbine Cycle Performance. *Appl. Sci.* **2018**, *8*, 535. [CrossRef]
3. Meng, Q.; Zhao, N.B.; Zheng, H.T.; Yang, J.L.; Qi, L. Numerical investigation of the effect of inlet mass flow rates on H2/air non-premixed rotating detonation wave. *Int. J. Hydrog. Energy* **2018**, *43*, 13618–13631. [CrossRef]
4. Sun, C.W.; Zheng, H.T.; Li, Z.M.; Zhao, N.B.; Qi, L.; Guo, H.B. Effects of Diverging Nozzle Downstream on Flow Field Parameters of Rotating Detonation Combustor. *Appl. Sci.* **2019**, *9*, 4259. [CrossRef]
5. Driscoll, R.; George, A.S.; Munday, D.; Gutmark, E.J. Optimization of a multiple pulse detonation engine-crossover system. *Appl. Eng.* **2016**, *96*, 463–472. [CrossRef]
6. Frolov, S.M.; Smetanyuk, V.A.; Gusev, P.A.; Koval, A.S.; Nabatnikov, S.A. How to utilize the kinetic energy of pulsed detonation products. *Appl. Therm. Eng.* **2019**, *147*, 728–734. [CrossRef]
7. Wang, Y.; Wang, K.; Fan, W.; He, J.N.; Zhang, Y.; Zhang, Q.B.; Yao, K.G. Experimental study on the wall temperature and heat transfer of a two-phase pulse detonation rocket engine. *Appl. Therm Eng.* **2017**, *114*, 387–393. [CrossRef]
8. Sorin, R.; Zitoun, R.; Desbordes, D. Optimization of the deflagration to detonation transition: Reduction of length and time of transition. *Shock Waves* **2006**, *15*, 137–145. [CrossRef]
9. Harris, P.; Farinaccio, R.; Stowe, R.; Higgins, A.; Thibault, P.; Laviolette, J. The effect of DDT distance on impulse in a detonation tube. In Proceedings of the 37th AIAA/ASME/SAE/ASEE Joint Propulsion Conference & Exhibit, AIAA 2001-3467, Salt Lake City, UT, USA, 8–11 July 2001.

10. Sorin, R.; Zitoun, R.; Khasainov, B.; Desbordes, D. Detonation diffraction through different geometries. *Shock Waves* **2009**, *19*, 11–23. [CrossRef]

11. Jackson, S.; Shepherd, J. Initiation systems for pulse detonation engines. In Proceedings of the 38th AIAA/ASME/SAE/ASEE Joint Propulsion Conference & Exhibit, AIAA 2002-3627, Indianapolis, IN, USA, 7–10 July 2002.

12. Kellenberger, M.; Ciccarelli, G. Advancements on the propagation mechanism of a detonation wave in an obstructed channel. *Combust. Flame* **2018**, *159*, 195–209. [CrossRef]

13. Li, J.; Zhang, P.; Yuan, L.; Pan, Z.; Zhu, Y. Flame propagation and detonation initiation distance of ethylene/oxygen in narrow gap. *Appl. Therm. Eng.* **2017**, *110*, 1274–1282. [CrossRef]

14. Chambers, J.; Ahmed, K. Turbulent flame augmentation using a fluidic jet for Deflagration-to-Detonation. *Fuel* **2017**, *199*, 616–626. [CrossRef]

15. Zhao, S.; Fan, Y.; Lv, H.; Jia, B. Effects of a jet turbulator upon flame acceleration in a detonation tube. *Appl. Therm. Eng.* **2017**, *115*, 33–40. [CrossRef]

16. Wang, F.; Kuthi, A.; Gundersen, M. Technology for transient plasma ignition. In Proceedings of the 43th AIAA Aerospace Sciences Meeting & Exhibit, AIAA 2005-951, Reno, NV, USA, 10–13 January 2005.

17. Xiao, H.; Oran, E.S. Shock focusing and detonation initiation at a flame front. *Combust. Flame* **2019**, *203*, 397–406. [CrossRef]

18. Shimada, H.; Kenmoku, Y.; Sato, H.; Hayashi, A.K. A new ignition system for pulse detonation engine. In Proceedings of the 42th AIAA/ASME/SAE/ASEE Joint Propulsion Conference & Exhibit, AIAA 2004-308, Reno, NV, USA, 5–8 January 2004.

19. Zhao, W.; Han, Q.X.; Zhang, Q. Experimental investigation on detonation initiation with a transversal hot jet. *Combust. Explos. Shock Waves* **2013**, *49*, 171–177. [CrossRef]

20. Brophy, C.; Sinibaldi, J.; Damphousse, P. Initiator performance for liquid-fueled pulse detonation engines. In Proceedings of the 40th AIAA Aerospace Sciences Meeting & Exhibit, AIAA 2002-472, Reno, NV, USA, 14–17 January 2002.

21. Brophy, C.; Werner, S.; Sinibaldi, J. Performance characterization of a valveless pulse detonation engine. In Proceedings of the 41th AIAA Aerospace Sciences Meeting & Exhibit, AIAA 2003-1344, Reno, NV, USA, 6–9 January 2003.

22. Ishii, K.; Tanaka, T. A study on jet initiation of detonation using multiple tubes. *Shock Waves* **2005**, *14*, 273–281. [CrossRef]

23. Ishii, K.; Akiyoshi, T.; Gonda, M.; Murayama, M. Effects of flame jet configurations on detonation initiation. *Shock Waves* **2009**, 239–244.

24. Saretto, S.R.; Lee, S.Y.; Conrad, C.; Brumberg, J.; Pal, S.; Stantoro, R.J. Predetonator to thrust tube detonation transition studies for multi-cycle PDE applications. In Proceedings of the 39th AIAA/ASME/SAE/ASEE Joint Propulsion Conference & Exhibit, AIAA 2003-4825, Huntsville, AL, USA, 20–23 July 2003.

25. Conrad, C.; Saretto, S.R.; Lee, S.Y.; Stantoro, R.J. Studies of Overdriven Detonation Wave Transition in a Gradual Area Expansion for PDE Applications. In Proceedings of the 40th AIAA/ASME/SAE/ASEE Joint Propulsion Conference & Exhibit, AIAA 2004-3397, Fort Lauderdale, FL, USA, 11–14 July 2004.

26. He, J.; Fan, W.; Chi, Y.; Zheng, j.; Zhang, W. A Comparative Study on Millimeter-scale Hot jet and Detonation Diffraction in a Rectangular Chamber. In Proceedings of the 21th AIAA International Space Planes and Hypersonics Technologies Conference, AIAA 2017-2149, Xiamen, China, 6–9 March 2017.

27. Wang, Z.; Zhang, Y.; Chen, X.; Liang, Z.; Zheng, L. Investigation of Hot Jet Effect on Detonation Initiation Characteristics. *Combust. Sci. Technol.* **2017**, *189*, 498–519. [CrossRef]

28. Kumar, K. Vented combustion of hydrogen-air mixtures in a large rectangular volume. In Proceedings of the 44th AIAA Aerospace Sciences Meeting & Exhibit, AIAA 2006-375, Reno, NV, USA, 9–12 January 2006.

29. Kindracki, J.; Kobiera, A.; Rarata, G.; Wolanski, P. Influence of ignition position and obstacles on explosion development in methane–air mixture in closed vessels. *J. Loss Prev. Process Ind.* **2007**, *20*, 551–561. [CrossRef]

30. Xiao, H.; Duan, Q.; Jiang, L.; Sun, J. Effects of ignition location on premixed hydrogen/air flame propagation in a closed combustion tube. *Int. J. Hydrog. Energy* **2014**, *39*, 8557–8563. [CrossRef]

31. Peng, Z.; Weng, C. Numerical Calculation of Effect of Plasma Ignition on DDT. *Combust Sci. Technol.* **2011**, 1.

32. Blanchard, R.; Arndt, D.; Grätz, R.; Scheider, S. Effect of ignition position on the run-up distance to DDT for hydrogen–air explosions. *J. Loss Prev. Process Ind.* **2011**, *24*, 194–199. [CrossRef]

33. Goodwin, G.B.; Houim, R.W.; Oran, E.S. Effect of decreasing blockage ratio on DDT in small channels with obstacles. *Combust. Flame* **2016**, *173*, 16–26. [CrossRef]

34. Perera, I.; Wijeyakulasuriya, S.; Nalim, R. Hot combustion torch jet ignition delay time for ethylene-air mixtures. In Proceedings of the 49th AIAA Aerospace Sciences Meeting including the New Horizons Forum and Aerospace Exposition, AIAA 2011-95, Orlando, FL, USA, 4–7 January 2011.

35. Luo, C.; Zanganeh, J.; Moghtaderi, B. A 3D numerical study of detonation wave propagation in various angled bending tubes. *Fire Safety J.* **2016**, *86*, 53–64. [CrossRef]

36. Zhang, Z.; Li, Z.; Wu, Y.; Bao, X. Numerical Studies of Multi-Cycle Detonation Induced by Shock Focusing. In Proceedings of the Asia-pacific International Symposium on Aerospace Technology, Toyama, Japan, 25–27 October 2016.

37. Qin, Y.; Yu, J.; Gao, G. Performance studies of annular pulse detonation engine ejectors. *J. Aerosp. Power* **2011**, *26*, 25–41.

38. Zhang, Z.; Li, Z.; Dong, G. Numerical studies of multi-cycle acetylene-air detonation induced by shock focusing. *Procedia Eng.* **2015**, *99*, 327–331. [CrossRef]

39. Petrova, M.; Varatharajan, B.; Williams, F. Theory of Propane Autoignition. In Proceedings of the 42th AIAA Aerospace Sciences Meeting & Exhibit, AIAA 2004-1325, Reno, NV, USA, 5–8 January 2004.

40. Goodwin, G.B.; Houim, R.W.; Oran, E.S. Shock transition to detonation in channels with obstacles. *Proc. Combust. Inst.* **2017**, *36*, 2717–2724. [CrossRef]

41. CEARUN. Available online: https://cearun.grc.nasa.gov (accessed on 5 February 2004).

42. Kuznetsov, M.; Alekseev, V.; Matsukov, I.; Dorofeev, S. DDT in a smooth tube filled with a hydrogen–oxygen mixture. *Shock Waves* **2005**, *14*, 205–215. [CrossRef]

43. Bychkov, V.; Fru, G.; Eriksson, L.E.; Petchenko, A. Flame acceleration in the early stages of burning in tubes. *Combust. Flame* **2007**, *150*, 263–276. [CrossRef]

44. Landau, L.D.; Lifshitz, E.M. *Course of Theoretical Physics*; Elsevier: Amsterdam, The Netherlands, 2013.

45. Hunt, J.C.R.; Wray, A.A.; Moin, P. Eddies, streams, and convergence zones in turbulent flows. Cent Turbul Res Report CTR-S88. 1988, pp. 193–208. Available online: https://www.researchgate.net/publication/234550074_Eddies_streams_and_convergence_zones_in_turbulent_flows (accessed on 29 October 2019).

46. Fan, J.; Zhang, Y.; Wang, D. Large-eddy simulation of three-dimensional vortical structures for an impinging transverse jet in the near region. *J. Hydrodyn.* **2007**, *19*, 314–321. [CrossRef]

47. Ahmed, U.; Prosser, R. Modelling Flame turbulence interaction in RANS simulation of premixed turbulent combustion. *Combust. Theory Model.* **2016**, *20*, 34–57. [CrossRef]

48. Ciccarelli, G.; Dorofeev, S. Flame acceleration and transition to detonation in ducts. *Prog. Energy Combust. Sci.* **2008**, *34*, 499–550. [CrossRef]

49. Gamezo, V.N.; Ogawa, T.; Oran, E.S. Flame acceleration and DDT in channels with obstacles: Effect of obstacle spacing. *Combust. Flame* **2008**, *155*, 302–315. [CrossRef]

50. Ott, J.D.; Oran, E.S.; Anderson, J.D. A mechanism for flame acceleration in narrow tubes. *AIAA J.* **2003**, *41*, 1391–1396. [CrossRef]

51. Kagan, L.; Sivashinsky, G. Parametric transition from deflagration to detonation: Runaway of fast flames. *Proc. Combust. Inst.* **2017**, *36*, 2709–2715. [CrossRef]

52. Wang, Z.; Qi, Y.; He, X.; Wang, J.; Shuai, S.; Law, C.K. Analysis of pre-ignition to super-knock: Hotspot-induced deflagration to detonation. *Fuel* **2015**, *144*, 222–227. [CrossRef]

53. Massa, L.; Austin, J.; Jackson, T. Triple point shear-layers in gaseous detonation waves. *J. Fluid Mech.* **2007**, *586*, 205–248. [CrossRef]

Article

A Steam Ejector Refrigeration System Powered by Engine Combustion Waste Heat: Part 1. Characterization of the Internal Flow Structure

Yu Han [1,2], Lixin Guo [1,*], Xiaodong Wang [1], Anthony Chun Yin Yuen [2], Cuiling Li [1], Ruifeng Cao [2], Hengrui Liu [2], Tim Bo Yuan Chen [2], Jiyuan Tu [3] and Guan Heng Yeoh [2,4]

1   School of Mechanical Engineering & Automation, Northeastern University, Shenyang 110819, China; yu.han4@student.unsw.edu.au (Y.H.); Xiaodongwang.neu@outlook.com (X.W.); clli@mail.neu.edu.cn (C.L.)
2   School of Mechanical and Manufacturing Engineering, University of New South Wales, Sydney, NSW 2052, Australia; c.y.yuen@unsw.edu.au (A.C.Y.Y.); ruifeng.cao@unsw.edu.au (R.C.); h.liu@unsw.edu.au (H.L.); timothy.chen@unsw.edu.au (T.B.Y.C.); g.yeoh@unsw.edu.au (G.H.Y.)
3   School of Engineering, RMIT University, Melbourne, Vic 3083, Australia; jiyuan.tu@rmit.edu.au
4   Australian Nuclear Science and Technology Organization (ANSTO), Locked Bag 2001, Kirrawee DC, NSW 2232, Australia
*   Correspondence: Lixinguo.neu@outlook.com; Tel.: +86-24-8368-7618

Received: 17 September 2019; Accepted: 8 October 2019; Published: 12 October 2019

Abstract: With the escalating production of automobiles, energy efficiency and environmental friendliness have always been a major concern in the automotive industry. In order to effectively lower the energy consumption of a vehicle, it is essential to develop air-conditioning systems that can make good use of combustion waste heat. Ejector refrigeration systems have become increasingly popular for this purpose due to their energy efficiency and ability to recycle waste heat. In this article, the elements affecting the performance of a typical ejector refrigeration system have been explored using both experimental and numerical approaches. For the first time, the internal flow structure was characterized by means of comprehensive numerical simulations. In essence, three major sections of the steam ejector were investigated. Two energy processes and the shock-mixing layer were defined and analyzed. The results indicated that the length of the choking zone directly affects the entertainment ratio under different primary fluid temperature. The optimum enterainment ratio was achieved with 138 °C primary fluid temperature. The shock-mixing layer was greatly affected by secondary fluid temperature. With increasing of back pressure, the normal shock gradually shifted from the diffuser towards the throat, while the shock train length remains lunchanged.

Keywords: combustion waste heat; experiment; steam ejector; shock-mixing layer; flow structure; operating parameter

---

1. Introduction

Energy conservation and recyclability are among the greatest concerns of modern research studies in automotive industry. Regarding all the energy consumed within a vehicle, the air-conditioning system typically takes up a larger part. With major breakthroughs in refrigeration system technologies in the usage of low-grade energy, steam ejector refrigeration systems have become one of the potential devices for effective energy conversion, which can be driven by industrial and automobile exhaust heat waste [1]. The applications of steam ejectors range from transportation systems, fire safety [2], petroleum, chemical industry and seawater desalination [3]. Throughout the last century, the automobile has accounted for the majority of transport worldwide. However, the fuel consumption against energy converting ratio is relatively low. The effective thermal efficiency of the engine is generally about

30%. The remaining 65% of the fuel combustion heat includes 10% lost for radiation, 30% lost to the exhaust and 30% lost by the cooling system. That is, the waste heat accounts for 30% [4]. The engine combustion waste heat has the characteristics of high temperature, fast flow rate and strong pressure. Its temperature can range from 200 °C to 600 °C [1]. This part of the heat not only exacerbates the air pollution but also causes the overheating of the water tank to affect the dynamic performance of the automobiles. To reduce waste heat or to improve the waste heat utilisation rate is an effective way to decline the air pollution and enhance the operating stability of the automobile. As an independent running individual, the recovery and utilisation of the waste heat are partly limited. In 1980, Hamner firstly proposed to apply the combustion waste heat of automobile engine to power a steam ejector refrigeration system [5]. From an application perspective, using the recycled waste heat as a power source for driving the automotive air conditioning system is a promising way to save energy [6]. Comparing the compression cooling system of the automobile, the steam ejector refrigeration system has the potential to be used in automobiles, due to it has no moving parts, simple structure, small equipment area, durable operation and low cost [2].

The steam ejector refrigeration system consists of four major parts: the generator, the steam ejector, the evaporator and the condenser. The water vapour generated from the generator enters the nozzle at a higher temperature. After it flowed through the nozzle, the water vapour was reduced to a low pressure fluid at the entrance of the ejector. Then the water vapor in the evaporator was sucked into the ejector by the pressure difference. Though a series of flow changes, the two streams of fluid are mixed into a mixed fluid and together discharged into the condenser to continue the next refrigeration cycle [7,8]. There is a direct relationship between the performance of the refrigeration system and ejector working characteristics [9] and the efficiency of the steam ejector refrigeration system largely depends on the characteristics of the steam ejector [10,11]. Therefore, a comprehensive understanding of the factors affecting the performance of the steam ejector is essential to improve the performance of the whole refrigeration system. The disadvantage of the steam ejector is that its efficiency is relatively low [12].

As early as 1910, Leblanc proposed the ejector refrigeration cycle system [13]. The researchers have conducted extensive experimental studies and numerical simulations of the steam ejector [14,15]. Though the main research of the ejector was still focused on the thermodynamic analysis of the ejector [16,17], the numerical simulation of the steam ejector developed rapidly. For example, the research on the spontaneous condensation inside the ejector has made some progress in the past five years. The wet steam model applied in the optimization of the ejector efficiency was studied by Li [18]. A developed wet steam model in the non-equilibrium condensation phenomenon to understand the intricate feature of the steam condensation was demonstrated by Yang [19]. Zhang [20] proposed a modified wet steam model to evaluate the pumping performance and optimize the operating conditions of the steam ejector. Moreover, the study of the numerical simulation on the pumping performance of the ejector by changing the geometry structure and the operating parameters have been investigated by the researchers. The internal complex flow phenomena such as the shock wave [21], boundary layer separation [22] and shock boundary layer interaction [23] have been studied to improve the design of the ejector. Bartosiewicz [24] studied the flow separation phenomenon in detail by both experimental and numerical simulation methods. Zhu [25] analysed the shock wave structures and their effect on the pumping performance of the converging-diverging and the converging nozzle ejectors. The effect of the first shock wavelength on the ejector performance is studied by using an image processing method. The results showed that the ejector performance decreases with the increase of the shock wavelength at a given primary flow pressure. Zhu [26] proposed a shock circle model in predicting the ejector performance is simpler than many 1D modelling methods and can predict the performance of critical mode operation ejectors much more accurately. Numerical study of the internal flow and heat transfer in the steam ejector has been conducted [27,28], which provided valuable insights for geometry improvements for the ejector design. It can effectively reveal the internal flow and the mixing process of the ejector. The shock's position of the mixed fluid played a very important role in the ejector performance [29]. Nevertheless, the flow structure inside the steam ejector was still lack of full understanding.

In this work, a steam ejector refrigeration system driven by the combustion waste heat of the automobile engine was proposed. A steam ejector refrigeration system with a cooling capacity of 12.5 kW was studied experimentally. The characteristics of the internal flow structure in the steam ejector are analyzed by using both experimental and numerical approaches. The whole flow region inside the ejector was firstly divided into three zones via static pressure, namely: inducing zone, choking zone and discharging zone. By identifying these three zones, the flow structures such as the shock train, choking position and the normal shock wave can be identified with more indication. The shock-mixing layer is defined as where the separation boundary of the primary and secondary fluids. The effects of the energy exchange and mixing progress on the pumping performance of the ejector were analyzed and discussed. By discussing the characteristic of the internal flow structure, the influence of the operating condition, i.e., the primary fluid temperature (134 °C–146 °C), the secondary fluid temperature (5 °C–20 °C) and back pressure (3000 Pa–5000 Pa) on the pumping performance under was comprehensively investigated.

2. Experiment Instrumentation

The steam ejector refrigeration system is mainly composed of four components: a generator, a steam ejector, an evaporator, and a condenser. Other experimental components include a storage tank, a water ring vacuum pump, an evaporative condenser, pressure regulating valves, vacuum gauge and pipes, as displayed in Figure 1. The steam generator generates saturated steam with a certain pressure and temperature, then the saturated steam passes through a steam-pressure regulating valve and reaches the steam storage tank. The high-speed primary fluid from the steam storage tank entrained the vapour from the evaporator through the ejector to achieve the purpose of refrigeration and then discharges into the condenser for the next cooling cycle. A water ring pump is installed at the outlet of the water-cooled condenser and supplies a stable back pressure for the steam ejector as the back-pressure pump. It is used to cool the high-temperature steam in the condenser to prevent the pump body from being corroded and stabilize the gas pressure in the condenser. The purpose of installing evaporative condenser is to make the condensed water from the condenser cool down rapidly so that it can enter the next cooking cycle quickly and improve the cycling efficiency of the refrigeration system. The steam ejector outlet is equipped with a back pressure regulating valve to regulate the back pressure at different levels. The secondary fluid is evaporated from the evaporator, and its mass flow rate is measured by the vortex shedding mass flow meter. The measurement and control of the primary fluid pressure are mainly achieved by means of a pressure transmitter, a PID controller and a pressure regulating valve. The pressure transmitter not only measures and displays the pressure, as a pressure gauge but also converts the pressure in the steam storage tank into an electrical signal sent to the PID controller. The PID operator automatically acquires a given signal from the system and valve position feedback signal and adjust the output of the corresponding control according to the difference between the pressure and the electrical signal. Then the pressure regulating valve will change the mass flow of primary fluid after accepting the DC signal from the PID controller. Therefore, the experimental primary fluid pressure can maintain at a given value.

A T-type thermocouple temperature control meter measures the temperature of the water in the evaporator which constantly adjusts to the temperature to reach the preset temperature, making sure the experiment can work under set conditions. 15 small holes are evenly opened on the ejector wall to measure the static wall pressure. The vacuum valve is connected to each hole through the vacuum silicon tubes. The pressure is carried through the valve and measured by the film vacuum gauge in turn. The error of the pressure transmitter of the full-scale is 0.5%. The instrument used to test the wall pressure is a film vacuum gauge with an error of 0.25% of the full scale. The mass flow rate of the secondary fluid is measured by a vortex flowmeter with the uncertainty of less than 0.33% and with the accuracy of 1%. The T-type thermocouple temperature control meter is used to measure the temperature with an error of ± 1 °C and an error of 0.5% of full scale. Besides, energy balance analysis was performed on the evaporator, generator and condenser during equipment commissioning to ensure measurement accuracy.

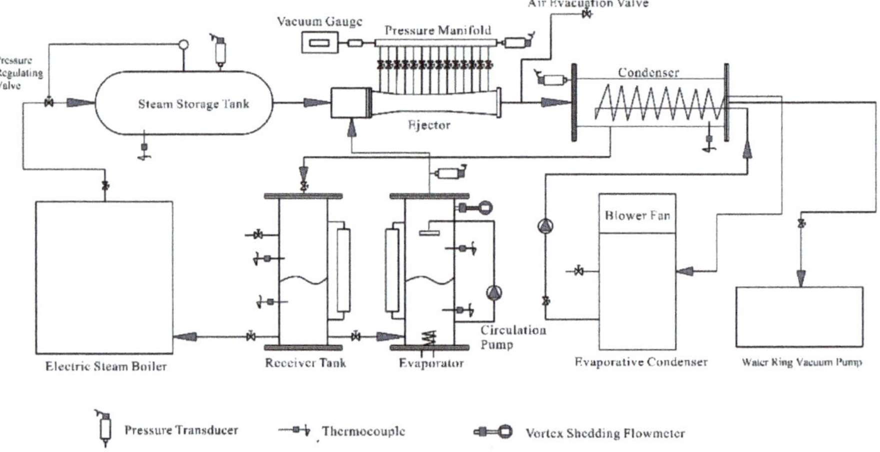

Figure 1. The experimental steam ejector refrigeration system.

The cooling capacity of the experimental system is 12.5 kW. The maximum pressure and the saturated temperature of steam provided by the generator are 0.44 MPa and 148 °C, respectively. The cooling water circulating flow rate is 0.34–0.9 m³/h. The geometric parameters of the steam ejector are shown in Table 1. The operating conditions of the steam ejector are shown in Table 2.

Table 1. Geometrical parameters of the steam ejector.

Geometrical Parameters	Value
Diameter of the nozzle inlet	12 mm
Diameter of the nozzle outlet	11 mm
Diameter of the nozzle throat	2.5 mm
Expand the angle of the nozzle	10°
Nozzle exit position	10 mm
Diameter of mixing chamber inlet	70 mm
Diameter of throat	28 mm
Length of mixing chamber	122.2 mm
Length of throat	90 mm
Length of diffuser	210 mm

Table 2. Flow conditions of the steam ejector.

Boundary Conditions	Value
Primary fluid temperature	134 °C~146 °C
Secondary fluid temperature	5 °C~20 °C
Back pressure	3000 Pa~5000 Pa

3. Numerical Analysis of Flow Behaviour inside the Steam Ejector

The coefficient of performance (*COP*) is evaluated as the ratio of the refrigeration effect to the energy input of the system as given in Equation (1) [30]:

$$COP = \frac{Q_e}{Q_g + W_p} \tag{1}$$

The *COP* can be rewritten after a series of calculations by neglecting pump work using Equation (2) [31]:

$$COP = \frac{\dot{m}_p}{\dot{m}_a} \times \frac{\Delta h_e}{\Delta h_b} = E_m \times \frac{\Delta h_e}{\Delta h_b} \tag{2}$$

Therefore, it is obvious that the *COP* of the whole system depends entirely on the entrainment ratio, and the value of *COP* will be closer to the value of the entrainment ratio. Entrainment ratio (E_m) is a common index for evaluating the pumping performance and reflects the working capability of the steam ejector which is defined as the mass flow rate ratio of the secondary fluid to the primary fluid's mass flow rate [32]. Based on the existing experimental data, Chou [18] established a 1-D model of the ejector with R113 and R141b as the primary fluid:

$$E_m = \left(\frac{P_2}{P_1}\right) \cdot \frac{1}{\left(\frac{T_2}{T_1}\right)^{1/2}} \cdot K_p \cdot K_f \cdot K_s \cdot \varphi \tag{3}$$

K_p is correlative with the structure of the nozzle, K_f is correlative with the effective area in the ejector, K_s is correlative with the aspect ratio and the exit Mach number of the primary nozzle, φ is a specific correction factor when the geometry structure of the ejector operating system is fixed. When the geometry dimensions of the nozzle are constant, the operating parameters K_p, K_s of the primary fluid are invariant, the entrainment ratio of the ejector is varied with the primary fluid pressure and secondary fluid pressure. The maximum E_m of the ejector will be obtained when the ejector only operated at the double-choked flow mode under certain operating conditions [33,34]. Previous researches have shown that the shock-mixing layer [35], effective area [36], shock train [37] and the choking flow [38] in the throat section of the ejector are key factors in dominating the pumping efficiency of the ejector [39,40]. However, the research of the complex flow characteristics inside the ejector is limited by the present existing research methods.

3.1. Numerical Simulation Method

The CFD simulation software ANSYS Fluent version 19.2 was employed in this numerical study. Previously, the research work studied by Pianthong and Seehanam [41] has already demonstrated that the simulation results by the two-dimensional axisymmetric model and three-dimensional model were almost identical. Therefore, a two-dimensional axisymmetric model can be made as an assumption to perform the analysis of this study. The segregated solver is applied in ANSYS Fluent and the pressure is modelled by pressure velocity coupling approaches, which is typical for a wide range heat fluid interaction problems [42,43]. Huang [44] divided the ejector operating mode into three parts. When the back pressure of the ejector was below the critical value, the ejector was in the double-choked flow mode and is operating normally. The ejector is in unchoked flow and reversed flow modes when it was greater than the critical value. In practice, the spontaneous condensation of water vapour exists inside the steam ejector. A few study results have pointed out that the ideal gas cannot truly reflect the flow characteristics of the ejector [45]. The wet steam model can better predict the actual flow conditions and the simulation results are closer to the experimental values [46,47]. Nevertheless, the research of comparing the CFD simulation values with the experimental values studied by Moore et al. [48], it was concluded that the simulated value of the ideal gas model is closer to the experimental value and can better reflect the fluid flow characteristics than the wet steam model in the double-choked flow mode. Garcia del Valle [38] pointed out that there was no significant difference between the ideal gas and the real gas in the linear axisymmetric supersonic ejector. The ideal gas can be used to replace the real gas in numerical analysis reflect the flow property inside the ejector. Aphornratana [49] also proved that the ideal gas model could provide similar simulation results to the real gas in a steam ejector with lower operating pressure. To save computational cost and to improve simulation efficiency, the ideal gas model is selected for simulation analysis for thermal fluidic flows [50,51]. The fluid flow inside the steam ejector is governed by the compressible steady-state flow conservation equations including the mass conservation equation, the momentum conservation equation and the energy conservation equation, which can be written as follows:

The continuity equation:

$$\frac{\partial \rho}{\partial t} + \frac{\partial}{\partial x_i}(\rho u_i) = 0 \tag{4}$$

The momentum equation:

$$\frac{\partial}{\partial t}(\rho u_i) + \frac{\partial}{\partial x_j}(\rho u_i u_j) = -\frac{\partial P}{\partial x_i} + \frac{\partial \tau_{ij}}{\partial x_j} \tag{5}$$

The energy equation:

$$\frac{\partial}{\partial t}(\rho E) + \frac{\partial}{\partial x_i}(u_i(\rho E + P)) = \vec{\nabla}\cdot\left(\alpha_{eff}\frac{\partial T}{\partial x_i}\right) + \vec{\nabla}\cdot\left(u_j(\tau_{ij})\right) \tag{6}$$

where:

$$\tau_{ij} = \mu_{eff}\left(\frac{\partial u_i}{\partial x_j} + \frac{\partial u_j}{\partial x_i}\right) - \frac{2}{3}\mu_{eff}\frac{\partial u_k}{\partial x_k}\delta_{ij} \tag{7}$$

with:

$$\rho = \frac{P}{RT} \tag{8}$$

where τ_{ij} is the stress tensor, E is the total energy, α_{eff} is the effective thermal conductivity, and μ_{eff} is the effective molecular dynamic viscosity.

The density-based model is employed for the simulation. The turbulence model was a realizable k-ε model [52,53] with which we can obtain higher accuracy of simulation results based on our previous research. The two inlets were set as pressure-inlets, and the outlet was set as a pressure-outlet. The wall-boundary was assumed to be a no-slip and adiabatic, using Enhanced Wall Treatment for near-wall treatment with y plus = 0.835 (<1) [54,55]. The second-order upwind format was used to calculate all the convective terms. The convergence criterion for residual of all dependent variables and the mass imbalance value was set to 1×10^{-6} and 1×10^{-7}, respectively.

The independence of the grid has been analysed by obtaining a resulting mesh composed of 51,203 quadrilateral elements. Three meshes-course grid, medium grid and fine grid were used to evaluate the grid independency; three grid units are 29,375, 51,203 and 74,512, respectively. As can be seen from Figure 2, the computational accuracy of the grid with 51,203 units and 74,512 units has a similar variable trend and no further improvement compared with the grid of 29,375 units. Therefore, the simulation model with 51,203 was chosen for reducing the computational time and saving costs in the later simulations. The grid was refined based on the adaptive technology of the Mach number gradient. The quadrilateral structure grid is displayed in Figure 3.

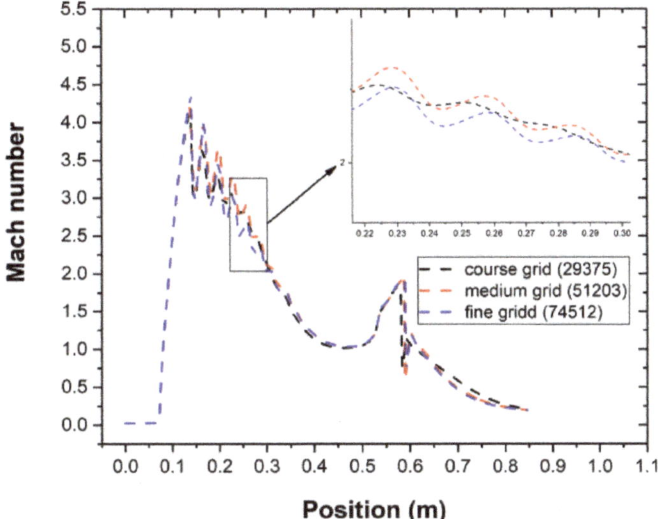

Figure 2. The Mach number with different grid density.

Figure 3. The structural grid of the computational domain.

3.2. CFD Model Validation

The static pressure distribution along the wall of the steam ejector was used as the reference data for validating the numerical simulation model. The comparison was conducted when the primary fluid pressure is 0.34 MPa, and the secondary fluid pressure is 1710 Pa with different back pressures. As displayed in Figure 4, the static pressure distribution of the experimental values and simulation values is close. Though in the mixing chamber of the ejector, there is a certain difference between the two. The cause may one is that the simulation assumes that the fluid is ideal gas and the actual situation is based on it but not all. The other is that the sealing property of the experimental equipment and its manufacturing error which it is still controlled within an acceptable range. In the throat and divergent sections, the agreement between the two is good, and the distribution trend of the wall pressure is relatively consistent. The total average deviation of experimental and CFD simulation is less than 8%, which is the allowable error range in the engineering industry. Thus, the feasibility and accuracy of the simulation model can be verified. During the experiment, a thermocouple is used to measure the temperature of the vapour in the evaporator. Owing to the influence of the inlet temperature and the ambient temperature on the thermocouple, the temperature of the steam is in a dynamic change near the set temperature.

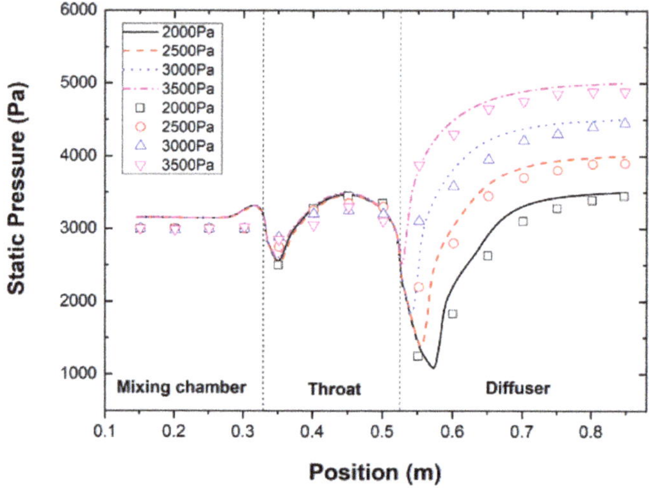

Figure 4. Wall pressure distribution under different back pressure.

3.3. Analysis of the Flow Structure inside the Steam Ejector

Research into the flow characteristics inside the steam ejector is still at the stage of theoretical analysis due to the lack of effective experimental methods. With the advantages of the computational fluid dynamics (CFD) technology, a detailed understanding of the internal flow structure of ejector is conducted in this part. The entire flow region is divided, and a new method of judging the flow characteristics inside the steam ejector is proposed. The simulation is underway in the condition of the primary fluid pressure, the secondary fluid pressure and the backpressure are 0.34 MPa, 1710 Pa and 3500 Pa, respectively.

3.3.1. The Division of the Entire Flow Region

The Mach number contour of the static pressure is shown in Figure 5. The fluid flow characteristics of the shock train, the diamond wave, the primary jet core, choking position, shock-mixing layer, the normal shock wave and oblique shock waves can be observed. The flow region is divided into three zones as indicated by the static pressure distribution along the ejector axis in Figure 6. The first flow area is the "inducing zone". In this zone, the supersonic primary fluid forms a primary jet core after spraying from the nozzle [10]. A low-pressure region was generated at the nozzle outlet. The secondary fluid was pumped into the mixing chamber driven by the pressure difference. Then the secondary fluid was accelerated by the viscous transport capacity of a high-speed shock-mixing layer and reached sonic at the end of the zone. Meanwhile, the intensity of the diamond wave [56] decreased until it disappeared. Under the combined action of the pressure difference and viscous transport capacity of shock-mixing layer, the purpose of pumping and accelerating the secondary fluid has been accomplished.

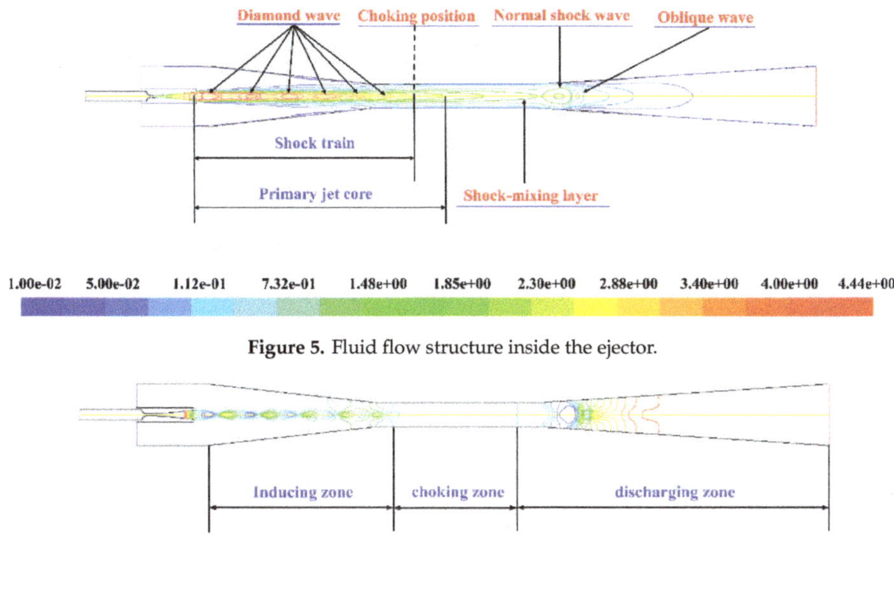

Figure 5. Fluid flow structure inside the ejector.

Figure 6. Flow zones distribution in the steam ejector.

The second flow area is the "chocking zone". It can be seen from Figure 6 that the static pressure was constant, and the velocity shows no obvious changes in this zone. One reason is that the energy of the primary jet core had been partly consumed in the previous flow. The other reason was that the secondary fluid flowed in a fixed cross-sectional area (i.e., the effective area [57]). The kinetic and internal energy was continuously added with the increasing of the mass flow rate of the secondary fluid. The two streams began to mix and are at co-velocity at the end of this zone. The mixed fluid was choked and exceeded sonic. Besides, the size of the effective area played an important role in determining the pumping performance of the ejector.

The third flow area is the "discharging zone" where the momentum of the primary fluid decreased while the secondary fluid's momentum correspondingly increased. However, the primary fluid is still in the supersonic state. A normal shock wave which resulted in the velocity dropping and the pressure rising abruptly was generated to make the mixed fluid overcome the backpressure to discharge from the diffuser section. Meanwhile, the shock-mixing layer was broken as the normal shock wave appeared. The velocity of the two streams completely mixed after a series of oblique shocks. In this zone, the back pressure was the main influencing factor that directly determined the position and intensity of the normal shock wave and oblique shock.

3.3.2. The Shock-Mixing Layer

With the aim of deeply understanding the structure and characteristics of the fluid flow inside the ejector, it is necessary to analyse and discuss the shock-mixing layer of which is the key factor affecting the mixing progress of the two streams. The position and the shape of the shock-mixing layer can be reflected through the powerful visualisation function of the simulation method. The secondary fluid did not mix with the primary fluid immediately after entering the mixing chamber as shown in Figure 7a. The shock-mixing layer with a large velocity gradient between the primary fluid and the secondary fluid is captured. It was the boundary layer of the momentum exchange and mixing between the two streams. It was mainly that the high-speed primary fluid transferred its kinetic and internal energy to the secondary fluid through the viscous transport capacity of the shock-mixing layer. Also, the velocity of the primary fluid decreases gradually through the restraint function of the shock-mixing layer. The borderline of the shock-mixing layer was displayed in Figure 7b. The length range of the shock-mixing layer is from the position of the nozzle outlet to the front of the normal shock wave where the two streams were reaching the co-velocity.

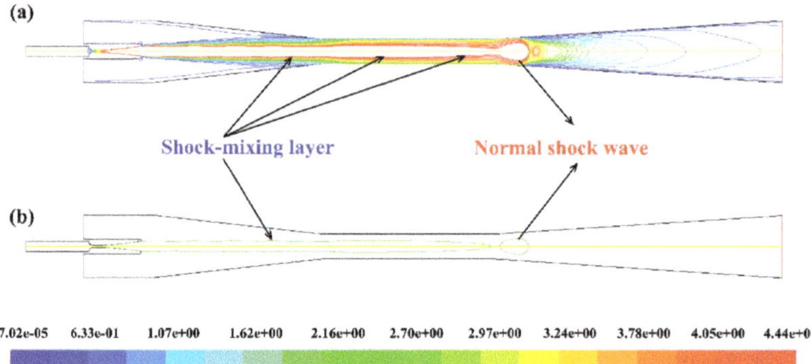

Figure 7. The location and range of the shock-mixing layer (a) the contour of Mach number (b) the borderline.

As indicated in Figure 8, the flow process in the ejector can be divided into the energy exchanging process and the mixing process. In Figure 8, the energy and momentum transmissions of the two streams in front of the normal shock wave were conducted in the exchanging progress and the shock-mixing layer was broken, and the two streams began to mix after the normal shock wave was generated in the mixing progress. However, before the beginning of the mixing process of the primary fluid and the secondary fluid, the shock-mixing layer with a certain thickness can be regarded as a wall surface separating the two fluids. The flow duct of the secondary fluid was a converging-diverging structure like that of the primary nozzle. It was consisting of mixing channel, throat and diffuser as displayed in Figure 7a. The only difference between the actual and the imaginary converging-diverging structure was that the inner wall was a moving wall to provide the maximum energy to accelerate the secondary fluid. The secondary fluid was accelerated from stagnation state to supersonic state relying on continuously absorbing the kinetic energy from the moving wall in the energy exchange process. The position where the secondary fluid reaches the sonic velocity is the choking position and mostly is located at the minimum cross-section of the mixing chamber. In the end, the mixed fluid flow completely mixed after flowing through a series of oblique shock waves and the mixed fluid overcame the back pressure to discharge from the ejector. In summary, the whole flow progress is divided by the position of the normal shock wave, the area before the normal shock wave is the energy exchange process region while the area behind the normal shock wave is the mixing process region.

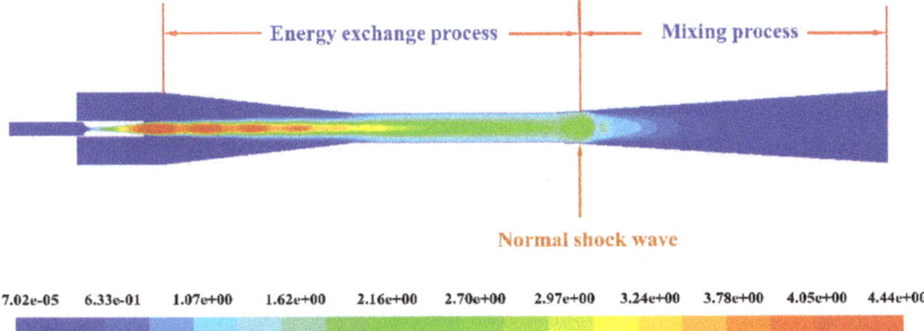

Figure 8. Flow structure partition under the influence of the shock-mixing layer.

4. Results and Discussion

4.1. The Influence of the Primary Fluid Temperature

As can be seen from the variation of the entrainment ratio with the primary fluid temperature in Figure 9. The operating condition was the secondary fluid temperature, and the back pressure is 15 °C and 3000 Pa, respectively, which the primary fluid temperature varies from 134 °C to 146 °C. The maximum value of the entrainment ratio is 0.69 as the primary fluid temperature is 138 °C. The entrainment ratio increases with the increasing of the primary fluid temperature when it is less than 138 °C. On the contrary, when the temperature is higher than 138 °C, the entrainment ratio gradual declines. In the condition of the constant geometry of the ejector, there exists a primary fluid temperature which optimizes the entrainment ratio. The static pressure lines of the flow structure are as shown in Figure 10 that the length of the shock train and the shock-mixing layer are almost the same. It indicates that these two flow characteristics are less affected by the primary fluid temperature. There is no significant difference inside the mixing chamber and diffuser sections when the temperature changes. Though the Mach number changes a lot at the inlet of the throat section, the shape of the primary fluid jet core is also similar. The local peak value of the normal shock wave increases with the temperature increasing.

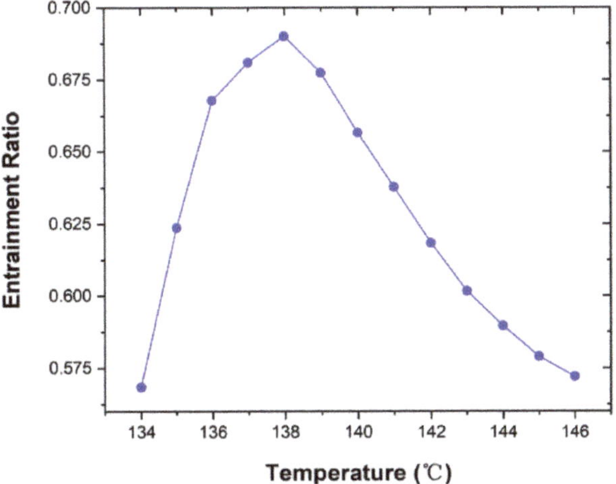

Figure 9. The variation of the entrainment ratio with the primary fluid pressure.

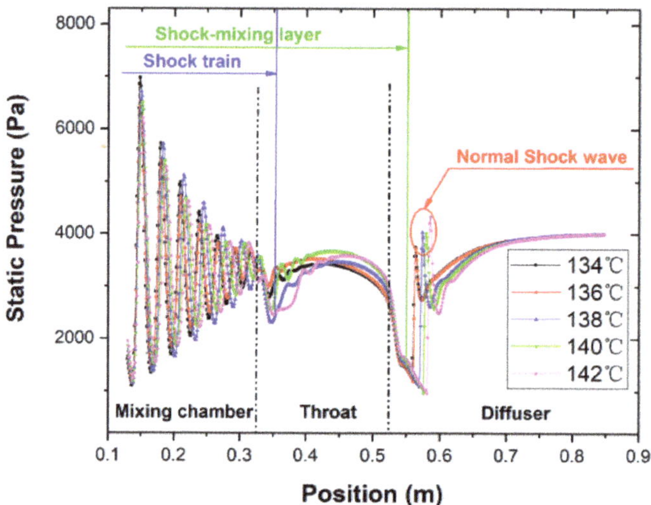

Figure 10. Comparison of static pressure distribution along the axis under different primary fluid pressure.

With the increase of the primary fluid temperature, the position of the normal shock wave gradually moves from the outlet of the throat section to the inlet of the diffuser section and the intensity of the normal shock wave is enhanced (see Figures 10 and 11). Thus, the ability to resist the interference of the back pressure is significantly improved. Only when the normal shock wave is generated in the diffuser section, the ejector will work in double-choking flow and operates in a normal working state [45]. As mentioned before, the choking zone is an area that the static pressure and velocity were constant. As can be seen from Figure 11, the length of the choking zone changes with the primary fluid temperature. The length of the choking zone reaches the maximum value of 86 mm when the temperature is 138 °C. As the length of the choking zone increases, the inducing zone which is in front of the choking zone is almost the same while the discharge zone after choking zone is shortened. Compared with Figure 9, it is found that the length of the choking zone is a positive linear correlation with the entrainment ratio and the length of the choking zone directly affects the efficiency of the ejector. As the aspect of the energy transmission, the inducing zone and the choking zone were the main areas where the kinetic and the internal energy exchanges between the primary fluid and the secondary fluid. The two zones are including in the energy exchange progress. The results illustrate that the pumping performance of the ejector would be directly determined by the length the energy exchange progress. The longer length of the energy exchange progress, the higher the efficiency of the ejector would be. The length of the choking zone should be increased as the length of the discharging zone should be reduced to improve the pumping performance of the ejector.

When the primary fluid temperature is too high, there will be a velocity reflux flow at the position of the nozzle outlet and a part of the kinetic energy would be lost as shown in the dotted line in Figure 11. It causes the kinetic energy of the primary fluid entering the choking zone reduced, and the entrainment ratio decreased. Although the length of the choking zone is much longer at 146 °C than at 134 °C, the entrainment ratio of the two is similar, which are 0.572 and 0.5684, respectively. Similarly, the length of the choking zone at 144 °C is also the same when the primary fluid temperature is at 136 °C, but the entrainment ratios are different.

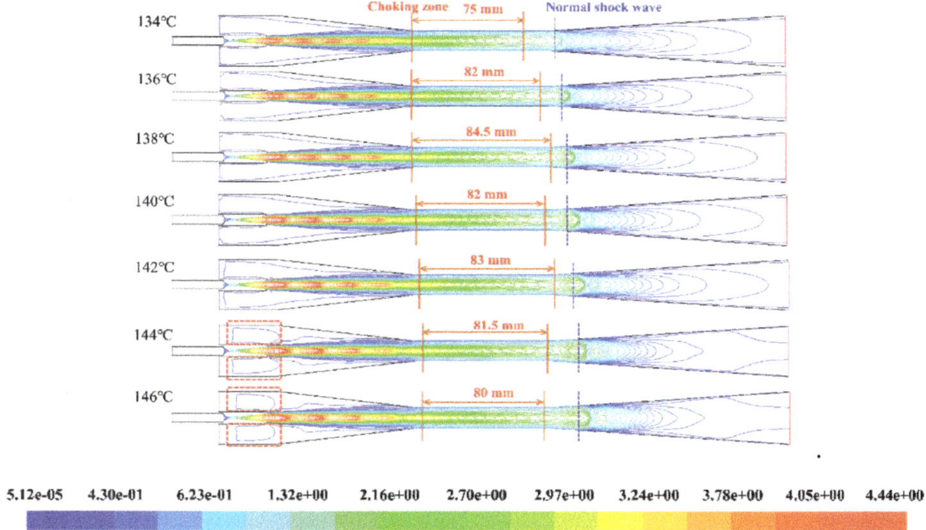

Figure 11. The contours of Mach number under various primary fluid temperature.

4.2. The Influence of the Secondary Fluid Temperature

The Mach number contour of the flow field is shown in Figure 12, as the primary fluid temperature and back pressures of the steam ejector are 138 °C and 3500 Pa, respectively. With the increase of secondary fluid pressure, the length of the shock-mixing layer firstly increases and then decreases. The position of the normal shock moves downstream, and the position of the shock train locates in the mixing chamber section while the choke occurs in the throat section. It shows that the secondary fluid pressure has a significant effect on the flow characteristics of the ejector.

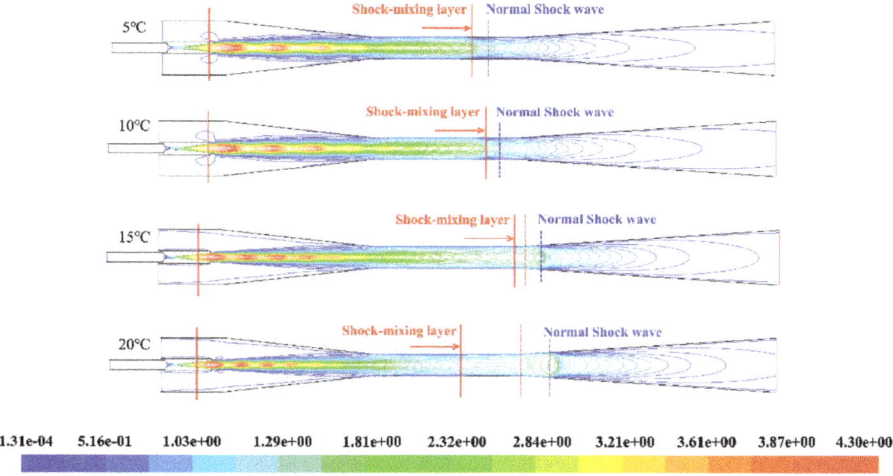

Figure 12. The contours of Mach number for various secondary fluid temperature.

Otherwise, the length of the primary jet core and the shock-mixing layer increases with the increasing of the secondary fluid temperature when it is less than 20 °C. As in the analysis above, the shock-mixing layer is the main momentum and energy exchange region, where the degree of energy and momentum transfer between the two streams are determined. The primary and secondary fluids will not be mixed until the shock-mixing layer disappears. The longer the length of the shock-mixing layer is, the more degree energy and momentum transmission would be. On the other hand, with the secondary fluid temperature increases, the position of the normal shock wave gradually moves downstream indicating that the ability of ejector to resist back pressure disturbance is enhanced and the ejector will operate more stably. As indicated by the variation of the entrainment ratio with the secondary fluid temperature in Figure 13, the increasing of the length of the shock-mixing layer causes the normal shock wave moving downstream and its intensity then the entrainment ratio increases except the temperature of the secondary fluid is at 20 °C. The entrainment ratio is not only influenced by the length of the shock-mixing layer but also relates to the area from the end of shock mixing layer to the beginning of the normal shock wave where the energy and momentum exchanging progress finished and the two streams starts to mix. In this area, the ability of the primary fluid pumping the secondary fluid is continuously enhanced and the mass flow rate of the secondary fluid is also increased resulting in the pumping performance of the ejector will be better and better.

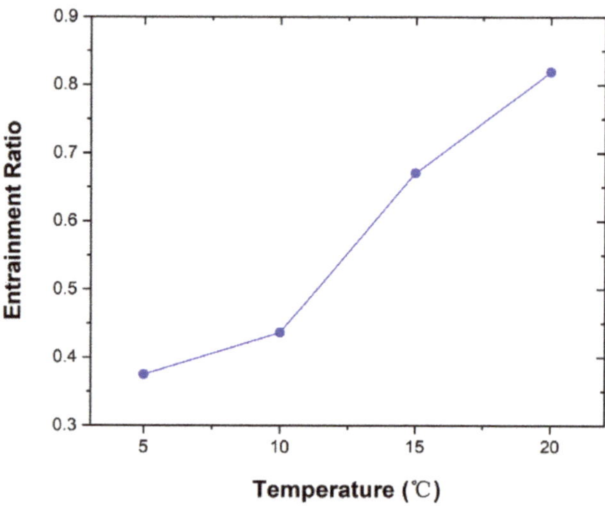

Figure 13. The variation of the entrainment ratio with different secondary fluid temperature.

The ratio of the primary fluid pressure to the secondary fluid pressure is generally referred as "expansion ratio", which is a dimensionless parameter to analyse the flow behaviour and to reflect the efficiency of a steam ejector [15]. As the value of expansion ratio increases, it means more secondary fluid is entrained given the same amount of primary fluid is being consumed. The pumping performance of the ejector will be greatly improved. Comparing Figure 12 with Figure 13, the steam ejector should be operated at relatively higher secondary fluid pressure to maintain the Mach number in a higher value and to pump more mass flow rate of the secondary fluid. The higher of the secondary fluid pressure is, the mixing progress will be, and the larger secondary fluid mass flow rate will be obtained.

4.3. The Influence of the Back Pressure

Figure 14 displays the flow structure inside the steam ejector under different back pressure, when the temperature of the primary and the secondary fluids are 138 °C and 20 °C, respectively. The existence of the normal shock wave prevents the disturbance caused by the back pressure spreading from the diffuser section to the throat section. The increase of the distance between the normal shock wave and the outlet of the ejector leads to reduce the ability to resist the back pressure disturbances.

It will be more difficult to form a choking flow which played an important role in affecting the pumping efficiency of the ejector [45]. As can be seen in Figure 15, when the back pressure is less than 4600 Pa, the entrainment ratios are always the same value of 0.546. The boundary layer separation occurs in the mixing chamber section, the normal shock wave disappeared when the back pressure is at 5000 Pa, and the entrainment ratio rapidly drops to 0.27 as shown in Figures 14 and 15. When the back pressure is more than 4600 Pa, the entrainment ratio decreases gradually with the increase of the back pressure until it reaches 0 at 7300 Pa. It shows that the critical back pressure is 4600 Pa. When the back pressure is higher than the critical value, the backpressure is so high that the primary fluid is kept from flowing into the diffuser section, and the choking flow would not be formed due to the largest primary jet core does not move into the mixing chamber. The energy exchange between the primary fluid and the secondary fluid is too short of exchanging momentum and energy with each other enough. The ideal flow passage of the converging-diverging structure of the secondary fluid flow passage is not formed where the secondary fluid cannot be further accelerated in the mixing chamber section to be pumped. Then, the flow rate of the secondary fluid is gradually reduced resulting the entrainment ratio decline (in Figure 14).

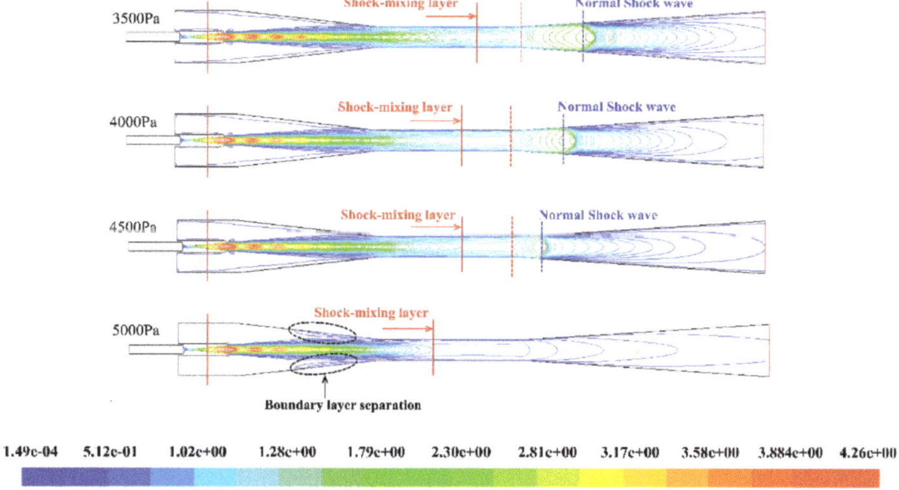

Figure 14. Contours of Mach number for various back pressure.

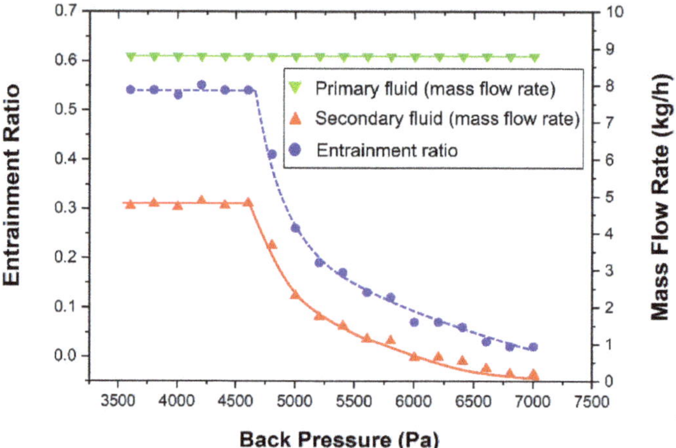

Figure 15. Comparison of entrainment ratio and the mass flow rate of two streams under different back pressure.

In Figures 14 and 16, the length of the shock train is relatively similar, and the flow structure inside the mixing chamber and the throat sections remains the same when the back pressure is less than the critical value (4600 Pa). Also, the entrainment ratio remains constant as the length of shock-mixing layer, and the energy exchange progress is nearly the same. Simultaneously, the energy and momentum transfer process between the primary fluid and the secondary fluid and the velocity distribution in this region are substantially the same. When the back pressure exceeds the critical value, there will be no normal shock exists in the diffuser section. The disturbance caused by the back pressure will propagate to the upstream to destroy the choking flow. Once the choking phenomenon cannot be generated, the mass flow rate of the secondary fluid will sharply drop until zero. When the back pressure is higher than the critical back pressure, the ejector will in the reverse flow mode resulting in mass flow rate of the secondary reduced and the entrainment ratio decreases as shown in Figure 15.

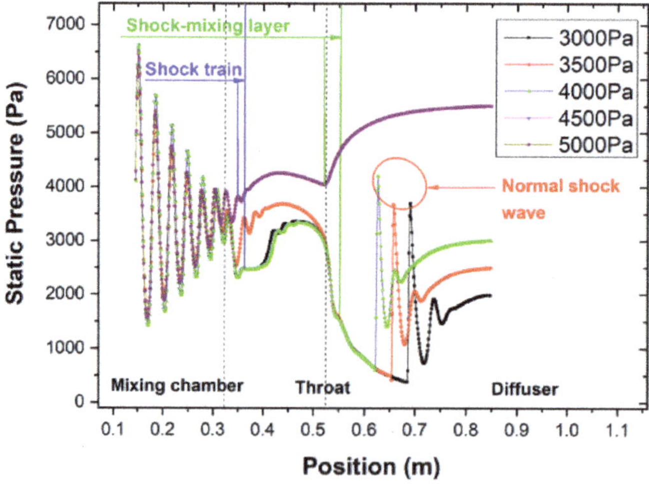

Figure 16. Comparison of static pressure distribution along the ejector axis by different back pressure.

5. Conclusions

The use of steam ejector refrigeration technology to recover waste heat from engine combustion and convert it into useful energy is an effective energy-saving and environmental protection measure. A steam ejector refrigeration system driven by the engine combustion waste heat in the minibus with the cooling capacity of 12.5 kW was proposed. An experimental setup of the steam ejector refrigeration system was established. The computational fluid dynamics (CFD) approach was used to analyse the internal flow characteristic inside the steam ejector which cannot be fully accessed and studied through the experiment research. The influence of fluid flow characteristics on the pumping performance of the steam ejector was comprehensively studied. The reasons for the variation of the entrainment ratio under the operating parameters (i.e., the primary fluid temperature, the secondary fluid temperature and the back pressure) were also discussed. In summary, the following are the key findings from this study: (i) the entire flow area inside the steam ejector was divided into three regions: the inducing zone, the chocking zone and the discharging zone. Also, the energy exchange progress and the mixing progress are determined; (ii) the position and the shape of the shock-mixing layer which the boundary layer separating the primary fluid from the secondary fluid was defined, (iii) in the range of the primary fluid temperature from 134 °C to 138 °C, the entrainment ratio was positively correlated with the length of the choking zone. However, though the length of the choking zone increases, the pumping performance of the steam ejector would be reduced when the temperature of the primary fluid exceeds 138 °C; (iv) The shock-mixing layer only increased with the temperature until 15 °C, and then a significant drop happened at 20 °C. It indicates that the entrainment ratio increased from 5 °C to 20 °C, in which the improvement from 15 °C to 20 °C was marginal; and (v) with the increase of the back pressure, the normal shock moved from the diffuser section to the throat section. When the back pressure is less than the critical value (4600 Pa), the length of the shock-mixing layer is not significantly affected. As the back pressure exceeds the critical pressure, due to the disappearance of the normal shock wave, the choking condition cannot be maintained causing the pumping efficiency drops rapidly until the steam ejector lost its pumping capacity. Generally, this in-depth numerical and experimental study provided major insight structural design optimizations of steam ejectors.

Author Contributions: Conceptualization, Y.H. and L.G.; algorithm, Y.H. and X.W.; software, Y.H., R.C., T.B.Y.C. and H.L.; data curation, Y.H., C.L. and H.L.; formal analysis, Y.H. and A.C.Y.Y.; numerical simulations, Y.H., and R.C.; validation, A.C.Y.Y. and C.L.; writing—original draft preparation, Y.H. and X.W.; writing—review and editing, L.G. and A.C.Y.Y.; visualization, C.L. and A.C.Y.Y.; supervision, J.T. and G.H.Y.; funding acquisition, X.W., J.T. and G.H.Y.

Funding: This research was funded by the National Natural Science Foundation of China (51775098), the Australian Research Council (DP160101953) and the Australian Research Council Industrial Transformation Training Centre (ARC IC170100032) in the University of New South Wale. All financial and technical support are deeply appreciated by the authors.

Conflicts of Interest: The authors declare no conflict of interest.

References

1. Wang, L.; Cai, W.; Zhao, H.; Lin, C.; Yan, J. Experimentation and cycle performance prediction of hybrid A/C system using automobile exhaust waste heat. *Appl. Therm. Eng.* **2016**, *94*, 314–323. [CrossRef]

2. Li, A.; Yuen, A.C.Y.; Chen, T.B.Y.; Wang, C.; Liu, H.L.; Cao, R.F.; Yang, W.; Yeoh, G.H.; Timchenko, V. Computational Study of Wet Steam Flow to Optimize Steam Ejector Efficiency for Potential Fire Suppression Application. *Appl. Sci.* **2019**, *9*, 1486. [CrossRef]

3. Wu, H.; Liu, Z.; Han, B.; Li, Y. Numerical investigation of the influences of mixing chamber geometries on steam ejector performance. *Desalination* **2014**, *353*, 15–20. [CrossRef]

4. Sumeru, K.; Martin, L.; Ani, F.N.; Nasution, H. Energy savings in air conditioning system using ejector: An overview. *Appl. Mech. Mater.* **2014**, *493*, 93–98. [CrossRef]

5. Hamner, R.M. An Alternate Source of Cooling- the Ejector-Compression Heat-Pump. *Ashrae J.-Am. Soc. Heat. Refrig. Air-Cond. Eng.* **1980**, *22*, 62–66.

6. Passengers, I. Determination of the ejector dimensions of a bus air-conditioning system using analytical and numerical methods. *Appl. Therm. Eng.* **2015**, *90*, 110–119.

7. Chunnanond, K.; Aphornratana, S. Ejectors: Applications in refrigeration technology. *Renew. Sustain. Energy Rev.* **2004**, *8*, 129–155. [CrossRef]

8. Poirier, M.; Giguère, D.; Sapoundjiev, H. Experimental parametric investigation of vapor ejector for refrigeration applications. *Energy* **2018**, *162*, 1287–1300. [CrossRef]

9. Haghparast, P.; Sorin, M.V.; Nesreddine, H. The impact of internal ejector working characteristics and geometry on the performance of a refrigeration cycle. *Energy* **2018**, *162*, 728–743. [CrossRef]

10. Sriveerakul, T.; Aphornratana, S.; Chunnanond, K. Performance prediction of steam ejector using computational fluid dynamics: Part 1. Validation of the CFD results. *Int. J. Therm. Sci.* **2007**, *46*, 812–822. [CrossRef]

11. Wang, X.; Dong, J.; Li, A.; Lei, H.; Tu, J. Numerical study of primary steam superheating effects on steam ejector flow and its pumping performance. *Energy* **2014**, *78*, 205–211. [CrossRef]

12. Riffat, S.B.; Jiang, L.; Gan, G. Recent development in ejector technology—A review. *Int. J. Ambient Energy* **2005**, *26*, 13–26. [CrossRef]

13. Stoecker, W.F. *Steam-Jet Refrigeration*; McGraw-Hill: Boston, MA, USA, 1958.

14. Chen, J.; Jarall, S.; Havtun, H.; Palm, B. A review on versatile ejector applications in refrigeration systems. *Renew. Sustain. Energy Rev.* **2015**, *49*, 67–90. [CrossRef]

15. Besagni, G.; Mereu, R.; Inzoli, F. Ejector refrigeration: A comprehensive review. *Renew. Sustain. Energy Rev.* **2016**, *53*, 373–407. [CrossRef]

16. Sanaye, S.; Emadi, M.; Refahi, A. Thermal and economic modeling and optimization of a novel combined ejector refrigeration cycle. *Int. J. Refrig.* **2019**, *98*, 480–493. [CrossRef]

17. Ahmadzadeh, A.; Salimpour, M.R.; Sedaghat, A. Thermal and exergoeconomic analysis of a novel solar driven combined power and ejector refrigeration (CPER) system. *Int. J. Refrig.* **2017**, *83*, 143–156. [CrossRef]

18. Li, A.; Yeoh, G.H.; Timchenko, V.; Yuen, A.C.Y.; Wang, X. Numerical Study of Condensation Effect on a Steam Ejector by Wet Steam Model. In Proceedings of the Thirteenth International Conference on Flow Dynamics (ICFD2016), Sendai, Japan, 10–12 October 2016; Volume 12, pp. 492–493.

19. Yang, Y.; Zhu, X.; Yan, Y.; Ding, H.; Wen, C. Performance of supersonic steam ejectors considering the nonequilibrium condensation phenomenon for e ffi cient energy utilisation. *Appl. Energy* **2019**, *242*, 157–167. [CrossRef]

20. Zhang, G.; Zhang, X.; Wang, D.; Jin, Z.; Qin, X. Performance evaluation and operation optimization of the steam ejector based on modified model. *Appl. Therm. Eng.* **2019**, *163*, 114388. [CrossRef]

21. Wang, X.; Dong, J.; Zhang, G.; Fu, Q.; Li, H.; Han, Y.; Tu, J. The primary pseudo-shock pattern of steam ejector and its influence on pumping efficiency based on CFD approach. *Energy* **2019**, *167*, 224–234. [CrossRef]

22. Han, Y.; Wang, X.; Sun, H.; Zhang, G.; Guo, L.; Tu, J. CFD simulation on the boundary layer separation in the steam ejector and its influence on the pumping performance. *Energy* **2019**, *167*, 469–483. [CrossRef]

23. Glushniova, A.V.; Saveliev, A.S.; Son, E.E.; Tereshonok, D.V. Shock wave-boundary layer interaction on the non-adiabatic ramp surface. *High Temp.* **2014**, *52*, 220–224. [CrossRef]

24. Bartosiewicz, Y.; Aidoun, Z.; Mercadier, Y. Numerical assessment of ejector operation for refrigeration applications based on CFD. *Appl. Therm. Eng.* **2006**, *26*, 604–612. [CrossRef]

25. Zhu, Y.; Jiang, P. Experimental and analytical studies on the shock wave length in convergent and convergent-divergent nozzle ejectors. *Energy Convers. Manag.* **2014**, *88*, 907–914. [CrossRef]

26. Zhu, Y.; Cai, W.; Wen, C.; Li, Y. Shock circle model for ejector performance evaluation. *Energy Convers. Manag.* **2007**, *48*, 2533–2541. [CrossRef]

27. Varga, S.; Oliveira, A.C.; Diaconu, B. Numerical assessment of steam ejector efficiencies using CFD. *Int. J. Refrig.* **2009**, *32*, 1203–1211. [CrossRef]

28. Jeong, H.; Utomo, T.; Ji, M.; Lee, Y.; Lee, G.; Chung, H. CFD analysis of flow phenomena inside thermo vapor compressor influenced by operating conditions and converging duct angles. *J. Mech. Sci. Technol.* **2009**, *23*, 2366–2375. [CrossRef]

29. Ruangtrakoon, N.; Thongtip, T.; Aphornratana, S.; Sriveerakul, T. CFD simulation on the effect of primary nozzle geometries for a steam ejector in refrigeration cycle. *Int. J. Therm. Sci.* **2013**, *63*, 133–145. [CrossRef]

30. Sankarlal, T.; Mani, A. Experimental studies on an ammonia ejector refrigeration system. *Int. Commun. Heat Mass Transf.* **2006**, *33*, 224–230. [CrossRef]

31. Ramesh, A.S.; Joseph Sekhar, S. Experimental Studies on the Effect of Suction Chamber Angle on the Entrainment of Passive Fluid in a Steam Ejector. *J. Fluids Eng.* **2017**, *140*, 011106. [CrossRef]

32. Ramesh, A.S.; Sekhar, S.J. Experimental and numerical investigations on the effect of suction chamber angle and nozzle exit position of a steam-jet ejector. *Energy* **2018**, *164*, 1097–1113. [CrossRef]

33. Fu, W.; Liu, Z.; Li, Y.; Wu, H.; Tang, Y. Numerical study for the influences of primary steam nozzle distance and mixing chamber throat diameter on steam ejector performance. *Int. J. Therm. Sci.* **2018**, *132*, 509–516. [CrossRef]

34. Bartosiewicz, Y.; Aidoun, Z.; Desevaux, P.; Mercadier, Y. Numerical and experimental investigations on supersonic ejectors. *Int. J. Heat Fluid Flow* **2005**, *26*, 56–70. [CrossRef]

35. Ariafar, K.; Buttsworth, D.; Al-Doori, G.; Sharifi, N. Mixing layer effects on the entrainment ratio in steam ejectors through ideal gas computational simulations. *Energy* **2016**, *95*, 380–392. [CrossRef]

36. Sharifi, N.; Sharifi, M. Reducing energy consumption of a steam ejector through experimental optimization of the nozzle geometry. *Energy* **2014**, *66*, 860–867. [CrossRef]

37. Matsuo, S.; Kanesaki, K.; Nagao, J.; Khan, M.T.I.; Setoguchi, T.; Kim, H.D. Effects of Supersonic Nozzle Geometry on Characteristics of Shock Wave Structure. *Open J. Fluid Dyn.* **2012**, *2*, 181–186. [CrossRef]

38. García Del Valle, J.; Sáiz Jabardo, J.M.; Castro Ruiz, F.; San José Alonso, J. A one dimensional model for the determination of an ejector entrainment ratio. *Int. J. Refrig.* **2012**, *35*, 772–784. [CrossRef]

39. Sriveerakul, T.; Aphornratana, S.; Chunnanond, K. Performance prediction of steam ejector using computational fluid dynamics: Part 2. Flow structure of a steam ejector influenced by operating pressures and geometries. *Int. J. Therm. Sci.* **2007**, *46*, 823–833. [CrossRef]

40. Pianthong, K.; Seehanam, W.; Behnia, M.; Sriveerakul, T.; Aphornratana, S. Investigation and improvement of ejector refrigeration system using computational fluid dynamics technique. *Energy Convers. Manag.* **2007**, *48*, 2556–2564. [CrossRef]

41. Eames, I.W.; Aphornratana, S.; Haider, H. A theoretical and experimental study of a small-scale steam jet refrigerator. *Int. J. Refrig.* **1995**, *18*, 378–386. [CrossRef]

42. Yuen, A.C.Y.; Yeoh, G.H.; Cheung, S.C.P.; Chan, Q.N.; Chen, T.B.Y.; Yang, W.; Lu, H. Numerical study of the development and angular speed of a small-scale fire whirl. *J. Comput. Sci.* **2018**, *27*, 21–34. [CrossRef]

43. Yuen, A.C.Y.; Chen, T.B.Y.; Yeoh, G.H.; Yang, W.; Cheung, S.C.P.; Cook, M.; Yu, B.; Chan, Q.N.; Yip, H.L. Establishing pyrolysis kinetics for the modelling of the flammability and burning characteristics of solid combustible materials. *J. Fire Sci.* **2018**, *36*, 494–517. [CrossRef]

44. Huang, B.J.; Chang, J.M.; Wang, C.P.; Petrenko, V.A. 1-D analysis of ejector performance. *Int. J. Refrig.* **1999**, *22*, 354–364. [CrossRef]

45. Selvaraju, A.; Mani, A. Analysis of a vapour ejector refrigeration system with environment friendly refrigerants. *Int. J. Therm. Sci.* **2004**, *43*, 915–921. [CrossRef]

46. Sharifi, N.; Boroomand, M.; Sharifi, M. Numerical assessment of steam nucleation on thermodynamic performance of steam ejectors. *Appl. Therm. Eng.* **2013**, *52*, 449–459. [CrossRef]

47. Sharifi, N.; Boroomand, M.; Kouhikamali, R. Wet steam flow energy analysis within thermo-compressors. *Energy* **2012**, *47*, 609–619. [CrossRef]

48. Ariafar, K.; Buttsworth, D.; Al-Doori, G.; Malpress, R. Effect of mixing on the performance of wet steam ejectors. *Energy* **2015**, *93*, 2030–2041. [CrossRef]

49. Aphornratana, S.; Eames, I.W. A small capacity steam-ejector refrigerator: Experimental investigation of a system using ejector with movable primary nozzle. *Int. J. Refrig.* **1997**, *20*, 352–358. [CrossRef]

50. Yuen, A.C.Y.; Yeoh, G.H.; Timchenko, V.; Chen, T.B.Y.; Chan, Q.N.; Wang, C.; Li, D.D. Comparison of detailed soot formation models for sooty and non-sooty flames in an under-ventilated ISO room. *Int. J. Heat Mass Transf.* **2017**, *115*, 717–729. [CrossRef]

51. Yuen, A.C.Y.; Yeoh, G.H.; Timchenko, V.; Cheung, S.C.P.; Barber, T.J. Importance of detailed chemical kinetics on combustion and soot modelling of ventilated and under-ventilated fires in compartment. *Int. J. Heat Mass Transf.* **2016**, *96*, 171–188. [CrossRef]

52. Shih, T.H.; Liou, W.W.; Shabbir, A.; Yang, Z.; Zhu, J. A new κ-ε eddy viscosity model for high reynolds number turbulent flows. *Comput. Fluids* **1995**, *24*, 227–238. [CrossRef]

53. Yang, X.; Long, X.; Yao, X. Numerical investigation on the mixing process in a steam ejector with different nozzle structures. *Int. J. Therm. Sci.* **2012**, *56*, 95–106. [CrossRef]

54. Yuen, A.C.Y.; Yeoh, G.H. Numerical Simulation of an Enclosure Fire in a Large Test Hall. *Comput. Therm. Sci.* **2013**, *5*, 459–471. [CrossRef]
55. Yuen, A.C.Y.; Yeoh, G.H.; Timchenko, V.; Cheung, S.C.P.; Chan, Q.N.; Chen, T. On the influences of key modelling constants of large eddy simulations for large-scale compartment fires predictions. *Int. J. Comput. Fluid Dyn.* **2017**, *31*, 324–337. [CrossRef]
56. Dvorak, V.; Safarik, P. Transonic instability in entrance part of mixing chamber of high-speed ejector. *J. Therm. Sci.* **2005**, *14*, 258–264. [CrossRef]
57. Monday, J.T.; Bagster, D.F. A New Ejector Theory Applied to Steam Jet Refrigeration. *Ind. Eng. Chem. Process Des. Dev.* **1977**, *16*, 442–449. [CrossRef]

*applied
sciences*

Article

A Steam Ejector Refrigeration System Powered by Engine Combustion Waste Heat: Part 2. Understanding the Nature of the Shock Wave Structure

Yu Han [1,2], Xiaodong Wang [1,*], Lixin Guo [1], Anthony Chun Yin Yuen [2], Hengrui Liu [2], Ruifeng Cao [2], Cheng Wang [2], Cuiling Li [1], Jiyuan Tu [3] and Guan Heng Yeoh [2,4]

1 School of Mechanical Engineering & Automation, Northeastern University, Shenyang 110819, China; yu.han4@student.unsw.edu.au (Y.H.); Lixinguo.neu@outlook.com (L.G.); clli@mail.neu.edu.cn (C.L.)
2 School of Mechanical and Manufacturing Engineering, University of New South Wales, NSW 2052, Australia; c.y.yuen@unsw.edu.au (A.C.Y.Y.); h.liu@unsw.edu.au (H.L.); ruifeng.cao@unsw.edu.au (R.C.); c.wang@unsw.edu.au (C.W.); g.yeoh@unsw.edu.au (G.H.Y.)
3 School of Engineering, RMIT University, Melbourne, VIC 3083, Australia; jiyuan.tu@rmit.edu.au
4 Australian Nuclear Science and Technology Organization (ANSTO), Locked Bag 2001, Kirrawee DC, NSW 2232, Australia
* Correspondence: xdwang@mail.neu.edu.cn; Tel.: +86-24-8368-7618

Received: 17 September 2019; Accepted: 16 October 2019; Published: 19 October 2019

Abstract: In general, engine fuel combustion generates 30% waste heat, which is disposed to the environment. The use of the steam ejector refrigeration to recycle the waste heat and transfer them to useful energy source could be an environmentally friendly solution to such an issue. The steam ejector is the main component of the ejector refrigeration system, which can operate at a low-temperature range. In this article, the internal shock wave structure of the ejector is comprehensively studied through the computation fluid dynamics (CFD) approach. The shock wave structure can be subdivided into two regions: firstly the pseudo-shock region consisting of shock train and co-velocity region; secondly the oblique-shock region composed of a single normal shock and a series of oblique shocks. The effect of the shock wave structure on both pumping performance and the critical back pressure were investigated. Numerical predictions indicated that the entrainment ratio is enhanced under two conditions including (i) a longer pseudo-shock region and (ii) when the normal shock wave occurs near the outlet. Furthermore, the system is stabilized as the back pressure and its disturbance is reduced. A critical range of the primary fluid pressure is investigated such that the pumping is effectively optimized.

Keywords: combustion waste heat; steam ejector; shock wave; entrainment ratio; pumping performance; critical back pressure

---

1. Introduction

The steam ejector refrigeration system has been well-recognized to be a very promising device in saving energy and reducing and environmental pollution [1]. The steam ejector refrigeration system was firstly proposed by Maurice Leblanc in 1910. Even since there was a gradual decline in research attention due to its low cooling efficiency [2]. As an important means of transportation in today's social development, the energy consumption and environmental pollution of automobiles have always been the concern of researchers. The traditional compression refrigeration systems may not meet the requirements of increasing awareness and pressure to protect the environment. Recently, owing to the low-grade energy (i.e., waste heat) from the engine combustion waste heat can be used as power consumption in the ejector refrigeration system, steam ejector refrigeration systems have regained its research interests in particularly on the optimization of energy conversion [3]. As the main component of the refrigeration system, the steam ejector is designed to be simply structured without any rotational parts, for the purpose of stable operation, long service life and low maintenance cost [4]. The ejector is mainly composed of a nozzle, a mixing section, a constant-area section and a subsonic diffuser section [5]. The efficiency of the ejector is affected by both the geometric and operating parameters. Nevertheless, the further improvement of the ejector pumping efficiency requires a deeper understanding of its internal flow structure and the fluid characteristics [6].

The research on the ejector is mainly carried out from two aspects: experiment and numerical simulation. Numerous experimental efforts on the ejector refrigeration system have been made to enhance the understanding of the flow structure, thus enhancing the design of the system. The choking of the secondary fluid in the mixing chamber had an important influence on the pumping performance of the ejector, and the maximum COP(the coefficient of performance) is obtained under the critical flow condition studied by Eames et al. [2,7,8]. These important conclusions were the theoretical foundations of future research on the steam ejector. Chen et al. [6] established a steam ejector refrigeration system to evaluate the performance of ejectors under different operating. The corresponding empirical equations were deduced according to the experimental results. Chunnanond et al. [9] measured the pressure along the wall of the ejector to examine the influence of operating and geometry parameters on the performance of the ejector. By analysing the experimental results, the expansion angle, effective area and shock wave were obtained to explain the flow characteristics of the two fluids and the mixing process inside the ejector. A solar-powered ejector refrigeration system with a movable spindle at the inlet of the primary nozzle was built to adjust the pumping performance of the ejector by changing the pressure of the boiler [10–12]. It was discovered that the position of the spindle greatly influences the entertainment ratio and the critical back pressure.

Similarly, the variable-area ejector refrigeration system was established by Chandra and Ahmed [13] in 2014. The results indicated that the variable-area ejector eliminated the loss of the total pressure caused by the shock wave and the efficiency of the system was improved. Ramesh and Sekhar [14,15] analysed the effect of suction chamber angle and different NXP (i.e., nozzle exit position) on the entrainment ratio of the steam ejector using the experimental method. The conclusion was that the entrainment ratio increased with the increasing of the suction chamber angle and decreased with a decline of the NXP. Li et al. [16] established an experimental device of the ejector refrigeration system using R134a as the refrigerant to investigate the influence of the operating parameters and the area ratio of the mixing chamber on the performance of the ejector. Dong et al. [1,17] conducted different experiments under three nozzle ejectors with a different cross-section area. A certain structure of a steam ejector was found to enable the ejector refrigeration system work in a heat source temperature between 40 °Cand 70 °C. Śmierciew et al. [18] established an experimental ejector refrigeration system driven by low-grade energy of 55 °C using HFO-1234ze(E), as the working fluid.

Computational fluid dynamics (CFD) method is widely used in the study of the steam ejector owing to its comprehensive analytical potential benefit by the visualization [19], as well as the advantages of prediction for the flow dynamic behavior [20]. It not only can predict and reflect the flow characteristics inside the ejector but also can estimate the pumping efficiency of the ejector. Besides, the complex and non-linear turbulent effect on the fluid structure can be characterized to evaluate the mixing process and the wall shearing behaviours [21–23]. The simulation study of geometric parameters were mainly in terms of nozzle structure [24], the area ratio of nozzle throat to constant-area section [25,26], the NXP (nozzle exit position) [27–29], diffusion angle [30] and the structure of mixing chamber [31]. The influence of five different nozzle geometry on the flow characteristics inside the ejector was studied by Yang et al. [32]. The results showed that the increase of ER can be achieved by increasing the exit perimeter and the special geometric structure. Ruangtrakoon et al. [33] discussed the effect of eight different primary nozzle structure on the ejector refrigeration performance. Riffat and

Omer [34] analysed the influence of the NXP on the performance of the ejector and obtained the optimal geometry of the ejector using the methanol as the working fluid. The variation of pumping performance with a variable primary nozzle geometry was discussed by Varga et al. [11]. Sriveerakul et al. [35,36], Yuan et al. [26] and Aly et al. [37] comprehensively analysed and discussed the variation of the pumping performance under different geometric and operating parameters. A better understanding of the mixing process of the working steam and the pumped steam under different back pressure was studies by Su and Agarwal [38]. According to the work of Dong et al. [27], a higher COP can be obtained by decreasing the temperature of primary steam and raising the temperature of secondary steam.

In Part 1, the flow structure characteristics of the entire fluid in the steam ejector were analysed meanwhile, the correctness and applicability of the CFD model were verified. Since the primary fluid stream drives additional flows via the secondary inlet, it significantly improves the overall momentum across the ejector and outlet thrust. With the utilization of pressurized steam as the primary fluid, such a system can be potentially applied for automobile cooling system. Nonetheless, the optimization of the pumping efficiency of the steam ejector has always been the major consideration in the design of ejector refrigeration systems. Different from other devices with one converging-diverging structure, there were two converging-diverging structures in a typical steam ejector, one was in the nozzle, and the other was in the subsonic diffuser. The primary fluid is accelerated into the supersonic fluid in the primary nozzle. The primary fluid is mixed with the suction steam and discharged into the condenser through the pressure difference in the subsonic diffuser. Due to the unique, plain structure of the ejector, the flow structure of the fluid in the ejector is also complex, such as transonic flow [15], choking flow [39], shock train [40], boundary layer separation [41], shock-mixing layer [42] and the pseudo shock [40,43], etc. Shock waves are generated during the transonic fluid flow. When a shock wave contacts with the boundary layer, a shock-mixing layer structure will be formed. Simultaneously, the shock wave would exert a strong pressure gradient on the boundary layer that can result in the boundary layer thicker to separate from the wall and to increase the fluid viscous dissipation. Though the shock wave is a common structure occurring in different fluid flows, the influence on the steam ejector cannot be ignored because the existence of the shock wave makes the flow structure of the ejector more complex and varied. However, there are few works focused on analysing the whole shock wave structure inside the ejector comprehensively. The better understanding of the shock wave structure would be helpful to improve the ejector efficiency and optimize the geometric structure of the ejector. In enlighten of the current knowledge gap, the shock wave structure of flow structure inside the steam ejector is entirely studied through the numerical simulation method in this article. The shock wave structure is identified and sub-divided into two regions based on physical observation of the flow via simulation results. Three types of the shock waves inside the steam ejector are classified and defined. The influence of the shock wave structure on the pumping performance of the steam ejector under different primary fluid pressure is discussed.

2. Numerical Algorithms

2.1. Governing Equations

The flow of the steam ejector is controlled by the conservation equation of the compressible steady-state asymmetric fluid flow [44]. Typically, the Navier-Stokes equation can be applied for thermally driven variable-density flow, as the turbulent interactions are conveniently accounted in the momentum and energy equations [45,46]. The total energy equation including viscous dissipation and the set coupled with the ideal gas law [47]. The thermodynamic and transport properties of steam remain unchanged during validation. The governing equations of the continuity equation, the momentum equation and the energy equation can be written as follows:

The continuity equation:

$$\frac{\partial \rho}{\partial t} + \frac{\partial}{\partial x_i}(\rho u_i) = 0 \tag{1}$$

The momentum equation:

$$\frac{\partial}{\partial t}(\rho u_i) + \frac{\partial}{\partial x_j}(\rho u_i u_j) = -\frac{\partial P}{\partial x_i} + \frac{\partial \tau_{ij}}{\partial x_j} \tag{2}$$

The energy equation:

$$\frac{\partial}{\partial t}(\rho E) + \frac{\partial}{\partial x_i}(u_i(\rho E + P)) = \vec{\nabla} \cdot (\alpha_{eff}\frac{\partial T}{\partial x_i}) + \vec{\nabla} \cdot (u_j(\tau_{ij})) \tag{3}$$

where

$$\tau_{ij} = \mu_{eff}(\frac{\partial u_i}{\partial x_j} + \frac{\partial u_j}{\partial x_i}) - \frac{2}{3}\mu_{eff}\frac{\partial u_k}{\partial x_k}\delta_{ij} \tag{4}$$

With

$$\rho = \frac{P}{RT} \tag{5}$$

where τ_{ij} is the stress tensor, E is the total energy, α_{eff} is the effective thermal conductivity, and μ_{eff} is the effective molecular dynamic viscosity.

2.2. Geometry and Mesh Approach

The geometry of the ejector is derived from the experimental system described in Part 1. In order to reduce the calculation cost and running time, a two-dimensional axisymmetric model is adopted in this work. The geometric parameters of the ejector are shown in Table 1.

Table 1. Geometrical parameters of the steam ejector.

Geometrical Parameters	Value
Diameter of the nozzle inlet	12 mm
Diameter of the nozzle outlet	11 mm
Diameter of the nozzle throat	2.5 mm
Expand the angle of the nozzle	10°
Nozzle exit position	10 mm
Diameter of the mixing chamber inlet	70 mm
Diameter of the throat	28 mm
Length of the mixing chamber	122.2 mm
Length of the throat	90 mm
Length of the subsonic diffuser	210 mm

The structured meshing of the internal flow field of the ejector is carried out. Encryption is carried out in the region with high local speed. The specific grid structure after adaptive technology is shown in Figure 1. The grid quality is above 0.9, and the maximum aspect ratio is 5:1. The grid quality is high enough to meet the requirements of numerical simulation. By comparing the Mach number distribution curves of the centerline under different grid densities in Figure 2, the variation trend of the medium grid and the fine grid is almost the same. The Mach number has reached mesh independence by using a medium grid. Therefore, the medium grid is adopted in the subsequent simulation process with the grid number of 47,562 units.

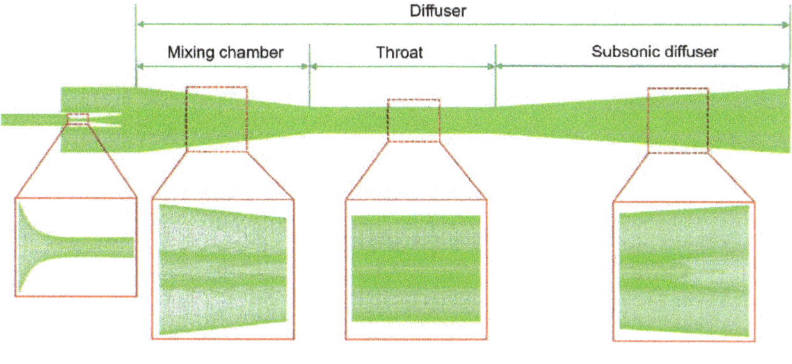

Figure 1. The mesh structure of the steam ejector.

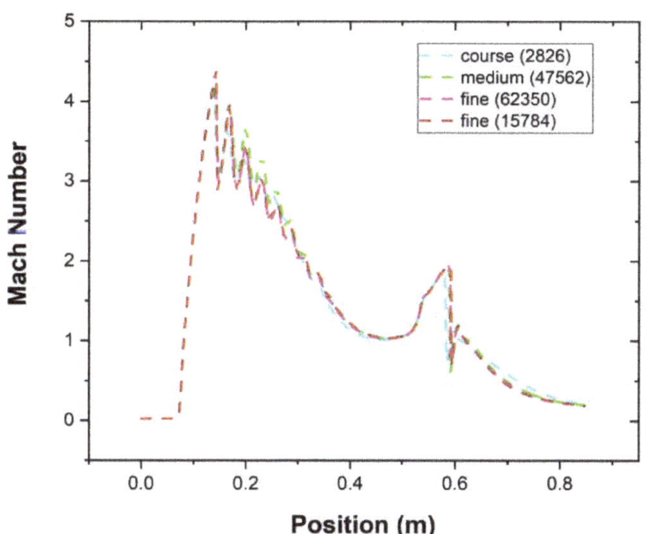

Figure 2. Grid independence verification.

2.3. Numerical Solution Procedure

The computational fluid dynamics software FLUENT19.2 based on ANSYS was used for simulation analysis. As mentioned in Part 1, with the application of the segregated solver and the pressure is simulated by pressure velocity coupling approaches [48,49]. The coupled implicit solver is used, and the diffusion term is discretized by the central difference scheme, to enhance the thermal and molecular diffusion for thermally driven reacting flows [45]. The field variables values are closed by Gauss-Seidel iteration method and subsequentially march in the time by multiple Runge-Kutta explicit schemes. All the convection terms are discretized using the second-order upwind scheme. The k-w SST turbulence model is applied for the consideration of turbulent interaction with the flow field, which can predict the complex flow in the ejector. The wall is considered as adiabatic isentropic flow with no-slip condition. To include the wall damping effect on turbulence, the near-wall treatment method uses the enhanced wall function with y < 0.893 [50,51]. The turbulence intensity at the entrance is 5%. The inlet and outlet are set as pressure inlet and pressure outlet, respectively. Besagni G et al. [41,52,53] have proved that the density of the vapour is not much different from the real gas model to the ideal gas model. Therefore, the ideal gas model is used to simplify the problem. The properties of water vapour are shown in Table 2. The operating conditions of the numerical simulations are displayed in Table 3.

Table 2. Properties of the working fluid used in the simulation.

Properties	Dynamic Viscosity	Thermal Conductive	Specific Heat Capacity	Molecular Weight
Value	1.34×10^{-5} $\text{kgm}^{-1}\text{s}^{-1}$	0.00261 $\text{Wm}^{-1}\text{K}^{-1}$	2014.00 J $\text{kg}^{-1}\text{K}^{-1}$	18.01534 kgkmol^{-1}

Table 3. The operating conditions of the numerical simulation.

Operating Condition	Value
Primary fluid pressure	360,000 Pa
Secondary fluid pressure	2330 Pa
Back pressure	3500 Pa

3. Results and Discussion

3.1. Structural Division of Shock Wave Structure in the Ejector

The efficiency of the ejector is mainly determined by the degree of the mixing process between the primary fluid and the secondary fluid, as discussed in Part 1. Another important aspect affecting the pumping performance of the ejector is the complex flow characteristics of the steam ejector produced. Among these flow characteristics, the shock wave structure is extremely complex and its structure is one of the factors to determine the pumping performance of the steam ejector [40]. However, there is still a less thorough understanding of the whole shock wave structures in detail. This section will analyse and discuss the shock waves inside the steam ejector.

The Mach number contour of the flow field and the static pressure distribution curve of the wall and the axis of the ejector are shown in Figure 3. The whole shock wave structure in the steam ejector can be clearly observed. The static pressure at the axis of the mixing chamber fluctuates up and down due to the existence of a series of shock waves in the mixing chamber and the throat sections. The zone ranges from the beginning point A of the shock wave to the point B where the two static pressure curves coincide is called the shock train region which is separated by the choking position of the secondary fluid. The range of this region is the length of the primary fluid jet core as described in Part 1. The pressure along the wall of the pseudo-shock region changes relatively smoothly, but the pressure along the axis continues to oscillate and its amplitude gradually decreases until the point B. The zone which ranges from the first coincidence point A to the second coincidence point C is called the co-velocity region. The pressures and the velocities are constant and the two streams begin to mix with each other. The shock train and the co-velocity region are collectively referred to as a pseudo shock region [43] while the oblique-shock region consisting of a single normal shock wave and a series of oblique shocks [40], and the range of which is from the position of the normal shock wave to the coincidence point C.

From the Mach number cloud diagram of the steam ejector in Figure 4a, it can be clearly seen that three types of shock waves exit in the whole flow field. One type is the shock train consisting of several diamond waves both in the mixing chamber and the throat sections as displayed in Part 1, the other type is a single inviscid normal shock wave in the subsonic diffuser section. The last type is the oblique shock following with the normal shock. Figure 4a displays the Mach number counter of the ejector when the primary fluid pressure, secondary fluid pressure and the back pressure are 0.36 MPa, 2330 Pa and 3500 Pa, respectively. As indicated by the Mach number distribution along the radial distance shown in Figure 4b, the larger change of the velocity gradient begins at 0.50m distance along the wall. When the distance is greater than 0.50 m on the X-axis, greater velocity fluctuation occurs near the wall showing that the normal shock wave starts to change into the oblique shock waves [40].

The oblique shock waves are generated after the 0.5 m, which corresponds to the Mach number contour in Figure 4a. The intensity of oblique shock wave reaches its maximum at 0.51 m and keeps constant between 0.52 m and 0.54 m. The results show that the bifurcation structure is generated at the root of the positive shock with the decrease of the Mach number values. Then, the oblique shock is formed. The energy of the primary jet core decreases which resulting in the separation degree of the boundary layer becomes stronger, and the mixing degree of the two streams is sufficiently affected. The efficiency of the steam ejector will decline with the non-fully mixing process between the two streams.

Figure 5 shows the Mach number distribution curve along the ejector axis. The Mach number values fluctuate continuously. There are six peak values and six trough values in the shock train region. However, there is only one inviscid normal shock wave in the subsonic diffuser section with only one peak value. It shows that the shock train is composed of the bifurcation shock waves causing the fluid oscillations. The concrete manifestation is that the static pressure along the axis drops first and then remains constant while the velocity increases first and then decreases, as shown in Figures 3 and 5. That indicates the shock train plays an important role in restoring pressure in the case of some energy loss. Therefore, the shock train area can be judged and defined according to the fluctuation range and oscillation amplitude of the pressure or the velocity.

In the subsonic diffuser section, a single normal shock wave appears near the outlet of the throat section, which is generally accompanied by the transonic behaviour and followed by a subsonic region. The normal shock wave can prevent the disturbance caused by the change of the back pressure propagating upstream. When the position of the normal shock wave is closer to the outlet of the subsonic diffuser, the critical back pressure is higher, the influence of the back pressure is less and the system operation will be more stable. However, a few oblique shock waves with small amplitude following by the normal shock wave are generated, which would reduce the energy exchange of the mixing process between the two streams [54]. The oblique shock region composing of the normal shock wave and a series of the bifurcated shock waves are different from the pseudo-shock region composed by the shock train and the co-velocity region. The former will prevent the disturbances caused by the back pressure from propagating upstream. The latter is an important area for mixing of the two streams. Moreover, the shock wave structures are the main components of the entire flow structure inside the ejector.

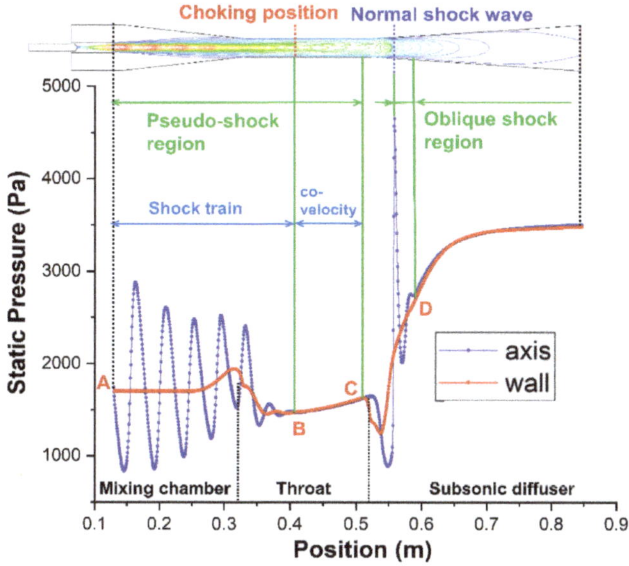

Figure 3. The structure of shock wave in the steam ejector.

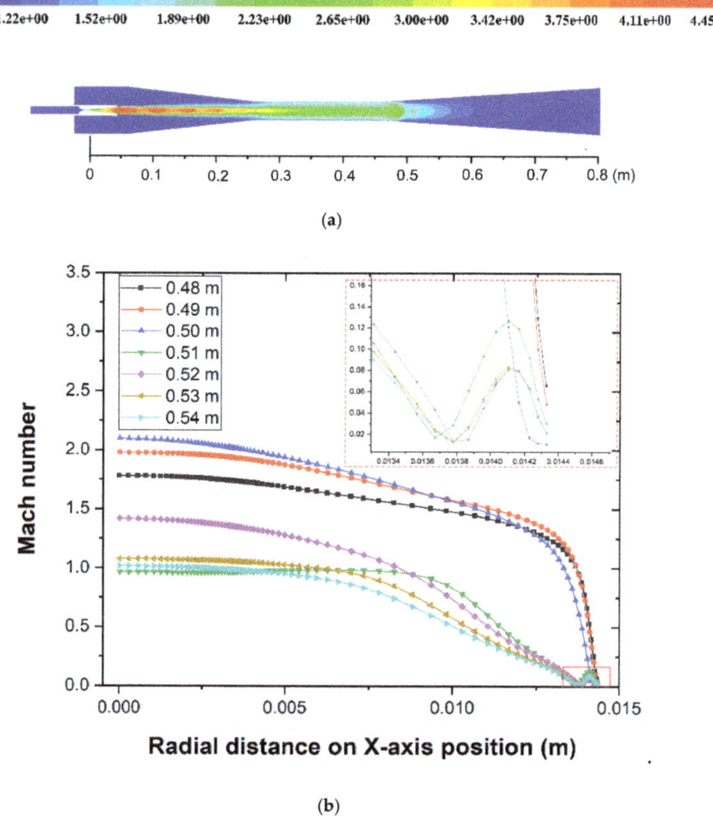

Figure 4. (**a**) The contour of Mach number at the primary fluid pressure; (**b**) Radial distance on different X-axis position.

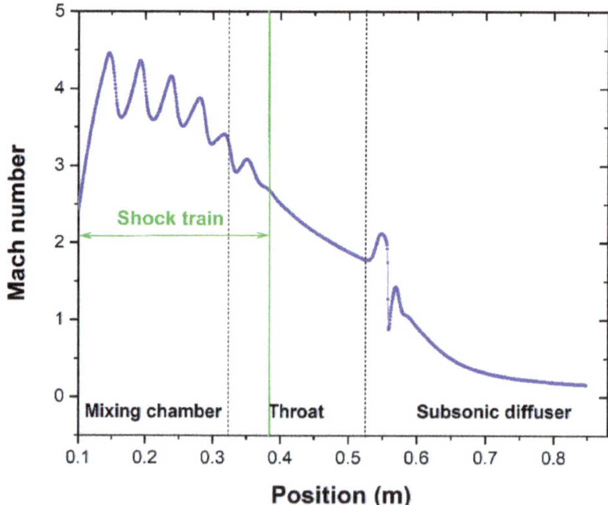

Figure 5. The distribution curve of Mach number at centerline.

3.2. Effect of the Shock Wave Structure on the Pumping Performance

Under the condition of the secondary fluid, pressure is 2330 Pa and the back pressure is 3500 Pa, the pumping performance of the steam ejector with changing of the primary fluid pressure is analysed. The contour of Mach number above 1 under different primary fluid pressure is shown in Figure 6. The structure and characteristics of the shock wave can be clearly observed. An oblique-shock region consisting of a normal shock wave and a series of bifurcated shock waves represented in the dashed box. The velocity of the mixed fluid suddenly drops from supersonic to subsonic due to the existence of the normal shock wave. To accurately capture the shock wave position, only fluids above 1 Mach number are displayed and the subsonic part would not appear. It can be seen that the higher the primary fluid pressure is, the closer the normal shock wave is to the outlet of the ejector. While the length of the oblique-shock in the subsonic diffuser section increases with the increasing of the primary fluid pressure. The flow region in the throat section where the fluid velocity is higher than 1 Mach number expands. The length of the pseudo-shock strain is elongated with the increase of the primary fluid pressure when the primary fluid pressure is less than 0.34 MPa. When the pressure is more than 0.34 MPa, the length of the pseudo-shock region maintains constant. It shows that the energy and momentum of the primary fluid mixing with the secondary fluid continuously increases when the primary fluid pressure is less than 0.34 MPa. However, the increase of the oblique shock region consumes a part of the energy and momentum of the mixed fluid causing the energy of the mixed fluid to be reduced when it flows through this region. The shock wave structure changes significantly from the inviscid normal shock to the bifurcation oblique shock with the primary fluid pressure increasing. When the primary fluid pressure is less than or equal to 0.34 MPa, the upstream Mach number values and the intensity of the shock wave decrease. At 0.34 MPa, the velocity at the center of the shock wave is the highest. When the primary fluid pressure is more than 0.34 MPa, the upstream Mach number values of the shock wave rises significantly, and the normal shock changes to be a bifurcated oblique shock.

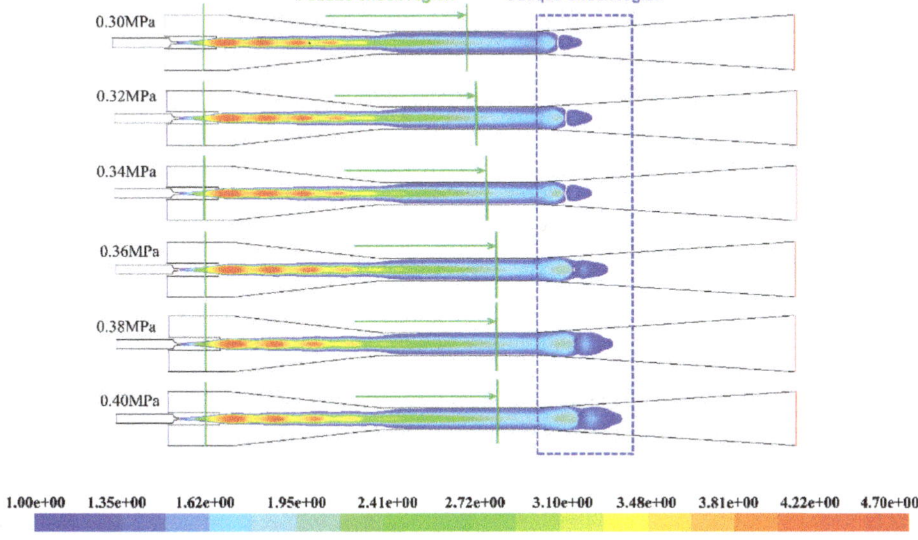

Figure 6. The influence of primary fluid pressure on the shock structure.

Figure 7 shows the Mach number distribution curve at the axis of the ejector to illustrate the influence of the oblique-shock region. The Mach number values of the fluid in the mixing chamber section is significantly higher than that in the subsonic diffuser section. When the Mach number in the subsonic diffuser section reaches 4.26, a shock train is generated. When the Mach number value is less than 1.53, a series of oblique shock waves are formed and accompanied by the transonic flow behaviour. As long as there is a normal shock wave in the subsonic diffuser section, the ejector is in the double-choked flow mode [55] with a better pumping performance. When the normal shock wave enters the throat section, the double-choked flow mode would be destroyed by the downstream back-pressure disturbance. The pumping performance of the ejector continuously decreases until it loses its function completely. It shows that the normal shock wave is of great significance in maintaining the pumping performance and the operational stability of the ejector.

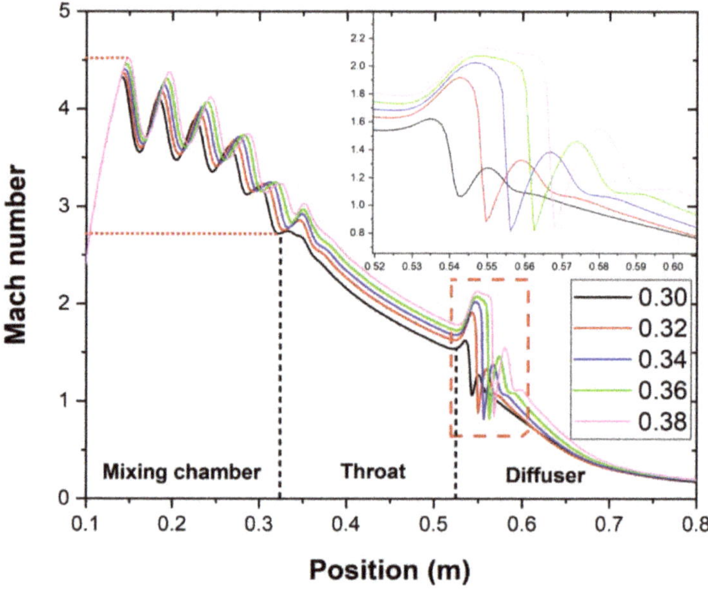

Figure 7. The curve of Mach number along the axis under different primary fluid pressure.

The variation of the entrainment ratio under different primary fluid pressure is shown in Figure 8. The experimental and simulation values of the entrainment ratio are in a good agreement. The entrainment ratio reaches the optimum value at 0.34 MPa. The entrainment ratio is more than 0.8 when the primary fluid pressure is at between 0.32 MPa and 0.36 MPa, while it is less than 0.8 when the primary pressure is outside of this range. As displayed in Figure 7, when the primary fluid pressure is between 0.32 MPa and 0.36 MPa, the Mach number values at the subsonic diffusion section is between 1.6 and 2.1. The peak Mach number value is 2.1 at 0.34 MPa. It indicates that when the primary fluid pressure is 0.34 MPa, the normal shock wave with more energy makes the mixing of two streams more completely and the mixed fluid with a higher pressure can effectively prevent the disturbance generated by the back. Simultaneously, the energy loss of the shock wave at the entrance of the subsonic diffuser section is the smallest when the primary pressure is at 0.34 MPa. As the primary fluid pressure is greater than 0.34 MPa, the front area of the normal shock wave becomes wider (in Figure 7), and the intensity of the normal shock wave becomes weak (in Figure 6). The velocity gradient generated at the wall surface will lead to the boundary layer separation. The partial energy and momentum of the mixed fluid will be lost, and the entrainment ratio will be reduced. When the primary fluid pressure is less than 0.34 MP, the mixing process of the two streams is not enough resulting in a smaller entrainment ratio because of the normal shock wave within less energy.

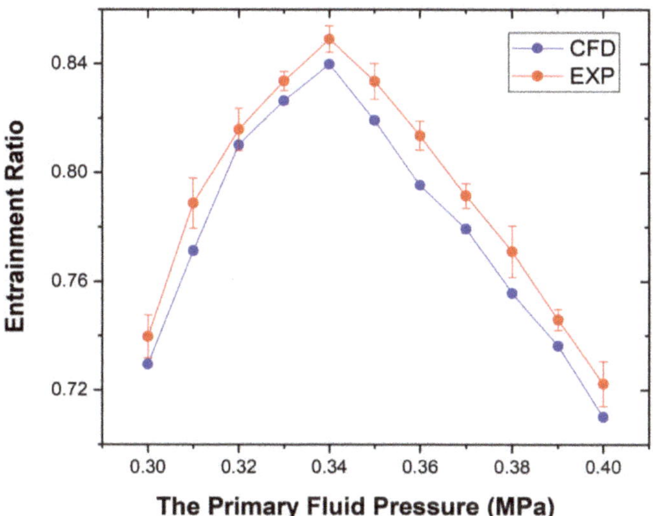

Figure 8. The variation of entrainment ratio with the primary fluid pressure.

3.3. Effect of the Shock Wave on the Critical Back Pressure

When the secondary fluid pressure and the back pressure are 1710 Pa and 3000 Pa, respectively, the variation curve of the entrainment ratio and the back pressure under different primary fluid pressure is shown in Figure 9. According to the previous analysis, the increasing of the primary fluid pressure will cause the normal shock wave moves downstream close to the outlet of the ejector. Meanwhile, the Mach number in the upstream region of the shock wave is also raised (as shown in Figure 7). The structure of the shock wave develops from the inviscid normal shock with lower intensity to the bifurcated oblique shock with higher intensity (as shown in Figure 6). The critical back pressure enlarges with the increasing of the primary fluid pressure, while the entrainment ratio declines. When the primary fluid pressure is 0.34 MPa, the critical back pressure is 3200 Pa; the critical back pressure is 6000 Pa as the primary fluid pressure is 0.46 MPa, increasing by nearly 47%. To ensure the stable operation of the experimental system, a higher primary fluid pressure should be selected as far as possible within the allowable range of the experimental equipment. When the primary fluid pressure is 0.34 MPa, the entrainment ratio is 0.84; the entrainment ratio is 0.44Pa, as the primary fluid pressure is 0.46 MPa, which decreases by nearly 48%. The critical back pressure is one of the main parameters to determine the working range of the ejector while the entrainment ratio is the parameter to determine the pumping efficiency of the ejector. In the case of ensuring the high pumping efficiency of the ejector, the primary fluid pressure can be appropriately increased to expand the operable range of the ejector.

As shown in Figure 10, when the primary fluid pressure is less than 0.40 MPa, the growth rate of critical back pressure is lower than the value when the primary fluid pressure is more than 0.40 MPa. The reason may be that the higher the primary fluid pressure is, the smaller the loss of momentum and kinetic energy produced by the mixed fluid. With the increase of the velocity and the intensity of the shock waves, the mixed fluid resisting the back pressure disturbance is improved. The analysis combined with Figures 6 and 9; the growth rate of the critical back pressure is related to the length of the pseudo-shock wave region. Moreover, the longer the pseudo-shock wave length results in the greater value of the critical back pressure. It indicates that the pseudo-shock wave region has a very significant influence on the critical back pressure. The growth rate of critical back pressure also is related to the primary fluid pressure. When the primary fluid pressure is kept in 0.32 MPa–0.40 MPa, the critical back pressure cure shows a slope of M1 = 0.0012. The slope of the critical back pressure increases to M2 = 0.0031 when the critical back pressure varies between 0.40 MPa–0.46 MPa. The growth rate of the critical back pressure at 0.40 MPa–0.46 MPa accelerates 61% comparing with the 0.32 MPa–0.40 MPa. Consequently, though the critical back pressure is not constantly increased with the primary fluid pressure at the same growth rate.

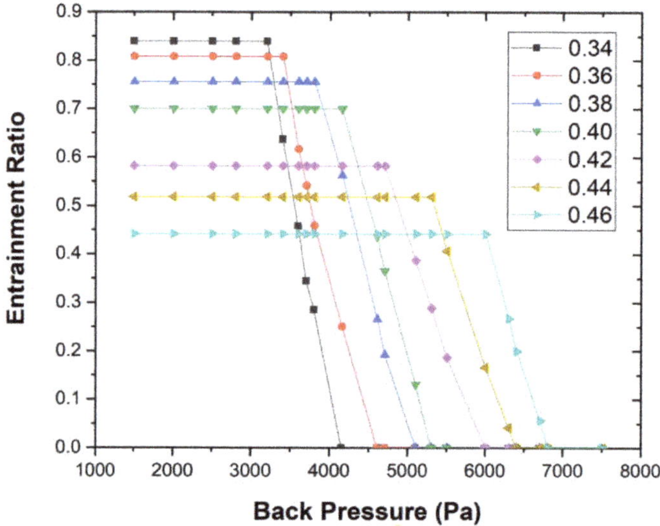

Figure 9. The relationship between entrainment ratio and back pressure under the primary fluid pressure.

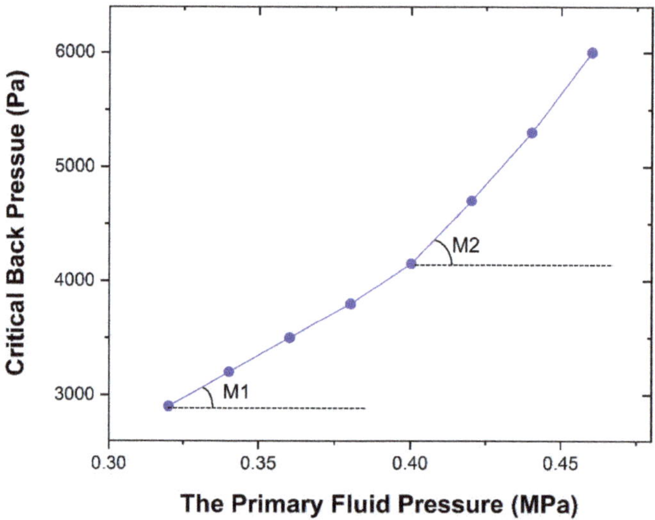

Figure 10. The influence of the primary fluid pressure on critical back pressure.

The disturbance caused by the back pressure propagates the inverse pressure gradient from the downstream to the upstream. When the reverse pressure value is lower than the critical back pressure, the shock wave will offset the disturbance caused by the reverse pressure value and decrease the energy loss. Simultaneously, with the increasing of the back pressure, the energy and intensity of the normal shock wave will decrease, the shock wave structure will be changed. Once the adverse pressure gradient value is higher than the critical back pressure value, the adverse pressure gradient will continue to move upstream and cause no normal shock wave to be generated. The inverse pressure gradient will directly enter the throat section and destroy the formation of choking flow where results in the flow passage of contraction structure not being produced, and the secondary fluid cannot be pumped into the mixing chamber. As a result, the pumping performance of the ejector will drop sharply until the ejector loses efficiency.

4. Conclusions

In this work, the engine combustion waste heat as the heat source is driven by the steam ejector refrigeration system, which can improve the energy utilization and save cost, as is proposed. The effect of the shock wave structure on the pumping performance of the steam ejector is thoroughly studied. The shock train and co-velocity region locating in the mixing chamber and throat sections is defined as the pseudo-shock region. While the other part is the oblique-shock region consisting of a single normal shock wave and a series of the oblique shock waves locating in the subsonic diffuser section. The shock train, the single normal shock wave and the oblique shock wave are the three types of shock waves, respectively. In conclusion, the main findings of this study are as follows: (i) the shock wave structure is thoroughly studied by the experimental-CFD method. An experimental system of the ejector refrigeration was established and the CFD model was verified. Based on the ideal gas model, the k-w SST turbulence model is chosen, as it can predict the complex flow in the ejector; (ii) the structure of the shock wave is divided into two regions: pseudo-shock region and the oblique-shock region. Three types of shock waves are determined as the shock train, the normal shock wave and the oblique shock wave, respectively; (iii) the length of the pseudo-shock region increases with the increasing of the primary fluid pressure when the pressure is less than 0.34 MPa, then maintains constant. Due to a part of the energy of the mixed fluid is consumed, the entrainment ratio and the pumping performance of the steam ejector decreases. The bifurcation shock waves increase, and the energy of the normal shock wave decreases with the increase of the primary fluid pressure when the primary fluid pressure is more than 0.34 MPa; (iv) as the primary fluid pressure increases, the entrainment ratio declines and the critical back pressure and the length of shock waves increasing of the steam ejector increase; (v) the idealized primary fluid pressure is between 0.32 MPa and 0.36 MPa and the idealized Mach number is between 1.6 and 2.1. The growth rate of the critical back pressure with the primary fluid pressure at 0.40 MPa–0.46 MPa accelerates 61% compared with the pressure at 0.32 MPa–0.40 MPa. With the utilization of CFD analysis, the critical operating primary fluid pressure of the steam ejector system was identified. It can enhance the effectiveness of converting combustion waste heat to recyclable and useful energy sources in a steam ejector refrigeration system. Future works can be carried out to further enhance the understanding of the shock structure via experimentation using high-speed photography techniques.

Author Contributions: Conceptualization, Y.H. and X.W.; algorithm, Y.H. and L.G.; software, Y.H., C.W., R.C. and H.L.; data curation, Y.H., A.C.Y.Y. and H.L.; formal analysis, Y.H. and A.C.Y.Y.; numerical simulations, Y.H. and X.W.; validation, A.C.Y.Y., C.L. and; writing—original draft preparation, Y.H. and X.W.; writing—review and editing, A.C.Y.Y., L.G., R.C. and H.L.; visualization, L.G. and R.C.; supervision, J.T. and G.H.Y.; funding acquisition, X.W., J.T. and G.H.Y.

Funding: This research was funded by the National Natural Science Foundation of China (51775098), the Australian Research Council (DP160101953) and the Australian Research Council Industrial Transformation Training Centre (ARC IC170100032) in the University of New South Wale. All financial and technical support are deeply appreciated by the authors.

Conflicts of Interest: The authors declare no conflict of interest.

References

1. Dong, J.; Yu, M.; Wang, W.; Song, H.; Li, C.; Pan, X. Experimental investigation on low-temperature thermal energy driven steam ejector refrigeration system for cooling application. *Appl. Therm. Eng.* **2017**, *123*, 167–176. [CrossRef]
2. Eames, I.W.; Aphornratana, S.; Haider, H. A theoretical and experimental study of a small-scale steam jet refrigerator. *Int. J. Refrig.* **1995**, *18*, 378–386. [CrossRef]

3. Besagni, G.; Mereu, R.; Inzoli, F. Ejector refrigeration: A comprehensive review. *Renew. Sustain. Energy Rev.* **2016**, *53*, 373–407. [CrossRef]

4. Wu, Y.; Zhao, H.; Zhang, C.; Wang, L.; Han, J. Optimization analysis of structure parameters of steam ejector based on CFD and orthogonal test. *Energy* **2018**, *151*, 79–93. [CrossRef]

5. Li, A.; Yuen, A.C.Y.; Chen, T.B.Y.; Wang, C.; Liu, H.; Cao, R.; Yang, W.; Yeoh, G.H.; Timchenko, V. Computational study of wet steam flow to optimize steam ejector efficiency for potential fire suppression application. *Appl. Sci.* **2019**, *9*, 1486. [CrossRef]

6. Chen, Y.M.; Sun, C.Y. Experimental Study of the Performance Characteristics of a Steam-Ejector Refrigeration System. *Exp. Therm. Fluid Sci.* **1997**, *15*, 384–394. [CrossRef]

7. Aphornratana, S.; Eames, I.W. A small capacity steam-ejector refrigerator: Experimental investigation of a system using ejector with movable primary nozzle. *Int. J. Refrig.* **1997**, *20*, 352–358. [CrossRef]

8. Eames, I.W.; Wu, S.; Worall, M.; Aphornratana, S. An experimental investigation of steam ejectors for applications in jet-pump refrigerators powered by low-grade heat. *Proc. Inst. Mech. Eng. Part A J. Power Energy* **1999**, *213*, 351–361. [CrossRef]

9. Chunnanond, K.; Aphornratana, S. An experimental investigation of a steam ejector refrigerator: The analysis of the pressure profile along the ejector. *Appl. Therm. Eng.* **2004**, *24*, 311–322. [CrossRef]

10. Ma, X.; Zhang, W.; Omer, S.A.; Riffat, S.B. Experimental investigation of a novel steam ejector refrigerator suitable for solar energy applications. *Appl. Therm. Eng.* **2010**, *30*, 1320–1325. [CrossRef]

11. Varga, S.; Oliveira, A.C.; Ma, X.; Omer, S.A.; Zhang, W.; Riffat, S.B. Comparison of CFD and experimental performance results of a variable area ratio steam ejector. *Int. J. Low-Carbon Technol.* **2011**, *6*, 119–124. [CrossRef]

12. Varga, S.; Oliveira, A.C.; Ma, X.; Omer, S.A.; Zhang, W.; Riffat, S.B. Experimental and numerical analysis of a variable area ratio steam ejector. *Int. J. Refrig.* **2011**, *34*, 1668–1675. [CrossRef]

13. Chandra, V.V.; Ahmed, M.R. Experimental and computational studies on a steam jet refrigeration system with constant area and variable area ejectors. *Energy Convers. Manag.* **2014**, *79*, 377–386. [CrossRef]

14. Ramesh, A.S.; Joseph Sekhar, S. Experimental Studies on the Effect of Suction Chamber Angle on the Entrainment of Passive Fluid in a Steam Ejector. *J. Fluids Eng.* **2017**, *140*, 011106. [CrossRef]

15. Ramesh, A.S.; Sekhar, S.J. Experimental and numerical investigations on the effect of suction chamber angle and nozzle exit position of a steam-jet ejector. *Energy* **2018**, *164*, 1097–1113. [CrossRef]

16. Li, F.; Li, R.; Li, X.; Tian, Q. Experimental investigation on a R134a ejector refrigeration system under overall modes. *Appl. Therm. Eng.* **2018**, *137*, 784–791. [CrossRef]

17. Dong, J.; Chen, X.; Wang, W.; Kang, C.; Ma, H. An experimental investigation of steam ejector refrigeration system powered by extra low temperature heat source. *Int. Commun. Heat Mass Transf.* **2017**, *81*, 250–256. [CrossRef]

18. Śmierciew, K.; Gagan, J.; Butrymowicz, D.; Łukaszuk, M.; Kubiczek, H. Experimental investigation of the first prototype ejector refrigeration system with HFO-1234ze(E). *Appl. Therm. Eng.* **2017**, *110*, 115–125. [CrossRef]

19. Wang, X.; Dong, J.; Li, A.; Lei, H.; Tu, J. Numerical study of primary steam superheating effects on steam ejector flow and its pumping performance. *Energy* **2014**, *78*, 205–211. [CrossRef]

20. Yuen, A.; Chen, T.; Yang, W.; Wang, C.; Li, A.; Yeoh, G.; Chan, Q.; Chan, M. Natural Ventilated Smoke Control Simulation Case Study Using Different Settings of Smoke Vents and Curtains in a Large Atrium. *Fire* **2019**, *2*, 7. [CrossRef]

21. Yuen, A.C.Y.; Yeoh, G.H.; Cheung, S.C.P.; Chan, Q.N.; Chen, T.B.Y.; Yang, W.; Lu, H. Numerical study of the development and angular speed of a small-scale fire whirl. *J. Comput. Sci.* **2018**, *27*, 21–34. [CrossRef]

22. Yuen, A.C.Y.; Yeoh, G.H.; Timchenko, V.; Cheung, S.C.P.; Chen, T. Study of three LES subgrid-scale turbulence models for predictions of heat and mass transfer in large-scale compartment fires. *Numer. Heat Transf. Part A Appl.* **2016**, *69*, 1223–1241. [CrossRef]

23. Chen, T.B.Y.; Yuen, A.C.Y.; Wang, C.; Yeoh, G.H.; Timchenko, V.; Cheung, S.C.P.; Chan, Q.N.; Yang, W. Predicting the fire spread rate of a sloped pine needle board utilizing pyrolysis modelling with detailed gas-phase combustion. *Int. J. Heat Mass Transf.* **2018**, *125*, 310–322. [CrossRef]

24. Fu, W.; Liu, Z.; Li, Y.; Wu, H.; Tang, Y. Numerical study for the influences of primary steam nozzle distance and mixing chamber throat diameter on steam ejector performance. *Int. J. Therm. Sci.* **2018**, *132*, 509–516. [CrossRef]

25. Varga, S.; Oliveira, A.C.; Diaconu, B. Numerical assessment of steam ejector efficiencies using CFD. *Int. J. Refrig.* **2009**, *32*, 1203–1211. [CrossRef]

26. Yuan, Y.; Tan, L.; Xu, Y.; Yuan, Y. Three-dimensional CFD modeling and simulation on the performance of steam ejector heat pump for dryer section of the paper machine. *Vacuum* **2015**, *122*, 168–178. [CrossRef]

27. Dong, J.; Wang, W.; Han, Z.; Ma, H.; Deng, Y.; Su, F.; Pan, X. Experimental investigation of the steam ejector in a single-effect thermal vapor compression desalination system driven by a low-temperature heat source. *Energies* **2018**, *11*, 2282. [CrossRef]

28. Zhang, K.; Zhu, X.; Ren, X.; Qiu, Q.; Shen, S. Numerical investigation on the effect of nozzle position for design of high performance ejector. *Appl. Therm. Eng.* **2017**, *126*, 594–601. [CrossRef]

29. Varga, S.; Oliveira, A.C.; Diaconu, B. Influence of geometrical factors on steam ejector performance—A numerical assessment. *Int. J. Refrig.* **2009**, *32*, 1694–1701. [CrossRef]

30. Li, X.; Wang, T.; Day, B. Numerical analysis of the performance of a thermal ejector in a steam evaporator. *Appl. Therm. Eng.* **2010**, *30*, 2708–2717. [CrossRef]

31. Wu, H.; Liu, Z.; Han, B.; Li, Y. Numerical investigation of the influences of mixing chamber geometries on steam ejector performance. *Desalination* **2014**, *353*, 15–20. [CrossRef]

32. Yang, X.; Long, X.; Yao, X. Numerical investigation on the mixing process in a steam ejector with different nozzle structures. *Int. J. Therm. Sci.* **2012**, *56*, 95–106. [CrossRef]

33. Ruangtrakoon, N.; Thongtip, T.; Aphornratana, S.; Sriveerakul, T. CFD simulation on the effect of primary nozzle geometries for a steam ejector in refrigeration cycle. *Int. J. Therm. Sci.* **2013**, *63*, 133–145. [CrossRef]

34. Riffat, S.B.; Omer, S.A. CFD modelling and experimental investigation of an ejector refrigeration system using methanol as the working fluid. *Int. J. Energy Res.* **2001**, *25*, 115–128. [CrossRef]

35. Sriveerakul, T.; Aphornratana, S.; Chunnanond, K. Performance prediction of steam ejector using computational fluid dynamics: Part 1. Validation of the CFD results. *Int. J. Therm. Sci.* **2007**, *46*, 812–822. [CrossRef]

36. Sriveerakul, T.; Aphornratana, S.; Chunnanond, K. Performance prediction of steam ejector using computational fluid dynamics: Part 2. Flow structure of a steam ejector influenced by operating pressures and geometries. *Int. J. Therm. Sci.* **2007**, *46*, 823–833. [CrossRef]

37. Aly, N.H.; Karameldin, A.; Shamloul, M.M. Modelling and simulation of steam jet ejectors. *Desalination* **1999**, *123*, 1–8. [CrossRef]

38. Su, L.; Agarwal, R.K. CFD Simulation of a Supersonic Steam Ejector for Refrigeration Application. In Proceedings of the ASME/JSME/KSME 2015 Joint Fluids Engineering Conference, Seoul, Korea, 26–31 July 2015.

39. Desevaux, P.; Mellal, A.; Alves de Sousa, Y. Visualization of secondary flow choking phenomena in a supersonic air ejector. *J. Vis.* **2004**, *7*, 249–256. [CrossRef]

40. Matsuo, K. Shock train and pseudo-shock phenomena in supersonic internal flows. *J. Therm. Sci.* **2003**, *12*, 204–208. [CrossRef]

41. Han, Y.; Wang, X.; Sun, H.; Zhang, G.; Guo, L.; Tu, J. CFD simulation on the boundary layer separation in the steam ejector and its influence on the pumping performance. *Energy* **2019**, *167*, 469–483. [CrossRef]

42. Ariafar, K.; Buttsworth, D.; Al-Doori, G.; Sharifi, N. Mixing layer effects on the entrainment ratio in steam ejectors through ideal gas computational simulations. *Energy* **2016**, *95*, 380–392. [CrossRef]

43. Wang, X.; Dong, J.; Zhang, G.; Fu, Q.; Li, H.; Han, Y.; Tu, J. The primary pseudo-shock pattern of steam ejector and its influence on pumping efficiency based on CFD approach. *Energy* **2019**, *167*, 224–234. [CrossRef]

44. Wang, X.; Dong, J.; Lei, H.; Li, A.; Wang, R.; Chen, W.; Tu, J. Modeling and simulation of steam-jet vacuum pump. *Zhenkong Kexue yu Jishu Xuebao/Vacuum Sci. Technol.* **2013**, *33*, 1069–1073.

45. Yuen, A.C.Y.; Yeoh, G.H.; Timchenko, V.; Chen, T.B.Y.; Chan, Q.N.; Wang, C.; Li, D.D. Comparison of detailed soot formation models for sooty and non-sooty flames in an under-ventilated ISO room. *Int. J. Heat Mass Transf.* **2017**, *115*, 717–729. [CrossRef]

46. Yuen, A.C.Y.; Yeoh, G.H.; Timchenko, V.; Cheung, S.C.P.; Barber, T.J. Importance of detailed chemical kinetics on combustion and soot modelling of ventilated and under-ventilated fires in compartment. *Int. J. Heat Mass Transf.* **2016**, *96*, 171–188. [CrossRef]

47. Li, A.; Yeoh, G.H.; Timchenko, V.; Yuen, A.C.Y.; Wang, X. Numerical Study of Condensation Effect on a Steam Ejector by Wet Steam Model. In Proceedings of the Thirteenth International Conference on Flow Dynamics (ICFD2016), Sendai, Japan, 10–12 October 2016; Volume 12, pp. 492–493.

48. Chen, T.B.Y.; Yuen, A.C.Y.; Yeoh, G.H.; Timchenko, V.; Cheung, S.C.P.; Chan, Q.N.; Yang, W.; Lu, H. Numerical study of fire spread using the level-set method with large eddy simulation incorporating detailed chemical kinetics gas-phase combustion model. *J. Comput. Sci.* **2018**, *24*, 8–23. [CrossRef]

49. Yuen, A.C.Y.; Chen, T.B.Y.; Yeoh, G.H.; Yang, W.; Cheung, S.C.P.; Cook, M.; Yu, B.; Chan, Q.N.; Yip, H.L. Establishing pyrolysis kinetics for the modelling of the flammability and burning characteristics of solid combustible materials. *J. Fire Sci.* **2018**, *36*, 494–517. [CrossRef]

50. Yuen, A.C.Y.; Yeoh, G.H. Numerical simulation of an enclosure fire in a large test hall. *Comput. Therm. Sci.* **2013**, *5*, 459–471. [CrossRef]

51. Yuen, A.C.Y.; Yeoh, G.H.; Timchenko, V.; Cheung, S.C.P.; Chan, Q.N.; Chen, T. On the influences of key modelling constants of large eddy simulations for large-scale compartment fires predictions. *Int. J. Comut. Fluid Dyn.* **2017**, *31*, 324–337. [CrossRef]

52. Besagni, G.; Mereu, R.; Inzoli, F. CFD study of ejector flow behavior in a blast furnace gas galvanizing plant. *J. Therm. Sci.* **2015**, *24*, 58–66. [CrossRef]

53. Besagni, G.; Mereu, R.; Chiesa, P.; Inzoli, F. An Integrated Lumped Parameter-CFD approach for off-design ejector performance evaluation. *Energy Convers. Manag.* **2015**, *105*, 697–715. [CrossRef]

54. Crocco, L. One-dimensional treatment of steady gas dynamics. *Fundam. Gas Dyn.* **1958**, *Chap-B*, 64–349.

55. Huang, B.J.; Chang, J.M.; Wang, C.P.; Petrenko, V.A. 1-D analysis of ejector performance. *Int. J. Refrig.* **1999**, *22*, 354–364. [CrossRef]

 *applied sciences*

Article

Simultaneous In-Cylinder Flow Measurement and Flame Imaging in a Realistic Operating Engine Environment Using High-Speed PIV

Atsushi Nishiyama *, Minh Khoi Le, Takashi Furui and Yuji Ikeda

Imagineering, Inc., 7-4-4 Minatojima-Minami, Chuo, Kobe 650-0047, Japan
* Correspondence: atsushi@imagineering.jp; Tel.: +81-78-386-8888

Received: 11 March 2019; Accepted: 18 April 2019; Published: 30 June 2019

Abstract: Among multiple factors that affect the quality of combustion, the intricate and complex interaction between in-cylinder flow/turbulent field and flame propagation is one of the most important. In this study, true simultaneous, crank-angle resolved imaging of the flame front propagation and the measurement of flow-field was achieved by the application of high-speed Particle Image Velocimetry (PIV). The technique was successfully implemented to avoid problems commonly associated with PIV in a combustion environment, such as interferences and reflections, avoided thanks to a number of adjustments and arrangements. All experiments were carried out inside a single-cylinder optical gasoline engine operated at 1200 rpm, using port fuel injection (PFI) with stoichiometric mixtures. It was found that the global vortex location of the tumble motion heavily influences the flame growth direction as well as the flame shape, mainly due to the tumble-induced flow across the ignition source. The flame propagation also influences the flow-field such that the pre-ignition flow can be maintained and the flow of unburned region surrounding the flame front will be enhanced.

Keywords: High speed PIV; flame propagation; planer laser tomography

1. Introduction

The current stringent emissions regulations as well as demands for better fuel economy in passenger vehicles require the combustion process in engines to be much more efficient while being cleaner overall. While other mode of powertrain such as electric or hybrid are developing, their drawbacks mean the combustion engines will still need to be improved. Researchers have proposed many approaches in order to achieve cleaner, more efficient combustion engines, including ultra-lean burning, high compression ratio, high exhaust gas recirculation (EGR), high intake charge pressure and others. Due to its strong effect on the overall combustion process and the complex in-cylinder turbulent environment [1–3], understanding the interaction of in-cylinder flow-field and flame development will be crucial in optimizing the effects of the aforementioned strategies.

Due to its importance and their intricate relationships, researchers have been working toward imaging both flow-field and flame imaging simultaneously. Flow-field measurement techniques have been quite well studied and developed, with diagnostics such as (Particle Image Velocimetry) PIV, (Laser Doppler Velocimetry) LDV and their variations providing useful information for engine researchers. LDV is a point-based measurement technique that provides the temporal flow velocity variations at a high sampling rate. Due to its setup, it is possible for a small probe of LDV to be made and applied to a production engine. For instance, special spark plug modified to contain LDV optics has been made and applied by engine researchers to provide flow-field information near the discharge location [4,5]. Despite its benefits, LDV cannot provide the important spatial information of in-cylinder flow, which is where PIV excels. In recent time, the advances in laser development has made powerful, high frequency laser available for researchers. Hence, PIV can now be applied to not just capture spatial characteristics of the in-cylinder flow but also its temporal evolution and development [6–8].

Most of the previous works about in-cylinder flow diagnostics of internal combustion engines were performed under motoring condition without combustion by using optical engines. However, the phenomenon in a practical engine is different from this condition. Figure 1 shows the phenomena of flow-field, flame propagation and its interaction in a spark ignition engine. During intake and compression stroke before ignition, the differences in flow-field between motoring and firing condition need to be discuss. For firing case, residual gas exists and its effect on flow-field, such as mean flow, turbulence intensity and scale, as well as cyclic variation should be investigated. After the ignition timing, one of the most interesting topics is the effect of flow-field on spark discharge, flame propagation speed, heat release and combustion duration. The large-scale flow will stretch the spark discharge and shift the propagating flame. The small-scale vortex will affect flame configuration and reaction zone thickness. The correlation between flow-field and flame propagation in the same cycle can provide the clear indication of the optimum velocity and turbulence distribution in-cylinder to realize faster flame propagation and combustion. As a result, detail analysis in these areas can provide valuable information for engine design to improve the thermal efficiency. Another point of interest is the effect of flame propagation on the flow-field in unburned mixture. Does the flame enhance or decay the turbulence of unburned mixture? Enhancement in the understanding of physical phenomena and accuracy of engine combustion simulations can be obtained from such results. Furthermore, heat loss is the one of the more important issues for thermal efficiency improvement and the information regarding on flame attachment to the wall as well as the corresponding velocity can help explain the heat transfer between flame wall (surface of piston top and cylinder head). The additional information in relation to the flame-wall interaction process can also be derived when using with advanced wall thermometry technique [9].

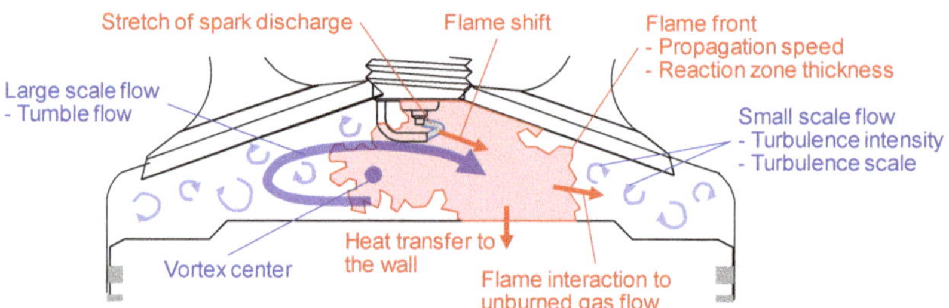

Figure 1. Flow-field, flame propagation, and its interaction in a spark ignition engine.

There are a number of diagnostics that one can use to observe the flame, ranging from simple natural emission based such as chemiluminescence or natural-flame luminosity to more complicated techniques such as tomography of hydroxyl planar laser-induced fluorescence (OH PLIF). Clark et al. investigated the effect of the timing of the first and second injection for an evenly split dual injection strategy in an optical engine [10,11]. Performance parameters derived from in-cylinder pressure data are analyzed alongside high-speed natural flame luminescence images in order to obtain relationships between engine output and the physical properties associated with flame propagation. They have been combined with PIV in attempts to observe flame and flow-field simultaneously and allow for interesting behaviors to be observed [12–14]. For instance, Mounaïm-Rousselle [13] used oil seeds for PIV that would be evaporated by the flame and hence, creating flame tomography effect to observe the flame. While this provide some information on the flame growth, it was not combined with flow-field and PIV results and the flow-field inside the burned region is missing. Other approaches involve the combination of laser diagnostic techniques: PIV and OH PLIF where there would be multiple cameras and lasers required to be implemented [12]. As OH PLIF is one of the best diagnostics for flame front, this set up has high accuracy and allows for derivation of more fundamental parameters such as turbulent burning velocity. However, OH PLIF is difficult to get enough signal intensity at high-speed to detect the change during the cycle.

In this study, true simultaneous, crank-angle resolved imaging of the flame front propagation, and the measurement of flow-field was achieved by the application of high-speed Particle Image Velocimetry (PIV). PIV measurement under combustion environment is quite challenging for multiple reasons, including the interferences of the natural luminosity from the flame, the interaction between the PIV seeding materials, and the high temperature flame, as well as complications from direct injection operations. All these issues decrease the optical clarity necessary for successful implementation of PIV. This is also why it is difficult to be able to observe both the flame and flow-field simultaneously in a reacting engine environment. Through careful optimization of a high-speed PIV setup and small modification of our engines, we have been able to observe simultaneously the flame and flow-field inside an operating engine environment. From the same captured PIV images, flow-field information as well as flame characteristics can be processed through a series of commercially-available and inhouse-developed software. The setup has been demonstrated to be able to capture the results at various engine operating conditions and provide valuable information on the relationship of flame development and in-cylinder flow [15]. In this report, we will discuss with more details on the setup, accuracy of the results and its potential errors. Focus will also be placed on the differences between the flow-field of non-reacting and reacting engine environment as well as the cyclic reaction on flame movement and combustion quality with respect to the flow-field.

2. Materials and Methods

2.1. Engine Specifications and Operating Conditions

The experiments were carried out in a single-cylinder optical spark ignition engine. Table 1 summarizes the specification of this engine and the selected operating condition for this study. The engine capacity is 500 cm^3 and has a compression ratio of 10.4 with a bore of 86 mm and stroke of 86 mm. Optical access into the combustion chamber of the engine is provided by replacing certain metal parts with quartz window: at the top of the piston (54 mm diameter) and in the pent-roof as shown in Figure 2a. The whole metal liner can be replaced by quartz for extended optical view into the chamber. To observe the bottom-view of the combustion chamber, a 45° mirror is placed in the hollow extended-piston. It also provides an optical-access path for laser for the observation of pent-roof area, which is the setup applied for this study. The experiments were carried out with the engine operating at 1200 rpm with port fuel injection and an absolute intake pressure of 60 kPa. The spark timing was set at −15° Crank-Angle after Top Dead Center (CA aTDC). Due to the focus of this study, only mixture at stoichiometry will be considered. In-cylinder pressure are monitored and recorded using a pressure transducer (Kistler, 6052C) and the engine is fired continuously during the PIV capturing process for synchronization purpose.

Table 1. Engine specifications and test conditions. PFI: port fuel injection; CA aTDC: Crank-Angle after Top Dead Center.

Engine Type	4 Stroke, Single Cylinder, PFI
Bore × stroke	86 × 86 mm
Displacement	500 cm^3
Compression ratio	10.4
Fuel injection	Port fuel injection
Fuel type	Regular gasoline
Injection pressure	300 kPa
Engine speed	1200 rpm
Equivalence ratio	1 (AFR 14.7)
Intake pressure	60 kPa
Spark timing	−15° CA aTDC

2.2. High-Speed PIV

Figure 2a also illustrates the laser setup for the application of high speed-PIV (HS-PIV) diagnostic in the optical engine. An Nd:YLF laser (Litron Lasers Ltd., LDY304) beam of 527 nm wavelength was converted to a 500 μm thin sheet through a series of optics and then introduced into the engine combustion chamber from the bottom by reflecting off the 45° mirror and then passing through the piston-top quartz window. The pulse width of the laser shot is 150 ns at 1 kHz repetition rate. An area of 37 mm × 54 mm in the middle of the side-view of the pent-roof, which includes the spark plug and its surrounding region was targeted as the measurement area for this study. Figure 2b displays the schematics of this measurement area (right) as well as the bottom-view of the fire deck to illustrate the relative position of the laser. The PIV images were captured using a high-speed camera (Vision Research Inc., Phantom v1611) at 1280 × 800 resolution and the laser was operated at 7.2 kHz (20 kHz maximum) for a temporal resolution of 1° CA in the current engine operating condition. The time interval for two images was set to 40 μs. A 527 nm bandpass filter with band width of 20 nm FWHM was utilized to isolate PIV signal from other interferences. The laser and camera operations were synchronized to engine rotation by using a pulse generator. The pulse generator was triggered by the engine rotating signal and the pulse generator generated the trigger signals for laser and camera. PIV data was captured for full crank angles of 19 consecutive fired cycles due to the high-speed camera memory limitation.

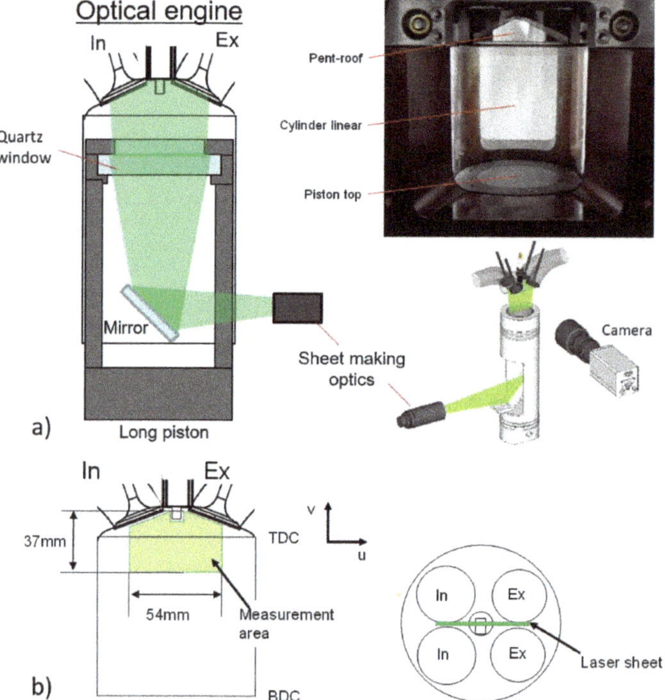

Figure 2. (a) Optical engine and Particle Image Velocimetry (PIV) set up, (b) PIV measurement area and position of the laser sheet with respect to the cylinder.

In this study, HS-PIV diagnostics was applied under engine-firing conditions which in turns introduces special challenges and difficulties. These will be described and discussed in the later sections. For optimum signal intensity with high traceability for kHz order, the material of the seeds need to be resilient to the harsh conditions of high pressure, high turbulence and high temperature inside a cylinder of a combustion engine while being inert and not affect the chemistry of combustion. As the practicability under the combustion condition was proven by previous LDV experiment at our laboratory [4], we selected a SiO_2 particle with 4 µm diameter as our tracer particles. The process on tuning for the best seeding amount will also be discussed in later sections. The particles in the tank were carried by the compressed air and seeded into the intake air flow upstream of the fuel injector of the optical engine to ensure the homogeneous mixing and trace to the intake air motion. Table 2 summaries the PIV setup used.

Table 2. PIV set up specifications.

Sampling frequency	7.2 kHz
Crank angle resolution	1°. CA
Image size	1280×800 (59×37 mm)
Spatial resolution of image	0.046×0.046 mm
Seeding particles	SiO_2
Measurement cycle number	19 (consecutive)
Interrogation area	32×32 px. (1.5×1.5 mm)
Overlap (Duplication degree of interrogation area)	50%
Spatial resolution for vector	0.74×0.74 mm

2.3. High-Speed Flame Tomography

Using the same HS-PIV setup, we have been able to optimize such that the burned gas region can also be identified and distinguished from the unburned region in the PIV images, which allows for flame front detection. Under combustion condition, the burned gas density and temperature will be drastically different to those of the unburned gas. The high temperature of the stoichiometric flame front can even lead to changes in the morphology of these solid seeds despite their resiliency to harsh environment. Moreover, the interaction between the PIV laser and combustion products and luminosity can also affect the laser signal intensity. As a result, these changes will be reflected in the seeding densities and scattering PIV laser intensities. Following these logics, the burned region should appear different to the unburned region. Hence, it is the gradient in intensity created that can be used for detection and identification of the burned region and hence, the flame front. To maximize the intensity gradient differences and avoid unwanted interferences from flame luminosity, a 527 nm bandpass-filter was placed in front of the camera. This pseudo flame tomography imaging was carried out at the same time as HS-PIV diagnostics under the same capturing procedures and operating condition.

Therefore, such HS-PIV setup allows for the simultaneous measurement of both flow-field and flame-front. The characteristics of both elements can be derived from the same captured HS-PIV image, maximizing the temporal accuracy when discussing the simultaneous relationship between flow-field and flame.

3. Results and Discussions

3.1. Considerations for Diagnostics Setup

First, the engine was installed with balancers and dampers to minimize vibrations and minimize any influences it can have on PIV arrangement. The camera and laser were also positioned on structures that well dampen from the engine vibrations. The results are confirmed to be free of vibration-induced errors which is normally seen as streaks, proving the effectiveness of this strategy. Secondly, to observe the side-view of the pent-roof area, the laser will be impinged on the cylinder head and cannot escape from the combustion cylinder, leading to possible reflections complications. As shown in Figure 3a, the reflection light can create smearing, scatters on the pent-roof and will decrease the clarity of a large portions of the measurement area. In this case, a layer of anti-reflection black paint (commercially available oily black ink) is applied onto the surface of the fire-deck and hence, the reflection light of the laser sheet is significantly reduced and the valid measurement area is maximized. However, it should be noted that the region around 1 mm away from the fire-deck of the pent-roof is still under the effect of reflection, especially at TDC.

While the stoichiometric combustion with port-fuel injection inside the S.I. engine does not produce as strong luminosity as other richer combustion seen in direct injection engine or diesel engine, the luminosity is still strong enough to create complications to the PIV signal. In fact, as seen in Figure 3b, the luminosity will be scattered on the seeds, especially inside the burned region. By adding a bandpass filter at the laser wavelength of 527 nm, it is possible to remove most of the scattered light from luminosity and only capture the dominating scattering signal from the PIV laser sheet. An example PIV image demonstrates the performance of this bandpass filter during combustion is also shown in Figure 3b.

a) Reduction of reflection light

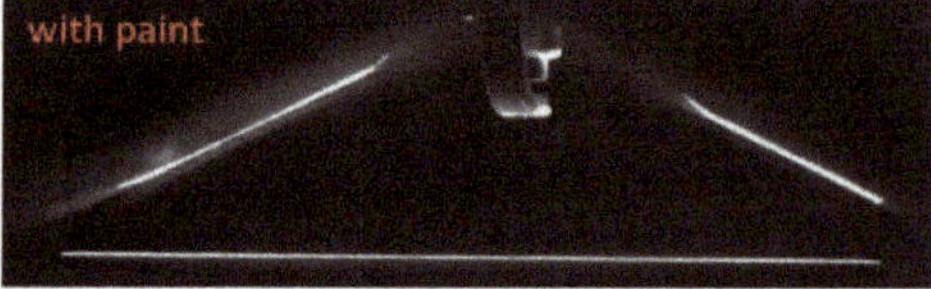

b) Elimination of flame emission
⇒ Bandpass filter (λ_{center}: 527 nm, Band width: 20 nm, Transmittance: 93 %)

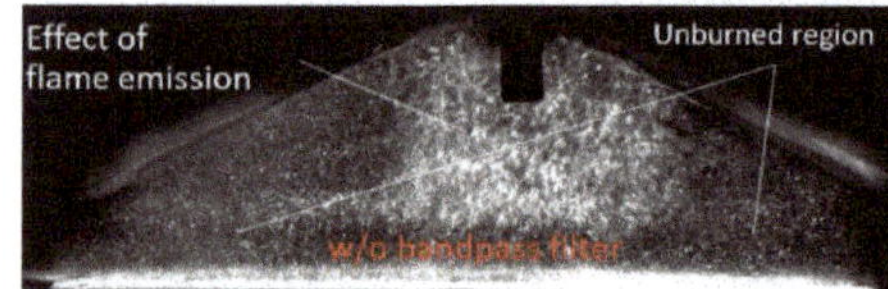

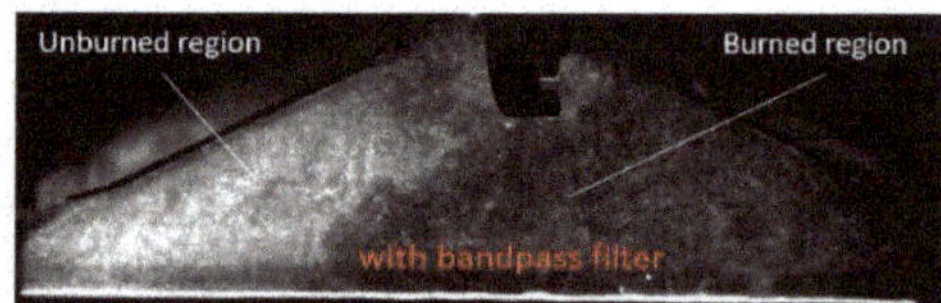

Figure 3. (**a**) Reducing reflection laser light during PIV measurement, (**b**) Avoiding the flame luminosity during combustion PIV measurement using band-pass filter.

For PIV measurement, selection of the seeding particle is crucial to achieve high quality data. As mentioned earlier, the seeds need to be inert (chemically and fluid-dynamically) and resistant to high temperature to be applicable for PIV during a combustion event in engine. We have chosen silica (SiO_2) particles in this work to ensure the flame will not burn all the particles such as the case when oil seeds were used [13]. However, to be able to distinguish between burned and unburned region from the flame analysis point of view while maintaining flow-field data from all regions, seeds materials will not be the only deciding factor. In fact, as shown in Figure 4a, depends on the amount of particles seeded into the cylinder during a firing cycle, we can either get an underexposed image where the laser scattering is not enough for PIV and flame front detection (top), an overexposed image where seeds amount was too much (middle) or only enough seeds to get PIV data but the flame cannot be detected bottom. The correct amount of seeds allows both PIV data and flame to be detected as shown in the bottom row of Figure 4a. To achieve this correct amount, the seed amount was controlled by adjusting the air flow rate for this seed supply. Once the image is confirmed, the same flow rate was kept for the rest of the experiment. Estimating the appropriate amount of seed particles in this experiment is about 1 mm^3 in bulk volume per cycle. Figure 4b displays the variation of the average intensity from PIV particle-scattering laser signal from BDC to TDC of a firing cycle and it demonstrated that the seeding amount can be used to monitor and measure the flow-field over the whole cycle and the intensity difference is not overwhelming so overseeding at the TDC has been avoided. It is worth noting that for this example, a spark timing of −30° CA aTDC is used and the dip in PIV signal intensity seen around −15° CA aTDC is most likely due to the lower density seen in the burned region.

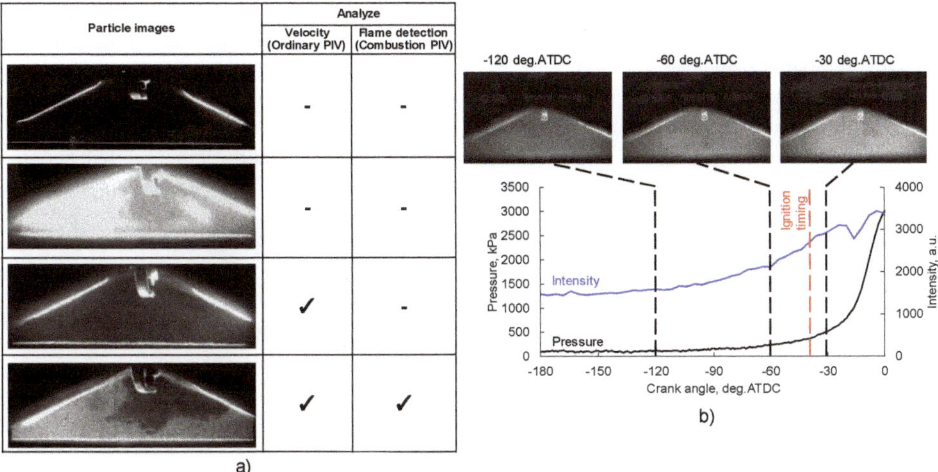

Figure 4. (**a**) The effects of difference seeds amount on the clarity and feasibility of the information that can be measured, (**b**) variation of PIV intensity from BDC to TDC of a firing cycle.

3.2. Data Processing and Validation

3.2.1. PIV Processing

The instantaneous velocity field is first derived from the captured PIV images using commercial software from Dantec Dynamics. The range of the interrogation areas for considerations are 16×16, 32×32 and 64×64 at 0.046 mm per pixels. The vector results of the TDC flow-field at different interrogation areas are shown in Figure 5. If the normalized vector residuals r_0 calculated by the following equation exceeds the detection threshold, that vector is defined as an error vector.

$$r_0 = \frac{|U_0 - U_m|}{r_m + \varepsilon} \tag{1}$$

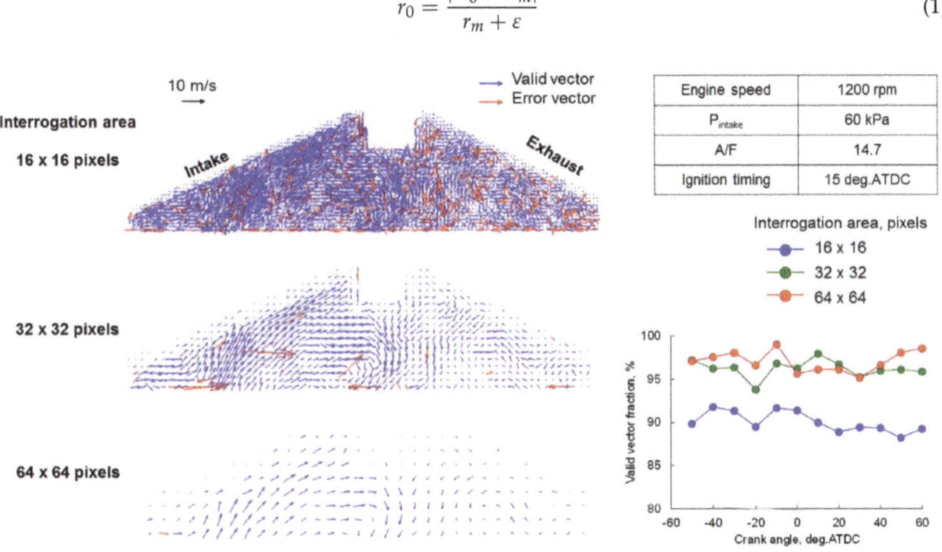

Figure 5. Validity of in-cylinder flow-field vectors from PIV processing with different interrogation areas: 16×16, 32×32, and 64×64 with images on left side and valid vectors % plot on the right side.

Here, U_0 is the displacement vector, U_m is the median vector calculated using neighborhood vectors, r_m is the median residual calculated using the neighborhood residuals and ε is the minimum normalization level. The normalized vector residuals r_0 was calculated using 7×7 neighborhood vectors and the detection threshold was set to 1.2. Both 32×32 and 64×64 areas allow for ~95% valid vectors while the 16×16 interrogation area has more than 87% valid vectors, with most of the error vectors coming from high scattered region near piston. While 87–90% valid vectors might be enough for interpretation of the flow-field behavior, for the rest of the discussion in the current articles, we will select 32×32.

The instantaneous velocity is further processed for mean velocity, velocity fluctuations and turbulence characteristics using in-house developed MATLAB software. The most common method to derive these specific flow-field characteristics from the instantaneous velocity is to use Fourier transformation to split it into the mean velocity ($\overline{u}$, $\overline{v}$) and the fluctuation of velocity (u', v'). The instantaneous velocity (u, v) and turbulence intensity are described as following equations.

$$u(\theta, n) = \overline{u}(\theta, n) + u'(\theta, n) \tag{2}$$

$$v(\theta, n) = \overline{v}(\theta, n) + v'(\theta, n) \tag{3}$$

$$k(\theta, n) = \frac{1}{2} \sqrt{[u'(\theta, n)]^2 + [v'(\theta, n)]^2} \tag{4}$$

Here, θ is crank angle and n is cycle number. While spatial filter is commonly used for PIV, high-speed PIV allows the use of temporal filter. Many studies have investigated the selection of the temporal cutoff frequency for turbulent velocity estimation in internal combustion engines, which is still a relatively arbitrary process. In this study, we selected four points in the important regions of the measurement area, with three points A, B, and C surrounding the spark plug and point D near the squish area to assess the derived temporal cutoff frequency at different spatial location. The power spectrum calculated from fast Fourier transform at the four selected measurement points is shown in Figure 6. The power spectrum was calculated from the instantaneous velocity of each 19 cycle, then the ensemble-averaged power spectrum was obtained. Around the frequency of 300 Hz, all power spectra have an inflection point which indicate that the variability characteristic of velocity changed at this frequency. For LDV measurement in the same optical engine under near engine operating condition, the inflection point in the power spectrum was also around 300 Hz [4]. Therefore, the cutoff frequency of this engine operating condition has been selected to be 300 Hz.

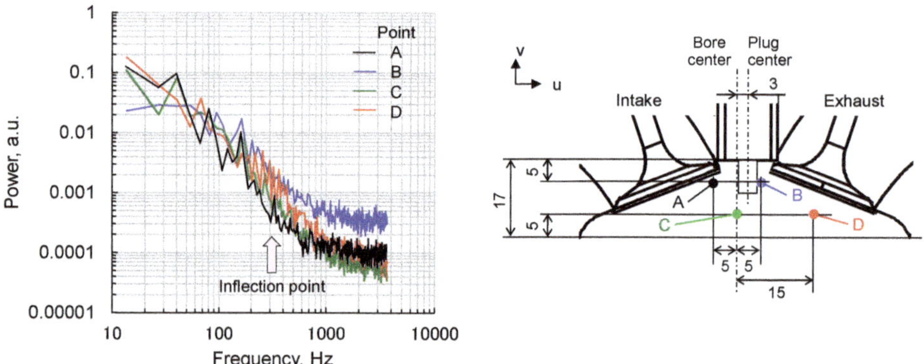

Figure 6. The power spectrum of the instantaneous velocity from PIV of different points in the measurement area.

Due to the existence of the flame during combustion PIV application inside operating engine, the propagation of the flame could affect the seed particles by agglomeration or melting/burning of the particles and hence the velocity measured from PIV processing. To assess this potential issue, we took a PIV image during the flame developing period and analyze the velocity across the flame front. Figure 7 displays the PIV image as well as the processed information for the velocity. The difference in intensity between the burned and unburned region is quite obvious: the top right of the figure shows the intensity along a line drawn horizontally across the burned and unburned region (the line is shown in dashed white line on the image top left of the figure) and the intensity is shown to be distinctively lower in the burned region compared to the unburned region. The intensity transition between burned and unburned region is also a sharp and steep rise which indicates the error in the detected flame front will be quite low (± 10 pixels, which is ± 0.5 mm). To estimate the error in velocity due to the flame front movement, an assessment of the velocity across the flame front at various interrogation area is carried out. As the interrogation area gets smaller, the less seed particles will be tracked and if the flame front affects the seeds movement, discrepancy with larger interrogation area that cover more than the flame front. The results were plotted as shown in bottom left; all the trends were similar with different interrogation area which see a smooth transition of velocity field across the flame front. This also suggests the unburned region contains enough valid particles for PIV analysis.

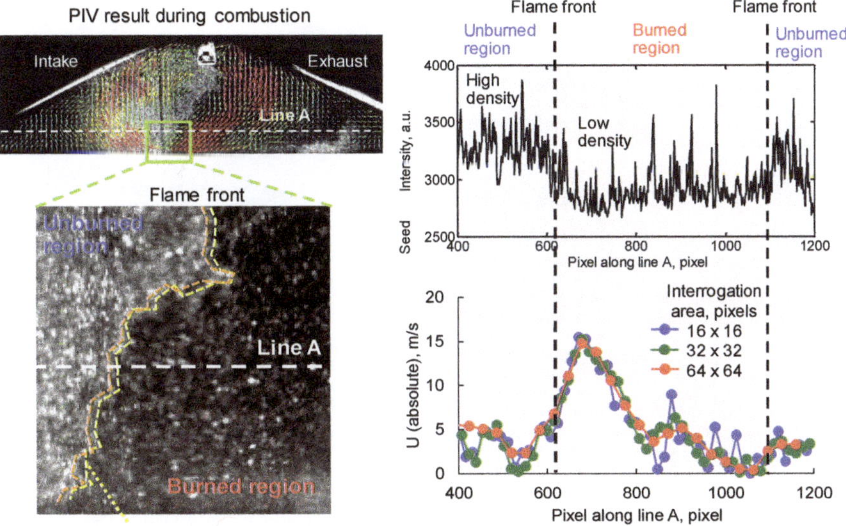

Figure 7. Effects of the flame front on derived velocity from combustion PIV at the flame front. PIV image and schematics of the analysis (**left**) and plots of the intensity across the measurement area on one line as well as the velocity calculated from different interrogation area (**right**).

3.2.2. Flame Analysis

The image processing steps in the algorithm for flame front detection and estimation is shown by an example at TDC (15° CA after Ignition Timing) in Figure 8. This flame front extraction process involves the use of both automated, in-house developed algorithm and manual visual inspections of the captured images. The region bounded by the piston and the pent-roof is selected as the area of interest. First, the images are inverted and background subtraction is applied. A combination of median and Gaussian filters is then applied to eliminate artefacts due to laser scattering effects from PIV seeds. An appropriate image intensity range and results of filtering is verified by visual inspections and binary conversion using most appropriate threshold is finally applied to extract the flame front. The derived flame front is once again confirmed via visual inspections to eliminate obvious errors due to laser intensity fluctuations. The resulted flame front is used to derive the projected flame area, the flame centroids in the vertical plane, the equivalent flame radius R and global flame-growth speed dR/dt. Here, the flame centroids are defined as the position of the centroids of the burned region. The equivalent flame radius R is defined as the radius of a circle having the same area as the burned area. The global flame-growth speed dR/dt is calculated from the difference of equivalent flame radius between two time points. The accuracy of the burned region detection and flame front derivation has an accuracy of around ±1 mm which correspond well to the 1.5 mm resolution of flow-field PIV data. An example of flame front boundary is found in the bottom of Figure 8. The flame front analysis will be limited to the early flame period from the first detected signal to when the flame completely fills the measurement area and pent roof.

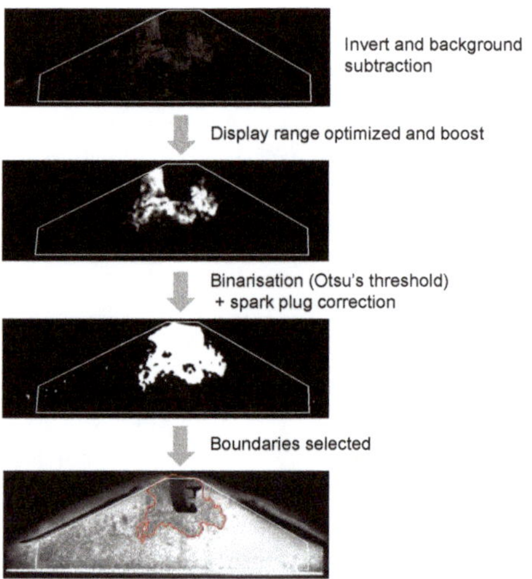

Figure 8. Image processing steps to extract the flame front from the high-speed PIV image.

3.3. Differences Between Firing and Motoring

Besides firing engine operation, high-speed PIV is also applied when the engine was operated in motoring mode without combustion. The PIV results were compared between firing and motoring to identify changes in the flow-field when there is a combustion event. One representative cycle out of the 19 tested cycles for each firing and motoring condition were chosen; the flow velocity of the two conditions were shown in Figure 9 and the turbulence intensity were shown in Figure 10. For brevity, images were shown in 10° CA timing increments, from −45° CA aTDC to 25° CAaTDC. From Figure 9, it can be seen that the flow-field between motoring and firing were not very different prior to the ignition timing (−15° CA aTDC) and the differences can be attributed to cyclic variations. The vortices centers are located more centrally in the pent roof which makes the flow across the spark plug stronger. After ignition, the flame boundary in the firing cycle is overlaid on top of the images, which can be seen from −5° CA aTDC onward. During this phase, significant differences between and firing and motoring flow-field can be observed, suggested strong effect of the flame on the dynamics of the unburned region. While in the motoring cycle, the strength of the flow-field is significantly diminished, the existing tumble flow continues to exist and even enhanced by the propagating flame. The flame first propagates in the tumble flow direction and the flow-field is maintained in this direction. From 5° CA aTDC, the flame continues to grow with the flow: flame propagation can be seen toward the other side of the pent-roof and interestingly, the flow here is significantly enhanced. This could be because it is in the later stages of combustion where the flame has occupied a much larger volume of the combustion chamber and hence the pressure rise, the higher temperature can contribute to a more complicated flow-field. This is well reflected in the high turbulence intensity in the pent-roof seen in Figure 10. The flow-field maintain a high level of turbulence intensity throughout the flame development period and at the region where the flame enhanced flow is seen, a very high turbulence intensity is also found, especially close to the flame front. It is interesting that the turbulence did not decrease at any timing, even when the flame kernel just appears. This could be due to the immediate pressure and temperature rise.

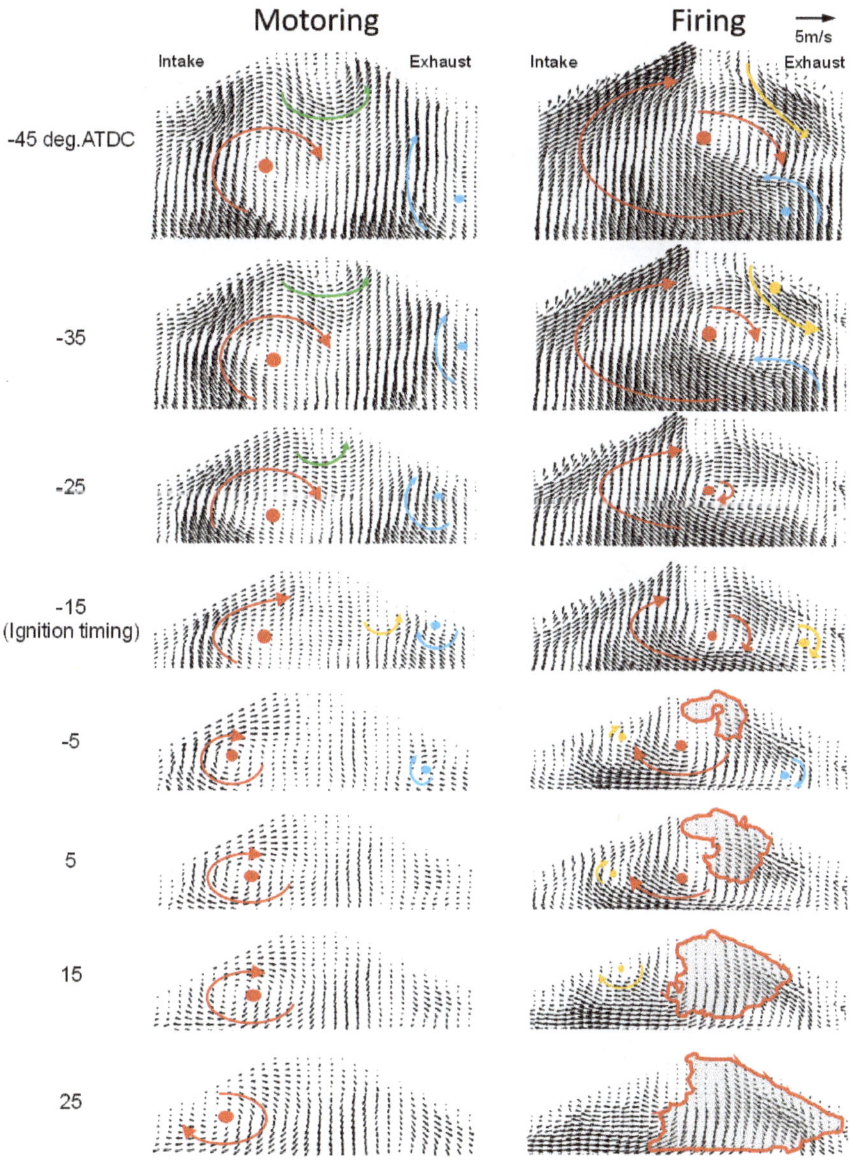

Figure 9. Mean in-cylinder flow velocity of a selected motoring cycle (**left**) and a selected firing cycles (**right**).

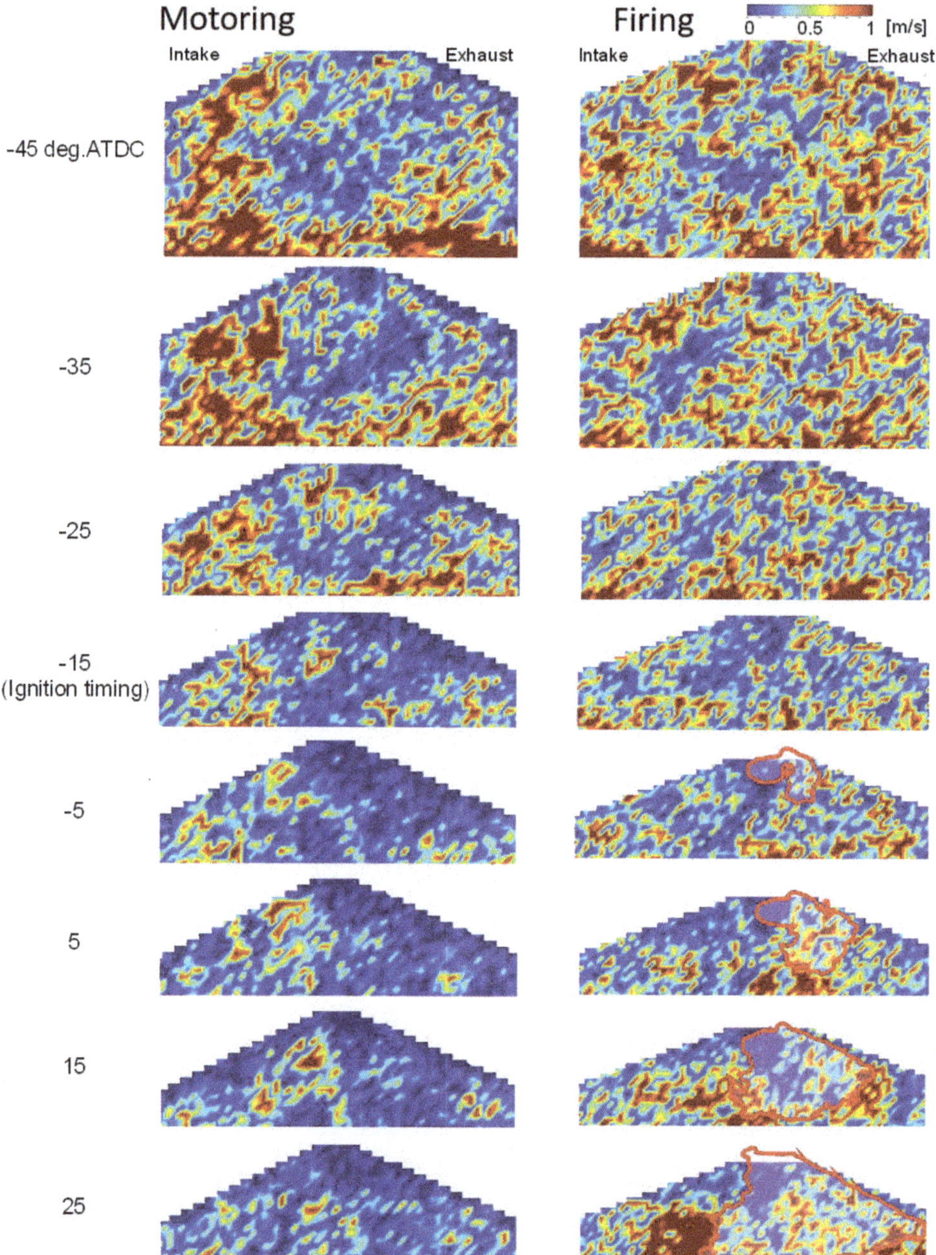

Figure 10. Turbulence intensity of a selected motoring cycle (**left**) and a selected firing cycles (**right**).

Figure 11 shows the velocity the flow at certain locations as the flame approaches and passes through them. In this case, 5 points at the same vertical distance of 2.5 mm to the TDC location of the piston were selected. Each point is 10 mm away from each other. These points were chosen as they allow the long distance away from each other while still provide a decent data size for analysis. Also, as they are close to the piston top in the period of investigation, only flow velocity in horizontal direction is considered. The colored dots represent the time when the flame front passes through the points. The flame crosses point D first followed by C, E, B and A which reflects its movement is influenced by the tumble direction. From the velocity plot, it is quite noticeable that as the flame passes by, the velocity of the flow achieves a local maximum. This likely suggests that as the flame approaches the flow is enhanced but quickly lost its energy as the mixture undergoes combustion.

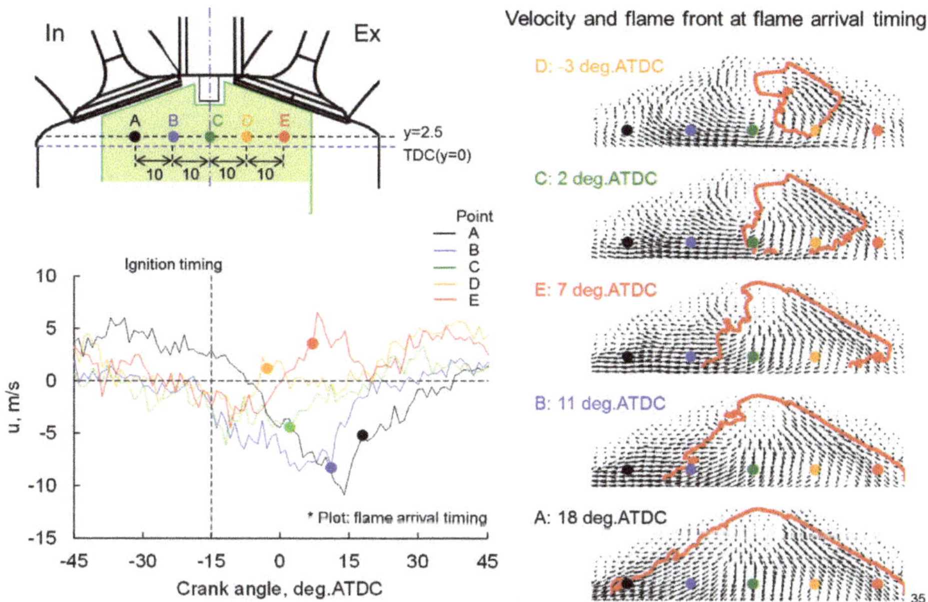

Figure 11. Velocity of the in-cylinder mixture flow at many different points as the flame front approaches and passes through those selected locations.

3.4. Connection to Cyclic Variations and Combustion Performance

Figure 12 displays at the top the in-cylinder pressure and the burn durations of all 19 tested firing cycles. Four cycles were selected to display the cyclic variations in flow and flame behavior. These four cycles represent different combustion performance at the same operating condition. The peak pressure of cycle A is the highest because the first initial combustion (short combustion period). On the other hand, combustion pressure of cycle B is the lowest with long combustion period. In short, cycle A represents the best performing cycle, cycle B the worst performing cycles while cycles C and D are closely matching to the ensembled average. Also shown in Figure 12 is the instantaneous flow at the ignition timing of these four cycles and their flame kernel at TDC (15° CA aIT) and 10° CA aTDC (25° CA aIT) together with the flame centroids movement over the whole investigation period. The plot of the derived flame growth speeds for the 4 cycles is also shown. All cycles have the same main tumble direction (clockwise) and the flames in cycles A, B, and C develop and propagate with the tumble direction, which can be easily observed when the centroids are plotted. Cycle D, while having the same tumble direction, the tumble center is located very far toward the squish area side of the intake and that make the flow velocity very minimal near the spark plug. As a result, the flame propagates in quite a symmetrical fashion across the spark plug axis. The centroids of this flame actually move along the spark plug axis, interestingly. Clear pattern in the flame speed in correlation with burn duration is not well observed but it is hinted that the early higher flame speed of cycle A does correspond to a faster 10% burnt duration. The variations in flame development path does not seem to have any correlation with the burnt duration.

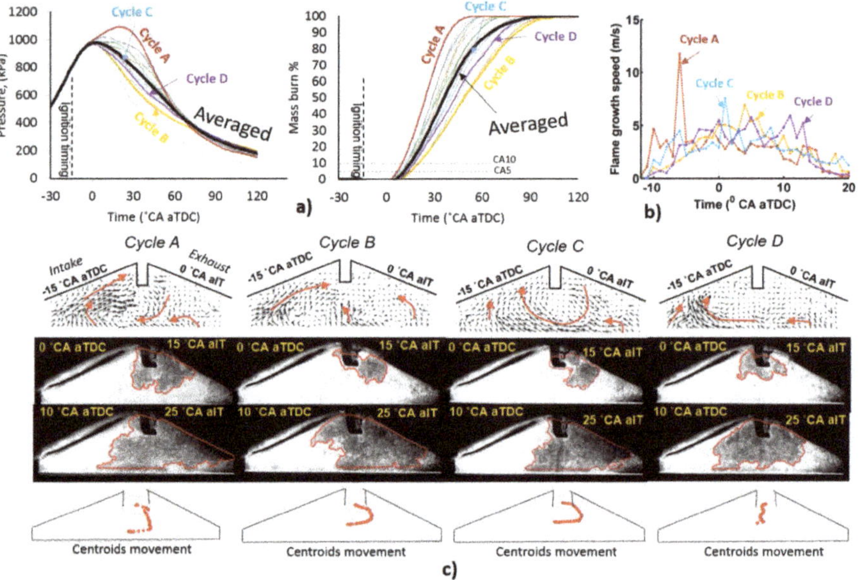

Figure 12. (**a**) In-cylinder pressure and burnt duration of 19 tested cycles with combustion PIV, (**b**) the flame velocity of the four selected cycles derived from the extracted flame front and (**c**) flow-field vectors at the ignition timing of the four selected cycles and the flame development at 0° and 10° CA aIT as well as the movement of the corresponding flame centroids over the whole inspected period.

Upon further analysis of the flow, the turbulence intensity near the spark plug appears to have stronger correlation with the burnt durations, with the velocity of the flow on the spark plug side that is downstream of the tumble flow results in a correlation factor of 0.56. This seems to agree with previous research about flow across the spark plug [2].

4. Conclusions

A high-speed PIV set up for optical spark-ignited engine has been modified and optimized to allow not only for the measurement of flow-field characteristics but also simultaneous flame front detection and imaging. Both properties of the flow-field such as flow velocity, turbulence intensity, and flame properties such as flame area, flame speed, and wrinkleness can be derived from the same image to explore their relationship in a true simultaneous fashion. Solutions and approach to avoid and overcome interferences and challenges in applying and developing this diagnostic technique are detailed in this paper. This include the process to minimize interferences from flame luminosity and laser reflection. The processing procedures are discussed and a few validation checks were performed to ensure the technique will yield good accuracy, especially near the flame front region. After the accuracy is confirmed, the technique was then applied during firing cycles of a stoichiometric operating optical engine as well as motoring cycles of the same engine. While prior to ignition the flow-field is similar between firing and motoring case, significant differences were observed after ignition timing. The flame propagates with the global tumble direction and help maintain this flow across the pent-roof area. It also enhances flow and turbulence intensity in flow region near the flame front as the combustion enters later stages and the flame occupies a larger portion of the pent-roof. In terms of cyclic variations, there seems to be no correlation between flame shape and propagation path with burning speed. However, there is a correlation between the flow across the spark plug with burning durations as well as flame propagation path. Additionally, this relatively simple high-speed PIV setup also has a lot of potential in combining with other diagnostics to capture other combustion species, mixture fractions as well as other important and influential processes which will help expanding the knowledge base for advance engine design and development.

Author Contributions: Conceptualization, Y.I.; Data curation, T.F.; Writing—original draft, M.K.L; Writing—review & editing, A.N.

Funding: This research received no external funding.

Conflicts of Interest: The authors declare no conflict of interest.

References

1. Nogawa, S.; Nataka, K.; Mohammadi, A. Effect of ignition system on combustion. In Proceedings of the 1st International Conference: Advanced Ignition System for Gasoline Engines, Berlin, Germany, 12–13 November 2012.

2. Le Coz, J. *Cycle-to-Cycle Correlations between Flow Field and Combustion Initiation in an S.I. Engine*; SAE Technical Paper 920517; SAE International: Warrendale, PA, USA, 1992; p. 15.

3. Abdel-Gayed, R.G.; Bradley, D.; Lawes, M. Turbulent Burning Velocities: A General Correlation in Terms of Straining Rates. *Proc. R. Soc. Lond. A.* **1987**, *414*, 389. [CrossRef]

4. Nishiyama, A.; Furui, T.; Ikeda, Y. Development of M12 Type Plug-in LDV Probe for Evaluation of Flow Field near Spark Location in SI Engine. In Proceedings of the 17th International Symposium on Applications of Laser Techniques of Fluid Mechanics, Lisbon, Portugal, 7–10 July 2014.

5. Ikeda, Y.; Nishiyama, A.; Furui, T.; Wachi, Y. PIV/LDV combination for optimum turbulence generation scheme to spark ignition in wide operating condition of SI engine. In Proceedings of the 16th International Symposium on Applications of Laser Techniques to Fluid Mechanics, Lisbon, Portugal, 9–12 July 2012.

6. Okura, Y.; Segawa, M.; Onimaru, H.; Urata, Y.; Tanahashi, M. Analysis of in-cylinder flow for a boosted GDI engine using high-speed particle image velocimetry. In Proceedings of the 17th International Symposium on Applications of Laser Techniques to Fluid Mechanics, Lisbon, Portugal, 7–10 July 2014.

7. Sick, V. High speed imaging in fundamental and applied combustion research. *Proc Comb. Inst.* **2013**, *34*, 3509–3530. [CrossRef]

8. Voisine, M.; Thomas, L.; Boree, J.; Rey, P. Spatio-temporal structure and cycle to cycle variations of an in-cylinder tumbling flow. *Exp. Fluids* **2011**, *50*, 1393–1407. [CrossRef]

9. Chi, T.; Zhai, G.; Kook, S.; Chan, Q.; Hawkes, E. Application of LED-based thermographic phosphorescent technique to diesel combustion chamber walls in a pre-burn-type optical constant-volume vessel. *Exp. Fluids* **2019**, *60*, 58. [CrossRef]

10. Clark, L.; Kook, S.; Chan, Q.; Hawkes, E. *Multiple Injection Strategy Investigation for Well-Mixed Operation in an Optical Wall-Guided Spark-Ignition Direct-Injection (WG-SIDI) Engine through Flame Shape Analysis*; SAE Technical Paper 2016-01-2162; SAE International: Warrendale, PA, USA, 2016; p. 12.

11. Clark, L.; Kook, S.; Chan, Q.; Hawkes, E. *Influence of Injection Timing for Split-Injection Strategies on Well-Mixed High-Load Combustion Performance in an Optically Accessible Spark-Ignition Direct-Injection (SIDI) Engine*; SAE Technical Paper 2017-01-0657; SAE International: Warrendale, PA, USA, 2017; p. 11.

12. Peterson, B.; Reuss, D.L.; Sick, V. High-speed imaging analysis of misfires in a spray-guided direct injection engine. *Proc. Comb. Inst.* **2011**, *33*, 3089–3096. [CrossRef]

13. Mounaïm-Rousselle, C.; Landry, L.; Halter, F.; Foucher, F. Experimental characteristics of turbulent premixed flame in a boosted spark-ignition engine. *Proc. Comb. Inst.* **2013**, *34*, 2941–2949. [CrossRef]

14. Aleiferis, P.G.; Behringer, M.K. Flame front analysis of ethanol, butanol, iso-octane and gasoline in a spark-ignition engine using laser tomography and integral length scale measurement. *Combust. Flame* **2015**, *162*, 4371–4674. [CrossRef]

15. Le, M.K.; Furui, T.; Nishiyama, A.; Ikeda, Y. Application of High-Speed PIV Diagnostics for Simultaneous Investigation of Flow Field and Spark Ignited Flame inside an Optical SI Engine. *SAE Int. J. Eng.* **2017**, *10*, 917–927. [CrossRef]

Article

CO_2 Emission of Electric and Gasoline Vehicles under Various Road Conditions for China, Japan, Europe and World Average—Prediction through Year 2040

Xue Dong [1], Bin Wang [2], Ho Lung Yip [3] and Qing Nian Chan [3,*]

1   China-UK Low Carbon College, Shanghai Jiao Tong University, Shanghai 200240, China;
    xue.dong@sjtu.edu.cn
2   State Key Laboratory of Ocean Engineering, School of Naval Architecture, Ocean & Civil Engineering,
    Shanghai Jiao Tong University, Shanghai 200240, China; wangbin1981525@sjtu.edu.cn
3   School of Mechanical and Manufacturing Engineering, University of New South Wales,
    Sydney, NSW 2052, Australia; h.l.yip@unsw.edu.au
*   Correspondence: qing.chan@unsw.edu.au

Received: 29 March 2019; Accepted: 29 May 2019; Published: 4 June 2019

Abstract: Many countries are making strategic plans to replace conventional vehicles (CVs) with electric vehicles (EVs), with the motivation to curb the growth of atmospheric CO_2 concentration. While previous publications have mainly employed social-economic based models to predict CO_2 emission trends from vehicles over the years, they do not account for the dynamics of engine and motor efficiency under different driving conditions. Therefore, this study utilized an experimentally validated vehicle dynamic model to simulate the consumption of gasoline and electricity for CVs and EVs, respectively, under eight driving cycles for different countries/regions. The CO_2 emissions of CVs and EVs through 2040 were then calculated, based on the assumptions of the improvement of engine efficiency and composition of power supply chain over the years. Results reveal that, assuming that the current projections and assumptions remain valid, China would have the highest CO_2 emission for EVs, followed by Japan, world average and the EU, mainly determined by the share of fossil fuels in the power grid. As for the influence of road conditions, the CO_2 emission of CVs was found to be always higher than that of EVs for all countries/regions over the years. The difference is around 10–20% under highway conditions, and as high as 50–60% in crowded urban driving conditions.

Keywords: CO_2 emission; gasoline vehicles; electric vehicles; driving cycles; vehicle dynamic model

---

1. Introduction

An increasing level of anthropogenic greenhouse emissions is considered to have significant impacts on the climate and marine environment [1]. As the adverse effects of climate change are leaving some countries vulnerable in terms of quality of life, physical infrastructure and biological diversity, etc. [2], the international community has been increasingly active in response to climate challenges in recent years [3]. For example, the Paris Agreement aims to control the increase in the global average temperature to well below 2°C above pre-industrial levels [4]. In order to achieve this long-term temperature goal, parties of the Paris Agreement aim to reach global peak greenhouse gas emissions as soon as possible, so as to achieve a balance between anthropogenic emissions of greenhouse gases by sources and removals by sinks in the second half of this century [5]. Of all the greenhouse gases, CO_2 is the major concern as it is the most abundant composition. Atmospheric CO_2 stabilization target at 450 ppm would be needed to forestall coral reef bleaching, thermohaline circulation shutdown, and sea level rise from disintegration of the West Antarctic Ice Sheet [6]. However, achieving such target would require significant effort, and even stabilizing atmospheric CO_2 concentration at 550 ppm requires a carbon-free energy usage of 15 TW by mid-21st century [6]. The current global annual energy usage is around 20 TW, with 82% of the energy consumption being supplied from non-carbon-neutral sources [7]. It can therefore be anticipated that, without effective measures to limit CO_2 emissions, it will be challenging to achieve the emission targets set by the Paris Agreement.

1.1. The Contribution of Road Traffic Emissions to Global Greenhouse Gas Emissions

Transportation has become a significant contributor to the global energy crisis and high greenhouse gas emissions. In 2006, 60.5% of the world's petroleum was consumed in cars, and this figure is anticipated to reach 62% by 2020 [8]. High consumption rate of fossil fuels is accompanied with a considerable amount of CO_2 emission. Specifically from 2001 to 2011, CO_2 emissions increased by 13%, and 25% of this increase is attributed to vehicle emissions [8]. Given that internal combustion engine is still the major powertrain system for on-road vehicles, many countries are making strategic plans to replace the conventional vehicles (CVs) with electric vehicles (EVs) and fuel cell vehicles with lower tailpipe emissions, in an effort to curb CO_2 emissions. For instance, the US government enacted energy policy provisions to support the development of hydrogen energy and fuel cell technologies in 2003. In 2009, the US also announced plans to achieve 1 million plug-in hybrid vehicles on the road by 2015 [8]. Similarly, Germany issued the national electro-mobility development plan in 2009, and it declared plans to have at least 1 million EVs by 2020 and 5 million by 2030 [9], with the investment of 5 billion euros ($\sim$5.6 billion USD) in hydrogen energy and fuel cell technology development from 2007 to 2017 [8]. In addition, the Japanese government planned to build 8500 hydrogen refueling stations by 2030 [8] and it is estimated that the country's fuel cell vehicles will account for 20% of the total vehicles on the road by that time [8]. Meanwhile, China became the country with the highest CO_2 emissions and energy consumption in 2006 and in 2009, respectively [10]. At the same time, China's transportation industry is still dominated by internal combustion engine vehicles (i.e., 62.39% of railway locomotives, 84.42% of civilian transport ships, and 98.93% of civilian motor vehicles) [8]. In order to alleviate these problems, China's central and local governments have taken extensive measures in recent years [11], including the introduction of more stringent fuel consumption regulations [12], subsidy scheme to accelerate penetration of the electric vehicle market [13], and restrictions on the purchase of vehicles and license plates, etc. [14]. Therefore, replacing CVs with EVs and FCVs seems to be an inevitable trend worldwide.

1.2. The CO2 Emission from CV, EV and HEV of Different Countries/Regions

The CO_2 emission of EV and hybrid electric vehicle (HEV), however, varies among countries/regions, due to the difference in the composition of power grid. According to Howey et al. [15], HEV showed the the lowest CO_2 emission, with more than half of the tested HEVs had an emission of 70 g CO_2/km, while that for most of the tested EVs were 70–110 g CO_2/km. The most efficient diesel engine vehicles, however, were reported to have the highest emission exceeding 110 g CO_2/km, despite the recent development of diesel fuel and injection strategy [16–18]. It is worth noting that these results were acquired based on the structure of British power grid, where CO_2 emission rate is 542 g CO_2/kWh. Similarly, Sullivan et al. [19] suggested that EV consumes less energy per kilometer compared with CV, but it was also identified that EV is not comparable with CV with regard to dynamic power and cruising range. In addition, Samaras and Meisterling [20] studied the life-cycle CO_2 emission of HEVs in the US, and their results revealed that HEV has lower emission than CV only when the power grid uses sufficient renewable energy. As coal-fired plants powered most of the grids in China, the CO_2 emission rate of EVs in China is higher than that of other developed countries. A previous study by Hao et al. [21] reported that the collective use of five control measures in China, including constraining vehicle registration, reducing vehicle travel, strengthening fuel consumption rate limits, vehicle downsizing and promoting EV penetration, will reduce energy consumption by 62.9% and 75.7%, by 2030 and 2050, respectively, when compared with the reference scenario with no measures taken in China. In the same study, it was also reported that similar potential in greenhouse gas mitigation can be realized. Likewise, Peng et al. [22] analyzed the CO_2-reduction potentials and emissions abatement costs from micro-vehicular and macro-industrial perspectives from 2010 to 2030, and found that the technologies with large emissions reduction potential are mainly available in plug-in hybrid electric vehicle (PHEV) and EV paths, which would be the main channels for reducing carbon emissions in the long run. While the carbon emission of CV, EV and HEV has been studied and reported by researchers from different countries, there is, however, no direct and fair comparison reported about the well-to-wheel CO_2 emission of CV and EV among different countries/regions, which this paper seeks to address.

1.3. The Effect of Road Conditions on Vehicle CO_2 Emission

Although previous social-economic based research model using macroscopic social-economic data, e.g., national-average sales, ownership and oil consumption, etc. provides valuable predictions on the emission trends over the years, results acquired from vehicle dynamic models are also desired because CO_2 emission is highly dependent on engine/motor efficiency, hybrid extent and road conditions, etc., which have to be obtained from vehicle dynamic models. Other institutes, such as Argonne National Laboratory, have previously developed the Powertrain Systems Analysis Toolkit (PSAT) to simulate vehicle efficiency and cycle implications over different of driving cycles [23]. Further work, however, is required to estimate CO_2 emissions of CVs and EVs in the future under the combined effects of driving cycles and energy utilizations, which is the motivation behind this study.

Driving cycles, which determine the relationship between vehicle speed and time, are produced by different countries and organizations to assess the performance of vehicles under real-life road conditions. For example, US06, high way fuel economy test (HWFET) cycle, urban dynamometer driving schedule (UDDS) cycle and New York City Cycle (NYCC) were formulated by the US Environmental Protection Agency [24], while new European driving cycle (NEDC) and JC08 are from Europe and Japan, respectively. Development and Reform Commission of China formulated urban test cycles with Chinese characteristics, including Chinese city driving cycle for urban road (CCUR) and Chinese city driving cycle for the expressway (CCEW) [25]. These driving cycles were originally used to measure exhaust emissions from CVs and are currently being employed to assist in the analysis, design and testing of EVs as well. More detailed information about these drive cycles will be provided in Section 2.4.

This current study aims to predict the well-to-wheel emission of gasoline fuelled CVs and EVs of different countries/regions under various driving conditions. The prediction is made from year 2018 through year 2040 because this is the time frame for which International Energy Outlook provided detailed information about energy utilization of different countries.

2. Methodology

2.1. CO_2 Emission of Electricity and Gasoline

In this work, the well-to-wheel, i.e., well-to-tank plus tank-to-wheel CO_2 emission of CV and EV was predicted. It is worth noting that, while this study accounts for the carbon emission from battery manufacturing and waste recovery, the CO_2 emission of car manufacturing and that is induced by the increased infrastructure for power supply are not considered. In this study, the well-to-wheel CO_2 emission was calculated using a vehicle dynamic model, which incorporates a spark-ignition direct-injection (SIDI) gasoline engine, a motor, battery and transmission system. The SIDI gasoline engine was chosen because it is the mainstream technology due to its improved fuel economy, more precise air-fuel ratio control and improved transient response [26–30].

The well-to-tank CO_2 emission of gasoline, i.e., carbon emission during the mining, storage and transportation process, is estimated to be 406.8 g/L [31]. The well-to-wheel CO_2 emission of gasoline vehicle is calculated to be 2.734 kg/L, with the CO_2 emission regulation and fuel consumption assumption of 128 g/km and 5.5 L/100 km, respectively. This emission level will be applied together with the fuel consumption calculated in the next section, to predict the CO_2 emission of gasoline vehicles over the years.

The CO_2 emission of EV mainly depends on the emission of power, as well as the carbon emission from battery manufacturing and waste recovery. According to Samaras and Meisterling [20], approximately 1510–1870 MJ of primary energy is required during the manufacturing process of 1 kWh of Li-ion battery storage capacity and its materials. Therefore, the average value of 1700 MJ or 10 g CO_2-eq/km was estimated as the carbon impacts of battery over the vehicle life-cycle from that of lithium-ion battery. The carbon emission of power is determined by the composition of the power grid because different forms of power generation result in different levels of CO_2 emission, as shown in Figure 1 [31].

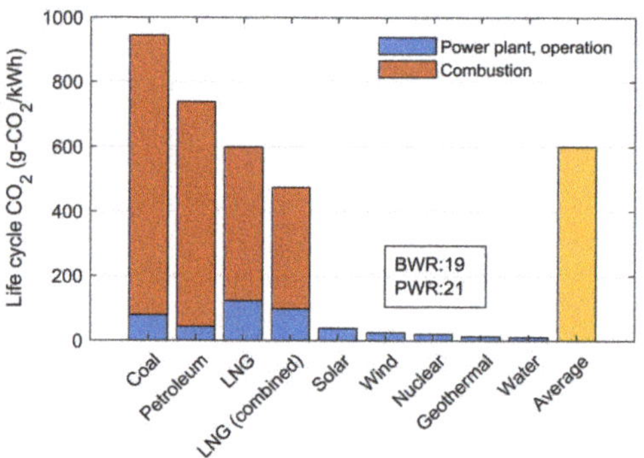

Figure 1. Life-cycle CO_2 emission from different forms of power generation [31]. BWR and PWR are abbreviations for boiling water reactor and pressurized water reactor, respectively.

Based on the life-cycle CO_2 emission of different forms of power generation shown in Figure 1, and the composition of power grid of different countries/regions reported in International Energy Outlook 2017 [7], the CO_2 emission of electricity (g/kWh) for the world average, EU, China and Japan can therefore be estimated, as shown in Figure 2. These countries/regions were chosen in this study as China and the EU are two extreme cases, representing the highest and lowest CO_2 emission per kWh of electricity generated, respectively, while that of Japan and World Average are in the middle. It can be seen from year 2018 to 2040 that the overall CO_2 emission of electricity is projected to decrease over the years for all countries/regions studied, as a result from the anticipated increase in the share of renewables with lower carbon emissions in all power supply chains. Of the individual countries/regions studied, China has the highest CO_2 emission per kWh of electricity generated due to its large share of coal-fired power, while the EU has the lowest over the years. The world average CO_2 emission level is in between the EU and China, whilst Japan has an emission level slightly higher than the world average over the 32-year projection. These data, when combined with the electricity consumption of EVs calculated in the following section, can be used to project the CO_2 emission of EVs for different countries/regions through 2040.

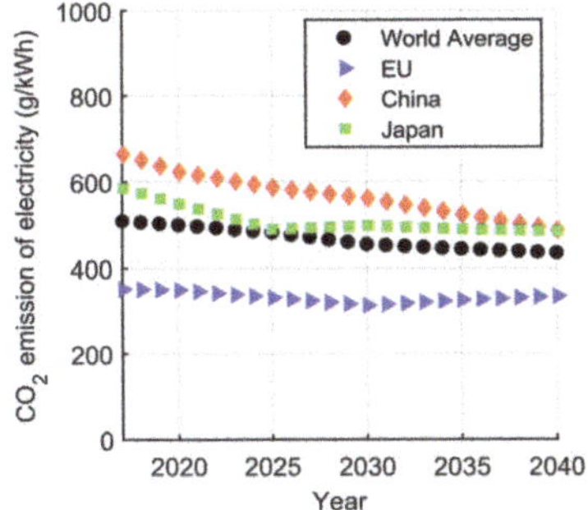

Figure 2. CO$_2$ emission per kWh of electricity generated for different countries/regions calculated according to International Energy Outlook 2017 [7].

2.2. Vehicle Dynamic Model

The vehicle dynamic model is usually employed to simulate the overall performance of vehicle during the design and optimizaiton process. In this work, a PHEV was studied because it is the most common type of HEV and it can run purely as a gasoline fuelled CV or a EV. As shown in Figure 3, the model consists of the power sources (engine and motor), the transmission systems (P2 clutch, Dual Clutch Transmission (DCT)), battery, vehicle and the driver module. The control strategy can be divided into three modules—driving mode identification module, required torque calculation module and driving mode actuation module. Together with the vehicle dynamic model, different driving modes can be achieved, which enables quantitative analysis on energy consumption under various driving modes, thus optimizing the energy management strategy. The vehicle model in the current study employs forward simulation. This means that the actuation starts from the driver, who can sense the speed of vehicles and adjust its speed accordingly. The action generates torque in the powertrain, and the torque is subsequently transmitted to the wheels. A flow diagram of the vehicle dynamic model employed in this study is presented in Figure 4, and the details of each submodule are discussed from Sections 2.2.1–2.2.7.

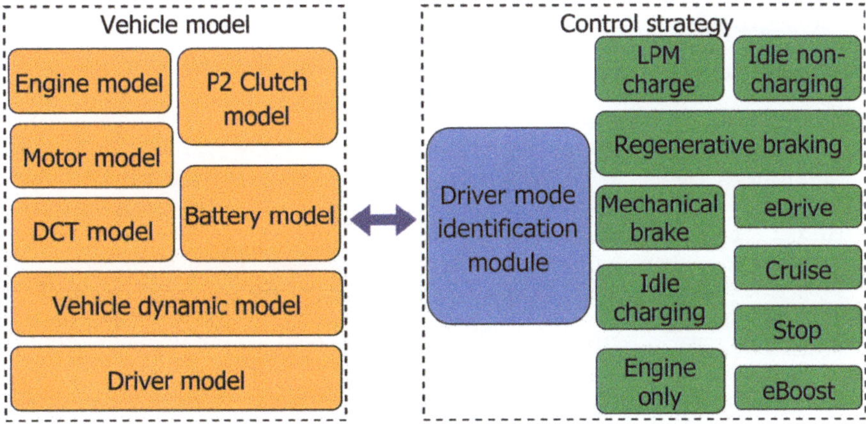

Figure 3. Structure of the vehicle dynamic model and its control strategy.

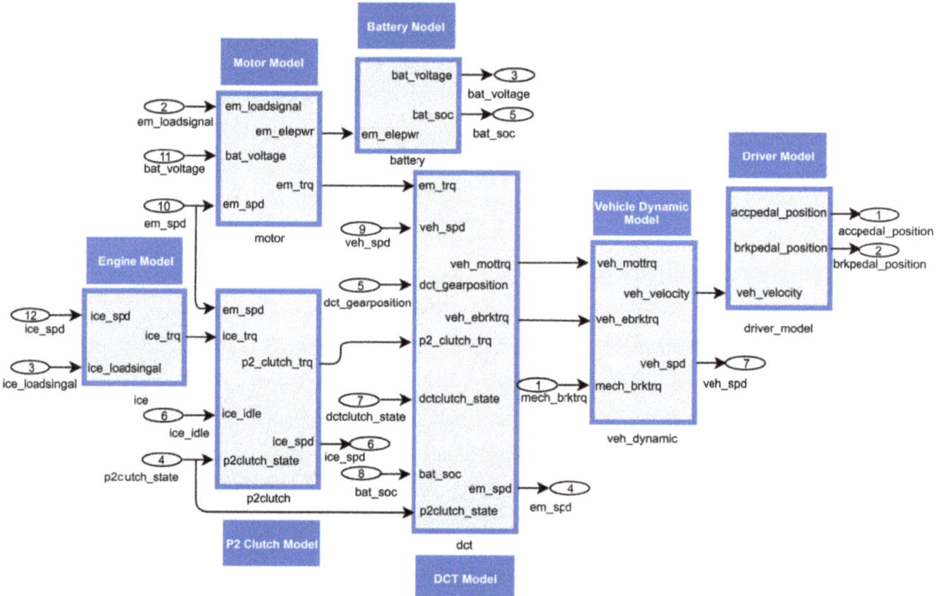

Figure 4. Flow diagram of the vehicle dynamic model of a hybrid electric vehicle employed in this study.

2.2.1. Engine Model

In the HEV dynamic model, only the input and output of the engine are considered. As such, the engine performance under different operating conditions can be acquired from the engine characteristic curve (shown later in Section 2.3). The input operating parameters include the engine speed and control signal, while the output signal is engine torque. Specifically, when the control signal is higher than zero, the fuel consumption rate is calculated based on the engine speed and torque; while the control signal equals to zero, the fuel consumption rate is determined by that in idle conditions.

2.2.2. Motor Module

The motor module is to convert the required vehicle speed and torque to the required battery power, and then generates the corresponding torque based on the battery capacity. The motor model was established based on the motor voltage, torque, power and operating characteristic equation. With torque and power being the outputs, the input is the speed connected to the transmission system and motor control signal. When the signal is positive, the motor is actuated, and the motoring torque was calculated based on the motor speed, control signal and voltage of the battery; whereas, when the control signal is negative, the transmission system drives the motor to regenerate electricity. The power P_m under motoring and regenerating condition follows Equation (1). The electricity transmission losses from both the motor control unit (MCU) and the motor itself were taken into account by experimentally testing the efficiency of the motor system (i.e., motor body + motor controller) under different working modes. The data acquired was then applied in the model. An equivalent battery charging efficiency of 98% was also used to account for the losses from battery charging and conversion from AC to DC:

$$P_m = \begin{cases} \dfrac{T_m \omega_m}{\eta_m} & motoring, \\ T_m \omega_m \eta_m & regenerating, \end{cases} \tag{1}$$

where η_m is motor efficiency, ω_m is the motor rotating speed connected with the transmission system, and T_m is the motor torque.

2.2.3. P2 Clutch Model

The inputs to the clutch model include the engine torque, rotation speed connected to the transmission system, the working conditions of the P2 clutch, and the engine speed in idle conditions. The outputs are the transmission torque of the clutch as well as the engine speed. While the P2 clutch is engaged, the speed and transmission torque of the engine is the same as those of the motor; however, when the P2 clutch is disconnected, the engine remains in an idle condition, as such the transmission torque of the clutch is zero.

2.2.4. Battery Model

The battery model describes the mathematical relationship among the voltage, current, resistance, temperature and state of charge (SOC). The current study employed the internal resistance battery model, for which the input is the discharge power of the battery, and the model outputs are voltage and SOC.

2.2.5. DCT Model

In the DCT model, the input parameters include the motor torque, P2 clutch torque, rotating speed of the wheel, DCT gear position, DCT state, P2 clutch state, and SOC, while the output parameters include the torque transmitted to the wheel, brake recovery torque, and motor speed. When the transmission torque is positive, the vehicle is driven by the power source; otherwise, the vehicle is under brake state, where the recovery torque is calculated. When the DCT is engaged, the motor speed was determined by the wheel speed. However, while the DCT is disconnected, the motor idle speed was determined through SOC.

2.2.6. Vehicle Dynamic Model and Driver's Model

The inputs of the vehicle dynamic model include driving torque, mechanical braking torque and brake recovery torque, and the outputs of this module are vehicle speed and wheel rotating speed. The driver's model can be approximated as a vehicle speed controller. This model employed a proportional–integral–derivative (PID) controller, where the difference between the expected driving speed and the actual vehicle speed is converted to the acceleration or deceleration command. Specific for the current study, the instantaneous command was given based on the speed and torque required by the driving cycle.

2.2.7. Control Strategy

Ten working modes of the current HEV model and their specifications, including veh_stop, idle_nocharging, idle_charging, brk_disableebrking, brk_enableebrking, cruise, eDrive, low power mode (LPM) charging, eBoost and ice_alone, are listed in Table 1. Figure 5 shows the flow diagram of control strategy. From the perspective of energy management, the control strategy aims to provide optimized working condition for the engine. Therefore, baseline control strategy was used. The baseline control restricts the engine to work under its high efficiency regime by setting threshold of the operating conditions. In this strategy, the motor was flexibly utilized as an assistant power source whenever the power from engine itself is insufficient, thus achieving "peak clipping" for the engine.

Table 1. The working modes and their specifications of the HEV model.

Number	HEV Working Mode	Specifications
1	veh_stop	Vehicle stops with engine and motor turned off
2	idle_nocharging	Idle condition for engine, no charging to battery
3	idle_charging	Idle condition for engine, charging to battery
4	brk_disableebrking	Mechanical brake
5	brk_enableebrking	Wheels drive motor to regenerate power, with mechanical brake being the assistance
6	cruise	Neither engine nor motor provides torque, the vehicle decelerates due to resistance
7	eDrive	The required torque was provided by motor only
8	LPM charging	The required torque was provided by engine, the engine works in optimized conditions, while the motor is charging the battery
9	eBoost	The required torque was provided by both engine and motor, the engine is working in optimized conditions, while the motor is charging the battery
10	ice_alone	The required torque was provided by engine only

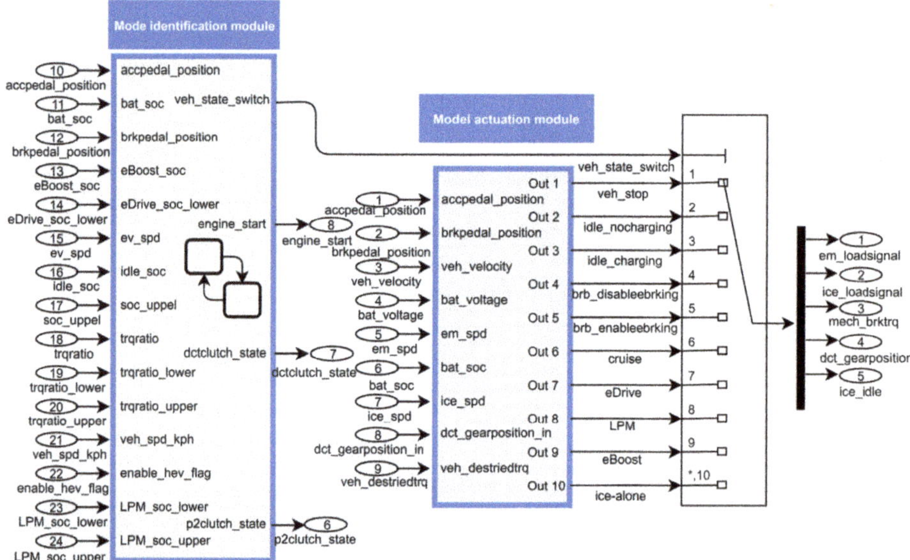

Figure 5. MATLAB/Simulink control strategy of the vehicle dynamic model presented in Figure 4.

2.3. Model Validation

The vehicle dynamic model was experimentally validated by the testing results of a PHEV SUV manufactured by the Shanghai Automotive Industry Cooperation (SAIC). The curb weight (dry weight) of vehicle is 2021 kg and the driving resistance coefficient A, B and C are 185.87 N, 1.0793 N/kph and 0.04899 N/kph^2, respectively. The vehicle also equipped with a lithium-ion battery with a maximum battery capacity of 8.8 kWh. The engine and motor characteristics are presented in Figure 6.

In this study, we calculated the fuel (gasoline) consumption of CVs under eight different driving cycles from year 2018 to 2040. Due to the development of advanced engine technology, the improvement of engine efficiency over the years needs to be considered. The current research follows the prediction from the Energy-saving and New Energy Vehicle Technology Roadmap released by China [32], and assumes a maximum engine efficiency of 38% for 2018, 40% for 2020, 50% for

2030 and 55% for 2040. Linear interpolation was employed to determine the engine efficiency for the years in between. In addition, the efficiency from the whole engine characteristic map (Figure 6a) was assumed to proportionally increase with the maximum efficiency. It should be noted that the engine characteristic map shown in Figure 6a has a maximum engine efficiency of 33% because it is not the most up-to-date engine model. In contrast, the efficiency of the motor is assumed to be constant over the years because the current motor efficiency is considered to be high enough with limited space for improvement. Therefore, the decrease of CO_2 emission for EV over the years is purely due to the reduced CO_2 emission from the power grid.

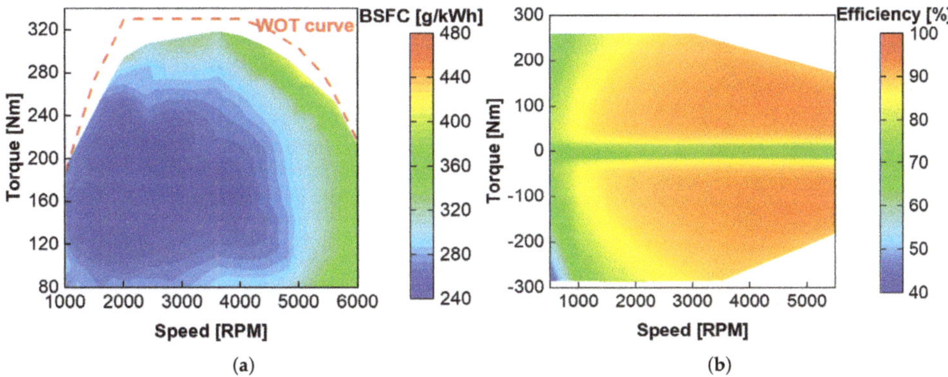

Figure 6. (a) engine and (b) motor characteristics of the hybrid electric vehicle employed in the current study.

The experiments follow the NEDC, with details shown in the following section. For the model validation, two conditions were studied—CV and P2 HEV. Under CV condition, both start and end SOC is 70, and the working modes experienced are 1, 2, 4 and 10 from Table 1. The battery did not discharge when the vehicle was driven by engine. The SOC of the vehicle therefore remained unchanged when operated in CV mode. However, for the P2 HEV condition, the start and end SOC are 63.0 and 64.5, respectively, and the working modes experienced include 1, 2, 5, 7, 8, 10. The results show that the deviation between the simulated and tested gasoline consumption (L/100 km) is within 3% under both CV and P2 HEV conditions. Considering that the HEV used in this study can be operated as pure EV or CV, in two extreme cases, this implies that our model is capable of capturing characteristics of the engine and motor modes of the vehicle. Therefore, the current vehicle dynamic model is considered to be accurate and valid to carry on the following analysis on CO_2 emission of CV and EV under various driving cycles for different countries. The starting SOC for the EV mode used in the model is 100% and a single cycle was run. The experimental conditions and specifications are summarized in Table 2.

Table 2. Conditions and specifications for model validation.

	CV	P2 HEV
Start SOC	70.0	63.0
End SOC	70.0	64.5
Working modes experienced	1, 2, 4, 10	1, 2, 5, 7, 8, 10
Gasoline consumption (L/100km)	9.733	6.975
Deviation between the simulation and experiments	< 3%	< 3%

2.4. Description of Driving Cycles

Figure 7 presents the dynamic characteristics of eight most commonly used driving cycles, including US06, HWFET, NEDC, UDDS, JC08, NYCC, CCUR and CCEW. Among them, US06 and HWFET are both used to indicate high-speed driving conditions. It can be seen from Figure 7a,b that US06 has greater vehicle acceleration/deceleration and peak vehicle speed. However, HWFET can be considered as a relatively stable highway operating condition with a maximum speed of no more than 96 km/h [33].

NEDC is a synthetic driving cycle that includes four repeating urban driving cycles and an extra-urban driving cycle. It is considered as the most robust driving cycle [33], i.e., vehicles based under the NEDC have the least degradation in performance than other driving cycles. However, the NEDC has relatively less dynamic changes compared with the other cycles described. In addition, in China's national standards GB/T 18386-2005 and GB/T 19233-2008, the NEDC is used to measure the mileage, emission level and fuel consumption of vehicles (including CV and EV).

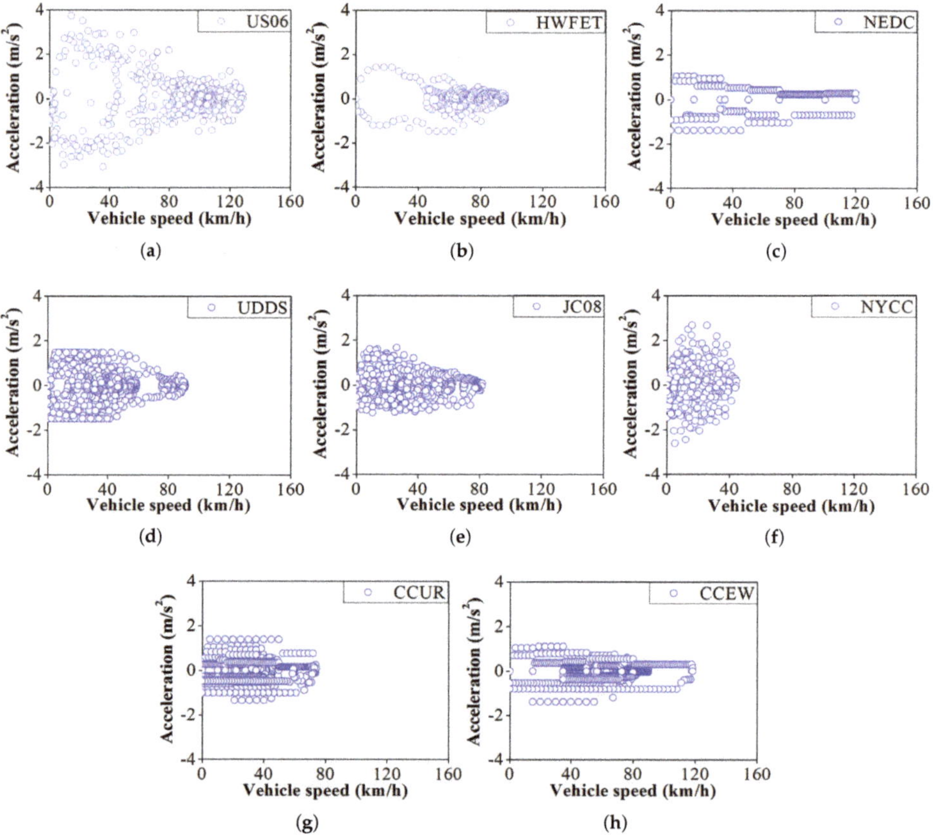

Figure 7. The dynamic characteristics of driving cycles based on acceleration and speed for: (a) US06; (b) HWFET; (c) NEDC; (d) UDDS; (e) JC08; (f) NYCC; (g) CCUR; and (h) CCEW.

In the US, the UDDS standard cycle is commonly used to evaluate the performance of light vehicles in urban conditions [24]. The widely-known FTP-75 driving cycle comes from UDDS, by adding the first 505 s of UDDS to its end. The JC08 driving cycle is a new fuel consumption and emissions test evaluation standard cycle implemented in Japan in 2015. It mainly reflects the characteristics of crowded urban driving conditions. Compared with the previously used Japan 10–15 mode driving cycle, JC08 has longer mileage and more stringent acceleration/deceleration.

NYCC represents the driving conditions in urban centers with frequent starts and stops. It can be found from Figure 7f that the peak speed is the lowest among all driving cycles, but the maximum acceleration/deceleration is large, only slightly smaller than that of US06. In contrast with HWFET, NYCC represents the low speed/high acceleration driving conditions, while HWFET cycle stands for the high speed/low acceleration driving conditions.

Finally, this study also selected two urban circulation conditions with Chinese traffic characteristics, i.e., CCUR and CCEW. Expressway is defined as conditions with a vehicle speed of more than 60 km/h and no traffic lights, for instance the city ring of Beijing and Shanghai, etc. [25], while the speed on urban roads is significantly lower than that of expressways.

3. Results and Discussion

In this study, the CO_2 emissions (kg/100 km) of CV (fueled by gasoline) and EV of different countries/regions from year 2018 to 2040 were predicted. Figure 8 presents the CO_2 emission from CV and EV of different countries under NEDC. It was chosen for comparing different countries' CO_2 emission because it is considered to have the least degradation in performance. It should be noted that the emission of CV is universal across the globe in this study, as it was calculated based vehicle fuel consumption. It can be seen in Figure 8 that, from year 2018 to 2040, CV has higher CO_2 emission than EV. For all the countries studied, China has the highest CO_2 emission for EV, around 50% of that from CV in Year 2018 and 66% of that from CV in Year 2040. However, carbon emission of EV from the EU remains at the lowest level (around 7 kg/100 km) in this time frame. Meanwhile, CO_2 emission of EV from Japan and the world average is in between that from China and the EU, and their differences are completely determined by the CO_2 emission of electricity (shown in Figure 2).

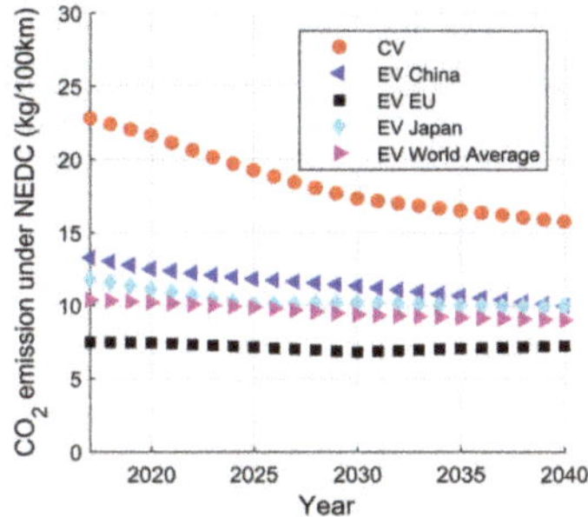

Figure 8. CO_2 emission of CV and EV under New European Driving Cycle for different countries from year 2018 to 2040. The EV model has electricity regeneration during braking.

Figure 9 presents the simulated results under eight different driving cycles. Since EV emission of different countries follow the exact same trend with their electricity emissions (shown in Figure 2), the following figures only show the trend for China and the EU as the two extreme representatives. Although the results are dependent on driving cycles (road conditions), there are some common trends for all types of driving cycles studied—(1) The CO_2 emission decreases over the years for both CVs and EVs, due to the increased engine efficiency and increased fraction of renewable energy in the power grid, respectively; (2) For all countries/regions, CV has higher CO_2 emission than EV from year 2018 to 2040; (3) China has the highest CO_2 emission for EVs, attributed to its higher CO_2 emission of electricity, followed by Japan, the world average and the EU.

Comparing Figure 9a,b with the other subfigures, it can be observed that the CO_2 emissions of CV (red curve) under China-fast and HWFET cycle are significantly lower, and closer to the emission of EV than other driving cycles. Specifically, under China Fast condition, the CO_2 emission of CV is about 23% and 17% higher than that of China EV with no electricity regeneration during braking (ebraking) in year 2018 and 2040, respectively. For HWFET, the CO_2 emission of CV is 16% higher than that of China EV with no ebraking in 2018, and this figure decreases to 9% by 2040. As described in the previous section, HWFET represents relatively stable highway operating conditions in the US, while China-fast represents the driving conditions in expressways with no traffic lights in China. Under these conditions, the engine operates at high speed, medium load (torque), where the brake specific fuel consumption is the lowest (engine efficiency highest), as shown in Figure 6a. This explains the lower CO_2 emission of CV under highway than other conditions.

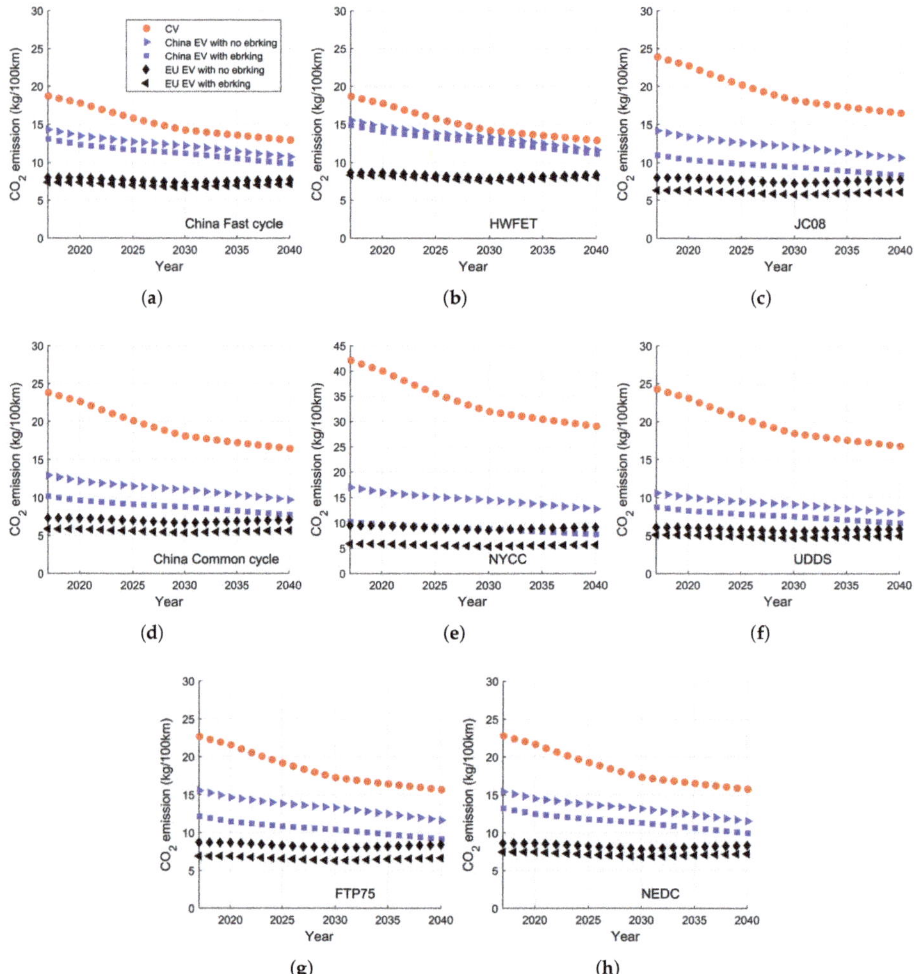

Figure 9. CO_2 emission from conventional vehicles fueled by gasoline and electric vehicles from China and the EU, predicted through 2040 under different driving conditions, including (**a**) China Fast cycle; (**b**) HWFET; (**c**) JC08; (**d**) China Common cycle; (**e**) NYCC; (**f**) UDDS; (**g**) FTP75 and (**h**) NEDC. ebraking means electricity regeneration during braking.

In contrast, in Figure 9c–g, the CO_2 emission of CV is a lot higher than that of EV. Among them, the most significant difference was observed in NYCC and UDDS cycle. For NYCC, the emission of

CV is 59% higher than that of China EV with no ebraking in 2018, and this difference is only slightly reduced to 56% in 2040, while that, for the UDDS cycle is 56% and 52%, respectively. This is because NYCC represents the driving conditions in urban centers with frequent starts and stops. UDDS is also used to evaluate the performance of light vehicles in urban conditions. On the other hand, JC08, China Common and FTP-75 cycle are standards for crowded urban driving conditions in Japan, China and the US, respectively. It can be seen from Figure 7a that, under these driving conditions, the vehicles maintain a low speed most of the time with frequent and significant acceleration and deceleration, which prohibits the engine from operating at its high efficiency regime (Figure 6a). Therefore, the CO_2 emission of CV is a lot higher than that of EV under these crowded urban conditions. The carbon emission of NEDC (Figure 9h) is similar to that under FTP-75 (Figure 9g). Additionally, it is found that the difference between EV with and without ebraking is larger in these urban driving conditions. This is because the frequent and substantial deceleration in the urban area recovers more electricity than other driving conditions. Therefore, it is suggested to activate ebraking for urban driving conditions in order to reduce carbon emission. Although it was found that EVs generally have less well-to-wheel CO_2 emissions across difference driving cycles as well as countries/regions than CVs, it should be noted that EV growth can lead to potential reshaping of the electricity load curve as well as increasing total electricity demand, which can place new strains grid in near to midterm, and in the long run, respectively [34]. While the impact of power-demand of the EVs on the grid as well as its corresponding emission characteristics is important and warrants additional research, the topic itself is complex and is beyond the scope of this study.

4. Conclusions

This work predicted the well-to-wheel CO_2 emission of gasoline vehicle and electric vehicles for China, Europe, Japan and World Average through year 2040 under eight different driving conditions. An experimentally validated vehicle dynamic model was employed to calculate the gasoline and electricity consumption of CV and EV. The key conclusions are as follows:

1. The CO_2 emission decreases over the years for both CVs and EVs, due to the increased engine efficiency and increased fraction of renewable energy in the power grid, respectively.
2. For all the countries/regions studied, CVs have higher CO_2 emission than EVs from year 2018 to 2040.
3. For all the countries/regions studied, China has the highest CO_2 emission for EVs, followed by Japan, the world average, and the EU. Their differences are determined by CO_2 emission of electricity supply chain.
4. The difference of CO_2 emission from CVs and EVs is smaller under highway conditions, compared with those under urban driving conditions. For instance, the CO_2 emission from CVs is 23% higher than that of EVs in China in year 2018, and decreases to 17% in year 2040 for China-fast driving conditions.
5. For urban driving conditions, e.g., New York City Cycle, the carbon emission gap between CVs and EVs in China is 59% and 56% in year 2018 and year 2040, respectively. However, electricity regeneration during braking is found to be effective in reducing carbon emission for EVs under urban driving conditions.

Author Contributions: Conceptualization, X.D. and B.W.; Formal analysis, B.W. and Q.N.C.; Methodology, X.D. and B.W.; Resources, H.L.Y. and Q.N.C.; Validation, X.D. and B.W.; Visualization, X.D. and H.L.Y.; Writing—original draft, X.D.; Writing—review and editing, H.L.Y. and Q.N.C. All authors have read and agreed with the final manuscript.

Funding: This research was funded by China Postdoctoral Science Foundation Grant No. 2016M600313, as well as Talent Plan Research Supporting Funds from Shanghai Jiao Tong University.

Conflicts of Interest: The authors declare no conflict of interest.

Abbreviations

The following abbreviations are used in this manuscript:

CVs	Conventional vehicles
EVs	Electric vehicles
HEV	Hybrid electric vehicle
PHEV	Plug-in hybrid electric vehicle
HWFET	High way fuel economy test
UDDS	Urban dynamometer driving schedule
NYCC	New York city cycle
NEDC	New European driving cycle
CCUR	Chinese city driving cycle for urban road
CCEW	Chinese city driving cycle for express way
DCT	Dual clutch transmission
SOC	State of charge
ebraking	Electricity regeneration during braking

References

1. O'Neill, B.C.; Oppenheimer, M. Dangerous climate impacts and the kyoto protocol. *Science* **2008**, *296*, 1971–1972. [CrossRef]
2. Klein, R.J. Identifying countries that are particularly vulnerable to the adverse effects of climate change: An academic or a political challenge? *CCRC* **2009**, *3*, 284–291. [CrossRef]
3. Takeshita, T. Assessing the co-benefits of CO_2 mitigation on air pollutants emissions from road vehicles. *Appl. Energy* **2012**, *97*, 225–237. [CrossRef]
4. Rogelj, J.; Elzen, M.D.; Höhne, N.; Fransen, T.; Fekete, H.; Winkler, H.; Schaeffer, R.; Sha, F.; Riahi, K.; Meinshausen, M. Paris Agreement climate proposals need a boost to keep warming well below 2 °C. *Nature* **2016**, *534*, 631–639. [CrossRef]
5. United Nations. *Paris Agreement*; United Nations: New York, NY, USA, 2015.
6. Hoffert, M.I.; Caldeira, K.; Benford, G.; Criswell, D.R.; Green, C.; Herzog, H.; Jain, A.K.; Kheshgi, H.S.; Lackner, K.S.; Lewis, J.S.; et al. Advanced technology paths to global climate stability: Energy for a greenhouse planet. *Science* **2002**, *398*, 981–987. [CrossRef]
7. *International Energy Outlook 2017*; U.S. Energy Information Administration: Washington, DC, USA , 2017.
8. Wang, D.; Zamel, N.; Jiao, K.; Zhou, Y.; Yu, S.; Du, Q.; Yin, Y. Life cycle analysis of internal combustion engine, electric and fuel cell vehicles for China. *Energy* **2013**, *59*, 402–412. [CrossRef]
9. Bitonto, S.; Rico, T. *Electromobility in Germany: Vision 2020 and Beyond*; Gremany Trade & Invest: Berlin, Germany, 2015.
10. Lin, B.; Ouyang, X. Energy demand in China: Comparison of characteristics between the US and China in rapid urbanization stage. *Energy Convers. Manag.* **2014**, *79*, 128–139. [CrossRef]
11. Hu, X.; Chang, S.; Li, J.; Qin, Y. Energy for sustainable road transportation in China: Challenges, initiatives and policy implications. *Energy* **2010**, *35*, 4289–4301. [CrossRef]
12. Wang, Z.; Jin, Y.; Wang, M.; Wu, W. New fuel consumption standards for Chinese passenger vehicles and their effects on reductions of oil use and CO_2 emissions of the Chinese passenger vehicle fleet. *Energy Policy* **2010**, *38*, 5242–5250. [CrossRef]
13. Hao, H.; Ou, X.; Du, J.; Wang, H.; Ouyang, M. China's electric vehicle subsidy scheme: Rationale and impacts. *Energy Policy* **2014**, *73*, 722–732. [CrossRef]
14. Hao, H.; Wang, H.; Ouyang, M. Comparison of policies on vehicle ownership and use between Beijing and Shanghai and their impacts on fuel consumption by passenger vehicles. *Energy Policy* **2011**, *39*, 1016–1021. [CrossRef]
15. Howey, D.A.; Martinez-Botas, R.F.; Cussons, B.; Lytton, L. Comparative measurements of the energy consumption of 51 electric, hybrid and internal combustion engine vehicles. *Transp. Res. Part D* **2011**, *16*, 459–464. [CrossRef]

16. Chan, Q.N.; Fattah, I.M.R.; Zhai, G.; Yip, H.L.; Chen, T.B.Y.; Yuen, A.C.Y.; Yang, W.; Wehrfritz, A.; Dong, X.; Kook, S.; et al. Color-ratio pyrometry methods for flame–wall impingement study. *J. Energy Inst.* **2018**, in press. [CrossRef]

17. Ming, C.; Fattah, I.M.R.; Chan, Q.N.; Pham, P.X.; Medwell, P.R.; Kook, S.; Yeoh, G.H.; Hawkes, E.R.; Masri, A.R. Combustion characterization of waste cooking oil and canola oil based biodiesels under simulated engine conditions. *Fuel* **2018**, *224*, 167–177. [CrossRef]

18. Fattah, I.M.R.; Ming, C.; Chan, Q.N.; Wehrfritz, A.; Pham, P.X.; Yang, W.; Kook, S.; Medwell, P.R.; Yeoh, G.H.; Hawkes, E.R.; et al. Spray and combustion investigation of post injections under low-temperature combustion conditions with biodiesel. *Energy Fuels* **2018**, *8727–8742*, 167–177. [CrossRef]

19. Sullivan, J.L.; Hu, J. *Life Cycle Energy Analysis for Automobiles*; SAE Technical Paper No. 951829; SAE International: Warrendale, PA, USA, 1995.

20. Samaras, C.; Meisterling, K. Life cycle assessment of greenhouse gas emissions from plug-in hybrid vehicles: Implications for policy. *Environ. Sci. Technol.* **2008**, *42*, 3170–3176. [CrossRef]

21. Hao, H.; Wang, H.; Ouyang, M. Fuel conservation and GHG (greenhouse gas) emissions mitigation scenarios for China's passenger vehicle fleet. *Energy* **2011**, *36*, 6520–6528. [CrossRef]

22. Peng, B.; Fan, Y.; Xu, J. Integrated assessment of energy efficiency technologies and CO_2 abatement cost curves in China's road passenger car sector. *Energy Convers. Manag.* **2016**, *109*, 195–212. [CrossRef]

23. Plotkin, S.; Singh, M.; Patterson, P.; Ward, J.; Wood, F.; Kydes, N.; Holte, J.; Moore, J.; Miller, G.; Das, S.; et al. *Multi-Path Transportation Futures Study: Vehicle Characterization and Scenario Analyses: Main Text and Appendices A, B, C, D, and F*; Argonne National Laboratory: Argonne, IL, USA, 2009.

24. Dynamometer Drive Schedules. Available online: https://www.epa.gov/vehicle-and-fuel-emissions-testing/dynamometer-drive-schedules (accessed on 15 February 2019).

25. *City Driving Cycle for Vehicle Testing*; National Development and Reform Commission of the People's Republic of China: Beijing, China, 2006.

26. Bao, Y.; Chan, Q.N.; Kook, S.; Hawkes, E. Spray penetrations of ethanol, gasoline and iso-octane in an optically accessible spark-ignition direct-injection engine. *SAE Int. J. Fuels Lubr.* **2014**, *7*, 1010–1026. [CrossRef]

27. Zhao, F.; Lai, M.C.; Harrington, D.L. Automotive spark-ignited direct-injection gasoline engines. *Prog. Energy Combust. Sci.* **1999**, *25*, 437–562. [CrossRef]

28. Chan, Q.N.; Bao, Y.; Kook, S. Effects of injection pressure on the structural transformation of flash-boiling sprays of gasoline and ethanol in a spark-ignition direct-injection (SIDI) engine. *Fuel* **2014**, *130*, 228–240. [CrossRef]

29. Yang, J.; Xu, M.; Hung, D.L.S.; Wu, Q.; Dong, X. Influence of swirl ratio on fuel distribution and cyclic variation under flash boiling conditions in a spark ignition direct injection gasoline engine. *Energy Convers. Manag.* **2017**, *138*, 565–576. [CrossRef]

30. Yang, J.; Dong, X.; Wu, Q.; Xu, M. Influence of flash boiling spray on the combustion characteristics of a spark-ignition direct-injection optical engine under cold start. *Combust. Flame* **2018**, *188*, 66–76. [CrossRef]

31. Hitomi, M. *Our Direction for ICE*; The Japan Society of Mechanical Engineers: Tokyo, Japan, 2017.

32. Weng, S. "Jieneng Yu Xin Nengyuan Qiche Jishu Luxiantu" Zhengshi Fabu. *Qiche Yu Peijian* **2016**, *46*, 30–31.

33. Whitefoot, J.W.; Ahn, K.; Papalambros, P.Y. The case for urban vehicles: Powertrain optimization of a power-split hybrid for fuel economy on multiple drive cycles. In Proceedings of the ASME 2010 International Design Engineering Technical Conferences & Computers and Information in Engineering Conference, Montreal, QC, Canada, 15–18 August 2010.

34. *The Potential Impact of Electric Vehicles on Global Energy Systems*; McKinsey & Co.: New York, NY, USA, 2018.

Article

The Relationship between In-Cylinder Flow-Field near Spark Plug Areas, the Spark Behavior, and the Combustion Performance inside an Optical S.I. Engine

Atsushi Nishiyama *, Minh Khoi Le, Takashi Furui and Yuji Ikeda

Imagineering, Inc., 7-4-4 Minatojima-Minami, Chuo, Kobe 650-0047, Japan; minh@imagineering.jp (M.K.L.); furui@imagineering.jp (T.F.); yuji@imagineering.jp (Y.I.)
* Correspondence: atsushi@imagineering.jp; Tel.: +81-78-386-8888

Received: 11 March 2019; Accepted: 8 April 2019; Published: 14 April 2019

Abstract: The stringent regulations that were placed on gasoline vehicles demand significant improvement of the powertrain unit, not only to become cleaner but also more efficient. Therefore, there is a strong need to understand the complex in-cylinder processes that will have a direct effect on the combustion quality. This study applied multiple high-speed optical imaging to investigate the interaction between the in-cylinder flow, the spark, the flame, and combustion performance. These individual elements have been studied closely in the literature but the combined effect is not well understood. Simultaneous imaging of in-cylinder flow and flame tomography using high-speed Particle Image Velocimetry (PIV), as well as simultaneous high-speed spark imaging, were applied to port-injected optical gasoline imaging. The captured images were processed using in-house MATLAB algorithms and the deduced data shows a trend that higher in-cylinder flow velocity near the spark will increase the stretch distance of the spark and decrease the ignition delay. However, these do not have much effect on the combustion duration, and it is the flow-field in the entire area surrounding the flame development that will influence how fast the combustion and flame growth will occur.

Keywords: flow-field; flame-front; gasoline engine; flame propagation; turbulence

1. Introduction

With the stringent restriction in emissions for passenger vehicles, as well as customers' demands for better fuel economy and efficiency, car manufacturers have to continuously invest more and more into research and development, especially for the powertrain system. While electric vehicles and electrification appears to be the future of automotive technology, difficult challenges, including infrastructures, energy density, and storage, mean these technologies will not be widespread globally in the coming years. Therefore, internal combustion engines will need to be improved significantly in these regards. For gasoline vehicles, many different technologies and strategies have been shown to be quite promising in terms of improving the efficiency of the engines. Namely, these include ultra-lean-burning, gasoline compression ignition, ultra-dilution, and high forced charge intake. Due to their substantial involvement and dependence on the in-cylinder flow and ignition, it is vital that the relationship between the flow, the ignition plasma, and the flame is well understood, so that these strategies can be optimized for maximal effect on efficiency.

Strong tumble flow and high turbulence level inside the combustion chamber have been demonstrated to be able to enhance the flame propagation rate and reduce the burnt duration [1–4]. Le Coz et al. revealed that the wrinkling accelerates the combustion process only later, during the propagation phase [1]. Li et al. showed that the presence of a strong tumble was necessary to obtain good fuel stratification [2]. Aleiferis et al. suggested that on the tumble plane of flow, a high convection velocity was preferable up to 30° crank angle (CA) after ignition timing [3]. Le et al. showed that the flame propagation is influenced by the general flow field and large-scale eddies, in which it can enhance, diminish, and guide the flame propagation [4]. However, in challenging conditions, such as lean burning, most of the misfire cases occur during the early flame kernel development stage before the free flame propagation period [5,6]. The most crucial factor during this period is the ignition process. The early kernel development period must be short to have the highest success rate for the associated combustion, and previous literature has indicated that to achieve this, the ignition source must deliver enough energy to the unburned mixtures, either thermally or kinetically through the supply of additional radicals [7]. While other ignition methods, such as microwave RF plasma, corona discharge, etc., can enhance ignition by having a bigger ignition volume and longer plasma duration (longer supply of energy) [8–10], the conventional spark plug produces a relatively short and high thermal plasma, and relies mostly on delivering energy to the unburned mixtures via thermal enhancement pathways. Due to the shape of the conventional spark plug, the spark volume is normally confined and early kernels are very susceptible to heat loss to the electrodes, especially ground electrodes, which in turn, can significantly inhibit the growth of the kernel leading to a slower early burnt duration [11]. Therefore, if the early flame kernel can avoid contact with the electrodes, the likelihood of non-misfiring will increase. Researchers have reported that when the spark is stretched away from the ground electrodes of J-type plugs, the early combustion period is shortened and the lean limit can be extended [12–15]. Herweg et al. evaluated a one-dimensional, time dependent model that described flame kernel development in spark ignition engines which explicitly accounts for all fundamental properties of the ignition system, of the combustible mixture, and of the flow field [12]. Smith et al. investigated the impact of intake air dilution with nitrogen, spark plug orientation, ignition system dwell time, and fuel injector targeting on ignition stability in an optical spray-guided spark ignition direct injection engine and identified the potential reasons for misfires [13]. Nishio et al. showed that it is possible to decrease heat loss and improve ignitability with the addition of a flow guide plate to promote a gas flow stream to the spark gap [14]. Pischinger indicated that under conditions where heat losses and electrical energy are relevant for the initial flame kernel growth, the flow velocity was identified as the most dominant single parameter affecting the initial kernel growth [15]. This enhancement mechanism of the stretch the spark plasma creates most likely is due to the increase in ignition volume, as well as less heat loss to the metal electrodes.

In a realistic engine environment, the in-cylinder flow surrounding the spark plug has been shown to have a correlation with the early combustion period, with stronger flow more likely to result in faster combustion. Optical studies in optical engine has also shown that the flow across the spark plug causes the plasma arc to stretch [13,16]. Moreover, when the spark plug orientation is fixed to match the flow or when the spark plug geometry is optimized for flow interaction, an improvement in lean limit is observed [14]. Hence, it is very likely that only certain characteristics of the flow surrounding the spark plug have some positive correlation with the stretch of the spark plasma, while others might have a negligible effect. In fact, having an understanding on such specific correlation is quite important for combustion performance. As the flow velocity and turbulence intensity reach a very high level, detrimental effects on the early kernel stability is prominent. An increase in misfire rate is attributed to high turbulence level restricting early growth of the kernel and blowing off of the spark plasma channel, which cuts off the energy supply to the mixtures [16].

However, despite the research effort, there is still a gap in the knowledge relating to the exact correlation between flow-field and discharge stretch length. This is mainly the result of the high difficulty in imaging the spark channel and flow-field simultaneously in a realistic operating engine environment. Besides the existing difficulties that all engine optical diagnostics face, such as vibration or interferences, in spark ignition (S.I.) engines with a pent-roof, the spark plug is commonly placed in locations that are difficult for the penetration of a laser in diagnostics, such as particle image velocimetry (PIV) or laser induced fluorescence (LIF). Furthermore, to have the best understanding possible of this ignition process, not only do the spark and the flow-field need to be measured but the resulting kernel should also be visualized in the same cycle. This means the diagnostics have to be carried out in a reactive environment. Previously, we have demonstrated our application of high-speed PIV to visualize both the in-cylinder flow-field and flame front in an optical engine [4,17]. A newly developed high-speed PIV technique that allows for time-resolved measurement of both flame front and flow field simultaneously was applied to observe the flame structure and its relationship to the engine flow and turbulence [4]. Moreover, the diagnostics were performed for both the vertical plane and the horizontal plane of the combustion chamber with a particular focus on the pent-roof area [17]. This diagnostic technique is very meaningful for investigating the interaction between flow field and flame propagation for each cycle and for discussing the cyclic variation.

In this study, to attempt to expand the understanding of the ignition process, particularly the interaction between flow-field and spark, we added high-speed imaging of the spark channel together with high resolution, high-speed combustion PIV. Both imaging techniques are carried out simultaneously in order to interpret the correct relationship during this ignition process and the focus area is the pent-roof region surrounding the spark plug. Commonly used performance parameters, such as burnt duration, are also measured and calculated. How the flow-field affects the spark channel behavior, and in turn, the flame front and combustion, will be explored in this study.

2. Materials and Methods

2.1. Engine Specifications and Operating Conditions

This relationship of the spark, in-cylinder flow-field, and flame kernel is observed inside an operating engine condition of a single-cylinder optical spark ignition engine. Table 1 summarizes the specification of this engine and the selected operating conditions. The engine capacity is 500 cm^3 with a bore of 86 mm and stroke of 86 mm and has a compression ratio of 10.4. The top of the piston (54 mm diameter) and the pent-roof liner is replaced by a quartz window to provide an optical access into the combustion chamber of the engine, as shown in Figure 1a. Reflection from a 45° mirror placed in the hollow extended-piston provides an optical-access path for the laser sheet for the observation of the pent-roof area in the vertical plane, which is the desired configuration for this study. This optical view of the vertical plane in the pent-roof via the pent-roof liner windows is also used for simultaneous high-speed spark imaging, which will be described in detail in the following section. The experiments were carried out with the engine operating at two different speeds, 1200 rpm and 1700 rpm in order to generate different magnitudes of flow-field near the spark plug. The engine operated with a port fuel injection and an absolute intake pressure of 60 kPa; the air fuel ratio (A/F) was kept at slightly lean A/F of 18. The spark timing was set at −40° Crank-Angle after Top Dead Center (CA aTDC). In-cylinder pressures were monitored and recorded using a Kistler pressure transducer and the engine was fired continuously during the PIV/spark capturing process. Nineteen cycles were captured for each condition.

Table 1. Engine Specifications and operating conditions.

Engine Type	4 Stroke, Single Cylinder, port fuel injection (PFI)
Displacement	500 cc
Bore × stroke	86 × 86 mm
Compression ratio	10.4
Number of valves	Intake: 2, Exhaust: 2
Engine speed	1200 and 1700 rpm
Intake pressure	60 kPa
Air fuel ratio (A/F)	18
Spark timing	40 deg. before top dead center (BTDC)

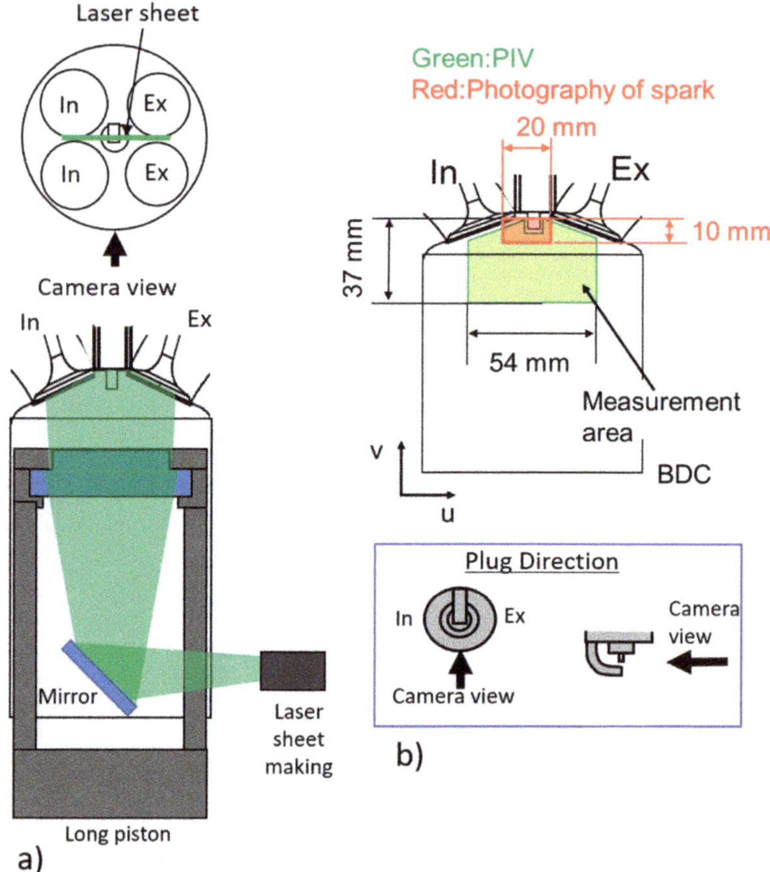

Figure 1. (**a**) Optical engine and PIV set up, (**b**) PIV measurement area, and position of the laser sheet with respect to the cylinder and plug direction with respect to the camera.

2.2. High-Speed PIV

The high-speed combustion PIV (HS-PIV) set up used in this study is the same as the one described in detail in a previous study [17]. The PIV signal of the pent-roof area is provided by inserting an 527 nm Nd:YLF laser sheet via a piston top quartz window using the reflection off the 45° mirror placed in the middle of the extended piston, as shown in Figure 1. Figure 1b displays the schematics of the measurement area, which contains an illustration of the spark plug in both top-view (left) and front-view (right). The measurement area is 37 mm × 54 mm, in which the spark-plug and its surrounding region is placed in the middle and the spark plug is orientated such that the camera view of the spark-gap is un-obstructed. Moreover, the ground electrodes are setup perpendicular to the tumble flow that has been observed previously in the same engine [17]. This ensures the spark will interact with the tumble flow without any obstruction. The PIV images were captured using a high-speed camera at 1280 × 800 resolution and a PIV time resolution of 1° CA and 2° CA corresponds to when the engine operates at 1200 rpm and 1700 rpm, respectively. For a higher resolution, a 150 mm macro lens was equipped to the camera.

While other PIV works under combustion conditioned has used liquid seeds [18,19], the PIV details are lost in the burned gas region as the seeds evaporated under high temperature. To retain the measurability of the flow field throughout the cycle in both burned and unburned region, burning resistant solid tracing particles SiO_2 at various sizes were used, similar to a previous study in the same engine [17]. This is in contrast with when liquid particles were used [18,20] which causes the PIV details to be lost in the burned gas region via seed evaporation. The seeding particles were introduced into the intake air flow upstream of the intake system of the optical engine to ensure the homogeneous mixing and trace to the intake air motion. The seeding amount was adjusted for the best signal quality. Figure 2 (top left) displays a sample result image during the combustion period of the cycle. A 527 nm bandpass filter was placed just in front of the camera lens to isolate PIV signal from other interferences and a beam splitter was used to divert part of the view into the pent-roof for simultaneous spark imaging. PIV data was captured for 19 consecutive fired cycles. Additional details on the setup when simultaneous spark imaging was involved will be discussed more in a later section detailing the high-speed spark imaging.

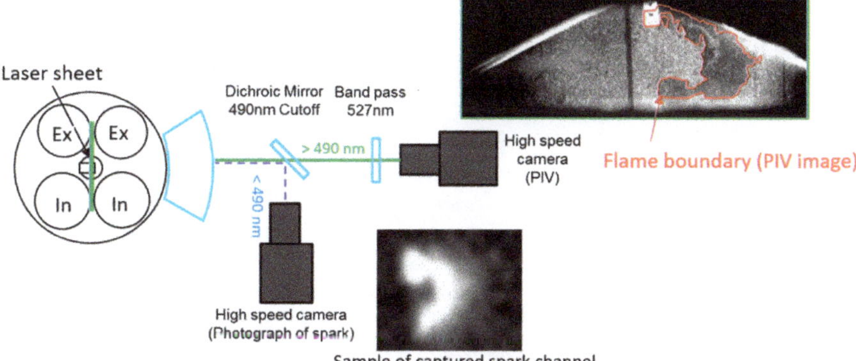

Figure 2. Simultaneous setup for PIV/flame imaging and high-speed spark imaging, together with their sample raw images.

2.3. High-Speed Flame Tomography

High-speed flame tomography imaging on the same laser plane can also be captured simultaneously with high-speed (HS)-PIV. This setup allows the flow and flame data to be analyzed and deduced from the same dataset, which essentially means that it eliminates potential errors as a result of misalignment, timing delays, or line-of-sight integration. This drastically improves the accuracy of the observed interactions between flow and flame. As seen in our previous work [17], the burned and unburned region can effectively be distinguished and identified in the PIV image by intensity gradient from the seed density, and hence, the planar flame front can be imaged. Advantageously, suitable seeds and seeding methods will still allow the PIV data in both the burned and unburned region to be measured.

In our works, the special in-house developed seed mixture, seeding procedures, and optimized laser fluence enables this simultaneous high-speed PIV and flame tomography. The PIV image is put through an in-house developed algorithm to determine the flame-front boundary (a sample is also shown in Figure 2). Each of the presented images in this paper will contain the flame boundaries and processed PIV flow-field results.

2.4. Spark High-Speed Imaging

To accomplish the goal set out in this investigation, PIV and flame imaging was carried out simultaneously with high speed spark imaging. As previously described, the PIV and flame data were captured in one dataset using one camera, and hence, the spark can be imaged using another camera. As illustrated in Figure 2, the two cameras for these diagnostics were positioned perpendicular to each other with the PIV/flame camera directly in-line with the pent-roof window. The light emissions from the combustion chamber through the pent-roof side window was split by a 45° dichroic mirror with a cut-off at 490 nm and then captured by the cameras. The dichroic mirror was orientated such that light above the cut-off of 490 nm was captured by the PIV/flame camera, whereas the signal below 490 nm was used for imaging the spark. This was due to the fact that the signal for PIV/flame was mainly the scattering of the laser light at 527 nm, while the spark emission was broadband with strong UV emissions. Moreover, this split also separated the interference of the laser signal from the high-speed spark imaging data. To be able to focus on the spark tip location for the spark imaging, a 200 mm macro lens was used together with 35 mm extension tubes. For the spark imaging, the area of interest was 20 × 10 mm around the spark tip. The camera was capturing at 50,000 frames per second to ensure all the spark movement was captured. It is also worth noting that the spark plug was installed in such a way that the J plug was behind the line of sight for the cameras, as this ensured the spark was not blocked. The spark plug used in this case was a small-tip iridium performance spark plug. The dwell time was around 3 ms, which translated to roughly 50 mJ ignition energy.

3. Results and Discussions

3.1. In-Cylinder Pressure and Mass Burned Fractions

Figure 3 displays the in-cylinder pressure and mass fraction burned (MFB) of the two tested operating conditions, 1200 rpm—A/F18 and 1700 rpm—A/F18 for all 40 tested cycles. The 1700 rpm case displays a shorter ignition delay and a shorter combustion duration. Moreover, the spread of both the pressure and MFB suggests that the combustion event is a lot more stable at 1700 rpm. For the analysis and discussion of the optical diagnostics results, three representative cycles were selected. These chosen cycles are the ones with highest, average, and lowest CA50, and hence, the observation of the flow-spark-flame interaction from optical diagnostics for them can provide clues to the cyclic variations. Images of the optical diagnostics will focus on the ignition period—just before the spark until the appearance of the early flame kernel.

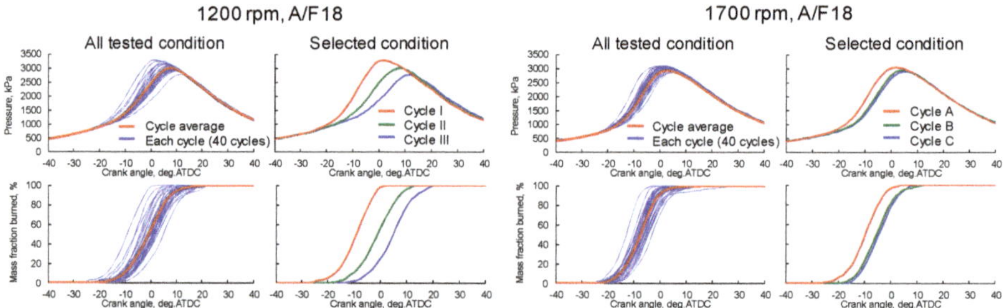

Figure 3. In-cylinder pressure traces (**top**) and mass fraction burnt (**bottom**) of all tested cases and selected cases for the two operating conditions (1200 rpm and 1700 rpm).

3.2. Operating Conditions 1200 rpm—A/F18

Figure 4 displays the images of combined in-cylinder flow from high resolution PIV and spark development from high-speed spark imaging near the spark plug. The blue area between the spark gap indicates the spark discharge path obtained from the spark high-speed imaging. The gray area indicates the burned area obtained from high-speed flame tomography. The earliest burning cycle, cycle I, is on the left-hand side, the average cycle, cycle II, is in the middle, and cycle III, the one with the longest ignition delay, is on the left of the figure. Cycle I also has a shorter combustion duration compared to cycle II and III. First, the in-cylinder flows of all the cycles appear to move clock-wise, from the intake valves to the exhaust valves which is the general tumble direction for this engine. For all cases, there exists a bulk flow perpendicular across the spark plug. It is quite noticeable that the earliest burning case has the fastest flow across the spark plug. Cycle III, however, only has a slightly slower flow across the plug compared to cycle II.

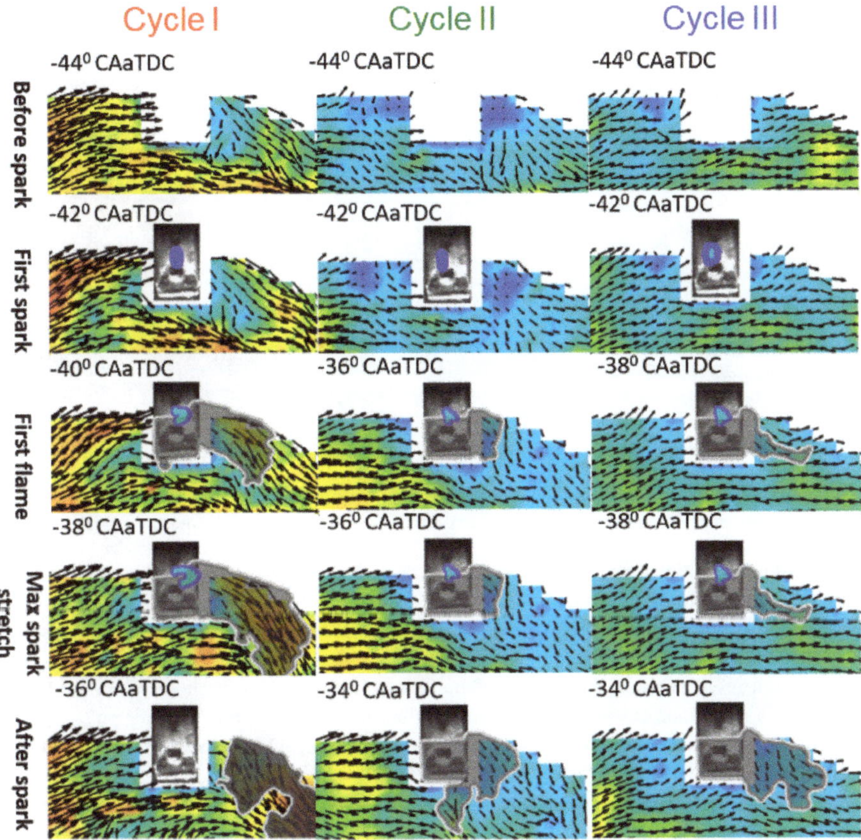

Figure 4. Near spark combined PIV-flame-spark images of three selected cycles (I, II, and III) at 1200 rpm condition.

Regarding the spark behavior, there is a general spark movement with the in-cylinder flow across the spark plug for all cycles. The spark also displays a stretching behavior observed in other work. For the earliest burning cycle I, with the strongest flow, a largest spark stretch is also observed. In fact, a break in the spark channel can also be seen when the spark is at its maximum stretch. However, for the two other cycles, the difference is not very noticeable. Hence, in order to get more details about the relationship between the flow and spark behavior, not only for cycle II and III but also for cycle I, we measured and compared the flow speed across the plug with the stretch distance. Figure 5 displays this plot. For the flow speed across the plug, velocity at a representative point is used. This chosen point is 1 mm away from the left side of the spark plug (before the flow crosses the plug), in-line with the spark gap. In particular for this study, only the horizontal velocity at the point will be considered. Furthermore, the time for the velocity to be considered is the instant just before the first spark. An illustration of the location of this point with regard to the spark plug is shown on the plot. Also, in the illustration is the measurement method to determine the spark stretch distance, which is the distance between the first spark discharge and the maximum stretch of this first discharge before the spark channel is broken off. The plot confirms the visual observation of the data, with cycle I having the longest stretch and flow followed by cycle II and then III. Hence, a trend of faster flow speed, longer spark stretch, and shorter ignition delay can be deduced from this data. It also appears that the length of this stretch is somewhat proportional to the strength of the flow.

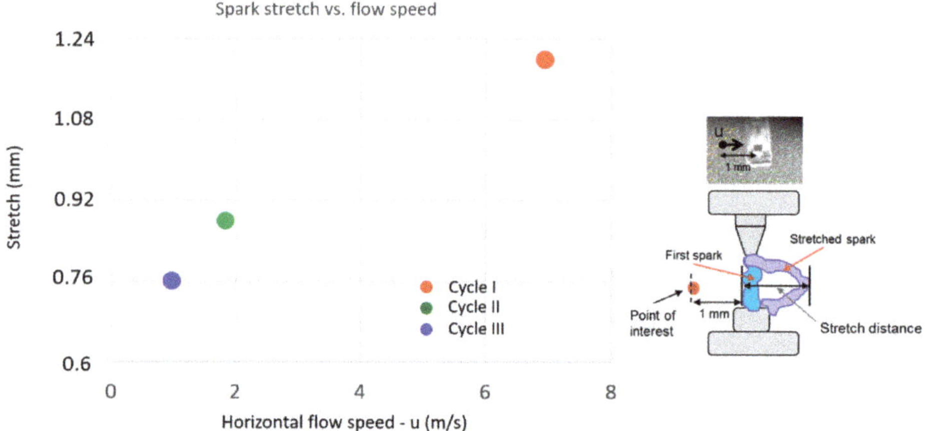

Figure 5. Flow speed and spark stretch relationship at 1200 rpm operating condition. Also shown is an illustration for the calculation of stretch distance and flow speed determination.

While there is a trend for the spark stretch, flow speed, and ignition delay, a slightly faster combustion duration in cycle I in comparison to cycle II and III will require additional information of the flame for a better understanding, due to the fact that the combustion duration of cycle II and III are very similar despite the difference in spark stretch and flow speed. Figure 6 displays images of flame development of these three cycles. The flames in these three cycles all develop toward the exhaust-valve side, which is with the direction of the bulk flow and the flow across the spark plug. Hence, it is reasonable to say that the direction of early flame kernel development is dependent on the flow direction across the spark plug. For all cases, the kernel appears while the spark channel is still presented. As the cycle I spark channel stretches much further than other cycles, the kernel also has extended much further with the flow at the point when the spark breaks and disappears. However, this effect is also due to the stronger overall tumble flow. As the flame develops, the kernel grows and continue to follow the bulk tumble flow direction. The flame also follows big vortices created by the flow and does not appear to be able to grow against the flow (as pointed out on the images with white arrows). Moreover, it is quite interesting that the growth of the flame also enhances the existing flow. Overall, it appears that the combustion duration is driven by the overall bulk flow strength rather than just the spark and flow near the spark region.

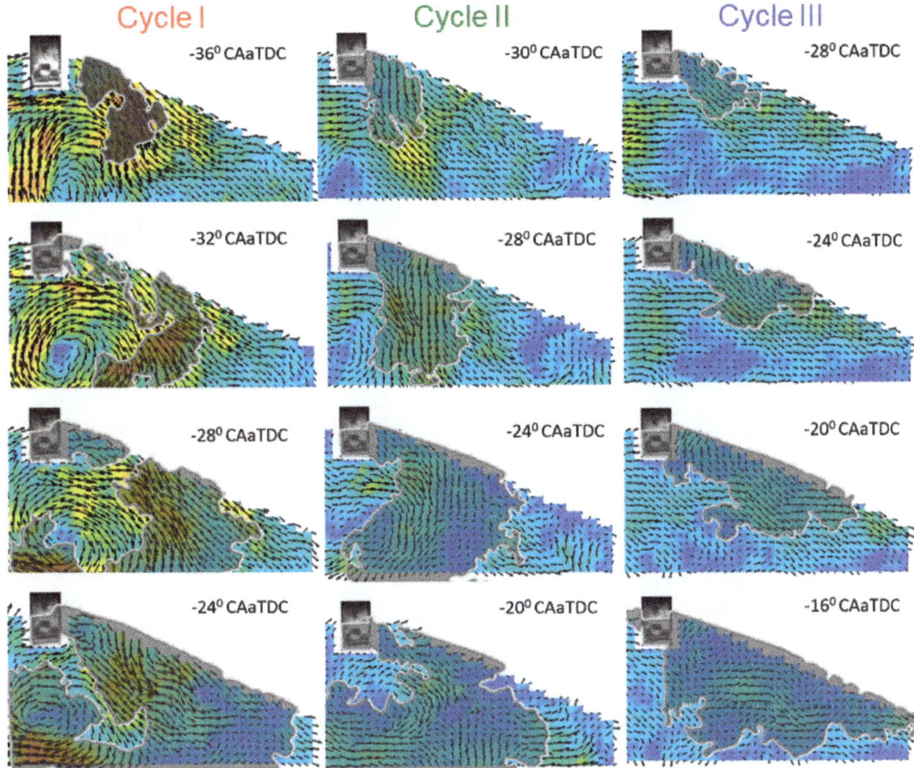

Figure 6. PIV-flame-development images after initial kernel development of three selected cycles (I, II, and III) at 1200 rpm condition.

3.3. Operating Conditions 1700 rpm—A/F18

Figure 7 displays the PIV and spark images at the near spark region of three selected cycles at 1700 rpm. Similar to Figure 4 of the 1200 rpm condition, these three chosen cycles have the earliest, average, and latest start of combustion; they will be referred to as cycle A, B, and C, respectively. First impression shows that the general strength of the flow is much higher than at 1200 rpm, as expected from the higher piston speed at 1700 rpm. The general flow direction still goes across the spark plug from intake to exhaust, which is the same in both operating conditions. At the beginning of the spark event, there is a flow across the spark plug for all cycles, however, it is quite weak in cycle B. It is also clearer in the 1700 rpm image that after the flow crosses the plug, it diminishes significantly.

For cycle A, the strong flow has a strong effect on the spark channel. The channel stretches out quite a bit, comparable to the cycle I of 1200 rpm condition. However, unlike cycle I, the spark channel in cycle A actually breaks off and restrikes close to the end of the spark event. It is likely that the stronger flow in this case overcomes the ignition energy required to sustain the spark channel. These breaks in the spark channel will affect the combustion event negatively as it momentarily stops the supply of energy to grow the flame kernels [11]. As shown on the flow speed versus spark stretch in Figure 8, higher flow speed leads to longer spark stretch. Cycle C displays a somewhat longer stretch compared to cycle B, which can be explained by the lower flow speed near the spark plug. However, the proportion of this relationship in this case is not similar to that of the 1200 rpm condition. This is understandable, as the flow that will affect the stretch behavior of the spark cannot be completely represented two-dimensionally. Hence, when looking at the spark stretch distance, the velocity of the flow should be used only for qualitative information.

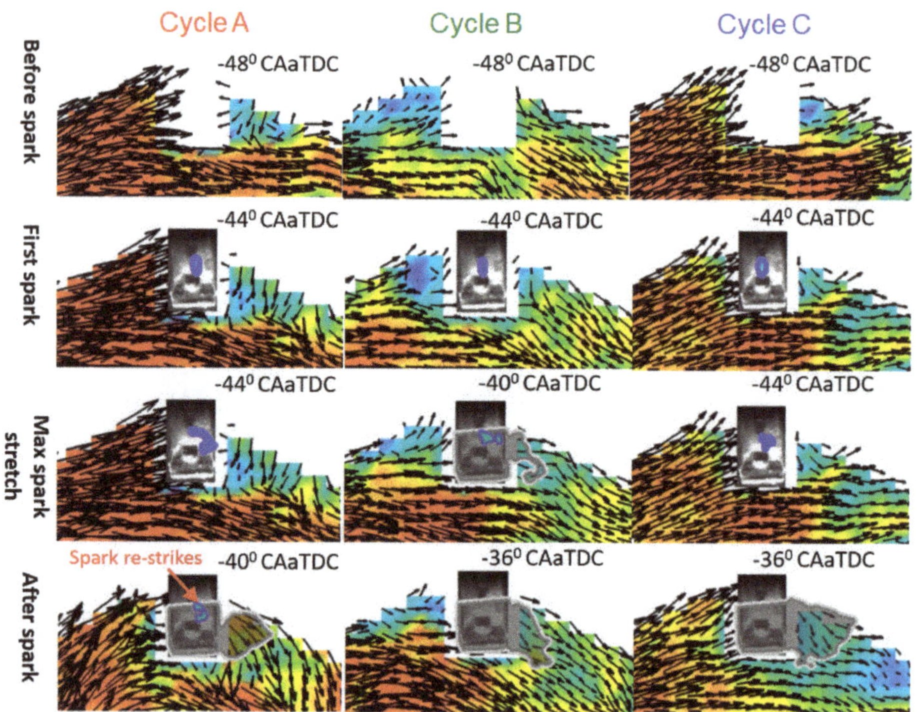

Figure 7. Near spark combined PIV-flame-spark images of three selected cycles (A, B, and C) at 1700 rpm condition.

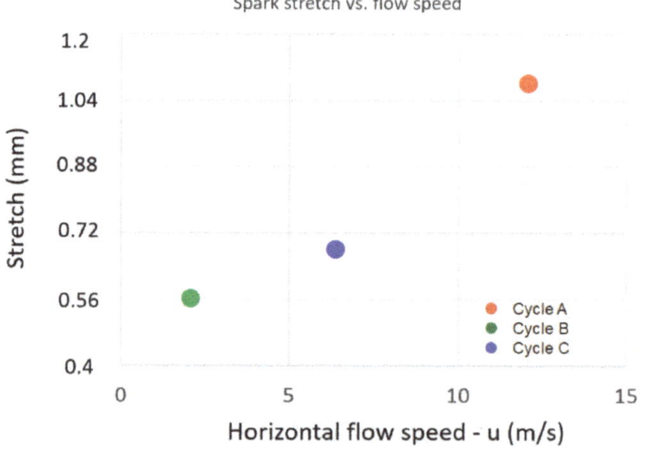

Figure 8. Flow speed and spark stretch relationship of selected cycles at 1700 rpm condition.

Despite having a longer spark stretch, cycle C has a slower start for the combustion compared to cycle B. However, the difference between the ignition delay of these two cycles is very small.

Upon closer inspection, at the start of the flame kernel formation, the flow near the spark plug where the flame will develop is weaker in cycle C. This might negate the effect of a small spark stretch advantage. Interestingly, all three cycles have very similar combustion duration. As shown in Figure 9, the flames in all cycles develop toward the intake side, following the tumble bulk flow. While they all have different shapes and grow paths, the overall flow condition seems to be somewhat similar. This could be why the combustion duration is the same across the investigated cycles. This further suggests that it is the bulk flow in the combustion chamber that will influence the combustion duration, while the spark stretch and flow near spark plug will influence the start of the combustion.

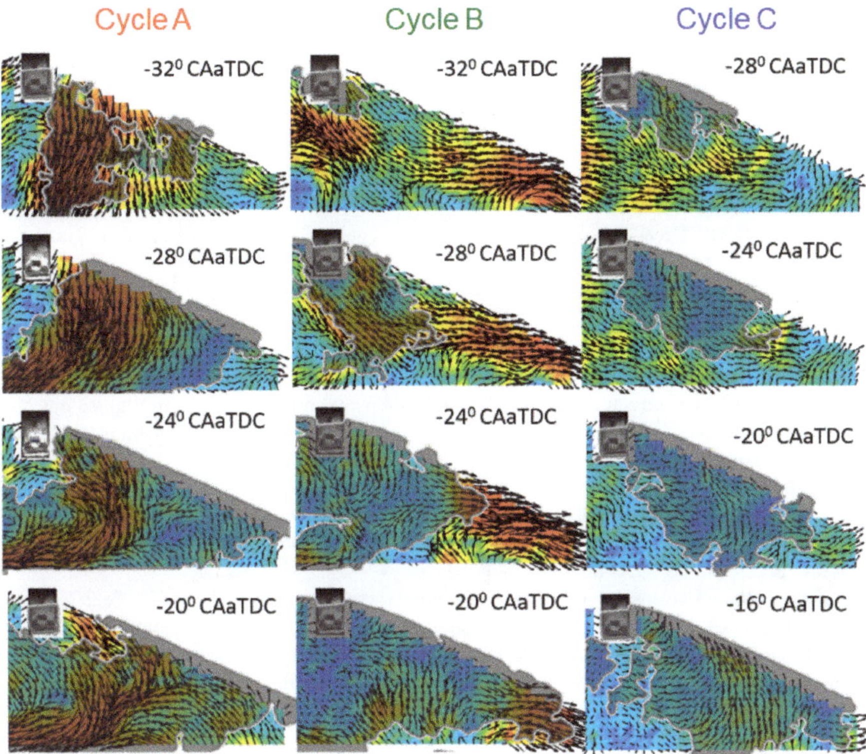

Figure 9. PIV-flame-development images after initial kernel development of three selected cycles (A, B, and C) at 1700 rpm condition.

4. Conclusions

Simultaneous high-speed PIV, flame tomography, and high-speed spark imaging is applied in an optical port-injected gasoline engine at two different operating conditions: 1200 rpm—A/F18 and 1700 rpm—A/F18. A dichroic mirror splits the signal to 2 cameras, with one setup for PIV and flame imaging, while the other is dedicated to high-speed spark imaging. The deduced results of the flow field, burned region, and spark imaging is combined and analyzed to understand the interaction between in-cylinder flow, spark behavior, flame development, and combustion performance. The following observations are made.

Higher flow velocity across the spark plug will result in a longer spark stretch distance. However, with a given ignition energy, the spark channel can only handle a certain flow speed limit before it breaks off and restrikes as the spark gap occurs.

A longer spark stretch and higher flow surrounding the spark plug (both sides, with and against the flow) shorten the ignition delay and allow an earlier start of combustion and flame kernel formation.

The flame will develop with the bulk in-cylinder flow and will follow the flow. It does not grow against the flow and will also enhance the existing flow in the same direction with its growth.

The combustion duration does not depend much on the spark stretch but mostly on the in-cylinder flow in the region of the combustion chamber where it is developing in.

Author Contributions: Conceptualization, Y.I.; methodology, F.T. and M.K.L.; software, M.K.L.; validation, M.K.L.; formal analysis, F.T. and M.K.L.; investigation, F.T.; resources, Y.I.; data curation, F.T.; writing—original draft preparation, M.K.L; writing—review and editing, A.N.; visualization, M.K.L.; supervision, Y.I.; project administration, A.N.

Funding: This research received no external funding.

References

1. Le Coz, J. Cycle-to-Cycle Correlations between Flow Field and Combustion Initiation in an S.I. Engine. *SAE Trans.* **1992**, *101*, 954–966.
2. Li, Y.; Zhao, H.; Ma, T. Flow optimization and fuel stratification for the stratified fuel spark ignition engine. *Int. J. Engine Res.* **2004**, *5*, 401–423. [CrossRef]
3. Aleiferis, P.G.; Taylor, A.M.K.P.; Ishii, K.; Urata, Y. The nature of early flame development in a lean-burn stratified-charge spark ignition engine. *Combust. Flame* **2004**, *136*, 283–302. [CrossRef]
4. Le, M.K.; Furui, T.; Nishiyama, A.; Ikeda, Y. The Interaction of Flow-Field and Turbulence on Flame Development using High-Speed Combustion PIV: High temporal and spatial resolved analysis of flame propagation and flame structure in a realistic combustion engine environment. In Proceedings of the 9th International Conference of Modeling and Diagnostics for Advanced Engine System (COMODIA 2017), Okayama, Japan, 25–28 July 2017.
5. Becker, H.; Monkhouse, P.B.; Wolfrum, J. Investigation of extinction in unsteady flames in turbulent combustion by 2D-LIF of OH radials and flamelet analysis. *Symp. Combust.* **1990**, *23*, 817–823. [CrossRef]
6. Peters, N. Local Quenching Due to Flame Stretch and Non-Premixed Turbulent Combustion. *Combust. Sci. Tech.* **1983**, *30*, 1–17. [CrossRef]
7. Ombrello, T.; Won, S.H.; Ju, Y.; Williams, S. Flame propagation enhancement by plasma excitation of oxygen. Part I: Effects O3. *Combust. Flame* **2010**, *157*, 1906–1915. [CrossRef]
8. Wolk, B.; DeFilippo, A.; Chen, J.Y.; Dibble, R.; Nishiyama, A.; Ikeda, Y. Enhancement of flame development by microwave-assisted spark ignition in constant volume combustion chamber. *Combust. Flame* **2013**, *160*, 1225–1234. [CrossRef]
9. Shiraishi, T.; Urushihara, T.; Gundersen, M.A. A trial of ignition innovation of gasoline engine by nanosecond pulsed low temperature plasma ignition. *J. Phys. Appl. Phys.* **2009**, *42*, 1–12. [CrossRef]
10. Kim, H.H.; Takashima, K.; Katsura, S.; Mizuno, A. Low-temperature NOx reduction processes using combined systems of pulsed corona discharge and catalysts. *J. Phys. Appl. Phys.* **2001**, *34*, 604–613. [CrossRef]
11. Kim, H.J.; Won, S.H.; Santner, J.; Chen, Z.; Ju, Y. Measurements of the critical initiation radius and unsteady propagation of n-decane/air premixed flames. *Proc. Combust. Inst.* **2013**, *34*, 929–936. [CrossRef]
12. Herweg, R.; Maly, R. A Fundamental Model for Flame Kernel Formation in S.I. Engines. *SAE Trans.* **1992**, *101*, 1947–1976.
13. Smith, J.D.; Sick, V. A Multi-Variable High-Speed Imaging Study of Ignition Instabilities in a Spray-Guided Direct-Injected Spark-Ignition Engine. *SAE Tech. Pap.* **2006**, 2006-01-1264. [CrossRef]
14. Nishio, N.; Aochi, T.; Yokoo, N.; Nakata, K.; Abe, Y.; Hanashi, K. Design of a High Ignitability Spark Plug with a Flow Guide Plate. *SAE Tech. Pap.* **2015**, 2015-01-0780. [CrossRef]
15. Pischinger, S. Effects of Spark Plug Design Parameters on Ignition and Flame Development in an SI-Engine. Ph.D. Thesis, Massachusetts Institute of Technology, Cambridge, MA, USA, 1989.
16. Suzuki, K.; Uehara, K.; Murase, E.; Nogawa, S. Ignition System for Gasoline Engines. In Proceedings of the 3rd International Conference, Berlin, Germany, 3–4 November 2016.

17. Le, M.K.; Furui, T.; Nishiyama, A.; Ikeda, Y. Application of High-Speed PIV Diagnostics for Simultaneous Investigation of Flow Field and Spark Ignited Flame inside an Optical SI Engine. *SAE Tech. Pap.* **2017**, 2017-01-0656. [CrossRef]

18. Mounaïm-Rousselle, C.; Landry, L.; Halter, F.; Foucher, F. Experimental characteristics of turbulent premixed flame in a boosted spark-ignition engine. *Proc. Comb. Inst.* **2013**, *34*, 2941–2949. [CrossRef]

19. Peterson, B.; Reuss, D.L.; Sick, V. High-speed imaging analysis of misfires in a spray-guided direct injection engine. *Proc. Comb. Inst.* **2011**, *33*, 3089–3096. [CrossRef]

20. Aleiferis, P.G.; Behringer, M.K. Flame front analysis of ethanol, butanol, iso-octane and gasoline in a spark-ignition engine using laser tomography and integral length scale measurement. *Combust. Flame* **2015**, *162*, 4371–4674. [CrossRef]

Review

Review of Advancement in Variable Valve Actuation of Internal Combustion Engines

Zheng Lou [1,*] and Guoming Zhu [2]

[1] LGD Technology, LLC, 11200 Fellows Creek Drive, Plymouth, MI 48170, USA
[2] Mechanical Engineering, Michigan State University, East Lansing, MI 48824, USA; zhug@egr.msu.edu
* Correspondence: gongda.lou@gmail.com

Received: 16 December 2019; Accepted: 22 January 2020; Published: 11 February 2020

Abstract: The increasing concerns of air pollution and energy usage led to the electrification of the vehicle powertrain system in recent years. On the other hand, internal combustion engines were the dominant vehicle power source for more than a century, and they will continue to be used in most vehicles for decades to come; thus, it is necessary to employ advanced technologies to replace traditional mechanical systems with mechatronic systems to meet the ever-increasing demand of continuously improving engine efficiency with reduced emissions, where engine intake and the exhaust valve system represent key subsystems that affect the engine combustion efficiency and emissions. This paper reviews variable engine valve systems, including hydraulic and electrical variable valve timing systems, hydraulic multistep lift systems, continuously variable lift and timing valve systems, lost-motion systems, and electro-magnetic, electro-hydraulic, and electro-pneumatic variable valve actuation systems.

Keywords: engine valve systems; continuously variable valve systems; engine valve system control; combustion optimization

1. Introduction

With growing concerns on energy security and global warming, there are global efforts to develop more efficient vehicles with lower regulated emissions, including hybrid electrical vehicles, electrical vehicles, and fuel cell vehicles. Hybrid electrical vehicles became a significant part of vehicle production because of their overall efficiency, and they still pose a significant cost penalty, resulting in a stagnant market penetration of 3.2% and 2.7% in 2013 and 2018, respectively, in the United States (US), for example [1]. Electrical vehicles reached a market penetration of 1.3% in the US in 2018 [1], still limited by high cost and concerns on well-to-wheel CO_2 emissions, charging speed, driving range, battery safety, and battery recycling issues. Fuel cell vehicles offer truly low overall emissions with driving range and fueling time comparable to vehicles powered by internal combustion engines. Hydrogen [2], the favorable fuel for fuel cells, can be generated from diverse sources, including natural gas, nuclear, coal, and renewable sources such as solar, wind, biomass, hydro, and geothermal sources. The development of fuel cell vehicles [3] is still in its infancy because of issues in technology maturity, cost, and performance, with only a few thousands of pilot vehicles on the road worldwide.

Internal combustion engines are believed to remain as a major part of vehicle powertrains in the foreseeable future, either standing alone or being part of highly electrified powertrains such as hybrid electrical vehicles, plug-in hybrid vehicles, and range extender vehicles, unless there is a major technology breakthrough in battery and/or fuel cell technology. For these reasons, it is imperative to continue advancement in more efficient and less polluting internal combustion engines.

1.1. Combustion and Need for Electronic Control of Gas Exchange

The combustion in an internal combustion engine involves three key components: in-cylinder fuel, air, and ignition. In gasoline engines, the fuel injection process evolved from a pure mechanical process such as via a carburetor to an electronically controlled process such as via port fuel injection or the recently adopted direction injection (see Figure 1). Spark ignition via a spark plug is completely electronically controlled for spark energy and timing. With its gaseous state, low density, and, thus, large volume, the electronic control of air exchange was a slow evolution process, from variable valve time (VVT) to more sophisticated systems such as discrete variable valve lift (DVVL), continuous variable valve lift (CVVL), cam-based variable valve actuation (VVA), and camless VVA, which are reviewed later in this article. There were also efforts to develop camless VVA to have complete electronic control of the air exchange process, which is a major enabler for advanced combustion such as homogeneous charge compression ignition (HCCI) to improve engine fuel economy and reduce emissions [4–7]. To improve engine fuel economy, engine downsizing techniques are widely used with the help of turbochargers. In this case, the VVT, VVL, or VVA system is able to improve engine system transient responses.

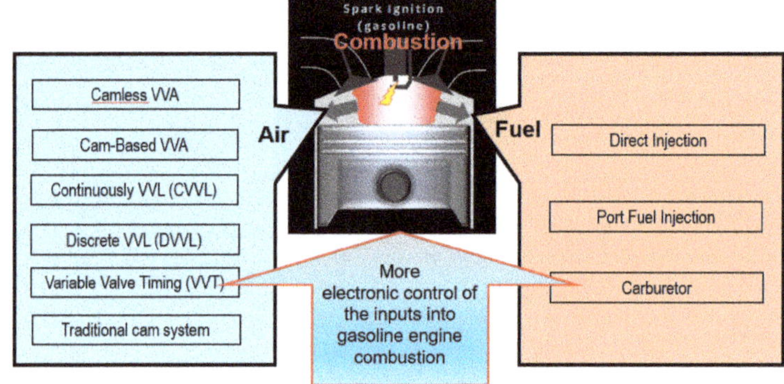

Figure 1. More electronic control of inputs into gasoline engine combustion.

In diesel engines, fuel injection evolved from pure mechanical pumping and injection systems to common-rail fuel injection via electronically controlled fuel pressure and injectors. Note that injection timing directly controls the ignition because of the compression ignition. There is less demand or development on controlling gas exchange in traditional diesel engines because of its compression ignition and lean combustion process. More sophisticate air charge management is needed for more advanced combustions such as the Miller cycle and PCCI. However, this review is limited to valve systems for gasoline engines.

1.2. Valve Lift, Valve Timing, and Valve Duration

The main function of an engine valve actuation system is to control the gas exchange into and out of a combustion chamber via intake and exhaust valves, respectively. The associated valve lift or travel is typically illustrated in a valve timing diagram (see Figure 2 for an example), where valve timing, valve lift, and valve duration are defined. A valve lift profile describes the valve lift as a function of camshaft angle between its opening and closing. The opening and closing points define the valve timing in the crank angle domain and, thus, the relationship between the lift profile and the rest of the engine components and events such as the piston movement and ignition. Valve lift is often optimized for minimal pumping loss. Often, intake and exhaust valve lifts are the same, but the diameter of the intake valve is larger than the exhaust one to ensure that fresh air can be easily charged into the cylinder. Note that variable valve lift has the potential of throttling the cylinder by reducing the intake valve lift to reduce the pumping losses associated with the conventional throttle. However, this requires very close control of lift to match changes in engine speed and load conditions, which is yet to be fully proven. The overlap between the intake valve opening and the exhaust valve closing is an important secondary parameter that has a major impact on combustion efficiency. The area under a lift profile represents the capacity for gas exchange. For the purpose of classifying valve actuation systems, the following definitions are provided:

- Valve lift refers to the amplitude, especially the peak value, of the valve lift profile.
- Valve timing refers to the phase shift in crank angle domain of the valve lift profile, especially the valve opening and closing events, such as EO, EC, IO, and IC.
- Valve duration refers to the duration when the valve is kept open, i.e., the span between the valve opening and closing events.

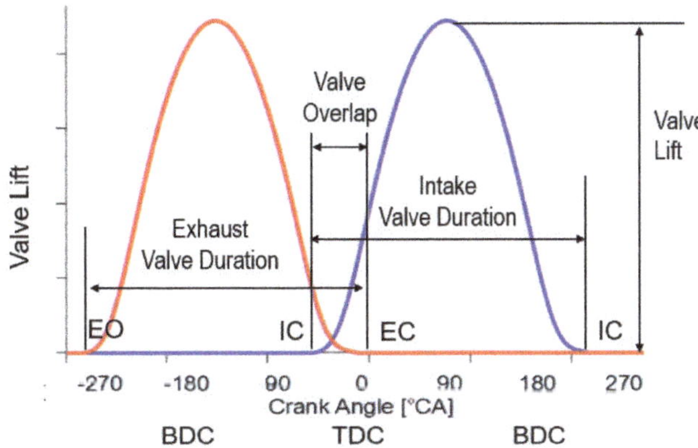

Figure 2. Regular valve timing diagram for naturally aspirated engines, defining valve lift, timing, and duration.

Modern engines are often equipped with multiple intake and exhaust valves, for example, two intake and two exhaust valves. The valve lift, timing, and duration can be optimized for each individual valve (for example, using VVA technology) to optimize in-cylinder mixing.

1.3. Classification of Valve Actuation Systems

Valve actuation systems are firstly classified into two large groups: cam-driven and camless systems. A cam-driven system utilizes cam lobes to actuate the valve lift, while a camless system does not include any cam mechanism and varies the valve lift using hydraulic, electro-magnetic, or pneumatic actuation to provide flexibility in control. Valve actuation systems are further classified based on the valve parameters being controlled.

Camless systems generally offer more control flexibility and capability, but they are yet to be implemented in production because of technical difficulties and commercial issues. For this reason, this review covers commercialized cam-based systems and only some development work in camless systems.

1.3.1. Cam-Based Valve Actuation Systems

A cam-based valvetrain system is based upon the traditional cam-system to drive the engine intake and exhaust valves with limited control over valve timing and/or lift, and it is now widely adopted in many new production engines. Cam-based systems include the following:

- **Variable valve timing:** Only the valve timing is independently controlled while the valve lift and duration remain the same. The VVT systems are also called cam (or camshaft) phasers. VVT systems are further classified into hydraulic (HVVT), mechanical (MVVT), and electrical (EVVT) types based on their respective cam phasing actuator designs. Many production VVT systems are HVVTs, using a device known as a variator that allows continuous adjustment of the cam timing, and EVVTs are getting popular for improving system response time at low temperature or engine start-up. However, the duration and lift cannot be adjusted.

- **Variable valve duration (VVD):** Only the valve duration is independently controlled.
- **Variable valve lift (VVL):** Only the valve lift is independently controlled. VVL systems further include discrete VVL (DVVL) and continuous VVL (CVVL) designs. A DVVL system includes a cam profile switching mechanism to activate one of two or three cam profiles or lobes, and a CVVL system includes a mechanism capable of continuous variation of the life profiles. In most, if not all, VVL systems, the lobes and mechanisms are designed such that the valve duration increases with the valve lift, which is a fixed relationship and not an independent control of the valve duration, although it serves the needs of a normal combustion. These VVL systems by themselves are, therefore, not classified as VVT, VVD, or VVA systems.
- **Cam-Based Variable Valve Actuation (VVA):** Cam-based VVA systems include (1) the VVL + VVT type, which is a combination of VVL (either DVVL or CVVL) and VVT, and (2) the lost-motion type (LMVVA).

Major cam-based valve actuation systems in production engines are listed in Table 1. Some of them are discussed in more detail in the later sections.

Table 1. Major cam-based valve actuation systems in production engines.

Classification	Company	System	Ind. Timing	Ind. Lift	Ind. Duration	Introd. Year, Comments & Refs
HVVT	Nissan	VTC/NVCS	2-stage			1987
HVVT	Toyota	VVT-i	Cont, Int			1996
HVVT	Mazda	S-VT	Variable, Int			1998
HVVT	Ford	Ti-VCT	2-stage, Both			2011
HVVT	Alfa Romeo	VCT	2-stage, Int			1980, 1st VVT, piston, [8]
HVVT	BMW	Single VANOS	2-stage & Cont, Int			1992, [9]
HVVT	BMW	Double VANOS	Cont, Both			1996, [9]
HVVT	Ford	VCT	2-stage, Int			
HVVT	GM	DCVCP	Cont, Both			
HVVT	Hyundai	CVVT	Cont, Both			
HVVT	Hyundai	VTVT	Variable, Both			
HVVT	Daihatsu	DVVT	Cont, Int			
HVVT	Ducati	DVT	Cont, Both			
HVVT	Nissan	CVVTCS/CVTC	Cont			
HVVT	Subaru	AVCS	VVT			
HVVT	Toyota	Dual VVT-i	Cont, Both			
HVVT	Toyota	VVT	2-stage			
MVVT	Porsche	VarioCam	Cont, Int			1992, 1st Cont VVT, [7,9]
EVVT	Toyota	VVT-iE	Cont, Both			2007, electric Int, hydraulic Exh
VVD	MG Rover	VVC			Cont, Int	1993, eccentric mechanism
VVD	Hyundai	CVVD			Cont, Int	2019, eccentric mechanism
DVVL	Honda	VTEC		2- & 3-lobe, Int		1989, [10]

Table 1. *Cont.*

Classification	Company	System	Ind. Timing	Ind. Lift	Ind. Duration	Introd. Year, Comments & Refs
DVVL	Audi	AVS		2-lobe, Both		2006, [11]
DVVL	Subaru	i-AVLS		2-lobe		2007, [12]
DVVL	Proton	CPS		2-lobe, Int		2016, [13]
DVVL	Yamaha	VVA		2-lobe		2017, motor cycle appl, [14]
DVVL + VVT	Mitsubishi	MIVEC	VVT, Both	2-lobe, Int		1992, [15]
DVVL + VVT	Nissan	VVL/VVL + VVT	VVT	2-lobe, Both		1997, [16]
DVVL + VVT	Porsche	VarioCam Plus	VVT	2-lobe, Int		1999, [4,17]
DVVL + VVT	Toyota	VVTL-i/VVT-iL	Cont	2-lobe		1999, [18]
DVVL + VVT	Honda	i-VTEC	Cont, Int	2-lobe, Int		2001, [19]
DVVL + VVT	Audi	AVS	Cont, Int	2-lobe, Int		2006, [11]
CVVL	Hyundai	CVVL		CVVL		2012, [20]
CVVL + VVT	Great Wall	CVVL + VVT	Cont, Both	CVVL		2018, [21]
CVVL + VVT	BMW	Valvetronic	Cont, Both	CVVL, Int		2001, [22]
CVVL + VVT	BMW and PSA	VTi	Cont, Ink	CVVL, Int		2002, [23]
CVVL + VVT	Nissan	VVEL + CVTC	Cont	CVVL		2007, [24,25]
CVVL + VVT	Toyota	Valvematic	Cont, Int	CVVL		2014, [26]
LMVVL	FCA	MultiAir	Cont	Cont		2009, [27,28]

Notes: Int = intake, Exh = exhaust, Both = both intake and exhaust, Cont = continuous.

1.3.2. Camless Valve Actuation Systems

Without the constraint from the cam mechanism, a camless system is capable of adjusting valve timing, duration, and lift independently to achieve more desired target levels that can be varied cycle-by-cycle. It also provides independent control of engine valves for each cylinder. For example, it is able to provide asymmetric opening for two intake valves for one cylinder, resulting in improved charge air and fuel mixing. It can also perform cylinder deactivation under low load conditions. It, thus, offers more control with greater benefits than conventional cam-based valve system. Camless systems include the following:

- Opposed solenoid electro-magnetic (or electromechanical) camless VVA (EMVVA) [29–34];
- Electro-hydraulic camless VVA (EHVVA) [35–37];
- Electro-pneumatic camless VVA (EPVVA) [38,39];
- Rotary motor EMVVA, also called intelligent valve actuation (IVA) system by Camcon [33,34].

Camless systems are a key technical enabler for other advanced engine technologies, such as air hybrid vehicles [40], HCCI [41], and high-efficiency diesel engines [42].

In the subsequent sections, various valve actuation systems are grouped and reviewed based on the valve actuation control parameters and structures.

2. Variable Valve Timing (VVT) System

The VVT technology was first applied to production engines by Alfa Romeo in 1980, and they are now widely used in most modern engines worldwide as shown in Table 1.

VVT systems were initially stand-alone systems for timing control, and gradually they were integrated with variable lift mechanisms to become part of a cam-based VVA system, i.e., VVL + VVT. They are often applied to intake valves only. Their control evolved from two-stage, i.e., two discrete positions, in the early days to continuous control ("Cont" in Table 1) in more recent engines.

2.1. Hydraulic VVT (HVVT)

An HVVT system includes a hydraulically actuated cam phaser or variator, the design of which evolved over the last 40 years.

The 1980 Alfa Romeo Spider 2.0 L had the first VVT system, which was an HVVT on the inlet camshaft. The design comes from a patented design (US Patent 4,231,330) by Alfa Romeo engineer Giampaolo Garcea [43]. The cam phaser is a cylinder containing a pressure chamber and a piston with helical splines. Alfa Romeo calls it mechanical VVT because of the helical splines, and it is classified as hydraulic VVT because of the hydraulic piston. Under oil pressure via a solenoid valve, the piston rotates slightly due to the helical splines and advances the inlet valve timing by 25° to increase engine valve overlap, which happens between 1500 and 2000 rpm and over 5000 rpm. Otherwise, the valve timing remains in its natural state.

Most phasers of the later HVVT systems use a rotary vane hydraulic motor, which is actuated by pressurized oil controlled by a solenoid valve. The cam phaser is operated either in two settings or, as in most of the more recent systems, continuously.

The BMW single VANOS system, when first introduced in 1992 on the BMW M50 engine, controlled the timing of the intake camshaft to one of two discrete positions. In 1998, infinitely variable single VANOS was introduced on the BMW M62 V8 engine. The double VANOS system, which appeared on the S50B32 engine in 1996, continuously adjusts the timing of the intake and exhaust camshafts [44]. The maximum range of phase timing relative to the sprocket is typically 60° [45].

In operation, the hydraulic VVT system can be vulnerable because of oil pressure fluctuation, oil quality, viscosity, and contamination. There is also a case where the phaser does not get enough oil because of a wear-induced leakage in the lubrication system [46]. At low temperatures, the system may not have adequate response time because of high oil viscosity, and the hydraulic VVT system cannot be activated and has to remain at its default lock position such that the cold-start performance and emissions cannot be improved [47]. For example, the camshaft phasing speed of the hydraulic VVT drops to about half of that of the electrical VVT and almost to zero when the operating temperature drops from 90 °C to 40 °C and −10 °C, respectively. For an engine cold start at −7 °C, the HC emissions are reduced by about one-third when replacing a hydraulic VVT with an electrical VVT.

2.2. Mechanical VVT (MVVT)

Porsche developed VarioCam, a mechanical VVT first used on the 1992 3.0 L engine in the Porsche 968 [1,8] It varies the timing of intake valves by adjusting the tension on the timing chain connecting the intake and exhaust camshafts. This mechanical design was not kept in a later version of VarioCam Plus, which uses a rotary vane hydraulic phaser as most HVVT systems do [1,4].

2.3. Electrical VVT (EVVT)

The issues associated with HVVT systems discussed above led to the development of EVVT systems, which was also made possible by recent advances in permanent magnetic motor technology and dramatically reduced motor drive cost. Toyota variable valve timing—intelligent electric (VVT-iE), for example, is a variation of dual VVT-i, by replacing the hydraulic cam phaser with an electric cam phaser for the intake camshaft timing [47]. The exhaust camshaft timing is still controlled using a hydraulic cam phaser. This technology was first introduced on the 2007MY Lexus LS 460 as a 1UR engine [18]. In operation, the electric motor in the cam phaser spins with the intake camshaft, running at the same speed to maintain camshaft timing. To advance or retard the camshaft timing, the actuator motor rotates slightly faster or slower, respectively, than the camshaft speed. The speed difference between the actuator motor and camshaft timing is used to operate a mechanism that varies the camshaft timing.

The performance of an EVVT is less dependent on engine oil temperature and pressure [47], thus providing better control precision and improving engine performance over a wider operational range. The control accuracy and fast response of a VVT system is more critical for advanced combustion such as HCCI, especially for the combustion mode transition control between spark ignition (SI) and HCCI combustion, where the engine cam timing needs to follow a desired trajectory to accurately control the engine charge and recompression process, as articulated by Ren and Zhu [48].

3. Variable Valve Duration (VVD) System

In 1993, MG Rover developed a 1.4 L K-series engine with a variable valve control (VVC) system, which was the first production continuous VVD (CVVD) system. It is based on an eccentric rotating disc to drive the inlet valves of every two cylinders. Since eccentric shape creates nonlinear rotation, the opening period of the valves can be varied by controlling the eccentric position of the disc. The basic concept was developed by Mitchell and it was published and patented back in 1973 [49]. In this design, the control is purely to vary the valve duration, with the valve lift fixed, thus differing from various DVVL or CVVL designs.

In 2019, Hyundai Motor Group announced that it developed CVVD technology to be in the Smartstream G1.6 T-GDi for future Hyundai and Kia vehicles [50–52]. Their design is based on a concept disclosed by Kim et al. [53]. It also involves utilization of some pins and slots to create eccentric alignment. With the duration variation in accordance to driving conditions, it is able to deliver a 4% increase in performance along with a 5% boost in fuel efficiency. The CVVD technology also helps reduce tailpipe emissions by 12% [50].

4. Discrete VVL (DVVL) and Associated VVA Systems

A discrete VVL (DVVL) system includes a cam profile switching mechanism to activate one of two or three cam profiles or lobes, and it becomes a cam-based VVA (or DVVL + VVT) system when a VVT mechanism is further incorporated. The applications of DVVL and DVVL + VVT systems include, but are not limited to, Honda, Audi, Subaru, Proton, Yamaha, Mitsubishi, Nissan, Porsche, Toyota, Honda, and Audi, as shown in Table 1.

In 1989, Honda launched, in Integra, the world's first commercial DVVL system in a motor vehicle engine called the variable valve timing and lift electronic control (VTEC) system [10,19,54]. VTEC uses two (or occasionally three) intake camshaft profiles, switched via hydraulically actuated rocker arm locking pins. In the system, the timing variation is fixed in the cam profiles and not independent of the lift variation. Honda then launched, in 2001, a cam-based VVA system called intelligent VTEC (i-VTEC) in high-output DOHC four-cylinder engines by adding continuous intake cam phasing (timing) to the traditional VTEC. VTEC controls are still limited to distinct low- and high-RPM profiles, but the intake camshaft is now capable of advancing between 25° and 50°, depending upon engine configuration [55].

In 1992, Mitsubishi launched the world's first cam-based VVA system, a DVVL + VVT system called the Mitsubishi innovative valve timing electronic control system (MIVEC). It has low-lift and high-lift cam profiles for low-speed and high-speed engine modes, respectively, which are switched via a locking pin mechanism. The low-lift cams and rocker arms, used to drive separate intake valves, are situated on two sides of a centrally located high-lift cam. Each intake valve is operated by a low-lift cam and rocker arm, while a T-lever between them engages the high-lift cam [15]. The VVT-i system from Toyota has a similar switching mechanism [18].

In 1999, Toyota launched the variable valve timing and lift intelligent system (VVTL-i or VVT-iL). The Toyota VVTL-i concept, including its a lift variation system via rocker arm locking pins, is similar to the Honda i-VTEC concept. Each cam has two lobes, one designed for lower-speed operation and another designed for high-speed operation, with higher lift and longer duration.

In 1999, Porsche launched its VarioCam Plus, a DVVL + VVT system on the intake side. The two-lobe valve-lift function is performed by electro-hydraulically controlled switchable tappets. Each of these 12 tappets consists of concentric lifters which can be locked together by a pin. The inner lifter and the outer ring element are actuated by a small cam lobe and a pair of larger-profile lobes, respectively. The timing of each valve is seamlessly varied by an electro-hydraulic rotary vane cam phaser [1,4]

In 2006, Audi launched, in the 1.8 L TFSI engine, the Audi valve lift system (AVS) [11]. It uses sliding electro-magnetic sleeves on the camshaft to vary the lift of the valves in two stages depending on load and engine speed. The system, thus, increases torque while also reducing fuel consumption. Two versions of the AVS system are available: (1) in the V6 engines in which AVS is used, it acts on the intake valves, regulating the amount of intake air so that the throttle can remain wide open for free breathing even at part load, thus reducing throttle losses and improving efficiency; (2) in the latest-generation 2.0 TFSI, the AVS varies the lift of the exhaust valves, thus reducing flushing losses in the combustion chamber and ensuring the optimal flow of the exhaust gas to the turbocharger.

5. Continuous VVL (CVVL) and Associated VVA Systems

A continuous VVL (CVVL) system includes a mechanism capable of changing the lift profile continuously, and it becomes a cam-based VVA (or CVVL + VVT) system when a VVT mechanism is further incorporated. The applications of the CVVL and CVVL + VVT systems include, but are not limited to, those by BMW, PSA, Hyundai, Nissan, and Toyota, as shown in Table 1.

In 2001, BMW launched the world's first CVVL + VVT system, as well as the first CVVL system called the Valvetronic system [45,56]. The Valvetronic system combines its double VANOS variable cam timing system for intake and exhaust valves with their CVVL system for lift control of the intake valve. The CVVL system includes an eccentric shaft moved by an electric stepper motor and the camshaft, with the camshaft being driven by the VANOS phaser. The valve lift can be varied from 0.18 mm to 9.9 mm. Later, in 2002, PSA Peugeot Citroën and BMW jointly developed a variable valve lift and timing injection (VTi) engine based on the Valvetronic concept [23].

In 2007, Toyota launched, in the Noah, the Valvematic system, which is essentially a combination of VVT-i and a continuously variable valve lift (CVVL) mechanism for the intake valve. This system is functionally similar to and structurally simpler and more compact than BMW Valvetronic. It varies intake valve lift in the range 0.9 mm to 10.9 mm, with a corresponding or coupled valve opening duration range of 106° to 260° in crank angle [26].

In 2007, Nissan launched a cam-based VVA system by combining its continuous variable valve timing control (CVTC) and variable valve event and lift (VVEL) systems, which are VVT and CVVL mechanisms, respectively [24,25]. It performs similarly to BMW's Valvetronic system but with desmodromic control of the output cam, allowing VVEL to operate at higher engine speeds. The Nissan VVEL system includes a rocker arm and two types of links that open the intake valves by transferring the rotational movement of a drive shaft with an eccentric cam to the output cam. The movement of the output cam is varied by rotating the control shaft with a direct current (DC) stepper motor and changing the fulcrums of the links.

In 2012, Hyundai launched a CVVL system [20] characterized by its compactness, i.e., no increase in engine height, using a unique six-linkage mechanism. In 2018, Great Wall became of the first Chinese OEMs launching a CVVL + VVT system [21].

6. Lost-Motion VVA (LMVVA)

Various lost-motion systems were disclosed in many patents, for example, US 4671221, US 5193494, US 5839400, US 6053136, US 6553950, US 6918364, US 681476, US 7819100, US 8578901, US 8820276, US 8776738, and US 9625050.

Fiat Powertrain Technologies and Schaeffler Group developed the only mass production systems branded as MultiAir and UniAir, respectively, which were first launched at the 2009 Geneva Motor

Show in the Alfa Romeo MiTo and were licensed in 2017 to Jaguar Land Rover for its Ingenium engine family.

In the MultiAir system, a solenoid valve controls the hydraulic pressure in a passageway connecting the intake valves and the camshaft [27]. The solenoid valve regulates the amount of oil that is pumped by the cam action either to the valve or a bypass reservoir. When pressurized, the hydraulic line behaves like a solid body and transmits the lift schedule imparted by intake cam directly to the intake valve in the full valve lift mode for max power. When the solenoid is disengaged, a spring takes over valve actuation, losing solid transmission of the motion from the cam in other three modes, which are the early intake valve closing mode, the late intake valve opening mode, and the multi-lift mode, thus leading to the name lost motion. This electro-hydraulic link allows independent LGD operation of the two components, resulting in certain control over the valve lift profiles. A closed solenoid keeps the hydraulic fluid pressurized, transmitting the intake cam profile to the valve in the normal fashion, while an open solenoid breaks the effective link between cam and valve, decoupling their profiles [27].

This system is not a full VVA system because the valve timing and duration are not independent of the lift in each of the three lost-motion modes although one has a choice to choose three different dependencies, i.e., variability or control flexibility, among these three modes. The intake valve opening event cannot be shifted ahead or left of that at maximum power, which may be necessary for certain EGR operations. Also, the intake valve closing event cannot be extended beyond or right of that at maximum power, which may be necessary for certain Miller cycles.

Jacobs Vehicle Systems Inc. also developed its own version of the lost-motion VVA system, with emphasis on diesel engine efficiency and after-treatment optimization [57]. It includes the capability of on-off control of secondary events for IEGR and engine braking, high load capacity for early exhaust opening and engine braking, and inherent protection against valve-to-piston contact.

More recently, there were efforts by Gongda Power [58] and Shandong University [59] to replace solenoid valves for individual actuators with motor-driven rotary valves common to a group of actuators, to achieve more stable and faster time response at low temperature and/or to devise an alternative hardware, but at the cost of losing independent controllability for individual actuators within a group. The Gongda Power system [58] uses two motor-driven rotary valves, instead of just one valve by Shandong University [59], to add control flexibility to achieve the Miller cycle by enabling much later intake valve closing to reduce pumping loss and lower air temperature. It can also incorporate a special cam lobe to achieve earlier exhaust valve opening and, thus, compression brake function for a diesel engine.

7. Electro-Magnetic VVA (EMVVA) Systems

7.1. Opposed Solenoid EMVVA

Most effort in camless VVA system development was devoted to EMVVA, actuated by a pair of opposed electromagnets and balanced by a pair of compression springs. It is capable of generating variable valve timing and duration, but with fixed lift operation.

The developers of this technology include Valeo [19,60–62], FEV [30,31,63–65], GM [29], Ford [66], Visteon [32], BMW [45], TRW [67], Siemens [68], MIT [69], Ibaraki University [70], LGD Technology [71], Instituto Motori of National Research Council of Italy [72], and Aura System [31].

Valeo acquired the related technologies from FEV, Sagem, and Johnson Control and developed them to a more mature system, which was marketed as smart valve actuation (SVA) [73] and later as e-Valve [60,62]. e-Valve claims to have reached the required maturity level for mass production [60].

The key issues and challenges, some of which may remain unresolved at this point, for EMVVAs in general include the following:

- Seating instability and the resulting noise and valve durability issues due to the highly nonlinear nature of the electro-magnetic latching force unique to the opposed solenoid design.

Chang et al. [69] incorporated a nonlinear spring or nonlinear mechanical transformer for better soft seating and/or low holding current.

- Need for an accurate, robust, and durable position sensor for each actuator [62].
- Limited or no capability to achieve a variable lift or low lift profile, necessary for some advanced combustions. Lou [71] proposed incorporating a hydraulic mechanism for enhanced capability.
- High incremental cost, which is a challenge for camless VVAs. A four-cylinder engine with electronic actuation on only the intake valves is expected to cost about €300 more to build [62].
- Electrical power consumption. Okada et al. [70] proposed a bias permanent magnet to reduce energy consumption and a seesaw architecture to improve performance and the fitness.

7.2. Rotary Motor EMVVA

Camcon Technology [33] is developing a camless engine for passenger vehicles based on their proprietary IVA system, which allows valve lift, timing, and duration to be independently and continuously controllable.

Different from earlier EMVVA systems using opposed solenoids, IVA employs a four-phase rotary actuator, i.e., a rotary motor, using a rotor which is extended to provide a separate camshaft for each individual poppet valve [34]. A desmodromic linkage connects this camshaft to the entirely conventional valve. The actuator is electronically synchronized with the crankshaft and drives the rotor through the required angular trajectory in order to provide the selected valve event, which is enabled via a non-contact absolute rotary encoder to determine the rotor position for each actuator.

Camcon collaborated with Jaguar Land Rover to fit the intake valve IVA onto an Ingenium 2.01 four-cylinder gasoline engine, with favorable test results in power consumption, lift repeatability, noise level, durability, and fuel economy [34,74].

Further development work is being carried out to achieve capability for higher engine speed and exhaust valve actuation [34]. Brunel University London is using a single-cylinder version of IVA technology called single-cylinder intelligent valve technology (SCI) to study future powertrain concepts and speed up OEM and tier 1 engine development [75]

7.3. Other EMVVAs

There are other kinds of EMVVAs. LaunchPoint Technologies, for example, developed a linear motor EMVVA, which includes a voice coil actuator, a position sensor, and a nonlinear energy storage mechanism [76]. The energy storage mechanism can both recover the valve's kinetic energy, thus reducing the system energy consumption, and help soft seating at the open and close. The low-power actuator is used only to catch and release the valve at the beginning or the end of the stroke. It is able to maintain repeatable performance with 1.63–3.82 ms switch times, 0.01–0.07 m/s seating velocity, and 1.33–3.15 J energy consumption per switch. No further report is available on the development since a news post in 2014 [76]. The need of a position sensor for its normal function may present cost and reliability issues in application.

8. Electro-Hydraulic VVA (EHVVA) Systems

In EHVVA systems, primary actuators are hydraulic actuators, such as a piston-cylinder mechanism, controlled by electro-hydraulic valves. Compared with an EMVVA, an EHVVA generally has higher power density but lower efficiency. The hydraulic fluid has high bulk modulus suitable for snubbing in the valve seating process, and its viscosity is highly sensitive to temperature, becoming too viscous for proper function at lower temperature. Some major EHVVA systems are listed below, which are also listed in Table 2 for comparison.

- Sturman Industries developed the hydraulic valve actuation (HVA) system. It includes two digital two-way pilot valves, a proportional valve, a hydraulic actuator with boost and drive pistons, and a position sensor necessary for closed-loop lift control [35,77]. The actuator is returned either hydraulically or by a return spring. It offers full control in valve timing, duration, and lift, and it was used in an experimental 15 L natural gas engine and as universal research modules for various research programs [78]. Its necessary use of a position sensor may incur high cost and reliability concerns for mass production.

- Lotus and Eaton jointly developed the active valve train (AVT) system. It includes one digital three-way pilot valve, one servo valve, one return spring, and a hydraulic actuator integrated with a position sensor needed for closed-loop lift control [79]. Like Sturman's HVA, the AVT system offers full control in valve timing, duration, and lift. It may also have cost and reliability issues associated with the position sensor.
- AVL and Bosch developed the electro-hydraulic valvetrain system (EHVS) system [80]. It includes two digital main valves, a hydraulic actuator with a two-stage differential piston drive, a pilot-controlled variable snubber for seating control, and no return spring. It uses an open-loop control and, thus, has no need for a position sensor, which offers substantial cost and reliability benefits but presents concerns in lift calibration and accuracy.
- Gongda Power developed the Gongda-VVA-2 (GD-VVA-2). It includes one digital three-way pilot valve, one digital three-way main valve, an actuator with one lift-control sleeve, two-step seating control, open-loop two-step lift control, and no position sensor [81–84]. The two-step lift control provides robust and accurate position control, which is delineated mechanically by the lift-control sleeve, without the need for an expensive and unreliable position sensor. It also has a two-level hydraulic damping mechanism for effective valve seating speed control over a wider temperature range. The two-step lift control does present certain functional compromise, which can be compensated for by its infinitely variable timing capability inherent in this and other EHVVAs. One GD-VV-2 prototype system passed 1000 h of durability testing on a test bench. There is also a proposal to incorporate some CVVL mechanisms into the base GD-VV-2 design, resulting in a full VVA system, still without the need for a position sensor for each engine valve [85].

Table 2. Some major electro-hydraulic VVA (EHVVA) systems.

Company	System	Design Features	Pros	Cons
Sturman	Hydraulic Valve Actuation (HVA)	Two digital 2-way pilot valves, a proportional valve, a return spring, and closed-loop control with a position sensor.	Full lift variability	High sensor cost and reliability concern
Lotus-Eaton	Active Valve Train (AVT)	One digital 3-way pilot valve, one servo valve, one return spring, and a hydraulic actuator integrated with a position sensor.	Full lift variability	High sensor cost and reliability concern
AVL-Bosch	Electro-hydraulic Valvetrain System (EHVS)	Two digital main valves, a hydraulic actuator with a two-stage differential piston drive and a pilot controlled variable snubber for seating control, no return spring, open-loop control without a position sensor	Full lift variability and low cost	Lift accuracy concern
Gongda Power	Gongda VVA-2 (GD-VVA-2)	One digital 3-way pilot valve, one digital 3-way main valve, an actuator with one lift-control-sleeve, 2-step seating control, open-loop 2-step lift control without position sensor.	Accurate lift and low cost	2-step lift

There are many other studies on EHVVA systems. One major effort is to minimize the energy consumption by the VVA system itself by using some kind of pendulum mechanism, similar to the compression spring pendulum used in the EMVVA system. Some examples are as follows:

- Ford developed an EHVVA system that has a unique hydraulic pendulum design, i.e., some fluid spring pendulum [7,86], which tries to convert the kinetic energy into hydraulic pressure or potential energy during both the opening and the closing stroke. The system includes a high-pressure and a lower-pressure switch valve and a couple of check valves, and it requires close monitoring and feedback on the engine valve position. However, the fluid spring may be difficult to manage because of the high bulk modulus of a typical hydraulic fluid. Additionally, the fluid bulk modulus is highly variable under the influence of the entrapped air.
- Gongda Power developed the LGD-VVA-1 system that consists of a two-spring actuation, a bypass passage, and an electro-hydraulic latch-release mechanism [36,37]. The two-spring pendulum system is used to provide efficient conversion between the moving mass kinetic energy and the spring potential energy for reduced energy consumption. Its latch-release mechanism can also compensate for the lost frictional energy during the pendulum motion. Prototypes of the system were bench- and engine-tested. This system, at least with its limited prototype design, presents some challenge in packaging because of its total height, considering adding two springs to the necessary hydraulic mechanism.
- DaimlerChrysler developed various designs using a two-spring pendulum with a hydraulic latching (US Patent Nos. 4930464, 5595148, 5765515, 5809950, 6167853, 6491007, and 6601552). However, the designs do not have an effective latching mechanism that can add energy to the pendulum to compensate for the frictional loss and cylinder air pressure, and there is no mechanism to change valve lift.

9. Electro-Pneumatic VVA (EPVVA) Systems

There were several studies and developments in electro-pneumatic VVA (EPVVA) systems [38, 39,87–89]. As a work medium, air in an electro-pneumatic system is better than hydraulic fluid in an electro-hydraulic system in terms of the insensitivity of its viscosity to the system temperature. Air leakage also does not impose pollution problem. However, Watson and Wakeman [88] found the following issues with the pneumatic actuator:

- Noise issues associated with air exhaust, choking, and hard valve seating associated with a pure pneumatic actuator design.
- Repeatability issues in lift control because of air flexibility.
- Sizing issues, at least for their particular design, due to the peak air pressure limit.

The most serious development of an EPVVA system was carried out by the Swedish company Freevalve AB, formerly Cargine and a sister company to Koenigsegg Automotive AB, which developed an EPVVA system branded as Freevalve on an existing SAAB car engine [89,90]. The Freevalve technology also appeared in the Qoros 1.6 L four-cylinder engine [89,91]. The Freevalve system includes pneumatic valve actuators for opening, springs for valve closing, and position sensors for feedback control. An oil damping mechanism must be incorporated, as shown in Reference [38], to help resolve the seating issue, and the technology is, therefore, also called an electro-hydraulic-pneumatic actuator [89]. The claimed benefits include up to a 30% increase in horsepower and torque, up to a 30% improvement in fuel economy, and a 50% reduction in overall emissions, based on a report [90] in 2013.

Ma et al. [38] proposed an adaptive lift control scheme for an early version of the Freevalve technology to improve the intake valve lift repeatability. A control-oriented electro-pneumatic valve model was developed and used for adaptive parameter identification, and a closed-loop control scheme of valve lift was developed, utilizing the identified parameters in real-time. The main control techniques used in the process include model reference adaptation and the MIT rule [92]. The resulting maximum steady-state lift errors were less than 0.4 mm at high valve lift and less than 1.3 mm at low valve lift, which is still not accurate enough for commercial application.

10. Valve Profile Tracking of Camless VVA Systems

In an engine without a camshaft, the accuracy and fidelity of the electronic control of valve profile are critical to achieve the desired engine performance. The valve profile tracking includes the following basic control objectives for most camless applications [81]:

1) Valve timing control for optimum combustion phase and valve collision avoidance.
2) Valve lift control.
3) Profile area (integration of valve lift profile over time or crank angle) control for accurate air exchange.
4) Engine valve soft seating for noise control and extending durability.

As noted by Li et al. [81], the overall valve duration control and the valve transition response (rising and falling slopes) control studied in literature can be classified into valve timing control and profile area control, respectively. For traditional cam-based engine valves, the above four properties are guaranteed by proper design of the cam profile. For camless VVAs, these control objectives can be achieved either partially or simultaneously, depending on the specific VVA system design and its application.

Adaptive peak lift control was employed by Levin et al. [93] for an EMVVA and by Ma et al. [94] for an EPVVA to achieve proper valve lift repeatability.

Feedforward control was used for valve timing control to compensate for the valve-opening or valve-closing delays for EHVVA and EPVVA by Liao et al. [95] and Ma et al. [38], respectively.

Soft seating control is a challenge for the EMVVA because of nonlinear magnetic force, and it is one of the most studied subjects in the field. Peterson et al. [96] studied guaranteed valve response using extreme seeking control. Tai and Tsao [97] used a combination of a feedforward linear–quadratic regulator and repetitive learning control to reduce cycle-to-cycle variations. These control designs [96,97] were intended for single or multiple control objectives. Others dealt with the overall valve profile control as a single tracking problem. Wang and Tsao [98], for example, applied a combination of model reference control and repetitive control to achieve asymptotic profile tracking. Eyabi and Washington [99] applied the sliding mode control to achieve repeatable tracking performance with guaranteed seating velocity. In addition, there were studies associated with the application of EMVVA systems in the combustion mode transition between SI and HCCI combustions [63], stratified lean combustion [69], and turbulent jet ignition [64].

For an EHVVA system, Sun and Kuo [100] and Gillella et al. [101] proved the effectiveness of robust repetitive control and time-varying internal-model-based control, respectively, in tracking the desired valve profile under both steady-state and transient engine operations.

For the EHVVA system by Lou et al. [84], Li et al. [81] studied the profile tracking problem without the need for complicated control scheme because of the inherent robust nature of its lift control and seating-velocity control. However, the valve timing and profile area controls are still challenging because of the nonlinear and time-varying nature of the hydraulic system, including nonlinear flow dynamics and temperature-sensitive fluid viscosity [83,102]. A receding horizon linear–quadratic tracking (LQT) controller was designed along with a Kalman optimal state estimation, which was proven to be effective through both steady-state and transient validations.

11. Summary

As a summary, the engine valve system with active control can be mainly divided into three groups: variable valve timing (VVT), variable valve lift (VVL), and camless valve system. The authors believe that each valve system has its own application domain. For the variable valve timing system, the trend is to move to electrical VVT systems motivated by reducing engine cold-start emissions and significant cost reduction of electrical drive systems. VVA systems may be used for engines with advanced combustion modes such as spark-controlled compression ignition (SpCCI). Among VVA systems and compared with the combined VVT and VVL systems, the camless systems have higher cost and less maturity but have the ultimate control flexibility, which is needed as an enabler for more advanced combustion modes such as HCCI to further improve the engine performance with reduced emissions. The benefit of different valve technologies with respect to engine fuel economy is not readily discernable or available because a new engine is typically incorporated with multiple new technologies; some of them are summarized in Table 3 below, where the baseline is the conventional cam-based valve system.

Table 3. Fuel economy benefits.

Valve System Type	System and Fuel Economy and Other Key Benefits	Reference
HVVT	General: 3%–5% better FE	
HVVT	BMW double Vanos: up to 10% better FE	[9]
EVVT	General: 3%–5% better FE, especially with cold-start tailpipe emission reduction	
DVVL	Audi AVS system: up to 7% better FE	[103]
DVVL	GM intake valve lift control (IVLC): up to 4% better FE	[104]
DVVL + VVT	Honda i-VTEC: 13% better FE	[105]
CVVL + VVT	BMW Valvetronic: 10% better FE	[103]
CVVL + VVT	Toyota Valvematic: 6% better FE	[106]
LMVVA	Fiat MultiAir: 10% better FE	[107]
VVL + EVVT	General: enabling HCCI and 20% better FE	[108]
Camless VVA	General: enabling HCCI and 25% better FE	[108]

Author Contributions: Conceptualization, Z.L. and G.Z.; methodology, Z.L.; software, N.A.; validation, N.A.; formal analysis, N.A.; investigation, Z.L. and G.Z.; resources, Z.L. and G.Z.; data curation, N.A.; writing— Z.L. and G.Z.; writing—review and editing, Z.L. and G.Z.; visualization, Z.L.; supervision, Z.L.; project administration, N.A.; funding acquisition, N.A. All authors have read and agreed to the published version of the manuscript.

Funding: This research received no external funding.

Abbreviations

AVCS	Active valve control system
AVLS	Active valve lift system
AVS	Audi valve lift system
AVT	Lotus-Eaton active valve train
BMW	Bayerische Motoren Werke automotive group
CA	Crank angle
CPS	Cam profile switching system
CVTC	Nissan continuous variable valve timing control
CVVD	Continuous VVD
CVVL	Continuous VVL
CVVT	Continuous VVT
CVVTCS	Continuously variable valve timing control system
DCVCP	Double continuous variable cam phasing
DOHC	double overhead camshaft
DVT	Discrete valve timing
DVVL	Discrete VVL
DVVT	Discrete VVT
EC	Exhaust closing
EHVS	AVL-Bosch electro-hydraulic valvetrain system
EHVVA	Electro-hydraulic VVA
EC	Exhaust closing
EGR	Exhaust gas recirculation
EO	Exhaust opening
EPVVA	Electro-pneumatic VVA
EVVT	Electrical VVT
FCA	Fiat Chrysler Automobiles

FE	Fuel economy
FEV	Forschungsgesellschaft für Energietechnik und Verbrennungsmotoren
GD-VVA-2	Gongda VVA-2
GM	General Motors
HC	Hydrocarbon
HCCI	Homogenous charge compression ignition
HVVT	Hydraulic VVT
IC	Intake closing
IEGR	Internal exhaust gas recirculation
IO	Intake opening
IVA	Camcon intelligent valve actuation
i-VTEC	Honda intelligent VTEC
LMVVA	Lost-motion VVA
MG	Morris Garages
MIT	Massachusetts Institute of Technology
MIVEC	Mitsubishi innovative valve timing electronic control
MVVT	Mechanical VVT
NVCS	Nissan valve control system
OEM	Original equipment manufacture
PCCI	Premixed charge compression ignition
PSA	Peugeot Société Anonyme
SpCCI	Spark-controlled compression ignition
SVA	Valeo smart valve actuation, also e-Valve
TFSI	Turbo fuel stratified injection
TRW	Thompson Ramo Wooldridge
VANOS	German words for variable camshaft timing
VTC	Valve timing control
VTEC	Honda variable valve timing and lift electronic control
VTVT	Variable timing valve train
VVA	Variable valve actuation
VVC	Variable valve control
VVD	Variable valve duration
VVEL	Variable valve event and lift
VVL	Variable valve lift
VVT	Variable valve time
VVT-iE	Toyota variable valve timing intelligent electric
VVTL-i or VVT-iL	Toyota variable valve timing and lift intelligent

References

1. Hybrid-Electric. Plug-in Hybrid-Electric and Electric Vehicle Sales. Available online: https://www.bts.gov/content/gasoline-hybrid-and-electric-vehicle-sales (accessed on 27 January 2020).
2. Alternative Fuels Data Center. Available online: https://afdc.energy.gov/fuels/hydrogen_basics.html (accessed on 27 January 2020).
3. US DOE. Fuel Cell Electric Vehicles. Available online: https://afdc.energy.gov/vehicles/fuel_cell.html (accessed on 3 February 2020).
4. Brüstle, C.; Schwarzenthal, D. VarioCam Plus-A Highlight of the Porsche 911 Turbo Engine. *SAE Tech. Paper* **2001**. [CrossRef]
5. Duesmann, M. Innovative Valve Train Systems, Spectrum: Technology Highlights and R&D Activities at FEV. 2002, p. 3. Available online: https://www.fev.com/fileadmin/user_upload/Media/Spectrum/en/spectrum19.pdf (accessed on 27 January 2020).
6. Tai, C.; Tsao, T.; Schörn, N.; Levin, M. Increasing Torque Output from a Turbodiesel with Camless Valvetrain. *SAE Tech. Paper* **2002**. [CrossRef]

7.   Schechter, M.; Levin, M. Camless Engine. *SAE Tech. Paper* **1996**. [CrossRef]
8.   Wikipeda, Variable valve timing. Available online: https://en.wikipedia.org/wiki/Variable_valve_timing (accessed on 27 January 2020).
9.   BMW Service. BMW Product Information Vanos. 2005. Available online: http://v12.dyndns.org/BMW/BMW%20Product%20info%20Vanos.pdf (accessed on 27 January 2020).
10.   2020 Honda, The VTEC Engine/1989. Available online: https://global.honda/heritage/episodes/1989vtecengine.html (accessed on 23 November 2019).
11.   Audi Techonology Portal. Audi Valvelift System. Available online: https://www.audi-technology-portal.de/en/drivetrain/engine-efficiency-technologies/audi-valvelift-system_en (accessed on 23 November 2019).
12.   Heidbrink, S. i-Active Valve Lift System. Available online: https://web.archive.org/web/20120624171722/http://drive2.subaru.com/Spring07_whatmakes.htm (accessed on 27 December 2019).
13.   Wikipedia, CamPro engine. Available online: https://en.wikipedia.org/wiki/CamPro_engine#Campro_CPS_and_VIM_engine (accessed on 27 January 2020).
14.   CanAndBike Team. 2017 Yamaha R15 Gets Variable Valve Timing. Available online: https://auto.ndtv.com/news/2017-yamaha-r15-gets-variable-valve-timing-1676599 (accessed on 27 January 2020).
15.   Wikipedia, MIVEC. Available online: https://en.wikipedia.org/wiki/MIVEC (accessed on 27 January 2020).
16.   Wikipedia, Nissan VVL engine. Available online: https://en.wikipedia.org/wiki/Nissan_VVL_engine (accessed on 27 January 2020).
17.   Wikipedia, VarioCam. Available online: https://en.wikipedia.org/wiki/VarioCam (accessed on 27 January 2020).
18.   Wikipedia, VVT-i. Available online: https://en.wikipedia.org/wiki/VVT-i (accessed on 27 January 2020).
19.   Valeo, Valeo Electromagnetic Valve actuation. Available online: https://www.slideshare.net/ValeoGroup/valeo-electromagnetic-valve-actuation (accessed on 27 January 2020).
20.   Ha, K.; Han, D.; Kim, W. Development of Continuously Variable Valve Lift Engine. *SAE Tech. Paper* **2010**, 1187. [CrossRef]
21.   Liu, T.; Yin, J.; Sun, X.W. Test for Continuously Variable Valve Lift Mechanism. *Intern. Combust. Engine Powerpl.* **2018**. [CrossRef]
22.   Witzenburg, G. It's All about Flow: Automakers Choose from a Wide Variety of Engine Technology. Automotive Industries. 2003. Available online: https://www.britannica.com/technology/automotive-industry (accessed on 27 January 2020).
23.   Wikipedia, VTi engine. Available online: https://en.wikipedia.org/wiki/VTi_Engine (accessed on 27 January 2020).
24.   Wikipedia, Variable Valve Event and Lift. Available online: https://en.wikipedia.org/wiki/Variable_Valve_Event_and_Lift (accessed on 27 January 2020).
25.   Wikipedia, Nissan VQ3VHR. Available online: https://www.engine-specs.net/nissan/vq37vhr.html (accessed on 27 January 2020).
26.   Eugenio, 77, Toyota Valvematic system. Available online: https://toyota-club.net/files/faq/12-11-03_faq_valvematic_eng.htm (accessed on 24 November 2019).
27.   Steven, A. *Inside Fiat's innovative MultiAir system*; SAE International: Warrendale, PA, USA, October 2010.
28.   Wikipedia, MultiAir. Available online: https://en.wikipedia.org/wiki/MultiAir (accessed on 27 January 2020).
29.   Theobald, M.; Lequesne, B.; Henry, R. Control of Engine Load via Electromagnetic Valve Actuators. *SAE Tech. Paper* **1994**. [CrossRef]
30.   Boie, C.; Kemper, H.; Kather, L.; Corde, G. Method for Controlling An Electromagnetic Actuator for Achieving a Gas Exchange Valve on a Reciprocating Internal Combustion Engine. US Patent 6340008, December 2000.
31.   Schneider, L.E. Electromagnetic Valve Actuator with Mechanical End Position Clamp or Latch. US Patent 6267351, 31 July 2001.
32.   Haskara, I.; Mianzo, L.; Kokotovic, V. Method of Controlling an Electromagnetic Valve Actuator. US Patent 6644253, 11 November 2003.
33.   Camcon Website. Available online: https://www.camcon-automotive.com/ (accessed on 27 January 2020).
34.   Stone, R.; Kelly, D.; Geddes, J.; Jenkinson, S. Intelligent Valve Actuation-A Radical New Electro-Magnetic Poppet Valve Arrangement. In Proceedings of the 26th Aachen Colloquium Automobile and Engine Technology, Germany, 9 October 2017; pp. 445–468.
35.   Sturman, O. Hydraulic Actuator for an Internal Combustion Engine. US Patent 5638781, 17 June 1997.

36. Lou, Z. Camless Variable Valve Actuation Designs with Two-Spring Pendulum and Electrohydraulic Latching. *SAE Tech. Paper* **2007**. [CrossRef]
37. Lou, Z.; Deng, Q.; Wen, S.; Zhang, Y.; Yu, M.; Sun, M.; Zhu, G. Progress in Camless Variable Valve Actuation with Two-Spring Pendulum and Electrohydraulic Latching. *SAE Int. J. Engines* **2013**, *6*, 319–326. [CrossRef]
38. Ma, J.; Zhu, G.; Schock, H. Adaptive control of a pneumatic valve actuator for an internal combustion engine. *IEEE Trans. Control Syst. Technol.* **2011**, *19*, 730–743. [CrossRef]
39. Ma, J.; Zhu, G.; Schock, H. A dynamic model of an electro-pneumatic valve actuator for internal combustion engines. *ASME J. Dyn. Syst. Meas. Control* **2010**, *132*. [CrossRef]
40. Tai, C.; Tsao, T.; Levin, M.; Barta, G.; Schechter, M.M. Using Camless Valvetrain for Air Hybrid Optimization. *SAE Tech. Paper* **2003**. [CrossRef]
41. Lang, O.; Salber, W.; Hahn, J.; Pischinger, S.; Hortmann, K.; Bücker, C. Thermodynamical and Mechanical Approach towards a Variable Valve Train for the Controlled Auto Ignition Combustion Process. *SAE Tech. Paper* **2005**. [CrossRef]
42. Kitabatake, R.; Minato, A.; Inukai, N.; Shimazaki, N. Simultaneous Improvement of Fuel Consumption and Exhaust Emissions on a Multi-Cylinder Camless Engine. *SAE Int. J. Engines* **2011**, *4*, 1225–1234. [CrossRef]
43. Wikipedia, Variator (Variable Valve Timing). Available online: https://en.wikipedia.org/wiki/Variator(variable_valve_timing (accessed on 27 January 2020).
44. Wikipedia, VANOS. Available online: https://en.wikipedia.org/wiki/VANOS (accessed on 27 January 2020).
45. Flierl, R.; Kluting, M. The third generation of new fully variable valvetrain for throttle free load control. *SAE Tech. Paper* **2000**. [CrossRef]
46. Carley, L. The Inner Workings of Variable Valve Timing. Available online: https://www.enginebuildermag.com/2014/01/the-inner-workings-of-variable-valve-timing/ (accessed on 27 January 2020).
47. Hattori, M.; Inoue, T.; Mashiki, Z.; Takenaka, A.; Urushihata, H.; Morino, S.; Inohara, T. Development of Variable Valve Timing System Controlled by Electric Motor. *SAE Int. J. Engines* **2009**, *V1*, 985–990. [CrossRef]
48. Ren, Z.; Zhu, G.G. Modeling and Control of an Electric Variable Valve Timing System. *J. Dyn. Sys. Meas. Control.* **2014**, *136*. [CrossRef]
49. Rover K-series Variable Valve Control (VVC). Available online: http://www.sandsmuseum.com/cars/elise/thecar/engine/vvc2.pdf (accessed on 27 January 2020).
50. Hyundai's Continuously Variable Valve Duration (CVVD) Technology. Available online: https://www.team-bhp.com/forum/technical-stuff/210770-hyundais-continuously-variable-valve-duration-technology.html (accessed on 27 January 2020).
51. Hyundai Motor Group Unveils CVVD Engine Technology; +4% Performance, +5% Fuel Economy, −12% Emissions. Available online: https://www.greencarcongress.com/2019/07/201090703-cvvd.html (accessed on 27 January 2020).
52. Hyundai·Kia Motors, Hyundai's Breakthrough Engine that Answers a 133-year Challenge. Available online: https://news.hyundaimotorgroup.com/Article/hyundai-announces-breakthrough-engine-that-answers-a-133-year-challenge (accessed on 27 January 2020).
53. Kim, B.S.; Lee, S.H.; Choi, K.; Kim, J.S.; Kim, D.S.; Im, H.; Ha, K.P. Continuous Variable Valve Duration Apparatus. US Patent 8813704, 24 August 2014.
54. Inoue, K.; Nagakiro, K.; Ajiki, Y.; Kishi, N. A high power wide torque range efficient engine with a newly developed variable valve lift and timing mechanism. *SAE Tech. Paper* **1989**, *98*, 822–832.
55. Wikipedia, VTEC. Available online: https://en.wikipedia.org/wiki/VTEC (accessed on 27 January 2020).
56. Bimerfest. How it Works: BMW Valvetronic. Available online: https://www.bimmerfest.com/news/1262694/how-it-works-bmw-valvetronic/ (accessed on 27 January 2020).
57. Schwoerer, J.; Kumar, K.; Ruggiero, B.; Swanbon, B. Lost-Motion VVA Systems for Enabling Next Generation Diesel Engine Efficiency and After-Treatment Optimization. *SAE Tech. Paper* **2010**. [CrossRef]
58. Lou, Z. Engine Valve Actuation System. US Patent Number 9625050, 18 April 2017.
59. Xie, Z.F. Oil Control Device for Fully Variable Hydraulic Valve System of Internal Combustion Engine, WO2015006886A1. US Patent 9,995,188, 22 January 2015.
60. Frederic, A.; Picron, V.; Hobraiche, J.; Gelez, N.; Gouiran, S. ElectroMagnetic Valve Actuation System e-Valve: Convergence Point between Requirements of Fuel Economy and Cost Reduction. *SAE Tech. Paper* **2010**. [CrossRef]

61. Valeo, Valeo Presents New Smart Valve Actuation Technology-the camless Engine becomes a reality, Frankfurt, Germany. Available online: http://www.valeo.com.cn/cws-content/www.valeo.cn/medias//fichiers/journalistes/en/CP/camless-uk.pdf (accessed on 13 September 2005).

62. Vale, 2008, e-Valve: The Electromagnetic Valve Control System. Available online: https://www.valeo.com/wp-content/uploads/2016/11/press-kit-2008-paris-motor-show.pdf (accessed on 19 November 2019).

63. Pischinger, M.; Salber, W.; van der Staay, F.; Baumgarten, H.; Kemper, H. Benefits of the electromechanical valve train in vehicle operation. *SAE Tech. Paper* **2000**. [CrossRef]

64. Wolters, P.; Salber, W.; Geiger, J.; Duesmann, M.; Dilthey, J. Controlled auto ignition combustion process with an electromechanical valve train. *SAE Tech. Paper* **2003**. [CrossRef]

65. Salber, W.; Kemper, H.; van der Staay, F.; Esch, T. The electro-mechanical valve train – a system module for future poewertrain concepts. *MTZ Mot. Z.* **2000**, *61*, 12.

66. Wang, Y.; Megli, T.; Haghgooie, M.; Peterson, K.; Stefanopoulou, A.G. Modeling and Control of Electromechanical Valve Actuator. *SAE Tech. Paper* **2002**. [CrossRef]

67. Hartwig, C.; Josef, O.; Gebauer, K. Dedicated Intake Actuator for Electromagnetic Valve Trains. *SAE Tech. Paper* **2005**. [CrossRef]

68. Butzmann, S.; Melbert, J.; Koch, A. 2000 Sensorless control of electromagnetic actuators for variable valve train. *SAE Tech. Paper* **2000**. [CrossRef]

69. Chang, W.S.; Parlikar, T.; Kassakian, J.G.; Keim, T.A. An Electromechanical Valve Drive Incorporating a Nonlinear Mechanical Transformer. *SAE Tech. Paper* **2003**. [CrossRef]

70. Okada, Y.; Marumo, Y.; Konno, M. Electromagnetic Valve Actuator for Automobile Engines. *SAE Tech. Paper* **2004**. [CrossRef]

71. Lou, Z. Electromechanical Variable Valve Actuator with a Spring Controller, WO2007092468A3. US Patent 7,591,237, 16 August 2007.

72. Giglio, V.; Iorio, B.; Police, G.; di Gaeta, A. Analysis of Advantages and of Problems of Electromechanical Valve Actuators. *SAE Tech. Paper* **2002**. [CrossRef]

73. Abuelsamid, S. Valeo has customers for camless engine with 'smart valve actuation'. *Automot. News*, 12 December 2006.

74. Cropley, S. New engine valve tech gives petrols the efficiency of diesels. *AutoCar*. 24 May 2017. Available online: https://www.autocar.co.uk/car-news/industry/new-engine-valve-tech-gives-petrols-efficiencydiesels (accessed on 27 January 2020).

75. Green Car Congress Brunel to Use Camcon Single Cylinder IVT in Researching Future Powertrain Concepts. Available online: https://www.greencarcongress.com/2019/07/20190704-camcon.html (accessed on 7 July 2019).

76. LaunchPoint Technologies Inc, New VVT Valve Actuator Cuts Power Consumption by More Than 50%, and Electromechanical Valve Actuator for Variable Valve Timing. Available online: https://www.launchpnt.com/news/news/topic/electromechanical-valve (accessed on 27 January 2020).

77. Turner, C.; Babbitt, G.; Balton, C.; Raimao, M.; Giordano, D.D. Design and Control of a Two-stage Electro-hydraulic Valve Actuation System. *SAE Tech. Paper* **2004**, *1265*. [CrossRef]

78. Sturman Industries. Available online: https://sturmanindustries.com/Solutions/Products/HVACamless/tabid/172/Default (accessed on 3 December 2019).

79. Turner, J.W.G.; Kenchington, S.A.; Stretch, D.A. Production AVT Development: Lotus and Eaton's Electrohydraulic Closed-Loop Fully Variable Valve Train System. Available online: https://www.semanticscholar.org/paper/ (accessed on 27 January 2020).

80. Denger, D.; Mischker, K. The Electro-Hydraulic Valvetrain System EHVS-System and Potential. *SAE Tech. Paper* **2005**. [CrossRef]

81. Li, H.; Huang, Y.; Zhu, G.; Lou, Z. Profile Tracking for an Electro-Hydraulic Variable Valve Actuator Using Receding Horizon LQT. *IEEE/ASME Trans. Mechatron.* **2019**, *24*, 338–349. [CrossRef]

82. Li, H.; Huang, Y.; Zhu, G.; Lou, Z. Adaptive LQT Valve Timing Control for an Electro-Hydraulic Variable Valve Actuator. *IEEE Trans. Control Syst. Technol.* **2019**, *27*, 2182–2194. [CrossRef]

83. Li, H.; Huang, Y.; Zhu, G.; Lou, Z. Linear Parameter-Varying Model of an Electro-Hydraulic Variable Valve Actuator for Internal Combustion Engines. *J. Dyn. Sys. Meas. Control* **2017**, *140*. [CrossRef]

84. Lou, Z.; Wen, S.; Qian, J.; Xu, H.; Zhu, G.; Sun, M. Camless Variable Valve Actuator with Two Discrete Lifts. *SAE Tech. Paper* **2015**. [CrossRef]

85. Lou, Z.; Wen, S. Continuously Variable Lift Actuator. China Patent CN201410614962.6, 18 August 2017.

86. Ashhab, M.; Stefanopoulou, A.; Cook, J.; Levin, M. Camless Engine Control for a Robust Unthrottled Operation. *SAE Tech. Paper* **1998**. [CrossRef]

87. Richeson, W.E.; Erickson, F.L. 1989 Pneumatical Actuator with Solenoid Operated Control Valves. US Patent 4,873,948, 17 October 1989.

88. Watson, J.P.; Wakeman, R.J. Simulation of a Pneumatic Valve Actuation System for Internal Combustion Engine. *SAE Tech. Paper* **2005**. [CrossRef]

89. Freevalve, A.B. Freevalve Technology. Available online: http://www.freevalve.com/technology/freevalve-technology/ (accessed on 18 November 2019).

90. Ernst, K. Inside Koenigsegg Looks at Future Engine Technology: Video. 2 February 2013. Available online: www.motorauthority.com (accessed on 17 November 2019).

91. Koenigsegg. Freevalve technology unveiled at Beijing Motor Show in Qoros Qamfree concept car. 26 April 2016. Available online: https://www.koenigsegg.com/freevalve-technology-unveiled-at-beijing-motor-show-in-qoros-qamfree-concept-car/ (accessed on 17 November 2019).

92. Astrom, K.J.; Wittenmark, B. *Adaptive Control*, 2nd ed.; Addison-Wesley: Boston, MA, USA, 1995.

93. Levin, M.B.; Tai, C.; Tsao, T.C. Adaptive nonlinear feedforward control of an electrohydraulie camless valvetrain. In Proceedings of the 2000 American Control, Chicago, IL, USA, 28–30 June 2000; pp. 1001–1005.

94. Ma, J.; Zhu, G.M.; Schock, H.; Winkelman, J. Adaptive control of a pneumatic valve actuator for an internal combustion engine. In Proceedings of the 2007 American Control Conference, New York, NY, USA, 9–13 July 2007; pp. 767–774.

95. Liao, H.H.; Roelle, M.J.; Chen, J.S.; Park, S.; Gerdes, J.C. Implementation and analysis of a repetitive controller for an electro-hydraulic engine valve system. *IEEE Trans. Control Syst. Technol.* **2011**, *19*, 1102–1113. [CrossRef]

96. Peterson, K.S.; Stefanopoulou, A.G. Extremum seeking control for soft landing of an electromechanical valve actuator. *Automatica* **2004**, *40*, 1063–1069. [CrossRef]

97. Tai, C.; Tsao, T.C. Control of an electromechanical actuator for camless engines. In Proceedings of the 2003 American Control Conference, Denver, CO, USA, 4–6 June 2003; pp. 3113–3118.

98. Wang, J.; Tsao, T.C. Repetitive control of linear time varying systems with application to electronic cam motion control. In Proceedings of the 2004 American Control Conference, Boston, MA, USA, 30 June–2 July 2004; Volume 4, pp. 3794–3799.

99. Eyabi, P.; Washington, G. Design and control of an electromagnetic valve actuator. In Proceedings of the 2006 IEEE Conference on Computer Aided Control System Design, 2006 IEEE International Conference on Control Applications, 2006 IEEE International Symposium on Intelligent Control, Munich, Germany, 4–6 October 2006; Volume 16, pp. 1657–1662.

100. Sun, Z.; Kuo, T.W. Transient control of electro-hydraulic fully flexible engine valve actuation system. *IEEE Trans. Control Syst. Technol.* **2010**, *18*, 613–621. [CrossRef]

101. Gillella, P.K.; Song, X.; Sun, Z. Time-varying internal model-based control of a camless engine valve actuation system. *IEEE Trans. Control Syst. Technol.* **2014**, *22*, 1498–1510. [CrossRef]

102. Zhang, S.; Song, R.; Zhu, G.G.; Schock, H. Model-based control for mode transition between spark ignition and HCCI combustion. *J. Dyn. Syst., Meas. Control* **2017**, *139*, 41004–41010. [CrossRef]

103. National Academies Press. Appendix I: Variable Valve Lift Systems, Cost, Effectiveness, and Deployment of Fuel Economy Technologies for Light-Duty Vehicles. Available online: https://www.nap.edu/read/21744/chapter/21 (accessed on 27 January 2020).

104. Kelly Blue Book. 2014 Chevy Impala Gets Variable Valve Lift on Ecotec 4-cylinder. Kelly Blue Book. 17 September 2012. Available online: http://www.kbb.com/car-news/all-the-latest/2014-chevy-impalagets-variable-valve-lift-on-ecotec-4_cylinder/2000008572/ (accessed on 6 August 2013).

105. Noh, D.Y. Honda's New VTEC Offers More Power, Better Fuel Economy, Cleaner Emissions. Available online: https://www.autoblog.com/2006/09/25/hondas-new-vtec-offers-more-power-better-fuel-economy-cleaner/ (accessed on 14 January 2020).

106. Borge, J.L. Toyota Engineers Put a Shine into the 2014 Corolla. SAE International, Automotive Engineering Magazine. 9 September 2013. Available online: http://articles.sae.org/12444/ (accessed on 27 January 2020).

107. Murphy, T. Fiat Breathing Easy with MultiAir. WardAuto. 26 March 2010. Available online: http: //wardsauto.com/ar/fiat_breathing_multiair_100326 (accessed on 27 January 2020).
108. Najafabadi, M.I.; Aziz, N.A. Homogeneous Charge Compression Ignition Combustion: Challenges and Proposed Solutions. *J. Combust.* **2013**. [CrossRef]

 *applied sciences*

Review

A Review of Hydrogen Direct Injection for Internal Combustion Engines: Towards Carbon-Free Combustion

Ho Lung Yip [1], Aleš Srna [1], Anthony Chun Yin Yuen [1], Sanghoon Kook [1], Robert A. Taylor [1,2], Guan Heng Yeoh [1], Paul R. Medwell [3] and Qing Nian Chan [1,*]

[1] School of Mechanical and Manufacturing Engineering, University of New South Wales, Sydney 2052, Australia; h.l.yip@unsw.edu.au (H.L.Y.); a.srna@unsw.edu.au (A.S.); c.y.yuen@unsw.edu.au (A.C.Y.Y.); s.kook@unsw.edu.au (S.K.); robert.taylor@unsw.edu.au (R.A.T.); g.yeoh@unsw.edu.au (G.H.Y.)

[2] School of Photovoltaic and Renewable Energy Engineering, University of New South Wales, Sydney 2052, Australia

[3] School of Mechanical Engineering, The University of Adelaide, Adelaide 5005, Australia; paul.medwell@adelaide.edu.au

[*] Correspondence: qing.chan@unsw.edu.au

Received: 30 September 2019; Accepted: 6 November 2019; Published: 12 November 2019

Abstract: A paradigm shift towards the utilization of carbon-neutral and low emission fuels is necessary in the internal combustion engine industry to fulfil the carbon emission goals and future legislation requirements in many countries. Hydrogen as an energy carrier and main fuel is a promising option due to its carbon-free content, wide flammability limits and fast flame speeds. For spark-ignited internal combustion engines, utilizing hydrogen direct injection has been proven to achieve high engine power output and efficiency with low emissions. This review provides an overview of the current development and understanding of hydrogen use in internal combustion engines that are usually spark ignited, under various engine operation modes and strategies. This paper then proceeds to outline the gaps in current knowledge, along with better potential strategies and technologies that could be adopted for hydrogen direct injection in the context of compression-ignition engine applications—topics that have not yet been extensively explored to date with hydrogen but have shown advantages with compressed natural gas.

Keywords: hydrogen; internal combustion engine; compression ignition; dual-fuel engine; direct injection; high pressure gas jet; jet penetration

---

Contents

---

1. Introduction

Constrained carbon-emission budgets and increasingly stringent emission standards for vehicles around the globe have placed enormous pressure on manufacturers to develop less carbon-intensive fleets. Despite the present global domination of internal combustion engine (ICE) in the transportation section, a number of legislative strategies have been developed and adopted to promote a gradual replacement of ICE propulsion technology by fuel cell (FC) and battery-electric vehicles [1], unless there is a breakthrough in ICE technology to enable a significant reduction of harmful emissions and dependence on fossil fuel. Hydrogen has long been considered a future fuel in transportation powertrains, due to its ability to eliminate carbon-based emissions (e.g., CO, CO_2 and soot) and to achieve high energy efficiency [2]. Furthermore, hydrogen can be produced from renewable energy sources [3]. The first reported successful commercial application of hydrogen-powered vehicles dates back to the 1930s, with more than 1000 vehicles converted into hydrogen and flexible hydrogen/gasoline operation; however, technical details were reportedly destroyed because of war and can no longer be found [4]. The development of hydrogen-fueled ICEs stagnated since, with ongoing scientific research but limited practical applications. During the second half of the 20th century, hydrogen-fueled ICEs were mainly demonstration projects. Recently, effort towards decarbonization and tightening emission standards has facilitated several breakthroughs in the development of renewable hydrogen technologies, including: advanced methods and materials for hydrogen storage (e.g., high pressure storage, up to 700 bar), production (e.g., solar thermo-chemical processes) and usage (e.g., high pressure direct in-cylinder injection of gas). These developments have catalyzed the re-ignition of global interests towards incorporating hydrogen as an energy carrier in powertrains [5,6]. The Hydrogen Council, a global initiative for hydrogen energy composed of various energy and transportation companies, estimated that approximately 25% of the passenger vehicles and 20% of the non-electrified rail transport would be fueled with hydrogen by 2050, potentially reducing daily oil consumption for transportation use by up to 20% [7,8].

1.1. Hydrogen Application and Production

In the hydrogen economy—a future scenario where hydrogen represents the primary energy carrier—hydrogen has a wide range of applications other than transportation. For instance, many industrial processes require high-grade heat, which can utilize hydrogen combustion as a more efficient route [7]. Hydrogen is also an important reactant in the production of industrial feedstocks such as ammonia, methanol, polymers as well as in other refining processes, including fuel desulfurization, iron making and conversion of captured CO_2 from air or flue gas into useful chemicals [7]. Countries including the United Kingdom (UK), the United States (US), South Korea and some

European countries have developed infrastructure to use natural gas as power and heating sources in buildings. Such infrastructure brings additional benefit from a convenient switch to hydrogen-methane blend for further decarbonization [7]; nevertheless, a technology advancement is required to facilitate the increased share of hydrogen. The universal applicability of hydrogen for modern energy needs has boosted the investment and development of renewable hydrogen production and its related technology by many countries, with China being the largest importer and the US being the largest exporter, as of 2017 [9]. Australia [10,11], Japan [12] and Germany [13] have also devised strategic plans to become major players in the potential future hydrogen economy.

As of 2016, 96% of the total hydrogen production (i.e., $\sim$55 million tonnes per annum [7]) originated from fossil fuels [14]. Apart from the thermo-chemical conversion of coal and oil into hydrogen, steam methane reforming is the most widely adopted method for hydrogen production due to its cost effectiveness [15]. This is achieved by the chemical reaction between purified methane or natural gas and high temperature (i.e., 970–1120 K) and pressure (i.e., 3–25 bar) steam in the presence of a catalyst, which is typically nickel [6]. It should be noted that CO_2 is produced during the steam methane reforming process. Therefore, and due to the fossil feedstock, hydrogen produced through these chemical pathways is not considered renewable.

On the other hand, electrolysis of water, an electro-chemical process of splitting water into oxygen and hydrogen using electric current, can be considered renewable if the electricity is sourced renewably, such as using hydro, wind or solar power. This method currently accounts for only 4% of the world hydrogen production but it is predicted to expand rapidly to 22% by 2050 [16]. The research community has also been developing other renewable hydrogen production pathways, including photocatalytic hydrogen production, biomass and waste gasification and biological hydrogen production through biomass fermentation, etc. Some of these technologies are expected to mature and to enter commercial scale production by 2030 [17] and will boost the production of hydrogen through renewable routes. Therefore, green hydrogen will become more readily available and find widespread application in energy generation systems and powertrains.

1.2. The Potential for Hydrogen

With abundant land and renewable energy resources, a number of countries are advantageously poised to produce green hydrogen at low cost and to implement a hydrogen economy for domestic use or export to their neighbouring countries with high energy demands. This is because, unlike electricity, hydrogen can be stored and transported over long distances at lower cost. In particular for countries such as Australia and Chile, their geographic locations and well-established reputation as reliable conventional energy exporters could make them major hydrogen suppliers targeting their neighbours with net demands, such as China, Japan and Korea [18]. In fact, such collaboration is already underway with a joint project between Australia and Kawasaki Heavy Industries (Japan) to build a hydrogen supply chain between the two countries, wherein Japan is forecasted to import 900,000 tons of hydrogen by 2030 [19]. This strategic hydrogen export opportunity is estimated to add as much as AUD10 billion to Australia's economy by 2040 in a high hydrogen demand scenario [19]. European countries with a considerable production of renewable electricity, as well as mature infrastructure and transportation facilities for natural gas, such as the Netherlands, are also poised to benefit from an increased use of hydrogen energy [20]. The potentials to store excess electricity in the form of hydrogen and to adopt existing gas pipelines for hydrogen transportation as the production volume increases, place the Netherlands in an advantageous position as an energy exporter. The attractiveness is demonstrated by an estimated $\sim$EUR17.5–25 billion investment by 2025 to develop a carbon-free hydrogen economy in the Northern Netherlands [20].

These figures may have a significant growth potential, considering that industrial applications, for instance oil refining (33%), ammonia production (27%), methanol production (11%) and steel production (3%) currently dominates the use of hydrogen and drives more than 50% of the demand; while the use of hydrogen for power generation and propulsion only accounts for 1–2% of the total consumption [19,21]. For the transportation sector, the low utilization could be attributed to the lack of hydrogen distribution infrastructure and the slow development of hydrogen powertrain that can efficiently and cost-effectively convert hydrogen into power. There is, therefore, significant room for growth for increased hydrogen usage in the transportation sector, as technologies surrounding hydrogen transportation and usage continue to improve and mature.

1.3. Recent Developments of Hydrogen Applications in the Transportation Sector

Hydrogen powertrains mainly utilize two energy conversion technologies—hydrogen FC and ICE. An FC, which converts hydrogen into electrical energy used for propulsion, is presently the more commercialized approach. A FC vehicle is reported to have a tank-to-wheel efficiency in the range from 31% to 36%, with water vapor being the only emission [22,23]. Toyota, Honda and Hyundai have already commercialized FC vehicles in selected markets, with more than 6500 units sold as of June 2018 [24]. A prototype FC truck with a range of 480 km was recently unveiled by Toyota and Kenworth [25], demonstrating the potential to replace the ICE technology in future heavy-duty applications.

The advantages of hydrogen ICE compared to the FC technology include a higher tolerance to fuel impurities, flexibility to switch between fuels, reduction of rare materials usage and a more straightforward transition from conventional vehicles [26]. In addition, hydrogen ICE technology benefits from the reduced cost of using the existing mature manufacturing facilities and processes for conventional ICEs. The development of advanced hydrogen ICEs (e.g., direct injection (DI) and dual-fuel methods) is still in the conceptual stage. A majority of hydrogen-fueled ICE prototypes use port fuel injection (PFI) system, benefiting from a straightforward conversion from existing gasoline engines. A proof-of-concept light-duty truck and 'microbus' powered by such hydrogen-fueled ICE was presented by Tokyo City University [26]. It has to be noted that even with such a simple modification, engine brake thermal efficiency (BTE) of ~30–37% at medium load was reported [27], readily comparable with the current FC technology. Despite the competitive peak efficiency, hydrogen PFI suffers from a series of issues that will be discussed in Section 4, which can be tackled by an advanced DI fuel system as further discussed later in Sections 5 and 6. The potential of DI system has been already demonstrated by BMW and its partners in 2009. A hydrogen DI system with up to 300 bar injection pressure has been developed and integrated into a spark-ignition (SI) engine, achieving a maximum efficiency of 42%, paralleling diesel engines with turbocharging [28]. In light of the potential of hydrogen engine, Mazda has been developing a hydrogen rotary engine combining PFI and DI technology since 2006. This technology has been integrated into a sports-car model with an extended driving range of 649 km [29]. Furthermore, a series of works on experimental metal and optical engines as well as numerical simulations for hydrogen DI was funded by the U.S. Department of Energy from 2004 to 2011 [30–39]. A peak BTE of 45% in a hydrogen ICE was demonstrated at 2000 rpm and a high load condition of 13.5 bar brake mean effective pressure (BMEP). Most recently, the Australian Renewable Energy Agency (ARENA) has allocated more than AUD22 million to develop an effective renewable energy supply chain centered around hydrogen, including an investment in developing advanced hydrogen ICE technology and support for establishing a fundamental understanding of hydrogen combustion in engines [10].

1.4. Scope

The potential of more readily available renewable hydrogen is often associated with its potential to reduce the consumption of fossil fuels and harmful combustion emissions, when used in energy generation. While technologies for producing and storing hydrogen are under development, current bottleneck to efficiently utilize hydrogen needs to be overcome especially in the transportation section, as a large contributor to carbon emissions. Thus, this paper reviews the current development of using hydrogen in the transportation sector, with a focus on hydrogen combustion in ICEs. The properties of hydrogen and their implications on the use in ICEs will first be discussed in Section 2. Different injection and ignition methods will be compared from Sections 4 and 5 in terms of performance and emissions, with a focus on hydrogen DI. A novel ignition strategy utilizing a small amount of pilot diesel jet to auto-ignite the hydrogen DI jet is proposed and evaluated in Section 6 as a solution to current issues associated with hydrogen DI. This is also known as the dual-fuel hydrogen-diesel direct injection (H2DDI) mode. Due to the need of in-depth understanding of the underlying combustion mechanisms in this proposed combustion mode, studies of the behaviour of high pressure hydrogen jets will be reviewed to map out the current level of fundamental understanding in Section 7. It is followed by a review of available hydrogen DI hardware as well as hydrogen DI injector design considerations in Section 8, which are required to facilitate development in this field. Discussion of the necessary next steps for hydrogen ICE will be provided by reviewing and evaluating this information.

2. Hydrogen Properties and Their Implications on Use in Internal Combustion Engine

Hydrogen has unique physical and chemical properties, compared to the conventional fossil fuels widely used in the transportation sector, namely compressed natural gas (CNG), gasoline and diesel, as shown in Table 1 [4,27,40–42]. Engine performance with these fuels in different engine modes is commonly compared with hydrogen and will be discussed throughout the study. One of the many advantages of using hydrogen in ICE as a clean alternative fuel is its zero carbon content. This means that carbon-based emissions, mainly CO, CO_2 and soot, can be eliminated, leaving NO_x as the only harmful combustion byproduct. With a high specific energy density, hydrogen can provide nearly three times as much energy by mass compared with other fossil fuels, reflected in its lower heating value.

Table 1. Hydrogen properties compared with compressed natural gas (CNG), gasoline and diesel.

Property	Hydrogen	CNG	Gasoline	Diesel
Carbon content (mass%)	0	75 [e]	84	86
Lower heating value (MJ/kg)	119.7	45.8	44.8	42.5
Density [a,b] (kg/m^3)	0.089	0.72	730–780	830
Volumetric energy content [a,b] (MJ/m^3)	10.7	33.0	33×10^3	35×10^3
Molecular weight	2.016	16.043 [e]	~110	~170
Boiling point [a] (K)	20	111 [e]	298–488	453–633
Auto-ignition temperature (K)	858	813 [e]	~623	~523
Minimum ignition energy in air [a,d] (mJ)	0.02	0.29	0.24	0.24
Stoichiometric air/fuel mass ratio	34.5	17.2 [e]	14.7	14.5
Stoichiometric volume fraction in air (%)	29.53	9.48	~2 [f]	-
Quenching distance [a,c,d] (mm)	0.64	2.1 [e]	~2	-
Laminar flame speed in air [a,c,d] (m/s)	1.85	0.38	0.37–0.43	0.37–0.43 [g]
Diffusion coefficient in air [a,b] (m^2/s)	8.5×10^{-6}	1.9×10^{-6}	-	-
Flammability limits in air (vol%)	4–76	5.3–15	1–7.6	0.6–5.5
Adiabatic flame temperature [a,c,d] (K)	2480	2214	2580	~2300

[a] at 1 bar, [b] at 273 K, [c] at 298 K, [d] at stoichiometry, [e] methane, [f] vapor and [g] n-heptane.

There are, nonetheless, a number of drawbacks related to the very low density of hydrogen (i.e., low volumetric energy content (MJ/m^3)). At atmospheric pressure and 273 K, the density of hydrogen is an order of magnitude less than that of natural gas, due to the very low molecular weight of hydrogen. The low boiling point suggests that compressed hydrogen will be the most prominent storage option. This presents a significant challenge for the implementation of hydrogen ICE in on-road applications due to the limited vehicle space. Increasing storage pressure is required to increase hydrogen density and hence the volumetric energy content. For instance, hydrogen at 350 bar (i.e., a current standard supply pressure for hydrogen refuelling) and 273 K can increase the gas density to ~31 kg/m^3 or the volumetric energy content to ~3700 MJ/m^3.

With the highest auto-ignition temperature and Research Octane Number (RON $\geq$ 130) [43] relative to the common fuels, the resistance of hydrogen to knocking is expected to be high. However, its minimum ignition energy in air at stoichiometry is an order of magnitude less than that of hydrocarbon fuels, which indicates that hydrogen can be easily ignited by hot spots or residues in combustion chamber. This may lead to pre-ignition of fuel, which is characterized by combustion during the compression stroke prior to the intended ignition. This results in a loss of combustion phasing control, knocking and possibly mechanical engine failure. It should be noted that the global effect of pre-ignition and knock is nearly indistinguishable [27], since pre-ignition usually leads to knock. However, the underlying causes for the two phenomena are very different. A previous study [44] has shown that the Motor Octane Number (MON) of hydrogen is much less than its RON, compared to the typical 8–10 points decrease for the gasoline fuel; although the exact value of hydrogen MON was not clear. Nevertheless, MON is reported to be a more accurate knock resistance metric in hydrogen engine designs [43]. This explains the frequent report of knock in hydrogen engine applications.

The quenching distance of hydrogen is small compared to conventional hydrocarbon fuels. Consequently, higher temperature gradients near the combustion chamber walls can be expected, leading to increased combustion heat losses. When hydrogen is used in PFI engine applications, the short quenching distance along with the high laminar flame speed in air imply an increased propensity for flame backfiring into the intake manifold. This issue can be alleviated by modifying engine geometry, reducing the crevice volume, retuning of engine operating conditions and a complete removal of abnormal discharge and residual electric energy in the ignition system [45,46]. Also, a non-platinum cold-rated spark plug should be used to avoid pre-ignition and backfiring in SI engine [47]. This is because the platinum material in the spark plug can result in an undesired catalytic response with hydrogen and air. A cold-rated spark plug, on the other hand, can facilitate quick heat transfer to minimize hydrogen exposure to hot spots that may lead to engine knock and abnormal combustion. This is where hydrogen DI shows a great advantage as the backfiring can be completely avoided using the injection after the intake valve closing.

Nevertheless, the unique physical and thermo-chemical properties of hydrogen can facilitate the design of a highly efficient ICE. For instance, the dispersion of hydrogen is four times faster than that of CNG, which can be inferred by comparing their diffusion coefficients in air in Table 1. This can promote in-cylinder fuel and air mixing in ICEs. The stoichiometric hydrogen volume fraction corresponds to 29.53 vol%. Nevertheless, the wide flammability limit from 4–76 vol% hydrogen in air, alongside with the high flame speed, indicates that hydrogen ICE can operate considerably lean, thus improving thermal efficiency. The adiabatic flame temperature of hydrogen at stoichiometry is relatively high, which promotes NO_x formation. However, lean operations or a high level of exhaust gas recirculation (EGR) can be used to reduce NO_x emissions due to the wide flammability limits.

It should be noted that utilizing hydrogen in ICEs may induce other safety concerns related to the on-board fuel storage and delivery system. For example, hydrogen embrittlement is a common cause of material failure when high pressure hydrogen is used [48]. Also, the high diffusivity of hydrogen indicates that hydrogen poses a high risk of leakage. These issues require special measures in the design of the vehicle and the fuel delivery system but are not the main focus of this study. Moreover, these challenges are commonly shared with the FC option for powering vehicles.

3. Hydrogen Engine Combustion Modes

A general categorization of hydrogen ICE technology is presented in Figure 1. Broadly, engines can be divided into two main categories based on the fuel injection method—PFI and DI. The ignition methods of hydrogen PFI engines typically employ spark discharge, dual-fuel operation with pilot diesel DI or auto-ignition in the homogeneous charge compression ignition (HCCI) mode. The hydrogen DI studies commonly employ spark-assisted or hot-surface-assisted (i.e., glow plug) ignition. This injection mode also possesses the potential to employ dual-fuel mode, with hydrogen ignited by a high temperature environment created by the pilot diesel-fuel combustion, known as the H2DDI mode. This injection strategy has, however, been more frequently studied with CNG instead of hydrogen, as discussed later in Section 6.

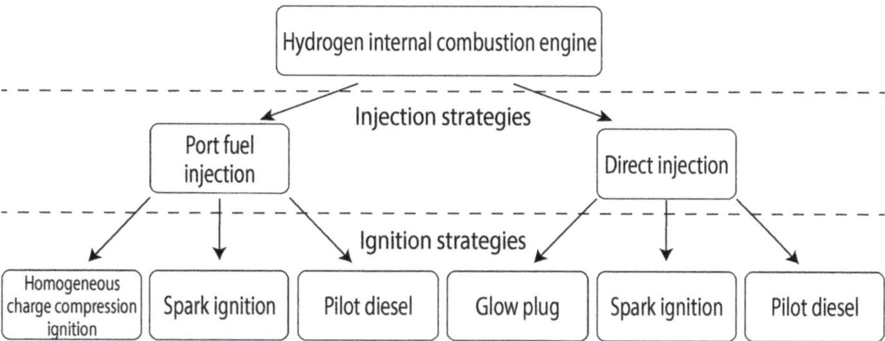

Figure 1. Categorization of hydrogen internal combustion engine (ICE) based on typical injection and ignition strategies.

4. Hydrogen Port Fuel Injection

PFI is a widespread fuel delivery strategy for SI ICEs—fuel is injected during the intake stroke into the intake port upstream of the intake valve. The modification of conventional PFI ICE to hydrogen involves a comparatively straightforward replacement of the injection system. However, as mentioned previously, this engine combustion mode operated with hydrogen fuel can suffer from a number of issues, such as pre-ignition, knock and backfiring due to the low minimum ignition energy and quenching distance of hydrogen. On the other hand, hydrogen displaces air in the intake and therefore limits engine power density. PFI also increases the work needed during the compression stroke compared with late hydrogen DI. These factors often lead to reduced power output and deteriorated efficiency of engines with hydrogen PFI. Combustion characteristics and engine performance of hydrogen PFI engines with different ignition strategies are discussed below.

4.1. Homogeneous Charge Compression Ignition

Due to the high diffusion coefficient of hydrogen, a homogeneous hydrogen-air mixture can be formed more readily than with other conventional fuels. HCCI mode with hydrogen fuel can be operated under very fuel-lean conditions, reducing NO_x formation while maintaining high engine efficiency [49]. Caton and Pruitt [50] found that NO_x emissions of hydrogen HCCI are nearly zero (i.e., ~1 ppm) and are one to three orders of magnitude less than that of conventional diesel operation at low load, compression ratio of 18 and intake temperature of 373 K. Nevertheless, the applicability of this combustion mode is limited by the high auto-ignition temperature of hydrogen—an unrealistic compression ratio of 42 would be required for HCCI combustion at equivalence ratio (ϕ) of 0.1, room temperature and cold start in a 0.5 L single-cylinder compression-ignition (CI) engine [51]. Control of the combustion phasing and peak pressure rise rate are some other challenges—high rate of heat release is commonly reported [51,52], which affects engine reliability. A previous study [53] in a 1.6 L single-cylinder CI engine reported that 10–90% of the total heat release occurs within ~3° crank angle (CA) at 1200 rpm, compression ratio of 17, ϕ of 0.2 and intake temperature of 390 K. Although a high engine indicated thermal efficiency (ITE) of 45% can be achieved with hydrogen HCCI, the achievable load was found to be limited at 3 bar indicated mean effective pressure (IMEP), which is half of the load limit in hydrogen SI PFI operation and 25% of the load limit in gasoline SI operation [53]. Increasing the engine load (i.e., rising ϕ) also causes higher NO_x emissions, due to advanced combustion phasing and thus increased peak temperature [50].

4.2. Spark-Ignited Port Fuel Injection

Hydrogen PFI SI engine is one of the most investigated modes of hydrogen ICE. The performance and control strategies of this engine mode have been reviewed in depth by Das [4] and White et al. [27].

In summary, running the engine at ultra-lean conditions (i.e., $\phi \leq 0.5$) can lower NO_x emissions to below 100 ppm without aftertreatment. Under stoichiometric condition using EGR and a three-way catalytic converter, a near zero engine-out NO_x emissions (i.e., less than 1 ppm) has been demonstrated [54,55]. Knocking, pre-ignition and backfiring are some of the well identified problems in this combustion mode [56], limiting the engine power output by forcing a very lean operation. Previous studies [27,57] have shown that the minimum ignition energy of hydrogen-air mixtures at atmospheric pressure increases exponentially with a decreasing ϕ from stoichiometry, which can alleviate the pre-ignition problem. The exact ϕ limit for engine operation depends on compression ratio, mixture temperature and engine speed, and so forth. Typically, the peak power output of hydrogen PFI SI engine decreases between 35% and 50%, compared to gasoline operation [58,59]. Despite the penalty in engine power, lean operation contributes to reduced combustion heat losses and increases the charge specific heat ratio, therefore improving engine BTE. A peak engine BTE of 38% at optimized compression ratio of 14.5 and $\phi = 0.55$ has been reported for a 2-L four-cylinder engine [58]. Nevertheless, the on-going development of more advanced engine technologies, for instance turbocharging with intake charge cooling and hydrogen DI, has the potential to alleviate the drawbacks of hydrogen PFI SI engine and even to improve engine BTE, as discussed later in Sections 5 and 6.

4.3. Pilot-Fuel-Ignited Engine with Port Hydrogen Injection

Ignition of a port-injected hydrogen charge can be triggered by a DI of pilot diesel fuel, as commonly employed in converted CI engines. One of the main advantages of this mode is its flexibility of hydrogen share in the total energy input. Usually, the substitution rate of diesel fuel with hydrogen is limited by the excessive rate of pressure rise and end-gas knocking of the well premixed charge. Also, at high hydrogen energy ratio, the increased ϕ raises the tendency for pre-ignition, similar to SI engine applications. As reviewed elsewhere, a majority of previous investigations were limited to a hydrogen energy share of $\sim$30–40% at low and medium loads and $\sim$6–25% at high load [60,61]. While a study by Santoso et al. [62] demonstrated a utilization of 97% hydrogen energy share in this engine mode is possible in a single-cylinder compression-ignition engine at 1.9 bar BMEP, a trade-off in the engine efficiency was also reported with increased hydrogen share. In terms of pollutant emissions, most studies show that this working mode cannot significantly alleviate the emissions of CO and particulate matter, while the emissions of CO_2 in first order decrease proportionally to the energy substitution rate. Nevertheless, over 50% of CO and smoke emission reduction relative to diesel operation has been reported at substitution rate of 46% [63]. The effect of diesel substitution with hydrogen on NO_x emissions is still not fully understood and contradictory engine testing results were reported [60]. For instance, Saravanan et al. [64] claimed that NO_x emissions can be lowered if hydrogen substitution is more than 30% of energy share, which was attributed to a reduced peak combustion temperature. To the contrary, Sandalci et al. [63] reported that increased NO_x emissions with hydrogen addition under all tested conditions, from 0–46% hydrogen energy fraction. This was attributed to the pathway of unburned hydrogen, oxidized to HO_2, boosting the conversion of NO to NO_2 [60]. This shows that the underlying mechanisms, governing the efficiency and pollutant formation in this combustion mode, are still not well understood. Undoubtedly, in the current state of technology, this combustion mode cannot fulfill the future emission legislation. The potential for the substitution of conventional fuel is also limited. Different injection strategies and charge cooling strategies should be further developed to improve NO_x emissions and pre-ignition problem leading to a higher hydrogen share, respectively, for this engine mode [60]. While hydrogen PFI is relatively straightforward to implement with the current engine technology and can be used in short term to promote hydrogen ICE development, more advanced combustion strategies are required to overcome the limitations of PFI in achieving high engine loads and reducing emissions.

5. Hydrogen Only Combustion with Direct Injection

A promising approach to improve hydrogen engine performance is to inject hydrogen directly into the cylinder during the compression stroke [65]. By doing so, the backfiring problem of PFI configuration can be avoided, since fuel injection occurs when the intake valves are already completely closed. Pre-ignition issue can also be avoided to a certain extent by reducing the exposure time of hydrogen mixture to hot-spots. The volumetric efficiency loss for PFI due to the displacement of air by hydrogen, as discussed above, is no longer an issue, if injection occurs after the inlet valves are closed. When injecting fuel late during the compression stroke, high injection pressure (i.e., ≥ 100 bar) is required to overcome the elevated in-cylinder pressure. Concurrently, higher injection pressure can increase fuel mass flow rate compared to typical low pressure PFI, which can provide higher energy input for the same injection duration, to drive high load operation. Therefore, a number of studies [38,65,66] demonstrated that high load hydrogen high pressure direct injection (HPDI) operation, under optimal operating conditions, can achieve similar efficiency as traditional diesel engines. Hydrogen HPDI also enables very flexible engine operation due to the many tuning parameters, for instance injection pressure, injection duration, ignition timing and injector orientation, which can be tuned to optimize the engine performance.

However, the high auto-ignition temperature of hydrogen still has to be overcome. Aleiferis and Rosati [43] investigated hydrogen DI HCCI mode in an optical engine with a compression ratio of 7.5 and it was reported that air intake preheating and high non-cooled internal EGR level are needed to auto-ignite hydrogen in this mode. It should be noted the mixture homogeneity may be affected by the late hydrogen DI after the intake valve is closed. Single-kernel flame propagation was also observed from OH laser induced fluorescence results for all equivalence ratios studied from 0.40 to 0.59, which is atypical to the multi-kernel fast combustion for hydrocarbon fuels in HCCI [43]. Authors proposed that coupling SI after the start of auto-ignition can establish a second flame front expanding towards the first one, similar to twin-spark engines, for better controlled auto-ignition. Glow plug and diesel pilot are also common ignition sources reported in literature. It is noted that hydrogen HPDI has been applied in modified SI as well as CI engines with a range of compression ratios. Typically, the conventional fuel injection system was replaced by a high pressure injection system for hydrogen DI. Additional engine modifications are required to accommodate the ignition assistance of choice.

5.1. Glow-Plug-Assisted Ignition

A glow plug is a device featuring an electrically heated surface protruding into the engine combustion chamber. It is a common equipment in diesel engines to assist engine cold start by increasing local charge temperature. When used in hydrogen DI ICE as conceptualized in Figure 2, the glow plugs need to operate continuously to ensure hydrogen ignition in every engine cycle. The required glow-plug's surface temperature in the range from 1200 to 1400 K was reported [67–69]. In 1979, Homan et al. [67] reported that glow plug is a more reliable ignition source for hydrogen ignition—a short and stable delay between the start of injection (SOI) and ignition of $10°–13°$ CA in a cooperative fuel research engine at 1240 rpm and compression ratio of 18 was reported. This ignition performance was compared to a multi-strike spark-plug ignition system (2.5 kHz strike rate), where a fluctuating delay of $0–25°$ CA between the injection and the initial pressure rise was observed. Homan et al. [67] attributed this fluctuation to a smaller surface area of the spark gap relative to the glow plug. A previous study [70] has shown that the glow-plug ignition method suffers from $\sim 10\%$ increased specific fuel consumption relative to diesel operation—for instance, at 5 bar IMEP and 1200 rpm the ITE decreased from approximately 47% to 42%. Although NO_x emissions in this combustion mode were lower than that in diesel operation, it was still significant, especially at high load (i.e., >500 ppm) [70]. Nevertheless, the above findings were from early research conducted decades ago. The durability of the glow plug due to the high surface temperature is questionable when it comes to commercial application and thus this technology is rarely used in recent engine development [71].

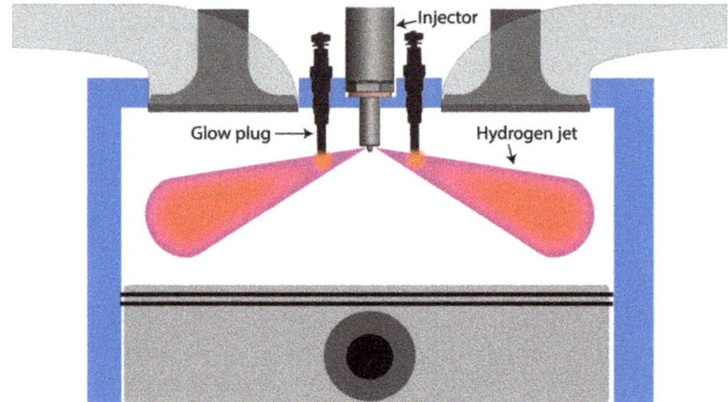

Figure 2. Schematic of the glow plug assisted ignition of hydrogen direct injection.

5.2. Spark-Assisted Ignition

A wealth of spark-assisted hydrogen DI combustion engine studies is available in literature, making this combustion concept the most investigated approach for hydrogen DI. The progress of hydrogen spark-assisted-ignition engines up to year 2013 was reviewed by Verhelst [26] and thus only some highlights and recent development are included in this review. The most common engine configuration is very similar to that of glow-plug-assisted ignition shown in Figure 2, with the glow plugs replaced by one or more spark plugs. In many research studies, the injector and spark plug arrangement was confined by the geometry of production cylinder head used for modification.

Wimmer et al. [40] converted a single-cylinder automotive size SI engine into hydrogen DI engine and achieved an ITE of 40% at low and medium loads, only slightly inferior to the state-of-art light-duty diesel engine at that time. A significant conclusion was that injection timing directly affects mixture homogeneity, which more significantly influences engine performance and emissions than ignition timing. Retarding SOI timing, from 120° to less than 65° CA before top dead center (BTDC), was found to improve overall engine efficiency. This is attributed to the more stratified mixture, which results in a faster initial flame development and improved combustion phasing, despite the increased local wall heat flux caused by the fuel-rich zone near spark plug and short hydrogen quenching distance. Furthermore, compression work during the compression stroke can be reduced as well since hydrogen is injected into an already compressed charge. The influence of injection timing on NO_x emissions is engine load dependent—at medium and low loads, NO_x emissions increase with delayed injection timing, attributed to a less homogeneous mixture leading to diffusion combustion and the formation of lean and hot zones. To the contrary, at high loads, NO_x emissions increase significantly for early injection, attributed to the global equivalence ratio becoming favorable for NO_x production. The effect of EGR on the NO_x emissions has also been investigated and a trade-off of efficiency was reported similar to the results from studies using hydrogen PFI. The validity of injection timing effects on the efficiency and NO_x emissions at low and medium load in SI DI engines by Wimmer et al. [40] has been confirmed by Kawamura et al. [72], Tanno et al. [73] and Takagi et al. [74], despite differences in engine geometry and operating condition (e.g., engine displacement, compression ratio, injection pressure, injection timing). In addition, under the low and medium loads, increasing the injection pressure can reduce NO_x emissions by promoting air entrainment and mixing [73,74] but at a cost of lower efficiency due to the increased engine wall heat losses associated with increased wall impingement arising from the longer jet penetration [73]. A more recent study by Takagi et al. [75] indicated that the injection angle has a strong impact on engine performance and needs to be optimized in order to reduce wall heat loss by ensuring best separation of the hydrogen jets from the chamber walls.

The hydrogen injection strategies discussed above mostly aimed at generating a relatively well mixed charge. To allow a larger degree of fuel-air mixture stratification, close-coupled injection and ignition strategies were proposed [76], namely the plume-head and plume-tail ignition. Both strategies utilize a late SOI—the plume-head ignition is triggered soon after the SOI, whereas the plume-tail ignition strategy triggers ignition of the jet just after the end of injection. The exact SOI timing and injection duration can be varied depending on engine requirement but the common SOI timings are after 50° and 20° CA BTDC for high and low load, respectively [76]. The results highlight the importance of ignition location within the jet. In a 1.05 L single-cylinder engine with CI combustion chamber geometry [76,77], at low load and 200 bar injection pressure, the plume-head ignition was reported to achieve the highest ITE of ~38% and lowest NO_x emissions below 500 ppm. The NO_x emissions for plume-tail ignition at the same load can exceed 1000 ppm. In contrast, at high load, NO_x emission levels can be less than 500 ppm for both ignition methods decrease with retarded injection timing and higher EGR. Nevertheless, the plume-tail ignition shows considerable advantages with respect to ITE at high load—at optimized injection and ignition timing, the plume-tail ignition can reach 48% ITE compared to that of 37% reported for the plume-head ignition. The governing mechanisms behind these combustion strategies were studied by Roy et al. [78,79] in a 0.31 L single-cylinder optical engine using 50 bar hydrogen injection pressure. It was found that the plume-head ignition has considerably lower peak cylinder pressure and its peak rate of heat release is approximately 60% lower than that of plume-tail ignition, indicating a diffusive combustion. Furthermore, the coefficient of variation of IMEP was found to be 24%, compared to 7% for plume-tail ignition, implying an unstable engine operation for the plume-head ignition strategy. Though not fully understood, local equivalence ratio measurements using spark-induced breakdown spectroscopy show that the mean local ϕ for plume-head ignition is relatively low, thus explaining the unstable combustion.

Overall, the literature suggested that the ideal in-cylinder fuel distribution before SI consists of a sufficiently fuel-rich region near the spark plug to ensure reliable and fast flame initialization but a fuel-lean mixture close to the wall to minimize wall heat losses [80]. The fuel and air mixing process of high pressure jets is understood to a certain extent, from a number of studies using flow visualization and simulation works [30,34,36,37]. In essence, the fuel concentration and flow field were visualized under non-reactive conditions in a motored engine by planar laser-induced fluorescence and particle image velocimetry measurement, respectively. When a hydrogen jet was injected towards the engine side wall with injection timing between 137–120.5° CA BTDC, it was observed to produce a wall-jet vortex after wall impingement. The jet was then redirected rapidly and slowed down by the air entrainment, followed by an upward push by the engine piston. It was also observed that a more stratified mixture can be formed with less tumble flow within the chamber. It is noted that injection angle, injection timing, injector nozzle number and piston geometry, etc, also have a significant impact on the final mixture formation.

Although the hydrogen DI SI engine concept mitigates several issues of the PFI configuration, the efficiency is still inferior to contemporary diesel CI engines. This can partially be attributed to the limitation of knocking and pre-ignition, which limit the compression ratio applicable for this combustion mode. Also, the combustion phasing might be sub-optimal since the number of combustion kernels in this combustion mode is usually limited by a single spark plug in the engine, leading to a slower early stage combustion. Therefore, the hydrogen jets from a multi-nozzle injector cannot be ignited simultaneously, limiting the combustion speed. A potential approach to mitigate this issue is seen in the concept of dual-fuel H2DDI, utilizing flame-kernels from pilot fuel auto-ignition to ignite the hydrogen jets.

6. Dual-Fuel High Pressure Direct Injection Compression-Ignition Engine

As discussed earlier, achieving hydrogen auto-ignition without any ignition assistance, such as the HCCI mode or diesel-like diffusion combustion, is challenging due to the high auto-ignition temperature of hydrogen. To circumvent this limitation, a small amount of diesel pilot fuel is injected into the combustion chamber as an ignition source for the high pressure gas jet. Although a similar injection strategy was proposed back in the 1980s [81], most of the studies used CNG as the primary fuel. Dual-fuel compression ignition with the DI of hydrogen has not been demonstrated to date. These studies typically inject a small amount of pilot diesel fuel prior to gas DI, in order to create a high temperature environment to assist the gaseous fuel ignition to achieve CI engine diffusion-like combustion. This novel combustion mode for hydrogen ICE can alleviate charge knocking and allows the engine to operate at a higher compression ratio to improve thermal efficiency up to levels that are comparable to contemporary CI engines. Since CNG has a comparable auto-ignition temperature to hydrogen, reviewing the results of CNG dual-fuel DI research can provide useful insights into the underlying mechanisms of this combustion mode, including the understanding of the gas jet interaction with the pilot diesel fuel. The effect of different operating parameters, for instance injection timing, injection pressure, interactions between the two fuels and ambient conditions will be discussed. The implications for changing the fuel from CNG to hydrogen are summarized later in this section. Dual-fuel DI engines are usually realized by employing an integrated concentric injector providing separate flow paths to independently admit both gas and diesel fuels from the same injector unit. Two separate injectors for the two fuels can also be used. As an alternative, earlier studies [82,83] investigated the possibility of external mixing of CNG and diesel before injection. These studies showed that mixing of CNG into diesel fuel lengthens the ignition delay, leading to an excessive pressure rise rate, especially at low load. Therefore, this approach did not attract attention in later studies and will not be discussed in depth.

In dual-fuel DI combustion mode, the nozzle and injector orientation was shown to play an essential role in improving the combustion process since it defines the interaction between the pilot fuel and gas jets. Figures 3 and 4 show the schematic of the axial cross section and top view of the dual-fuel DI jet orientation with a concentric injector, respectively [81]. The injection vertical angle is defined as the angle between the jet axis and the cylinder head in the axial cross-sectional plane, while the interlace angle is defined as the angle between the gas and diesel jet axis of a concentric injector on the top plane. It should be noted that the interlace angle is diverging from the concentric injector in Figure 4, however, depending on the injector configuration, it can be arranged as parallel or converging.

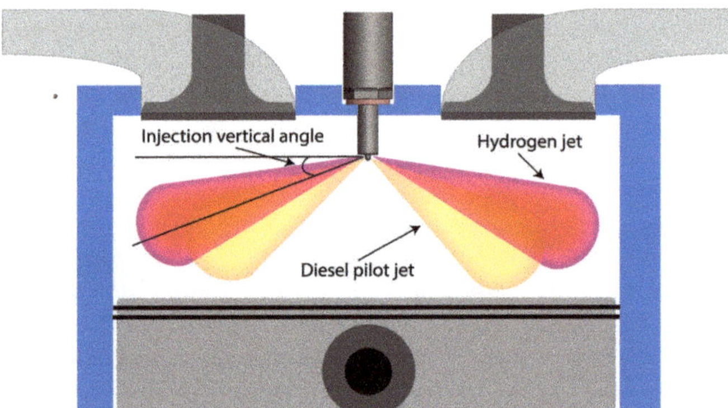

Figure 3. Schematic of the dual-fuel DI concentric injector jet configuration on the axial cross-sectional plane. The injection angle is defined as the angle between the jet axis and the horizontal axis in this plane. Reproduced from Trusca [81].

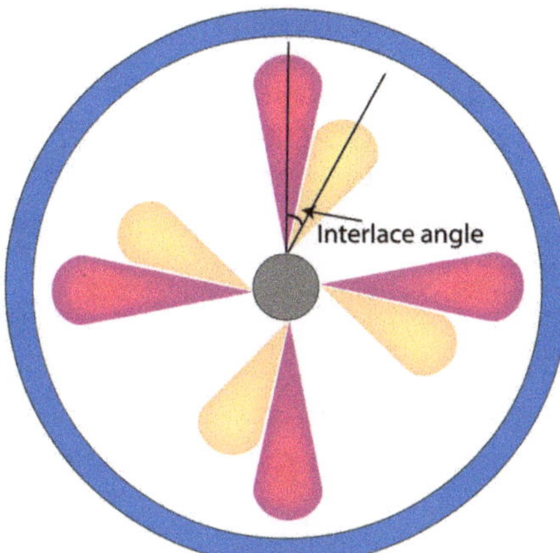

Figure 4. Schematic of the dual-fuel DI concentric injector jet configuration on the top-plane. The interlace angle is defined as the angle between the two jets' axes on this plane. Reproduced from Trusca [81].

Early investigations by Miyake et al. [83] in 1983 demonstrated a higher BTE of CNG-diesel dual-fuel DI engine than the contemporary diesel engines at the time. A modified large bore (420 mm) four-stroke single-cylinder diesel engine with two separate injectors was used. Diesel pilot jets were injected from the periphery of the cylinder in a radial direction and the gas injector was mounted in the center, perpendicular to the cylinder head. Using an injection pressure of CNG at 250 bar, only 5% of diesel fuel in the total energy input was required to achieve 85% and 100% of the engine full load.

Later studies demonstrated the significant influence of injection timing of diesel and gas jets on engine performance in the dual-fuel DI mode. Trusca [81] performed engine testing using a modified 1.2 L single-cylinder diesel engine, operated at 1200 rpm at low and medium loads, using an integrated concentric dual-fuel injector. Gas jet injection began just before the end of pilot diesel injection, which accounted for 5% of the total energy share. The relative SOI, defined as the delay between the start of diesel and gas injections, was approximately 10° CA. It was found that an early pilot diesel injection at ~25° CA BTDC can achieve the highest peak cylinder pressure, across various injection pressures and loads studied. Delayed injection resulted in a reduced peak cylinder pressure. This was attributed to a delayed pilot-fuel ignition, which leads to a later combustion phasing of the gaseous fuel. Nevertheless, the later combustion phasing may be compensated by a decrease in other engine losses, as engine efficiency was found to be insensitive to SOI timing for most cases at the same injection pressure and engine load [81]. A more recent optical investigation using schlieren imaging by Dai [84] in a constant-volume chamber under non-reactive conditions used two parallel injectors (Figure 5) to investigate the importance of relative SOI between the jets. Findings suggested that the gas jet should not be injected too early before the auto-ignition of diesel, as it was observed to rapidly mix with the diesel jet, which may lengthen diesel ignition delay due to the entrainment of colder gaseous fuel instead of hot oxidizer required for ignition. On the other hand, late gas injection may result in misfire of the gas jet, because the hot diesel combustion products of typically short pilot injections rapidly lean out and cool. This was later confirmed by Ishibashi and Tsuru [85] under reactive conditions using an optically-accessible rapid compression and expansion machine (RCEM) at 300 rpm to observe the interaction of diesel and gas jets from single-hole injectors with a converging arrangement. The effect of injection timing was demonstrated by altering the diesel injection timing at a fixed gas injection timing (i.e., 4°, 2°, 0° CA relative SOI with pilot diesel being injected first). An improved performance was observed at the medium temporal separation of injections (i.e., 2° CA relative SOI) of both fuels, with slightly reduced unburned hydrocarbon and NO_x emissions. When both fuels were injected simultaneously, both fuel jets merged and the diesel fuel rapidly mixed with the gaseous fuel before ignition occurred. This increased the ignition delay and induced a high peak heat release rate due to the large mass of accumulated combustible charge formed by the time of ignition.

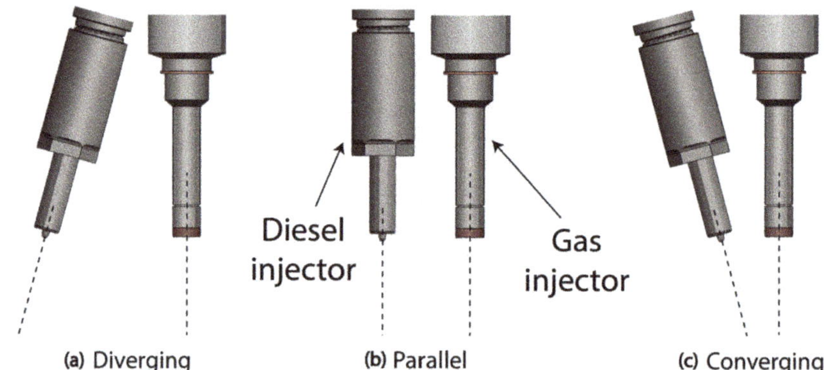

Figure 5. Two co-axial injector Dual-fuel DI configurations, with (a) diverging, (b) parallel and (c) converging nozzle orientation. The figure is not drawn to scale.

Douville [86] performed testing using the same engine and injector setup as Trusca [81] and investigated the effect of CNG injection pressure on engine efficiency and emissions. The injection pressure of CNG was varied from 100 to 140 bar. Over this range, the results at low and medium loads indicated only minor effect of injection pressure on thermal efficiency, similar to conventional diesel operation. Nevertheless, higher engine loads could be reached at higher injection pressures. On the other hand, an increase of NO_x emissions was detected with increasing gas injection pressure for all tested loads. This conclusion was confirmed in later studies for a larger range of injection pressure by Trusca [81] and McTaggart-Cowan et al. [87]. However, these studies reported faster combustion leading to a moderate enhancement of engine efficiency with an increasing injection pressure from 130 to 170 bar and from 280 to 400 bar, respectively. This is inconsistent with the results by Douville [86]. Nevertheless, no further increase in efficiency was observed at injection pressure exceeding 480 bar. At this optimized condition, a BMEP of 22.5 bar and BTE exceeding 40% were reached [87]. While these studies provide a good starting point for parametric analysis, the understanding of the influence of injection pressure on combustion and efficiency is limited and more work with optical diagnostics is needed for further understanding. The study by Dai [84] showed that both injection and ambient pressure affect gas jet penetration similarly to diesel jet. The penetration increases with higher injection pressure but decreases with higher ambient pressure.

The interaction between the pilot fuel and gas jet was studied optically under room conditions employing schlieren imaging by White [88]. A range of jet alignments was investigated as demonstrated in Figure 5 (i.e., diverging, parallel as well as converging). As expected, a converging configuration promotes interaction between the diesel and gas jets. However, in a converging arrangement with large angle between the jets, the gas jet was observed to pass through the diesel liquid jet and the development of a conical gas jet was not affected by the interaction with diesel jet. Under an assumption that a maximized overlapping of diesel and gas jets after the auto-ignition of diesel fuel will yield the best performance, a converging injector configuration with small angle (a few degrees) was recommended [88]. The effect of the dual-fuel jet interaction under engine-relevant reactive conditions was studied by Fink et al. [89,90] in a RCEM. Different injector orientations were used to change the degree of jet interaction—the gas injection angle was changed by rotating the injector while the diesel jet was fixed. The results confirmed non-reactive predictions that ignition is improved when the two jets overlap, especially at low ambient temperature. The combustion was found to be unstable when parallel jet arrangement was used. Although a large extent of gas jet overlapping with the pilot diesel lowers diesel combustion intensity, the ignition of the gas jet occurs earlier. The interlace angle, as shown in Figure 4, is another degree of freedom in dual-fuel DI configuration to induce various level of interactions between the diesel and gas jets but it is more relevant for concentric injectors. Similar to the injector vertical configuration, a previous simulation study [91] showed that when the interlace angle is reduced from 30° to 15°, the two jets have a larger overlapping area, leading to an increased combustion rate with a trade-off of NO_x emissions.

Fink et al. [90] investigated the effects of ambient pressure and temperature on dual-fuel combustion using shadowgraphy and OH-chemiluminescence in a RCEM. It was found that the ignition of gas jet becomes possible at a wider range of injector orientations and relative SOIs with an increasing ambient pressure and temperature (i.e., from 780 to 920 K). At a fixed slightly diverging injection configuration and a negative relative SOI (i.e., gas jet is being injected first), the ignition delays relative to their SOIs decrease for both jets with an increase in ambient temperature. However, the temporal difference between the diesel fuel auto-ignition and ignition of the gas jet increases. This was attributed to an earlier ignition of diesel fuel, which therefore occurrs closer to the injector orifice—at a larger distance from the gas jet. It should be noted that this observation may only be applicable to a specific injector configuration. The study also observed that heat release rates between different ambient temperatures show a similar profile at the same level of premixing but the peak heat release rate increased with higher ambient temperature.

Despite the wealth of studies, the interplay between injector configuration, ambient conditions, combustion premixing and heat release rate is not completely understood. Therefore, additional dedicated studies under a wider range of conditions and configurations both in engines and optically-accessible test rigs are needed. It should be noted all of the above studies used CNG as the gaseous fuel—the potential of dual-fuel DI combustion mode with hydrogen remains largely unexplored (i.e., H2DDI). However, the above studies demonstrated the potential to reduce the use of diesel significantly compared to diesel-ignited hydrogen PFI, where pre-ignition and knocking limit hydrogen to 6–25% of total energy share at high load as discussed above. One of the obvious advantages to use hydrogen over CNG is that carbon-based emissions will be reduced significantly. Although pilot fuel potentially forms soot, the close-coupled high velocity hydrogen jet could enhance mixing within the chamber, which can subsequently lead to enhanced soot oxidation and suppress soot formation processes, similar to diesel post-injection strategy [92,93]. The larger speed of sound and higher calorific value of hydrogen largely compensate for an order of magnitude lower density relative to CNG, therefore, requiring only ~20% longer injection duration at 249 bar injection pressure ratio and 353 K fuel temperature [94]. This is because the rate of fuel energy input is dependent on the fuel's energy content and injection mass flow rate, as will be discussed in the next section. A few studies [81,95] have investigated the effect of hydrogen-CNG blend in dual-fuel DI combustion using heavy-duty engines and integrated dual-fuel injectors. Gas injection duration was adjusted to achieve the same engine load as when operating on neat CNG. The results show good agreements that carbon-based emissions, for instance unburned hydrocarbon, CO and CO_2, decrease with a higher hydrogen share but with a trade-off of increasing NO_x emissions. At low load, a higher peak heat release rate was observed when a hydrogen-CNG blend was used, with the opposite trend at high load. This is because the combustion rate is limited by chemical reaction at low load and the addition of hydrogen can provide reactive species (e.g., H and OH) in the reaction zone to widen the mixture's flammability range, leading to enhanced combustion rate [95]. However, the early stage combustion process is limited by the availability of fuel at high load, due to the lower density of hydrogen. Gas ignition delay decreases with high hydrogen share for all loads, indicating an improvement on fuel ignitability. Furthermore, combustion stability can be significantly improved at low load with increasing hydrogen addition, attributable to a more complete consumption of fuel.

The hydrogen dual-fuel H2DDI combustion concept shows the potential to facilitate a broad market penetration of hydrogen fueled ICE in different applications. However, the underlying processes and implications for the engine performance are not well understood, therefore, additional research is required to facilitate its mass adoption. Among others, the ARENA in Australia along with a number of academic and industry partners are leading several projects to investigate the potential of dual-fuel H2DDI mode [96]. Studies in engine testing, optical diagnostics and numerical simulation will be produced in the next few years, to advance the understanding of the governing mechanisms and to optimize engine performance of this engine mode.

7. Non-Premixed Hydrogen Diffusion Combustion

An important intermediate step towards optimized hydrogen DI ICE performance and emissions is to understand the characteristics of hydrogen jets under non-reactive and reactive conditions. Fundamental studies of hydrogen jet combustion are limited; nevertheless, studies in constant-volume combustion chamber with only auto-ignition have identified some of the hydrogen jet features. Naber and Siebers [97] reported that injection pressure and nozzle size have a negligible effect on ignition delay, as it is more sensitive to ambient conditions. A decrease of ignition delay from ~5–0.2 ms was reported, with an increasing ambient temperature from 970 to 1200 K at a fixed initial fuel temperature, ambient density and O_2 concentration of 450 K, 20.5 kg/m^3 and 21 vol%, respectively. However, the study from Tsujimura et al. [98] found a decrease of ignition delay from 2 to 0.5 ms, when increasing the nozzle diameter from 0.3 to 1 mm, at 1000 K ambient temperature. They attributed this finding to the quantity of mixture and the turbulence length-scale generated during injection. However, in the absence of high-fidelity optical diagnostics and simulations results, the underlying mechanisms behind these findings are not well understood. Also, no literature information on the flame structure of a combusting hydrogen jet and its interaction with pilot-fuel jet under engine relevant conditions is available up to date. Future work with advanced optical diagnostic techniques measuring the fuel distribution, turbulence conditions and reactive species evolution would certainly provide useful insights into reactive hydrogen jet development.

Gas Jet Model

Since hydrogen jet combustion in non-premixed diffusion combustion mode relies significantly on high pressure jet characteristics, the characteristics of the gas jet need to be understood. The jet features are equally important, if hydrogen is burned in partially premixed combustion mode, as the hydrogen jet features affect fuel and air mixing as well as fuel stratification. Figure 6 shows a widely accepted turbulent transient gaseous jet model, known as the vortex ball model, proposed by Turner in 1962 [99,100]. If a round nozzle is used, the gaseous jet in an unconfined environment is assumed to have a conical shape with axisymmetric head vortex and some minor irregularities. Most of the air entrainment is expected to occur in the steady-state region, with Gaussian velocity profiles in the fully developed region upstream of the head vortex [100]. The jet penetration, defined as the axial distance between the jet tip and injector nozzle, alongside with the jet cone angle, were shown to have a strong influence on ambient air entrainment.

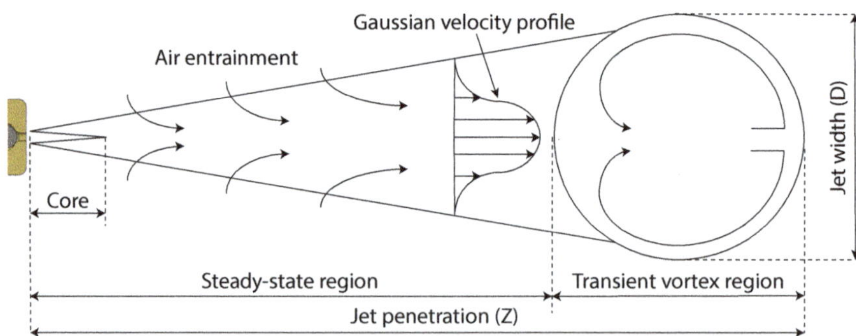

Figure 6. The vortex ball model of turbulent transient gaseous jet. Reproduced from Ouellette [100].

As discussed earlier, increasing the injection pressure increases fuel density in reservoir and fuel mass flow rate during the injection, which implies that HPDI of hydrogen has potential to achieve a high engine load. Usually, hydrogen in the jet can be described as an ideal gas and the flow is considered as an isentropic process, indicating no heat transfer and frictionless system [98]. The flow pattern in the core region depends on the pressure ratio across the nozzle—the jet can be considered as underexpanded when sonic condition is reached and flow is choked at nozzle exit. Literature distinguishes the moderately underexpanded and highly underexpanded jets, which occurs when the pressure ratio between the reservoir and ambient exceeds 2 and 3.85, respectively [101]. For a moderately underexpanded jet, a shock normal to the flow direction is formed with repeated diamond-shaped oblique shocks established in the core region [101]. On the other hand, for a highly underexpanded jet, the hydrogen gas will rapidly accelerate upon leaving the nozzle. The large difference in gas pressure after exiting the nozzle results in the formation of expansion waves in a process described by the Prandtl-Meyer expansion fan [102]. These waves grow until the boundary is generated to resist the jet expansion and to reflect the expansion waves as compression shock waves. A barrel-shaped shock with a Mach disc is formed downstream the injector nozzle due to the coalescence of compression shock waves, as shown in Figure 7 [100]. The diameter and distance of the Mach disc from the nozzle exit increase with the pressure ratio across the nozzle as well as nozzle diameter [98]. It is noted that the highly supersonic flow and shock boundary prevent any gas exchange within the barrel-shaped shock [102]. Therefore, the formation of a barrel shock can considerably affect the overall fuel and air mixing.

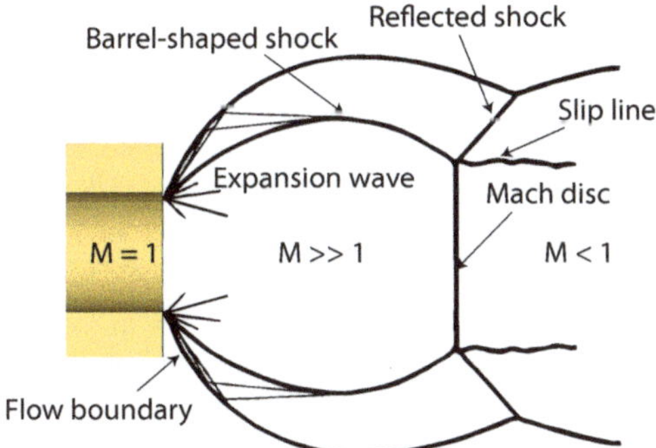

Figure 7. Schematic diagram of the highly underexpanded gaseous jet, with shock formation downstream of the injector nozzle. Reproduced from Ouellette [100].

However, the influence of combustion on the gas jet model under reactive engine conditions still has to be explored. It is also noted that in realistic applications, the hydrogen jet will inevitably interact with the engine chamber wall. Therefore, flame-wall interaction effect also needs to be considered, constituting a more complex scenario than free jet. Literature investigating hydrogen jet interaction with walls is scarce. Examples from liquid fuel jet research include optical investigations studying jet impingement on different wall geometries at a defined distance from the nozzle, for instance a plane flat wall or a confined wall [103,104].

8. Fuel System for High Pressure Hydrogen Injection

One of the major obstacles in commercializing hydrogen DI or dual-fuel DI engine is the lack of commercially available hydrogen injection hardware. An integrated two-fuel HPDI injector is marketed by the Westport Innovation Inc. This injector was used to inject CNG in the studies by Douville [86] and Trusca [81], and so forth. Despite this injector was mainly developed for CNG, a few studies [81,95] successfully used the injector for a hydrogen-CNG blend. This injector consists of two concentric spring-loaded needles, which allows a separate control of the gas and liquid fuel injections. The internal structure and more information about this injector can be found in the study by Trusca [81]. The Westport HPDI technology advanced to a tank-to-tip system, known as HPDI 2.0 and it is utilized commercially for heavy-duty truck applications in the US, Europe and Australia [87,105]. The injection pressure limit of the HPDI 2.0 system is around 600 bar for both diesel and CNG. Previous studies [89,90] employed a prototype injector by L'Orange from Germany to inject CNG at high pressure of up to 330 bar. However, no information about the applicability of this prototype for hydrogen injection is available. Furthermore, in an engine application, such injector will have to be combined with a second injector to deliver pilot diesel fuel.

The challenges of developing a hydrogen injector was documented in a publication by Welch et al. [94]. One of the main challenges is the low lubricity and viscosity of hydrogen. The low lubricity leads to increased friction wear. The low viscosity leads to reduced internal damping resistance of moving components, resulting in a stronger impact when moving parts reach their final positions, especially when the needle impacts the seat during the injector closing. This may also lead to resonant effects, part disassembly, material failures, wear and possible needle bouncing. An application of dry lubricants or low friction coating on surfaces could be used to alleviate this issue. Additionally, due to the high diffusivity of hydrogen, it can permeate through various materials. As a consequence, the epoxy material used in piezoelectric actuator can be delaminated, when the pressure is relieved, leading to an internal short circuit. Hydrogen is also known to cause embrittlement of common engineering steels, which reduces the injector durability [106].

Various prototype injectors for gas only hydrogen HPDI are documented in the literature. For example, Westport and its partners reported two generations of development of hydrogen only HPDI injector by modifying the design of CNG DI injectors. The first generation used solenoid actuation, which was upgraded to piezo-actuation in the second generation. The performance of these injectors was extensively tested by the Argonne National Laboratory [30–39]. Concurrently, the Japanese National Traffic Safety & Environment Laboratory (NTSEL) program developed an electrohydraulic-actuated injector to achieve hydrogen only HPDI and applied it in a number of studies within the program [72,74,75]. The schematic and operation principles of these injectors can be found in previous studies [80,94,107]. A brief summary of their characteristics is given below.

1. Electrohydraulic-actuated (NTSEL): This type of injector requires high-pressure hydraulic fluid (usually diesel fuel) for actuation. The injection pressure is limited to 200 bar. During the injection actuation, the electronically-triggered solenoid acts on the pilot-needle to relieve diesel pressure at the upper part of the injector, reducing hydraulic force that pushes the needle into its seat. Therefore, the high pressure hydrogen can lift the needle and the injection begins. In this design, the diesel pressure needs to be high enough to ensure needle sealing in closed position. It also provides lubrication to some of the injector moving parts. However, the long opening transient duration due to the inertia of hydraulic actuation system might be undesirable in some applications.

2. Solenoid-driven (Westport): The first generation Westport hydrogen DI technology is entirely driven by a solenoid. The direct solenoid actuation imposes an injection pressure limit, which is the lowest among the listed injectors at 150 bar. In addition, a serious durability issue was reported and attributed to the lack of needle motion control required to minimize the needle impact into the seat. Hoerbiger Valve TEC GmbH has also developed a similar solenoid-driven hydrogen DI injector but with a maximum injection pressure of 100 bar [72].

3. Piezo-driven (Westport): This second generation injector with maximum injection pressure of 250 bar is directly driven by a piezoelectric crystal, using analog voltage to proportionally control the needle displacement, enabling a very fast response time. It has a short opening transient duration of 0.5 ms, similar to their solenoid-driven design but only with 35% of that of the NTSEL's injector. Additionally, the injector lifetime is improved by the flexible control of the needle velocity, which can be decelerated at closing to reduce impact. Multiple injections can also be performed.

Many of the commercial injectors for gasoline direct injection (GDI) engines feature directly actuated needle and can therefore be operated with gaseous fuels. The durability due to the poor lubricity of gaseous fuels is the main concern; nevertheless, several researchers have successfully utilized GDI injectors for hydrogen DI injection, with no significant gas leakage reported [43,78,79,102]. Although most of these studies tested the injection pressure of hydrogen only up to 100 bar, the commercialization of higher GDI injection pressure injectors might raise possibilities to utilize a higher gas injection pressure. Due to the durability concerns, such modifications are only applicable for research purposes, where the injector does not need to be operated continuously for a long period of time. For fundamental jet research purposes, a typical multi-hole GDI injector can be modified to a single nozzle configuration to avoid complexities that can arise from jet-jet interaction. Figure 8 shows the author's design of a single-nozzle GDI injector modification, similar to those in References [102,108,109]. A part of the original injector tip is removed and a customized cap with single nozzle is threaded onto the injector shaft, with an o-ring to prevent hydrogen leakage. This approach is relatively cost effective and the nozzle geometry can be simply modified for different test requirement, even multiple nozzles can be featured. Material suitability suggests the use of Viton elastomer since hydrogen induces swelling in lower-grade rubber components [106]. Great care needs to be taken at the needle seat, as needle valve sealing is determined by surface finish—precision of less than 1 micron is recommended [107]. The clearance around the needle should be large enough to ensure sufficient flow rate and the nozzle inner wall should have a smooth surface finish to minimize flow disturbance. A spherical contact between the needle and inner cap surface might improve sealing integrity and prolong the injector lifetime. The cap should be made of softer materials (e.g., brass) than the needle, since the material can deform when in contact with the needle due to its ductility to ensure a mating face for sealing purposes [102]. Nevertheless, soft material will raise durability issues—Rogers [102] reported a lifetime of only several thousand cycles for the customized cap before the sealing surface starts rolling over and alters the nozzle flow behavior.

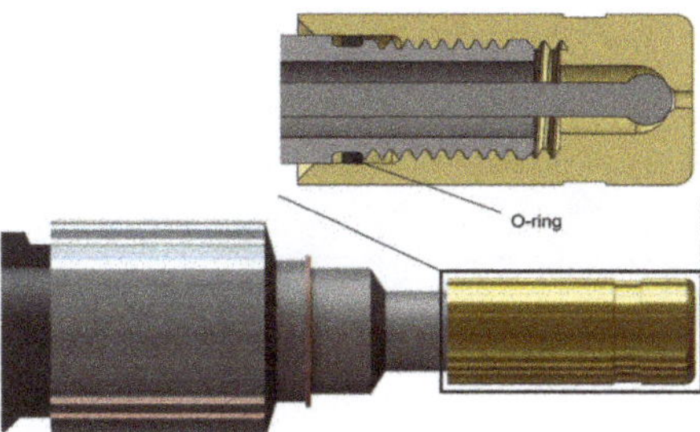

Figure 8. GDI injector modification design with a brass coaxial single-nozzle cap and its detailed section view.

8.1. Injector Design Considerations

The fuel injector design plays an important role in engine performance and has to be carefully considered. The injector characteristics, such as maximum injection pressure and nozzle size, directly affect the fuel injection rate and mixing and also govern the fuel injection quantity. Consequently, the injector configuration influences the heat loss through flame-wall interaction, and so forth.

8.1.1. Injection Rate

Under choked flow conditions, the maximum flow velocity and thus the maximum mass flow rate ($\dot{m}_{max}$) across the nozzle are limited by the speed of sound. By assuming that the enthalpy is only temperature dependent and incorporating the compressible flow theory under choked condition, the maximum mass flow rate can be approximated using Equations (1)–(3), with the subscripts o and superscript $*$ representing the infinite reservoir and speed of sound condition, respectively [98].

These equations should be applicable for different nozzle geometries, using the nozzle area at sonic condition. The validity of these equations for high pressure hydrogen injection was tested by measuring the mass flow using a straight single-nozzle electrohydraulic-actuated injector by Tsujimura et al. [98]. A range of pressure ratios and nozzle diameters was tested and the agreement between the measured and calculated average mass flow rates was better than 90% in most cases. The agreement becomes worse, if the injection duration is shorter than 5 ms, attributed to the needle opening transient.

$$\dot{m}_{max} = \rho^* u^* A^* \tag{1}$$

$$= \left[\rho_o \cdot \left(\frac{2}{\kappa + 1} \right)^{1/(\kappa-1)} \right] \left[\sqrt{ \frac{2\kappa}{\kappa + 1} \frac{P_o}{\rho_o} } \right] A^* \tag{2}$$

$$= P_o A^* \sqrt{ \frac{\kappa}{R_o T_o} } \cdot \left(\frac{2}{\kappa + 1} \right)^{(\kappa+1)/2(\kappa-1)} \tag{3}$$

where A = area

$\quad P$ = pressure

$\quad \rho$ = density

$\quad u$ = velocity

$\quad T$ = temperature

$\quad R$ = specific gas constant (i.e., universal gas constant/molar mass)

$\quad \kappa$ = specific heat ratio

Equation (3) was used to calculate the sensitivity of maximum injection rate to nozzle diameter and injection pressure (Figure 9a) and to estimate the theoretical minimum injection duration for engine operation (Figure 9b), assuming a single-hole injector. The fuel temperature was fixed at 298 K and a minimum injection pressure of 150 bar was postulated to ensure that a critical pressure ratio condition of 2 is reached for choked flow and underexpanded jet, assuming a back pressure representative of conditions around the top dead center in CI engines. The injection duration was derived from the nozzle mass flow using an injection quantity of $\sim$7 mg of hydrogen per injection, assuming the load and efficiency from a previous hydrogen DI SI study [73] in a 2.2 L four-cylinder engine (i.e., IMEP of 6 bar and ITE of $\sim$45% at 2000 rpm). The injection rate of hydrogen jet, as shown in Figure 9a, increases proportionally to injection pressure and with the square of nozzle diameter. The indicative minimum injection duration, as shown in Figure 9b, therefore decreases inversely proportional to higher injection pressure and quadratically with larger nozzle diameter. Despite the low density of hydrogen, with a nozzle diameter of 1 mm, the theoretical minimum injection duration as short as 5° CA can be achieved at injection pressure of 350 bar. At moderate injection pressure of 200 bar, the flow rate is sufficient to deliver the fuel mass within 10° CA. However, it should be noted that the calculation is just a theoretical indication using the maximum flow rate and the actual injection duration is expected to be longer due to the injector opening and closing transient [98]. Apart from delivering the required fuel quantity for a certain load, nozzle size and arrangement as well as the injection pressure need to be optimized for the engine geometry in order to achieve the best performance.

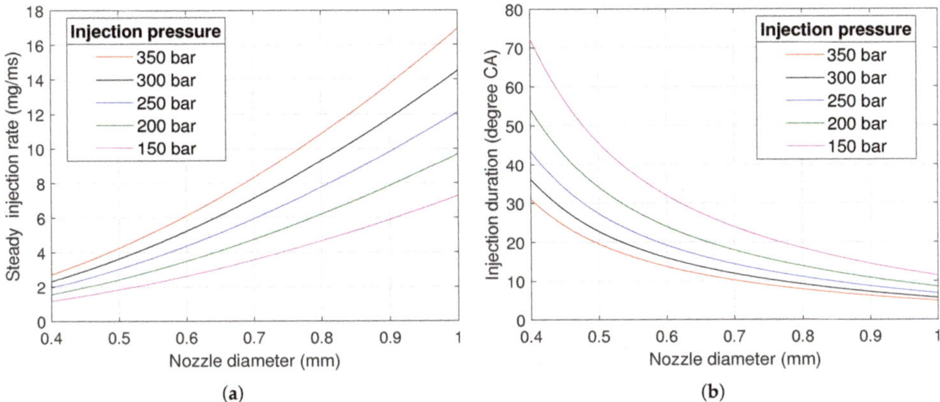

Figure 9. Theoretical (**a**) maximum steady injection rate and (**b**) minimum injection duration required for injecting 7 mg hydrogen at 2000 rpm at different injection pressures and nozzle diameters.

8.1.2. Axial Jet Penetration

Hill and Ouellette [110] developed a widely used jet tip penetration (Z_t) prediction model for a gaseous jet as a function of time assuming the conservation of momentum. The validity of this model has been demonstrated for highly underexpanded jets with pressure ratio of up to 70, unaffected by the barrel-shaped shock. This model, described by Equations (4) and (5) [110,111], assumes a constant momentum discharge rate and density during the injection, as well as self-similar jet velocity and mixing distribution. Since the underexpanded jet flow is choked, the nozzle exit velocity (subscript e) is assumed to be the sonic speed [112]. The subscript a represents the ambient condition and Γ is a constant dependent on jet cone angle, as shown in Equation (6) [110]. The sensitivity of Γ to jet cone angle is small. A 50% increase of s from 0.2 to 0.3 only decreases Γ from 3.04 to 2.89 (i.e., 5% decrease). Hill and Ouellette [110], therefore, proposed that a universal estimation of 3 for Γ can be used. A study [102] in a constant-volume chamber has also shown a small sensitivity of fully developed underexpanded jet spread angles to ambient conditions. Regardless of the fuel type and pressure ratios from 3 to 12, the jet spreading angles remained similar in the range of $22°$–$28°$.

$$Z_t = \Gamma \left(d_e \times u_e\right)^{1/2} \times \left(\frac{\pi}{4} \frac{P_o}{P_a} \frac{R_a}{R_o} \frac{T_a}{T_o} \left(\frac{2}{\kappa_o + 1}\right)^{\frac{1}{\kappa_o - 1}}\right)^{1/4} t^{1/2} \tag{4}$$

$$= \Gamma \left(d_e\right)^{1/2} \times \left(\frac{\kappa_o \pi}{4} \frac{P_o}{P_a} R_a T_a \left(\frac{2}{\kappa_o + 1}\right)^{\frac{\kappa_o}{\kappa_o - 1}}\right)^{1/4} t^{1/2} \tag{5}$$

$$\Gamma^4 + \frac{1.92(1 - s)^2}{\sqrt{\pi}(2 - s)s^3} \Gamma^2 - \frac{24}{\pi(2 - s)s^3} = 0 \tag{6}$$

where t = time after SOI

$\quad d$ = diameter

$\quad s$ = ratio of jet width to jet penetration distance (D_t / Z_t)

In reality, the effective pressure (P_{eff}) at the nozzle exit is less than the fuel supply reservoir pressure due to the high compressibility of gas flow [113], which is not considered in Equation (4). An estimation of the effective pressure depending on the reservoir pressure, ambient pressure and other gas properties was developed by Hajialimohammadi et al. [111] based on the shock tube diaphragm rupture process as described by Equation (7). The effective pressure ratio to the ambient pressure (P_{eff}/P_a) in Equation (7) can be approximated using the Newton-Raphson method. This effective pressure can then be substituted to Equation (4) to predict jet tip penetration accounting for the pressure loss at nozzle exit, as shown in Equation (8). Therefore, the jet penetration mainly changes with the injection pressure and nozzle exit diameter at an unchanged operating condition and depends on the gas properties. Table 2 shows a comparison of the effective pressure ratio for hydrogen, helium, methane and nitrogen, using a range of reservoir/ambient pressure ratios. A typical CI engine ambient condition around top dead center was assumed, with ambient density, pressure and temperature at 20.8 kg/m^3, 60 bar and 1000 K, respectively [104]. The reservoir temperature was set at 298 K. It is noted that the effective pressure still increases with the reservoir pressure but at a lower rate and methane has a higher pressure loss than hydrogen at the same condition. Also, the effective pressure ratio decreases with both increasing gas's molar mass and specific heat ratio, as seen from the comparison of different gases in Table 2.

$$1 + \left(\frac{u_o^*}{u_a^*}\right)\left(\frac{\kappa_a - 1}{\kappa_o - 1}\right) - \left[\left(\frac{u_o^*}{u_a^*}\right)\left(\frac{\kappa_a - 1}{\kappa_o - 1}\right)\right]\left(\frac{P_a}{P_o}\right)^{\frac{\kappa_o - 1}{2\kappa_o}}\left(\frac{P_{eff}}{P_a}\right)^{\frac{\kappa_o - 1}{2\kappa_o}} - \left(\frac{P_{eff}}{P_a}\right)^{\frac{\kappa_a - 1}{2\kappa_a}} = 0 \tag{7}$$

$$Z_t = \Gamma \left(d_e\right)^{1/2} \times \left(\frac{\kappa_o \pi}{4}\frac{P_{eff}}{P_a}R_a T_a \left(\frac{2}{\kappa_o + 1}\right)^{\frac{\kappa_o}{\kappa_o - 1}}\right)^{1/4} t^{1/2} \tag{8}$$

Table 2. Calculated effective pressure ratio for different reservoir pressure ratios for hydrogen, helium, nitrogen and methane.

Reservoir Pressure Ratio (P_o/P_a)	Effective Pressure Ratio (P_{eff}/P_a)			
	Hydrogen	Helium	Methane	Nitrogen
2	1.58	1.46	1.34	1.27
4	2.46	2.09	1.77	1.59
6	3.17	2.55	2.07	1.80

Accounting for the difference in effective pressure ratio, Table 2 suggests a faster penetration of hydrogen jet relative to nitrogen or methane. This is demonstrated in Figure 10, which compares the theoretical jet penetration of hydrogen to methane as conventional gaseous fuel in HPDI applications. Two comparisons are demonstrated—Figure 10a shows the jet penetration for a variation of injection pressures at a fixed nozzle diameter of 1 mm. Ambient condition was assumed the same as that in Table 2. A comparison with increased hydrogen injection pressure to match the energy flow of methane jet is presented as well. Figure 10b compares the jet penetration for a variation of nozzle diameters at a fixed injection pressure of 150 bar. Similarly, a comparison with increased hydrogen jet orifice diameter to match that of the methane jet's energy flow rate is presented. It is noted that hydrogen with 158 bar and 369 bar injection pressure provide the same theoretical energy flow rate as methane at 150 bar and 350 bar, respectively; nozzle diameters of hydrogen injection need to be increased to 1.03 mm and 1.54 mm for the same theoretical energy flow rate as methane injection with 1 mm and 1.5 mm nozzle diameters, respectively. Equation (8) and Figure 10 show a higher sensitivity of jet penetration to nozzle diameter than to injection pressure, as confirmed by schlieren imaging experiments [98]. In addition, hydrogen jet indeed penetrates faster than methane under the same conditions, which is mainly attributed to a difference in effective pressure ratio, as shown in Table 2. Therefore, the injection pressure of methane jet has to be nearly doubled to provide the same penetration as hydrogen jet. Previous experiments by Rogers [102] and large eddy simulation by Hamzehloo and Aleiferis [114] have validated the faster jet tip penetration of hydrogen than methane at the same pressure ratio. This needs to be considered when transferring current knowledge from CNG dual-fuel DI combustion to hydrogen. However, hydrogen combustion under engine conditions may affect the jet penetration behaviour, which needs to be further verified.

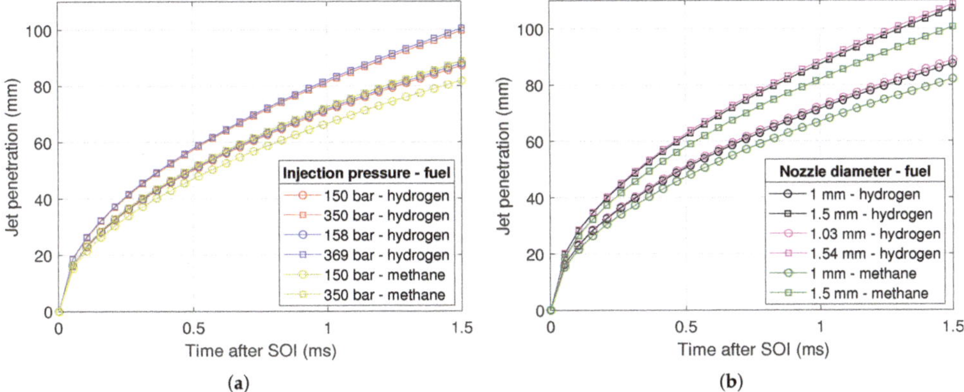

Figure 10. Jet tip penetration prediction for hydrogen and methane at (**a**) different injection pressures and 1 mm nozzle diameter and (**b**) different nozzle diameters and 150 bar injection pressure. The ambient density, pressure and temperature used in this model are 20.8 kg/m^3, 60 bar and 1000 K, respectively.

8.2. Fuel Delivery Strategies

Apart from the fuel injector, other parts of the on-board fuel delivery system also require further innovation. For instance, if the injection pressure were to be solely reliant on the storage pressure, the tank capacity can only be utilized until the instance when the storage pressure decreases below the set injection pressure. This would reduce the maximum achievable range of the vehicle. Integrating a hydrogen pump in the on-board fuel delivery system with a sufficiently small footprint could be a potential solution. A wide range of pneumatic-driven hydrogen booster pumps has recently been commercialized, with various output pressures as high as 1000 bar [115]. Despite the high cost of such a booster pump, it may still be a viable option to provide compressed hydrogen for research applications. However, a more cost-effective and robust tank-to-tip solution is required in the long run, before the hydrogen H2DDI technology can widely penetrate the transportation market.

9. Conclusions

This study reviews the current development of hydrogen internal combustion engines, with a focus on the hydrogen direct injection strategy. The port fuel injection engine configuration suffers many limitations, including pre-ignition, knocking, backfiring, low volumetric efficiency and compression loss problems. This limits the engine achievable load and efficiency. A potential solution to circumventing or alleviating these limitations is hydrogen direct injection. Despite being a promising strategy, the low engine compression ratio of typical SI engines limits the thermodynamic efficiency. In a compression-ignition engine, the spark ignition can be replaced by pilot fuel ignition, which induces multiple ignition kernels to promote fast combustion of gaseous fuel. This combustion mode, known as dual-fuel hydrogen-diesel direct injection, can potentially alleviate the power and compression ratio limitations reported for hydrogen applications in spark-ignition engines. When using green hydrogen, such combustion mode will greatly reduce the reliance on fossil fuels and thus reduce carbon-based emissions. However, hardware development for commercialization and research towards enhanced understanding of the mechanisms governing the engine efficiency and pollutant formation in such dual-fuel combustion mode is needed:

- Metal engine testing is required to prove the effectiveness of this combustion concept in terms of emissions and performance, and to investigate the effect of different operating parameters, for instance injector configuration and operation strategy.
- Fundamental optical and laser-based investigation as well as numerical simulations are needed to understand the governing mechanisms to facilitate engine performance optimization.
- For research purposes, this combustion mode can be studied by using single-fuel injectors—prototype injectors for hydrogen while using commercial diesel injectors to deliver pilot fuel. An integrated dual-fuel injector for hydrogen and pilot fuel may be needed as a long term solution.
- Further technological advancement towards a complete, compact, cost-efficient and robust on-board fuel delivery system is required for commercialization.

Author Contributions: Conceptualization, H.L.Y.; A.S. and Q.N.C.; resources, A.C.Y.Y.; S.K.; R.A.T. and G.H.Y.; writing—original draft preparation, H.L.Y.; A.S. and Q.N.C.; writing—review and editing, A.C.Y.Y.; S.K.; R.A.T.; G.H.Y. and P.R.M.; visualization, H.L.Y. and A.S.; supervision, G.H.Y. and Q.N.C.; project administration, A.S.; A.C.Y.Y.; S.K.; G.H.Y. and Q.N.C.; funding acquisition, S.K.; G.H.Y. and Q.N.C.

Funding: This research was funded by the Australian Renewable Energy Agency (ARENA) 2018/RND011 and Australian Research Council (ARC) Industrial Transformation Training Centre under grant number IC170100032.

Acknowledgments: The first author would like to acknowledge the financial support of UNSW Sydney through the University International Postgraduate Award (UIPA) scheme.

Conflicts of Interest: The authors declare no conflict of interest.

Abbreviations

ARENA	Australian Renewable & Energy Agency
BMEP	Brake mean effective pressure
BTDC	Before top dead center
BTE	Brake thermal efficiency
CA	Crank angle
CI	Compression ignition/compression ignited
CNG	Compressed natural gas
DI	Direct injection
EGR	Exhaust gas recirculation
FC	Fuel cell
GDI	Gasoline direct injection
HCCI	Homogeneous charge compression ignition
H2DDI	Hydrogen-diesel direct injection
HPDI	High pressure direct injection
ICE	Internal combustion engine
IMEP	Indicated mean effective pressure
ITE	Indicated thermal efficiency
MON	Motor Octane Number
NTSEL	National Traffic Safety & Environment Laboratory
PFI	Port fuel injection
RCEM	Rapid compression and expansion machine
RON	Research Octane Number
SI	Spark ignition/spark ignited
SOI	Start of injection

Nomenclature

A	Area
d	Diameter
D	Jet width
$\dot{m}_{max}$	Maximum mass flow rate
P	Pressure
R	Specific gas constant
s	Ratio of jet tip penetration to jet width
T	Temperature
t	Time after SOI
u	Velocity
Z	Jet tip penetration
ρ	Density
ϕ	Equivalence ratio
κ	Specific heat ratio

Subscripts

a	Ambient condition
o	Infinite fuel supply reservoir
eff	Effective
t	Time after SOI

Superscript

*	Sonic condition

References

1. Dong, X.; Wang, B.; Yip, H.L.; Chan, Q.N. CO_2 emission of electric and gasoline vehicles under various road conditions for China, Japan, Europe and world average—Prediction through year 2040. *Appl. Sci.* **2019**, *9*, 2295. [CrossRef]
2. Berry, G.D.; Pasternak, A.D.; Rambach, G.D.; Smith, J.R.; Schock, R.N. Hydrogen as a future transportation fuel. *Energy* **1996**, *21*, 289–303. [CrossRef]
3. Sharma, S.; Ghoshal, S.K. Hydrogen the future transportation fuel: From production to applications. *Renew. Sustain. Energy Rev.* **2015**, *43*, 1151–1158. [CrossRef]
4. Das, L.M. Hydrogen engines: A view of the past and a look into the future. *Int. J. Hydrogen Energy* **1990**, *15*, 425–443. [CrossRef]
5. Winkler-Goldstein, R.; Rastetter, A. Power to gas: The final breakthrough for the hydrogen economy? *Green* **2013**, *3*, 69–78. [CrossRef]
6. Abdalla, A.M.; Hossaina, S.; Nisfindy, O.B.; Azadd, A.T.; Dawoodb, M.; Azada, A.K. Hydrogen production, storage, transportation and key challenges with applications: A review. *Energy Convers. Manag.* **2018**, *165*, 602–627. [CrossRef]
7. *Hydrogen Scaling Up: A Sustainable Pathway for the Global Energy Transition*; Hydrogen Council: Belgium, 2017.
8. *Oil Market Report*; International Energy Agency: Paris, France, 2019.
9. The Observatory of Economic Complexity. Hydrogen Trade. Available online: https://atlas.media.mit.edu/en/profile/hs92/2804/ (accessed on 18 June 2019).
10. Bruce, S.; Temminghoff, M.; Hayward, J.; Schmidt, E.; Munnings, C.; Palfreyman, D.; Hartley, P. *National Hydrogen Roadmap: Pathways to an Economically Sustainable Industry*; The Commonwealth Scientific and Industrial Research Organization: Canberra, Australia, 2018.
11. Palmer, G. *Australia's Hydrogen Future*; Energy Transition Hub: Melbourne, Australia, 2018.
12. *Basic Hydrogen Strategy*; The Ministry of Economy, Trade and Industry: Tokyo, Japan, 2017.
13. Wegener, J. *Developments in Green Hydrogen Production and Use in Germany*; National Organization Hydrogen and Fuel Cell Technology: Berlin, Germany, 2019.

14. Banerjee, S.; Musa, M.N.; Jaafar, A.B. Economic assessment and prospect of hydrogen generated by OTEC as future fuel. *Int. J. Hydrogen Energy* **2017**, *42*, 26–37. [CrossRef]

15. Maggio, G.; Nicita, A.; Squadrito, G. How the hydrogen production from RES could change energy and fuel markets: A review of recent literature. *Int. J. Hydrogen Energy* **2019**, *44*, 11371–11384. [CrossRef]

16. Cornell, A. Hydrogen production by electrolysis. In Proceedings of the 1st International Conference on Electrolysis, Copenhagen, Denmark, 13–15 June 2017.

17. *Hydrogen Enabling a Zero Emission Europe: Technology Roadmaps Full Pack*; Hydrogen Europe: Brussels, Belgium, 2018.

18. *Hydrogen from Renewable Power: Technology Outlook for the Energy Transition*; International Renewable Energy Agency: Abu Dhabi, UAE, 2018.

19. *Opportunities for Australia from Hydrogen Exports*; ACIL Allen Consulting: Sydney, Australia, 2018.

20. *The Green Hydrogen Economy in the Northern Netherlands*; Noordelijke Innovation Board: Groningen, The Netherlands, 2017.

21. International Energy Agency. Hydrogen: A Key Part of a Clean and Secure Energy Future. Available online: https://www.iea.org/topics/hydrogen/demand/ (accessed on 19 October 2019).

22. Helmolt, R.V.; Eberle, U. Fuel cell vehicles: Status 2007. *J. Power Sources* **2007**, *165*, 833–843. [CrossRef]

23. Hänggi, S.; Elbert, P.; Bütler, T.; Cabalzar, U.; Teske, S.; Bach, C.; Onder, C. A review of synthetic fuels for passenger vehicles. *Energy Rep.* **2019**, *5*, 555–569. [CrossRef]

24. Manoharan, Y.; Hosseini, S.E.; Butler, B; Alzhahrani, H.; Senior, B.T.F; Ashuri, T; Krohn, J. Hydrogen fuel cell vehicles; Current status and future prospect. *Appl. Sci.* **2019**, *9*, 2296. [CrossRef]

25. Toyota and Kenworth unveil fuel cell heavy truck at Port of LA. *Fuel Cells Bull.* **2019**, *2019*, 3–4.

26. Verhelst, S. Recent progress in the use of hydrogen as a fuel for internal combustion engines. *Int. J. Hydrogen Energy* **2014**, *39*, 1071–1085. [CrossRef]

27. White, C.M.; Steeper, R.R.; Lutz, A.E. The hydrogen-fueled internal combustion engine: A technical review. *Int. J. Hydrogen Energy* **2006**, *31*, 1292–1305. [CrossRef]

28. *BMW Hydrogen Engine Reaches Top Level Efficiency*; BMW: Munich, Germany, 2009.

29. Ozcanli, M.; Bas, O.; Akar, M.A.; Yildizhan, S.; Serin, H. Recent studies on hydrogen usage in Wankel SI engine. *Int. J. Hydrogen Energy* **2018**, *43*, 18037–18045. [CrossRef]

30. Salazar, V.M.; Kaiser, S.A.; Halter, F. Optimizing precision and accuracy of quantitative PLIF of acetone as a Tracer for hydrogen fuel. *SAE Int. J. Fuels Lubr.* **2009**, *2*, 737–761. [CrossRef]

31. Kaiser, S.; White, C. PIV and PLIF to evaluate mixture formation in a direct-injection hydrogen-fuelled engine. *SAE Int. J. Engines* **2009**, *1*, 657–668. [CrossRef]

32. Wallner, T.; Scarcelli, R.; Nande, A.M.; Naber, J.D. Assessment of multiple injection strategies in a direct-injection hydrogen research engine. *SAE Int. J. Engines* **2009**, *2*, 1701–1709. [CrossRef]

33. Obermair, H.; Scarcelli, R.; Wallner, T. *Efficiency Improved Combustion System for Hydrogen Direct Injection Operation*; SAE Paper 2010-01-2170; SAE International: Warrendale, PA, USA, 2010. [CrossRef]

34. Scarcelli, R.; Wallner, T.; Salazar, V.M.; Kaiser, S.A. Modeling and experiments on mixture formation in a hydrogen direct-injection research engine. *SAE Int. J. Engines* **2010**, *2*, 530–541. [CrossRef]

35. Salazar, V.M.; Kaiser, S.A. An optical study of mixture preparation in a hydrogen-fueled engine with direct injection using different nozzle designs. *SAE Int. J. Engines* **2010**, *2*, 119–131. [CrossRef]

36. Scarcelli, R.; Wallner, T.; Matthias, N.; Salazar, V.; Kaiser, S. Mixture formation in direct injection hydrogen engines: CFD and optical analysis of single- and multi-hole nozzles. *SAE Int. J. Engines* **2011**, *2*, 2361–2375. [CrossRef]

37. Salazar, V.; Kaiser, S. *Interaction of Intake-Induced Flow and Injection Jet in a Direct-Injection Hydrogen-Fueled Engine Measured by PIV*; SAE Paper 2011-01-0673; SAE International: Warrendale, PA, USA, 2011. [CrossRef]

38. Matthias, N.S.; Wallner, T.; Scarcelli, R. A hydrogen direct injection engine concept that exceeds U.S. DOE light-duty efficiency targets. *SAE Int. J. Engines* **2012**, *5*, 838–849. [CrossRef]

39. Wallner, T.; Matthias, N.S.; Scarcelli, R.; Kwon, J.C. A Evaluation of the efficiency and the drive cycle emissions for a hydrogen direct-injection engine. *J. Automob. Eng.* **2013**, *227*, 99–109. [CrossRef]

40. Wimmer, A.; Wallner, T.; Ringler, J.; Gerbig, F. *H2-Direct Injection—A Highly Promising Combustion Concept*; SAE Paper 2005-01-0108; SAE International: Warrendale, PA, USA, 2005. [CrossRef]

41. Glaude, P.A.; Fournet, R.; Bounaceur, R.; Molière, M. Adiabatic flame temperature from biofuels and fossil fuels and derived effect on NO$_x$ emissions. *Fuel Process Technol.* **2010**, *91*, 229–235. [CrossRef]

42. Chong, C.T.; Hochgreb, S. Measurements of laminar flame speeds of liquid fuels: Jet-A1, diesel, palm methyl esters and blends using particle imaging velocimetry (PIV). *Proc. Combust. Inst.* **2011**, *33*, 979–986. [CrossRef]

43. Aleiferis, P.G.; Rosati, M.F. Controlled autoignition of hydrogen in a direct-injection optical engine. *Combust. Flame* **2012**, *159*, 2500–2515. [CrossRef]

44. Ingersoll, J.G. *Natural Gas Vehicles*; Fairmont Press: Lilburn, GA, USA, 1996.

45. Kondo, T.; Iio, S.; Hiruma, M. *A study on the Mechanism of Backfire in External Mixture Formation Hydrogen Engines—About Backfire Occurred by Cause of the Spark-Plug*; SAE Paper 971704; SAE International: Warrendale, PA, USA, 1997.

46. Lee, J.T.; Kim, Y.Y.; Lee, C.W.; Caton, J.A. An investigation of a cause of backfire and its control due to crevice volumes in a hydrogen fueled engine. *J. Eng. Gas Turbines Power* **2001**, *123*, 204–210. [CrossRef]

47. Huyskens, P.; Van Oost, S.; Goemaere, P.J.; Bertels, K.; Pecqueur, M. *The Technical Implementation of a Retrofit Hydrogen PFI System on a Passenger Car*; SAE Paper 2011-01-2004; SAE International: Warrendale, PA, USA, 2011. [CrossRef]

48. Zheng, J.; Liu, X.; Xu, P.; Liu, P.; Zhao, Y.; Yang, J. Development of high pressure gaseous hydrogen storage technologies. *Int. J. Hydrogen Energy* **2012**, *37*, 1048–1057. [CrossRef]

49. Antunes, J.G.; Mikalsen, R.; Roskilly, A.P. An investigation of hydrogen-fuelled HCCI engine performance and operation. *Int. J. Hydrogen Energy* **2008**, *33*, 5823–5828.

50. Caton, P.A.; Pruitt, J.T. Homogeneous charge compression ignition of hydrogen in a single-cylinder diesel engine. *Int. J. Engine Res.* **2009**, *10*, 45–63. [CrossRef]

51. Lee, K.J.; Kim, Y.R.; Byun, C.H.; Lee, J.T. Feasibility of compression ignition for hydrogen fueled engine with neat hydrogen-air pre-mixture by using high compression. *Int. J. Hydrogen Energy* **2013**, *38*, 255–264. [CrossRef]

52. Szwaja, S.; Grab-Rogalinski, K. Hydrogen combustion in a compression ignition diesel engine. *Int. J. Hydrogen Energy* **2009**, *34*, 4413–4421. [CrossRef]

53. Stenlåås, O.; Christensen, M.; Egnell, R.; Johansson, B.; Mauss, F. Hydrogen as homogeneous charge compression ignition engine fuel. *J. Fuels Lubr.* **2004**, *113*, 1317–1326.

54. Heffel, J.W. NO$_x$ emission and performance data for a hydrogen fueled internal combustion engine at 1500 rpm using exhaust gas recirculation. *Int. J. Hydrogen Energy* **2003**, *28*, 901–908. [CrossRef]

55. Heffel, J.W. NO$_x$ emission reduction in a hydrogen fueled internal combustion engine at 3000 rpm using exhaust gas recirculation. *Int. J. Hydrogen Energy* **2003**, *28*, 1285–1292. [CrossRef]

56. Mathur, H.B.; Das, L.M. Performance characteristics of a hydrogen fuelled S.I. engine using timed manifold injection. *Int. J. Hydrogen Energy* **1991**, *16*, 115–127. [CrossRef]

57. Lewis, B; von Elbe, G. *Combustion, Flames, and Explosions of Gases*; Academic Press: Orlando, FL, USA, 1987.

58. Tang, X.; Kabat, D.M.; Natkin, R.J.; Stockhausen, W.F.; Heffel, J. *Ford P2000 Hydrogen Engine Dynamometer Development*; SAE Paper 2002-01-0242; SAE International: Warrendale, PA, USA, 2002.

59. Stockhausen, W.F.; Natkin, R.J.; Kabat, D.M.; Reams, L.; Tang, X.; Hashemi, S.; Szwabowski, S.J.; Zanardelli, V.P. *Ford P2000 Hydrogen Engine Design and Vehicle Development Program*; SAE Paper 2002-01-0240; SAE International: Warrendale, PA, USA, 2002. [CrossRef]

60. Dimitriou, P.; Tsujimura, T. A review of hydrogen as a compression ignition engine fuel. *Int. J. Hydrogen Energy* **2017**, *42*, 24470–24486. [CrossRef]

61. Chintala, V.; Subramanian, K.A. A comprehensive review on utilization of hydrogen in a compression ignition engine under dual fuel mode. *Renew. Sustain. Energy Rev.* **2017**, *70*, 472–491. [CrossRef]

62. Santoso, W.B.; Bakar, R.A.; Nur, A. Combustion characteristics of diesel-hydrogen dual fuel engine at low load. *Energy Procedia* **2013**, *32*, 3–10. [CrossRef]

63. Sandalcı, T.; Karagöz, Y. Experimental investigation of the combustion characteristics, emissions and performance of hydrogen port fuel injection in a diesel engine. *Int. J. Hydrogen Energy* **2014**, *39*, 18480–18489. [CrossRef]

64. Saravanan, N.; Nagarajan, G. An experimental investigation of hydrogen-enriched air induction in a diesel engine system. *Int. J. Hydrogen Energy* **2008**, *33*, 1769–1775. [CrossRef]

65. Mohammadi, A.; Shioji, M.; Nakai, Y.; Ishikura, W.; Tabo, E. Performance and combustion characteristics of a direct injection SI hydrogen engine. *Int. J. Hydrogen Energy* **2007**, *32*, 296–304. [CrossRef]

66. Oikawa, M.; Ogasawara, Y.; Kondo, Y.; Sekine, K.; Takagi, Y.; Sato, Y. Optimization of hydrogen jet configuration by single hole nozzle and high speed laser shadowgraphy in high pressure direct injection hydrogen engines. *Int. J. Automot. Eng.* **2012**, *3*, 1–8.

67. Homan, H.S.; Reynolds, R.K.; De Boer, P.C.T; McLean, W.J. Hydrogen-fueled diesel engine without timed ignition. *Int. J. Hydrogen Energy* **1979**, *4*, 315–325. [CrossRef]

68. Furuhama, S.; Kobayashi, Y. Development of a hot-surface-ignition hydrogen injection two-stroke engine. *Int. J. Hydrogen Energy* **1984**, *9*, 205–213. [CrossRef]

69. Furuhama, S. Hydrogen engine systems for land vehicles. *Int. J. Hydrogen Energy* **1989**, *14*, 907–913. [CrossRef]

70. Welch, A.B.; Wallace, J.S. *Performance Characteristics of a Hydrogen-Fueled Diesel Engine with Ignition Assist*; SAE Paper 902070; SAE International: Warrendale, PA, USA, 1990. [CrossRef]

71. Wei, L. and Geng, P. A review on natural gas/diesel dual fuel combustion, emissions and performance. *Fuel Process. Technol.* **2016**, *142*, 264–278. [CrossRef]

72. Kawamura, A.; Sato, Y.; Naganuma, K.; Yamane, K.; Takagi, Y. *Development Project of a Multi-Cylinder DISI Hydrogen ICE System for Heavy Duty Vehicles*; SAE Paper 2010-01-2175; SAE International: Warrendale, PA, USA, 2010. [CrossRef]

73. Tanno, S.; Ito, Y.; Michikawauchi, R.; Nakamura, M.; Tomita, H. High-efficiency and low-NO$_x$ hydrogen combustion by high pressure direct injection. *SAE Int. J. Engines* **2010**, *3*, 259–268. [CrossRef]

74. Takagi, Y.; Mori, H.; Mihara, Y.; Kawahara, N.; Tomita, E. Improvement of thermal efficiency and reduction of NO$_x$ emissions by burning a controlled jet plume in high-pressure direct-injection hydrogen engines. *Int. J. Hydrogen Energy* **2017**, *42*, 26114–26122. [CrossRef]

75. Takagi, Y.; Oikawa, M.; Sato, R.; Kojiya, Y.; Mihara, Y. Near-zero emissions with high thermal efficiency realized by optimizing jet plume location relative to combustion chamber wall, jet geometry and injection timing in a direct-injection hydrogen engine. *Int. J. Hydrogen Energy* **2019**, *44*, 9456–9465. [CrossRef]

76. Kawamura, A.; Yanai, T.; Sato, Y.; Naganuma, K.; Yamane, K.; Takagi, Y. Summary and progress of the hydrogen ICE Truck development project. *SAE Int. J. Commer. Veh.* **2009**, *2*, 110–117. [CrossRef]

77. Naganuma, K.; Honda, T.; Yamane, K.; Takagi, Y.; Kawamura, A.; Yanai, T.; Sato, Y. Efficiency and emissions-optimized operating strategy of a high-pressure direct injection hydrogen engine for heavy-duty trucks. *SAE Int. J. Engines* **2010**, *2*, 132–140. [CrossRef]

78. Roy, M.K.; Kawahara, N.; Tomita, E.; Fujitani, T. High-pressure hydrogen jet and combustion characteristics in a direct-injection hydrogen engine. *SAE Int. J. Fuels Lubr.* **2012**, *5*, 1414–1425. [CrossRef]

79. Roy, M.K.; Kawahara, N.; Tomita, E.; Fujitani, T. Jet-guided combustion characteristics and local fuel concentration measurements in a hydrogen direct-injection spark-ignition engine. *Proc. Combust. Inst.* **2013**, *34*, 2977–2984. [CrossRef]

80. Verhelst, S.; Demuynck, J.; Sierens, R.; Scarcelli, R.; Matthias, N.S.; Wallner, T. *Update on the Progress of Hydrogen-Fueled Internal Combustion Engines*; Elsevier: Amsterdam, The Netherlands, 2013; pp. 381–400.

81. Trusca, B. High Pressure Direct Injection of Natural Gas and Hydrogen Fuel in a Diesel Engine. Master's Dissertation, The University of British Columbia, Vancouver, BC, Canada, 2000.

82. Hodgins, K.B.; Gunawan, H.; Hill, P.G. *Intensifier-Injector for Natural Gas Fueling of Diesel Engines*; SAE Paper 921553; SAE International: Warrendale, PA, USA, 1992. [CrossRef]

83. Miyake, M.; Biwa, T.; Endoh, Y.; Shimotsu, M.; Murakami, S.; Komoda, T. The development of high output, highly efficient gas burning diesel engines. In Proceedings of the CIMAC 1983, Paris, France, 1983.

84. Dai, L.M. Study of the Injection and Mixture Process and the Combustion Simulation of Natural Gas/Diesel Dual Fuel. Master's Dissertation, Jiangsu University, Zhenjiang, China, 2016.

85. Ishibashi, R; Tsuru, D. An optical investigation of combustion process of a direct high-pressure injection of natural gas. *J. Mar. Sci. Tech.* **2017**, *22*, 447–458. [CrossRef]

86. Douville, B. Performance, Emissions and Combustion Characteristics of Natural Gas Fueling of Diesel Engines. Master's Dissertation, The University of British Columbia, Kelowna, BC, Canada, 1994.

87. McTaggart-Cowan, G.; Mann, K.; Huang, J.; Singh, A.; Patychuk, B.; Zheng, Z.X.; Munshi, S. Direct injection of natural gas at up to 600 bar in a pilot-ignited heavy-duty engine. *SAE Int. J. Engines* **2015**, *8*, 981–996. [CrossRef]

88. White, T.R. Simultaneous Diesel and Natural Gas Injection for Dual-Fuelling Compression-Ignition Engines. Ph.D. Dissertation, University of New South Wales, Sydney, NSW, Australia, 2006.

89. Fink, G.; Jud, M.; Sattelmayer, T. Influence of the spatial and temporal interaction between diesel pilot and directly injected natural gas jet on ignition and combustion characteristics. *J. Eng. Gas Turb. Power.* **2018**, *140*, 102811. [CrossRef]

90. Fink, G.; Jud, M.; Sattelmayer, T. Fink. Fundamental study of diesel-piloted natural gas direct injection under different operating conditions. *J. Eng. Gas Turb. Power.* **2018**, *141*, 091006. [CrossRef]

91. Li, G.; Ouellette, P.; Dumitrescu, S.; Hill, P.G. Optimization study of pilot-ignited natural gas direct-injection in diesel engines. *SAE Int. J. Fuels Lubr.* **1999**, *108*, 1739–1748.

92. Yip, H.L.; Fattah, I.R.; Yuen, A.C.Y.; Yang, W.; Medwell, P.R.; Kook, S.; Yeoh, G.H.; Chan, Q.N. Flame-wall interaction effects on diesel post-injection combustion and soot formation processes. *Energy Fuels* **2019**, *33*, 7759–7769. [CrossRef]

93. Fattah, I.R.; Ming, C.; Chan, Q.N.; Wehrfritz, A.; Pham, P.X.; Yang, W.; Kook, S.; Medwell, P.R.; Yeoh, G.H.; Hawkes, E.R.; et al. Spray and combustion investigation of post injections under low-temperature combustion conditions with biodiesel. *Energy Fuels* **2018**, *32*, 8727–8742. [CrossRef]

94. Welch, A.; Mumford, D.; Munshi, S.; Holbery, J.; Boyer, B.; Younkins, M.; Jung, H. *Challenges in Developing Hydrogen Direct Injection Technology for Internal Combustion Engines*; SAE Paper 2008-01-2379; SAE International: Warrendale, PA, USA, 2008. [CrossRef]

95. McTaggart-Cowan, G.P.; Rogak, S.N.; Munshi, S.R.; Hill, P.G.; Bushe, W.K. Combustion in a heavy-duty direct-injection engine using hydrogen-methane blend fuels. *Int. J. Engine Res.* **2009**, *10*, 1–13. [CrossRef]

96. Australian Renewable Energy Agency. Enabling Efficient, Affordable & Robust Use of Renewable Hydrogen. Available online: https://arena.gov.au/projects/enabling-efficient-affordable-robust-renewable-hydrogen/ (accessed on 27 May 2019).

97. Naber, J.D.; Siebers, D.L. Hydrogen combustion under diesel engine conditions. *Int. J. Hydrogen Energy* **1998**, *23*, 363–371. [CrossRef]

98. Tsujimura, T.; Mikami, S.; Achiha, N.; Tokunaga, Y.; Senda, J.; Fujimoto, H. A study of direct injection diesel engine fueled with hydrogen. *SAE Int. J. Fuels Lubr.* **2003**, *112*, 390–405.

99. Turner, J.S. The 'starting plume' in neutral surroundings. *J. Fluid Mech.* **1962**, *13*, 356–368. [CrossRef]

100. Ouellette, P. Direct Injection of Natural Gas for Diesel Engine Fueling. Ph.D. Dissertation, University of British Columbia, Vancouver, BC, Canada, 1996.

101. Donaldson, C.D.; Snedeker, R.S. A study of free jet impingement. Part 1. Mean properties of free and impinging jets. *J. Fluid Mech.* **1971**, *45*, 281–319. [CrossRef]

102. Rogers, T. Mixture Preparation of Gaseous Fuels for Internal Combustion Engines Using Optical Diagnostics. Ph.D. Dissertation, RMIT University, Melbourne, Australia, 2014.

103. Chan, Q.N.; Fattah, I.R.; Zhai, G.; Yip, H.L.; Chen, T.B.Y.; Yuen, A.C.Y.; Yang, W.; Wehrfritz, A.; Dong, X.; Kook, S.; et al. Color-ratio pyrometry methods for flame-wall impingement study. *J. Energy Inst.* **2018**, *92*, 1968–1976. [CrossRef]

104. Fattah, I.M.R.; Yip, H.L.; Jiang, Z.; Yuen, A.C.Y.; Yang, W.; Medwell, P.R.; Kook, S.; Yeoh, G.H.; Chan, Q.N. Effects of flame-plane wall impingement on diesel combustion and soot processes. *Fuel* **2019**, *255*, 115726. [CrossRef]

105. Ouelette, P.; Goudie, D.; McTaggart-Cowan, G. Progress in the development of natural gas high pressure direct injection for Euro VI heavy-duty trucks. In *Internationaler Motorenkongress*; Springer: Berlin/Heidelberg, Germany, 2016; pp. 591–607.

106. Antunes, J.M.G.; Mikalsen, R.; Roskilly, A.P. An experimental study of a direct injection compression ignition hydrogen engine. *Int. J. Hydrogen Energy* **2009**, *34*, 6516–6522. [CrossRef]

107. Yamane, K.; Nogami, M.; Umemura, Y.; Oikawa, M.; Sato, Y.; Goto, Y. *Development of High Pressure H2 Gas Injectors, Capable of Injection at Large Injection Rate and High Response Using a Common-Rail Type Actuating System for a 4-Cylinder, 4.7-Liter Total Displacement, Spark Ignition Hydrogen Engine*; SAE Paper 2011-01-2005; SAE International: Warrendale, PA, USA, 2011. [CrossRef]

108. Baert, R.; Klaassen, A.; Doosje, E. Direct injection of high pressure gas: Scaling properties of pulsed turbulent jets. *SAE Int. J. Engines* **2010**, *3*, 383–395. [CrossRef]

109. Bovo, M.; Rojo, B. *Single Pulse Jet Impingement on Inclined Surface, Heat Transfer and Flow Field*; SAE Paper 2013-24-0003; SAE International: Warrendale, PA, USA, 2013. [CrossRef]

110. Hill, P.G.; Ouellette, P. Transient turbulent gaseous fuel jets for diesel engines. *J. Fluid Eng.* **1999**, *121*, 93–101. [CrossRef]

111. Hajialimohammadi, A.; Edgington-Mitchell, D.; Honnery, D.; Montazerin, N.; Abdullah, A.; Mirsalim, M.A. Ultra high speed investigation of gaseous jet injected by a single-hole injector and proposing of an analytical method for pressure loss prediction during transient injection. *Fuel* **2016**, *184*, 100–109. [CrossRef]

112. Rubas, P.J. An Experimental Investigation of the Injection Process in a Direct-Injected Natural Gas Engine Using PLIF and CARS. Ph.D. Dissertation, University of Illinois at Urbana-Champaign, Champaign, IL, USA, 1998.

113. Yu, J.; Vuorinen, V.; Hillamo, H.; Sarjovaara, T.; Kaario, O.; Larmi, M. *An Experimental Study on High Pressure PULSED jets for DI Gas Engine Using Planar Laser-Induced Fluorescence*; SAE Paper 2012-01-1655; SAE International: Warrendale, PA, USA, 2012. [CrossRef]

114. Hamzehloo, A.; Aleiferis, P.G. Large eddy simulation of highly turbulent under-expanded hydrogen and methane jets for gaseous-fuelled internal combustion engines. *Int. J. Hydrogen Energy* **2014**, *39*, 21275–21296. [CrossRef]

115. Haskel. Hydrogen. Available online: https://www.haskel.com/industries/hydrogen/ (accessed on 25 June 2019).

Review

Numerical Investigations of Micro-Scale Diffusion Combustion: A Brief Review

P. R. Resende [1], Mohsen Ayoobi [2,*] and Alexandre M. Afonso [3]

[1] INEGI—Institute of Science and Innovation in Mechanical and Industrial Engineering, 4200-465 Porto, Portugal
[2] Division of Engineering Technology, College of Engineering, Wayne State University, Detroit, MI 48202, USA
[3] CEFT, Department of Mechanical Engineering, University of Porto, 4200-465 Porto, Portugal
* Correspondence: mohsen.ayoobi@wayne.edu

Received: 28 June 2019; Accepted: 9 August 2019; Published: 15 August 2019

Abstract: With the increasing global concerns about the impacts of byproducts from the combustion of fossil fuels, researchers have made significant progress in seeking alternative fuels that have cleaner combustion characteristics. Such fuels are most suitable for addressing the increasing demands on combustion-based micro power generation systems due to their prominently higher energy density as compared to other energy resources such as batteries. This cultivates a great opportunity to develop portable power devices, which can be utilized in unmanned aerial vehicles (UAVs), micro satellite thrusters or micro chemical reactors and sensors. However, combustion at small scales—whether premixed or non-premixed (diffusion)—has its own challenges as the interplay of various physical phenomena needs to be understood comprehensively. This paper reviews the scientific progress that researchers have made over the past couple of decades for the numerical investigations of diffusion flames at micro scales. Specifically, the objective of this review is to provide insights on different numerical approaches in analyzing diffusion combustion at micro scales, where the importance of operating conditions, critical parameters and the conjugate heat transfer/heat re-circulation have been extensively analyzed. Comparing simulation results with experimental data, numerical approaches have been shown to perform differently in different conditions and careful consideration should be given to the selection of the numerical models depending on the specifics of the cases that are being modeled. Varying different parameters such as fuel type and mixture, inlet velocity, wall conductivity, and so forth, researchers have shown that at micro scales, diffusion combustion characteristics and flame dynamics are critically sensitive to the operating conditions, that is, it is possible to alter the flammability limits, control the flame stability/instability or change other flame characteristics such as flame shape and height, flame temperature, and so forth.

Keywords: micro-scale combustion; diffusion flames; micro power generation; numerical investigations

---

1. Introduction

With the recent technology advancements over the past few decades, the development of integrated devices at micro scales, known as Micro Electro-Mechanical Systems (MEMS), has been widely recognized and used in various industries. In order for such systems to work, an external power source is required, which is typically batteries. However, the low energy density of batteries and the weight limitations in MEMS devices have convinced researchers to seek alternative efforts in the power generation at small scales and take advantage of the extremely higher energy densities of hydrocarbon fuels through micro combustors [1]. However, combustion-related power generation in micro scales has its own challenges as it is not trivial to obtain a stable flame. This is essentially attributed to (a) the low residence time inside of the micro combustor to complete the combustion process and (b) the higher heat losses from the combustion chamber walls due to the larger surface to volume ratios [1,2].

Designing and conducting numerous micro-scale experiments are both time consuming and expensive. Therefore, it is more reasonable to direct most of the efforts towards numerical simulations and to limit experimental investigations to when it is necessary especially for validation purposes. Hence, the literature in combustion fields is very rich with numerical methodologies and simulations that have been conducted to investigate combustion physics at micro scales and seek ways to (a) develop more proficient and accurate numerical models, (b) capture different phenomena in micro-scale combustion and (c) optimize micro-scale power generation devices.

In the last two decades, significant efforts have been invested in the modeling of the micro-scale combustion devices, through which numerous methodologies have emerged and prominent conclusions have been drawn for different operating conditions such as different fuel types and phases, different geometries, different fuel/oxidizer compositions, and so forth. Currently, such numerical methodologies and simulation tools are widely adopted to conduct research in micro combustion. With that, most micro-combustion studies consist of numerical simulations to complement the experimental research, allowing a better understanding of the dynamics of the flames and the optimization of the corresponding devices.

Numerical simulations for both premixed and non-premixed (diffusion) combustion at micro scales have been evolving extensively as the topic is becoming more and more interesting to researchers. In premixed combustion, researchers have concluded that combustion characteristics and flame behavior are very sensitive to the dimensional and thermo-physical parameters [3–6]. Other than premixed combustion at micro scales, non-premixed or diffusion combustion at such scales has also been under great attention due to the fact that diffusion combustion is a more typical combustion regime to use in practical systems [7]. By comprehending the effects of the complex transport phenomena associated with the chemical kinetics of such flames, researchers can further investigate complex phenomena such as flame ignition and extinction and pollution formations and eventually improve the efficiency of combustion-oriented devices such as combustion engines [8].

In this review paper, the authors will present a succinct review of the progress that has been made in numerical investigations and methodologies on the topic of diffusion combustion at micro scales, where a summary of the state of the art will be provided to further direct researchers on adding to the body of knowledge in this field.

2. Governing Equations

The main governing equations used for the numerical simulations are the conservation equations for a continuous, compressible, multi-component and thermally-perfect gas mixture for laminar flows. It is noted that the numerical simulations for combustion at micro scales do not concern turbulence as the flow is usually laminar due to the small scales of the problems. The conservation equations of total mass, momentum, species mass fractions and energy are presented below, respectively [9].

$$\frac{\partial \rho}{\partial t} + \nabla(\rho \mathbf{u}) = 0 \tag{1}$$

$$\rho \left[\frac{\partial \mathbf{u}}{\partial t} + \nabla \cdot \mathbf{u}\mathbf{u} \right] = -\nabla p + \nabla \tau + \rho \mathbf{g} \tag{2}$$

$$\frac{\partial}{\partial t}(\rho Y_k) + \nabla(\rho Y_k \mathbf{u}) = -\nabla(\rho Y_k \mathbf{V_k}) + \dot{R}_k \qquad k = 1, \dots, N_C \tag{3}$$

$$\rho C_P \frac{\partial T}{\partial t} + \rho C_P \mathbf{u} \nabla T = -\nabla \mathbf{q} - \rho \sum_{k=1}^{N_C} C_{P,k} Y_k \mathbf{V_k} - \sum_{k=1}^{N_C} h_k \dot{\Omega}_k \tag{4}$$

Here, t, ρ, $\mathbf{u}$, τ, $\varnothing$ and $\mathbf{g}$ represent time, the mixture density, the pressure, the mixture velocity vector, the stress tensor, the fluid stress tensor, and the acceleration vector due to gravity, respectively.

N_C is the total number of species, T is the temperature, and Y_k, $\mathbf{V_k}$, $\dot{R}_k$, h_k and $\dot{\Omega}_k$ are the mass fraction, the diffusion velocity, the formation rate, the enthalpy and the net production rate of the kth species, respectively. C_p represents the specific heat coefficients. The gas mixture is assumed to be an ideal gas and its density is calculated through the equation of state.

The diffusion velocities of species k are calculated from Fick's law and the thermal diffusion effect:

$$\mathbf{V_k} = -\frac{D_k}{Y_k}\nabla Y_k - \frac{D_k\Theta_k}{X_k}\frac{1}{T}\nabla T \tag{5}$$

where, X_k, Θ_k, and D_k are the mole fraction, the thermal diffusion and the mixture-averaged diffusion coefficient of species k.

The heat flux vector, $\mathbf{q}$, is defined as:

$$\mathbf{q} = -\lambda\nabla T + \mathbf{q}_{rad} \tag{6}$$

where λ is the mixture thermal conductivity, $\mathbf{q}_{rad}$ is the radiative heat transfer. However, in some studies the radiative heat transfer is neglected, due to the lower impact at micro scale [2].

The radiative heat transfer can be modeled by assuming the optically thin radiation hypothesis, in which the self-absorption of radiation is neglected. The radiative heat transfer contribution in the energy equation is described as:

$$\nabla\mathbf{q}_{rad} = -4\sigma a_P(T^4 - T_{env}^4) \tag{7}$$

where, T_{env} is the environment temperature, σ is the Stefan-Boltzmann constant and a_P is the Planck mean absorption coefficient.

The initial numerical studies of micro diffusion combustion did not consider the wall thermal effects and, as a consequence, the numerical results close to the flame edge shifted away from those of the experiments, showing the important impact of the wall quenching in that region, and the importance of the thermal interactions between the solid and the fluid. For this reason, the numerical studies started to incorporate the energy equation for the solid region, using conjugate heat transfer condition for the fluid-solid interaction.

3. Numerical Studies of Micro-Diffusion Flames

The numerical studies of micro-diffusion flames have emerged with the aim of better understanding the physical phenomena of diffusion combustion at micro scales, where it is challenging to collect such comprehensive data through experimental investigations.

Nakamura et al. performed numerical simulations to investigate the impacts of the fuel velocity on micro flame structures in laminar methane-air combustion [10]. In their simulations, they used both one-step and detailed reaction mechanisms and were able to capture the flame heights, temperature and the methane and oxygen concentration distributions in an acceptable agreement with the experimental measurements for different inlet velocities. Through simulation results, they could capture a linear correlation between the flame height and the methane flow rate, that was consistent with the observations in the experimental data as represented in top plot of Figure 1. The contours illustrated in the bottom of this figure compare the simulation results (flow vectors (a), temperature contours (b), methane and oxygen concentrations (c) and reaction contours (d)) for two different methane inlet velocities. This work also reported on the limited buoyancy effect on the flow field (the order of 10^{-2} m/s), when compared with the minimum fuel velocity of 1 m/s, and a flame temperature difference of 700 K (temperature difference between the flame edge at the vicinity of the flame base and the adiabatic temperature). Such predictions were not as exact near the flame edge, where the impacts of wall quenching and detailed finite-rate chemical kinetics become more significant.

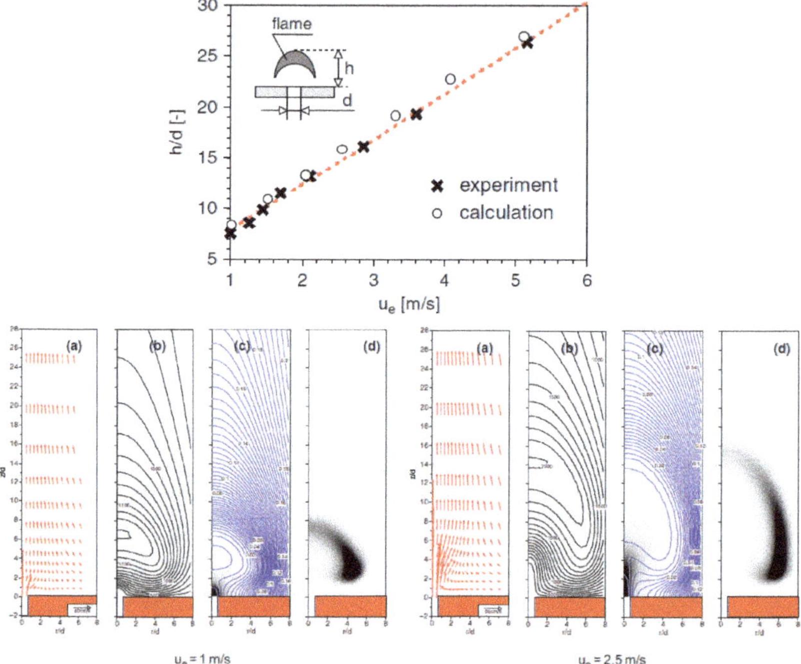

Figure 1. Comparison between experimental and numerical simulations of a micro diffusion flame using methane and air by Nakamura et al. [10]. **Top**: Simulation results against experimental data of the flame height as a function of the methane inlet velocity, u_e; **Bottom**: 2D profiles of the simulation results for $u_e = 1$ and 2 m/s: (**a**) flow vectors, (**b**) temperature contours, (**c**) methane and oxygen concentrations ratio, and (**d**) reaction contours.

Subsequently, Nakamura et al. conducted a numerical study to investigate the impacts of the burner size on the extinction behaviour of methane diffusion flames at micro scales [11]. They used a convective–diffusive heat and mass transport model, where the combustion reaction was assumed by a one-step, irreversible, exothermic reaction. In order to take the near-extinction effects into account, they selected and validated specific kinetics constants. Using the Damkohler principle, they concluded that "there is the minimum burner diameter for the micro flame to exist" [11]. As represented in the left diagram of Figure 2, the normalized flame height and radius (both are normalized by the burner diameter) and the scalar dissipation rate (SDR) at the maximum heat release rate location under the same Reynolds (Re) number, are plotted against the burner diameters. The SDR, or diffusive transport process rate, $D(dZ/dn)^2$, defined by the derivative of the mixture fraction, Z, based on inert gas normal to the flame surface, n, and multiplied by a non-dimensional diffusion coefficient, D, showed to be smaller than the chemical reaction rate (CRR) for $d > 0.1$ mm, and about the same order of the chemical reaction rate for $d = 0.1$ mm. It is shown that for $d < 0.1$ mm, the SDR, flame height and flame radius decrease rapidly with the increase of the burner diameter. However, for $d > 0.1$ mm, all three parameters experience a more smooth transition with the change of the burner diameter. Reducing the burner diameter further toward the limits, the SDR exceeded the chemical reaction rate and led to extinction. This work illustrates that the the crossover between the SDR and the CRR can be defined by a critical Damkohler (Da_{cr}) number, which is representative of the ratio of the diffusion time and the chemical time, i.e., the flame is controlled by diffusion or chemical process, in case of $Da >> Da_{cr}$ number or $Da << Da_{cr}$ number, respectively. Investigating the impacts of different inlet methane velocities in three different burner diameters ($d = 0.0375$, 0.05 and 0.075 mm) on the flame extinction behavior, they also showed the existence of a critical minimum flame height and a critical maximum heat release rate, at which the flame can experience extinction, regardless of the burner diameter as presented in the right diagram of Figure 2.

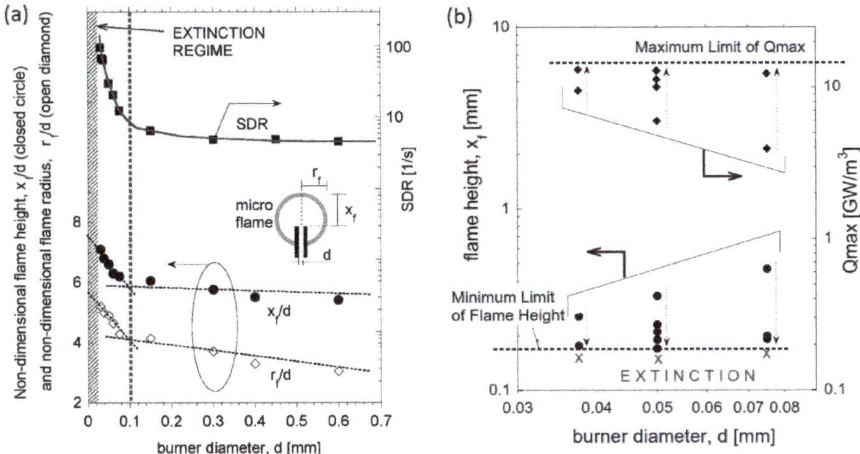

Figure 2. Numerical study of methane micro diffusion flame to analyze the burner size effects by Nakamura et al. [11]. (**a**): Size effects on flame shapes and scalar dissipation rate (SDR) evaluated at the location of maximum heat release rate; (**b**): Minimum flame height and corresponding maximum local heat release rate, Q_{max}.

Using a different fuel, Cheng et al. investigated the combustion characteristics of hydrogen diffusion flames at micro-scale, both experimentally and numerically [12]. In their experiments, they implemented non-intrusive techniques to measure all major species concentrations (O_2, H_2O, H_2, N_2) and, for the first time, the absolute hydroxyl radical concentration (OH), together with the temperature, at $Re = 30$ with inner diameters of 0.2 mm and 0.48 mm for the tube injector fuel (Figure 3). The OH images illustrated a spherical flame forming with reducing the tube diameter. Furthermore, flames were found to be in the convection–diffusion controlled regime due to the low Peclet (Pe) number, indicating the insignificant effect of Pe number on the flame shape. In the numerical simulations, they applied detailed transport and multi-step reaction mechanisms with 9 species and 21 reactions for hydrogen–air combustion with the following boundary conditions: non-slip, non-catalytic reactions, and constant wall temperature (300 K). These numerical simulations were, however, unable to correctly predict the flame structure for the domains with dominant air entrainment and thermal-diffusive effects. This was later addressed by Cheng et al. [13], where the burner was coupled to the computational domain to solve for the properties both inside and outside of the burner wall and to allow for taking the conjugate heat transfer and the back-diffusion of the species into account. More specifically, they obtained the wall temperature distribution by solving the energy balance of diffusive, convective and conductive fluxes into the gas flow and imposed that in their numerical calculations.

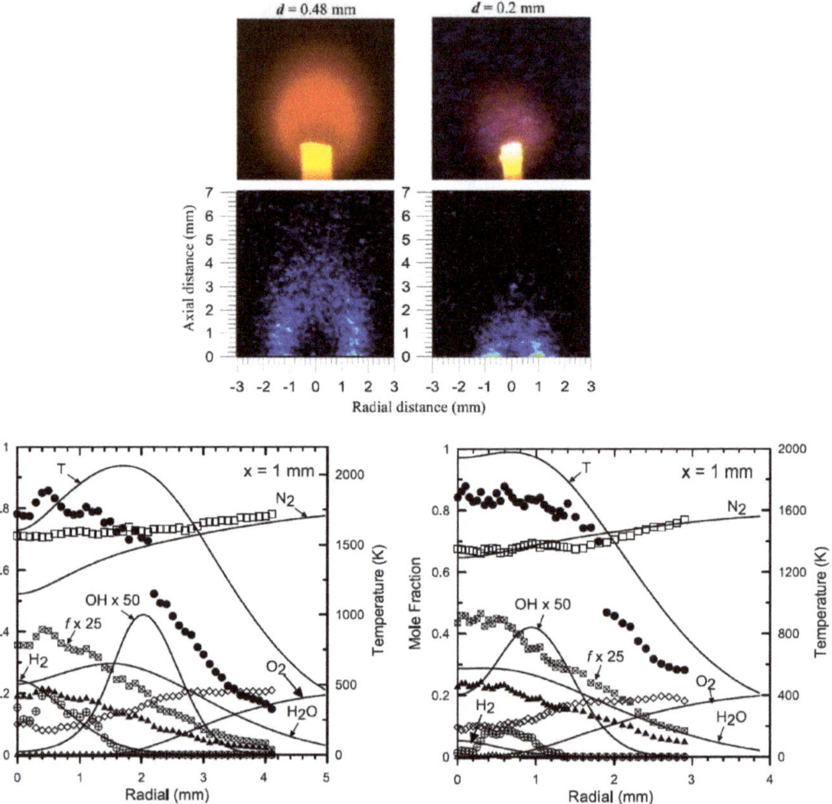

Figure 3. Experimental and numerical study of hydrogen micro-diffusion flame for different tube diameters by Cheng et al. [12]. **Top**: Photographs and single-pulse LIPF-OH (Laser Induced Predissociative Fluorescence - OH radical) (a pre-dissociation fluorescence technique to collect densities of OH radicals) images of the microscale hydrogen diffusion flames with tube diameters of 0.2 and 0.48 mm; **Bottom**: Comparison of measured mean temperature, major species, and OH mole fraction with calculated radial profile for $d = 0.48$ mm (**bottom-left**) and 0.2 mm (**bottom-right**) flame at $x = 1$ mm.

Moving forward with their previous work, Cheng et al. [8] conducted detailed measurements and assessments of laminar hydrogen jet flames through non-intrusive experimental techniques and compared the findings with computational results. They investigated non-premixed reactive flows with different Reynolds (Re) numbers ($Re = 30$ and $Re = 330$). As presented in Figure 4, it was observed that the impact of the thermal diffusion on the flame structure is more pronounced at the case with higher Re number. In addition to varying Re numbers, Cheng et al. examined the flame structure at $Re = 30$ with five different chemical kinetic mechanisms and did not observe any major differences among them. The effects of the burner wall on the flame structure were also studied, using a commercial CFD (Computational Fluid Dynamics) code, CFD-ACE [14], which was found to affect the flows with a lower Re number more considerably. It was also concluded that a variable wall temperature profile as a boundary condition would result in a better agreement between the computed and measured data. The complete measurements of laminar hydrogen jet diffusion flames in this work can be used as benchmarks for the development and validation of CFD codes.

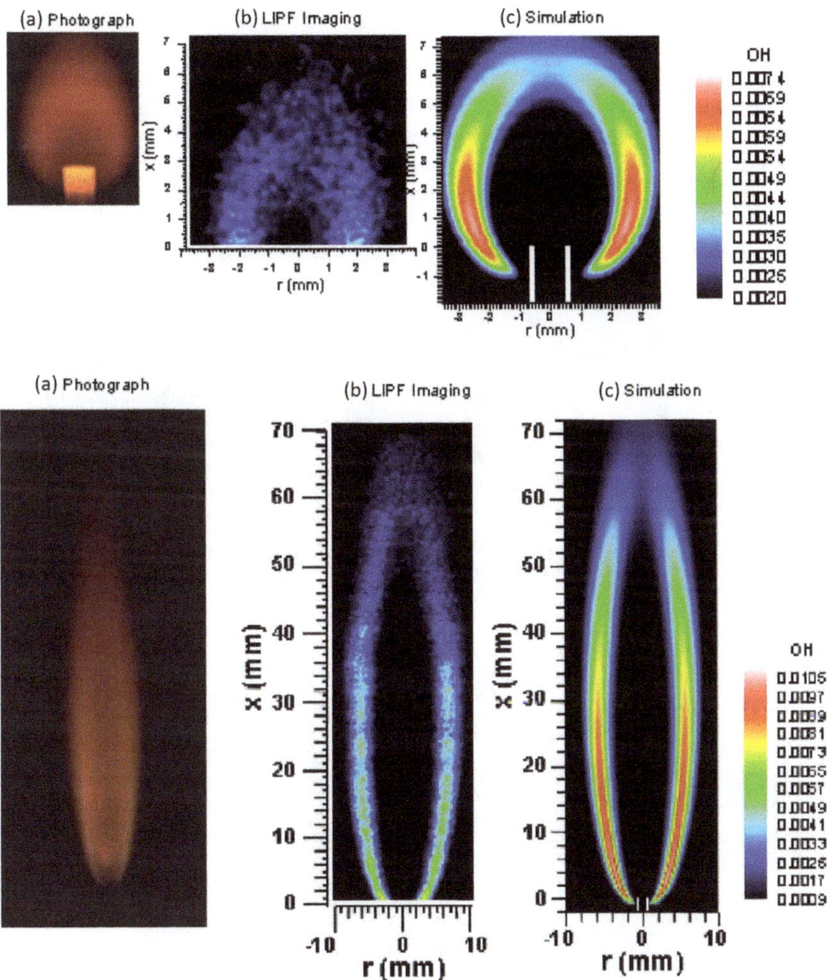

Figure 4. Experimental (**a** and **b** in both **top** and **bottom** figures) and numerical (**c** in both **top** and **bottom** figures) study of hydrogen micro-diffusion flame for different Reynolds numbers by Cheng et al. [8]. (**a**) Photograph, (**b**) single-pulse LIPF-OH image and (**c**) computed OH isopleths of the hydrogen jet diffusion flame, **Top:** $Re = 30$; **Bottom:** $Re = 330$.

Chen et al. continued the research that was initially pursued by Nakamura et al. [10] by conducting experimental and numerical analyses of the structure and stabilization mechanism of a micro jet methane diffusion flame at near extinction conditions, especially in the standoff region [13] (see Figure 5). The numerical simulations were conducted by the simulation tool CFD-ACE [14], where the energy balance of conductive, diffusive, and convective fluxes in the gas flow with thermal conduction at the wall are solved to compute the wall temperature profile. Similar to the previous work [8], the radiation heat loss was neglected and non-slip and non-catalytic reaction conditions were applied on the burner surface. The code was validated through initial simulations at different tube diameters (0.148 mm, 0.4 mm and 1.0 mm) and fuel exit velocities. The stoichiometric mixture fraction at the symmetric jet center line was common in all cases. The simulations were able to capture flame heights as measured in the experimental results, where the flame height increased with the increase of the exit velocity, for all tube diameters. In terms of the prediction of the micro-flame stabilization characteristics, CH mass fraction isopleths were also in good agreement with those of measurements (Figure 5). They also made a comparison between CH contours of adiabatic and variable wall temperature conditions and observed that, with the adiabatic wall, the flame tends to extend farther upstream to the port and anchor on the tube. However, in the case of variable wall temperature, the flame quenched on the tube wall resulting in a gap. Subsequently, the heat transfer through the wall led to accelerated fuel decomposition and chain-branching reactions. They also suggested that the role of the buoyancy becomes important in flames with larger tube diameters and fuel velocities. On the other hand, for lower tube diameters, $d = 0.186$ mm, and inlet velocities <3.68 m/s, the buoyancy effect becomes less important and the flame is controlled only by diffusion.

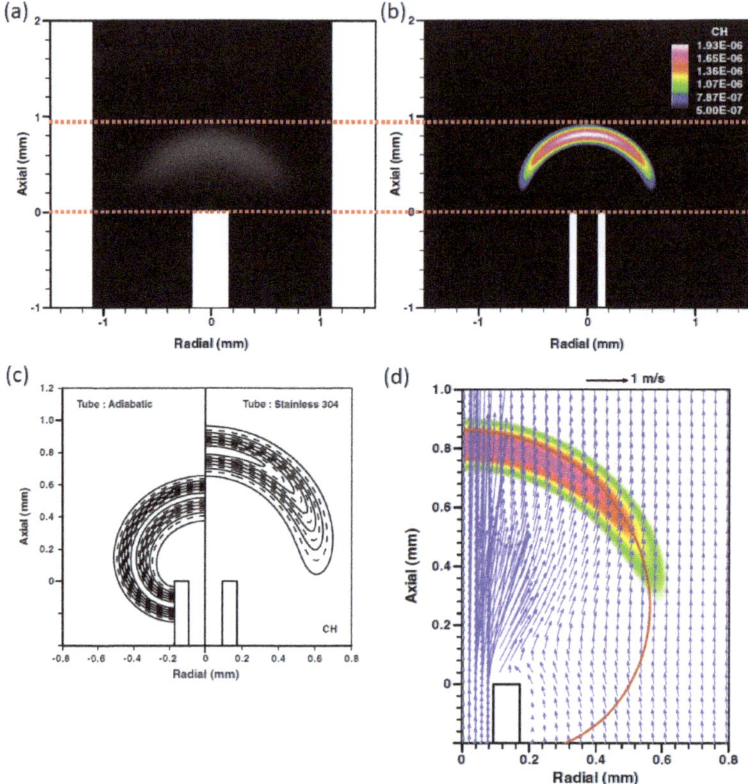

Figure 5. Numerical study of a microjet methane diffusion flame near extinction by Chen et al. [13]. **Top:** Comparison of the measured CH* image (**a**) with computed CH mass fraction isopleths (**b**); **Bottom:** Comparison of CH contours calculated with adiabatic and variable wall temperature conditions (**c**), and Computed 2-D velocity vector coupled with CH isopleths in the standoff region (**d**).

The design of micro combustors has always been the main concern in order to increase their efficiency and stability. One of the first attempts to stabilize and enhance a methane diffusion flame in micro combustors was made by Badra and Masri [15]. They simulated different 2D and 3D designs to optimize the mixing fractions of the methane and air streams and obtain the stability conditions for the flame. To do so, they examined various inlet temperatures, multiple chamber inner materials (changing the conductivity), and reducing the external heat losses. With varying the materials inside the chamber, it was verified that the material with higher thermal conductivity allowed for stronger and more stable flames. This is attributed the fact that more conductive materials allow for more heat transfer to the incoming mixtures, which results in increasing their temperature to 100–300 °C. On the other hand, for an adiabatic micro reactor with inlet mixtures at ambient temperature, they observed that the flame could not sustain. As shown in Figure 6, they also suggested that the optimal design to ensure suitable mixing would require a restriction at the inlet of the combustion chamber to help the incoming streams converge while introducing a minor pressure drop. It is noted that the simulations in this work were conducted using the commercial CFD tool ANSYS Fluent [16], where the authors took the thermal diffusion effects into account.

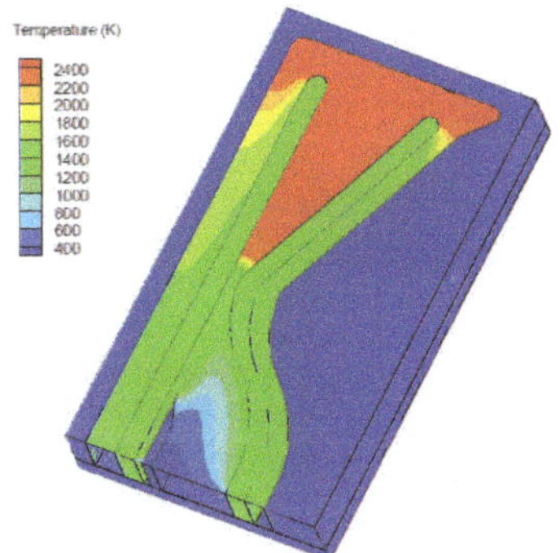

Figure 6. Numerical study of micro-combustor design for optimization of methane and air mixing by Badra and Masri [15]. 3D Temperature contours using aluminum as the conductive material.

To further understand the importance of conjugate heat transfer for the inlet mixture and the negative role of the external heat losses, Li et al. investigated the diffusion combustion of liquid heptane in a small tube with and without heat recirculation via both experimental and numerical approaches [17]. They demonstrated that it was possible to reach a stable flame even without heat recirculation. Analyzing the impacts of equivalence ratio and the external heat loss coefficient, they observed the reaction zone moving with the variations of equivalence ratio; i.e., the reaction zone moves farther upstream when decreasing the equivalence ratio. The flame position was found to be sensitive to the heat recirculation as well, where no heat recirculation (adiabatic) resulted in the flame formation closer to the tube exit as compared to the cases with the existence of heat recirculation (more details are presented in Figure 7). In terms of the flammability limits, the numerical simulations showed that with the increase of the external heat loss, the flammability limit falls down rapidly if there is no heat recirculation. On the other hand, even for higher values of external heat loss coefficient (e.g., 100 W/m^2-K), the burner can sustain at the presence of heat recirculation. It is noted that with the heat recirculation, the wall temperature of the inner tube reaches temperatures higher than the boiling point of the liquid n-heptane, which results in the pre-evaporation of the liquid fuel. Otherwise, the liquid fuel would be accumulated in the tube. Numerical simulations in this work were performed by the commercial CFD tool ANSYS Fluent [16], where the effects of solid and gas-phase conduction, solid radiation, solid-to-gas heat transfer, species diffusion, and chemistry were taken into account. To simulate the droplet vaporization, a vaporization and heating model in the ANSYS Fluent package was used. This model follows the mass transfer laws when the evaporation of the droplets takes place at the vaporization temperature. Also, the discrete phase model based on the Euler-Lagrange method was employed to simulate the droplets movement and the atomization of n-heptane.

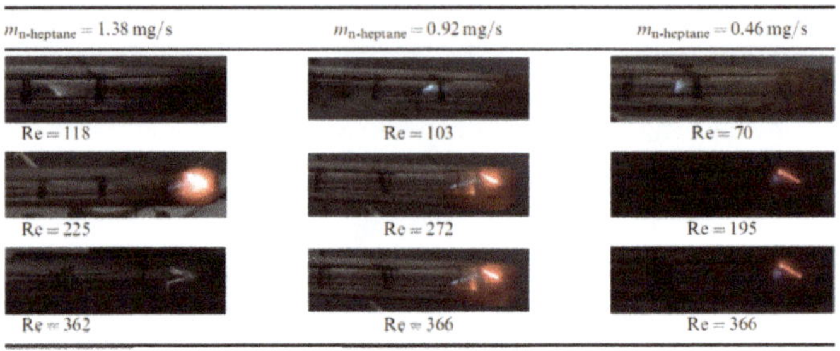

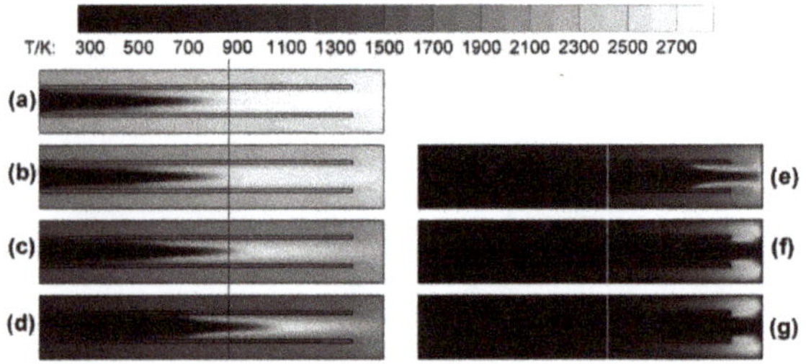

Figure 7. Experimental and numerical study of diffusion combustion in a micro channel using liquid n-heptane and air mixture, with and without heat recirculation, by Li et al. [17]. **Top**: Photos of flame at different Reynolds numbers (**b**); **Bottom**: Temperature contours for various heat loss coefficients: (*h* Unit: W/m^2 − K): (**a**) $h = 0$, (**b**) $h = 1$, (**c**) $h = 10$, (**d**) $h = 20$, (**e**) $h = 50$, (**f**) $h = 75$, (**g**) $h = 100$.

Xu et al. also studied a liquid fuel (ethanol) by performing experimental and numerical investigations of laminar flow with diffusion combustion in micro scale ceramic tubes of different inner diameters [18]. They employed numerical methods to capture micro flame characteristics as a way around the limitation of measurement tools in experiments. They designed several tests by varying the numerical conditions to examine the model capacity for predicting the micro flame characteristic correctly. As presented in Figure 8, Xu's group confirmed that the liquid droplet radiation heat transfer and boundary slip had significant impacts on the results and should not be overlooked in modeling. The numerical simulations were obtained by solving the typical governing equations of mass conservation, momentum, energy, and species conservation. In these equations, the following physical phenomena are taken into account: multi-component diffusion, wall heat diffusion, radiation between the ceramic tube and combustion gasses, and evaporation to simulate droplets combustion process. Comparisons between numerical and experimental results such as height and width of the flames and maximal temperature showed that the discrepancies become maximum if the liquid droplet radiation heat transfer is not taken into account. Consistent with the findings by Nakamura et al. [10], the flame height and width increased almost linearly with the volume flow rate of the fuel, but the gradient increases as the tube inner diameter decreases. In addition, the maximal temperature increases first and then decreases after a certain flow rate.

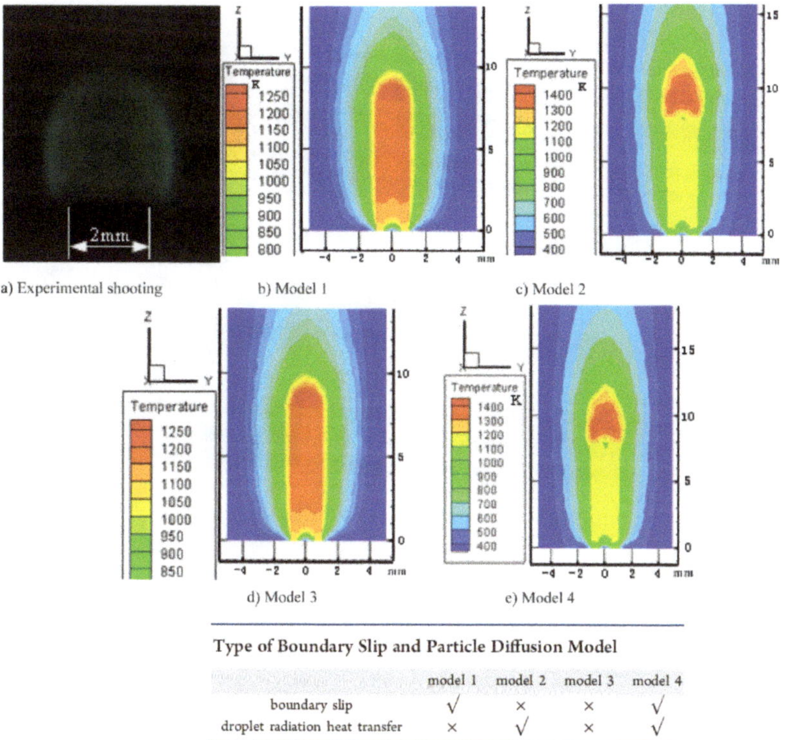

a) Experimental shooting   b) Model 1   c) Model 2

d) Model 3   e) Model 4

Type of Boundary Slip and Particle Diffusion Model

	model 1	model 2	model 3	model 4
boundary slip	√	×	×	√
droplet radiation heat transfer	×	√	×	√

Figure 8. Experimental and numerical study of a liquid ethanol micro-combustion in ceramic tubes with different inner diameters by Xu et al. [18]. Flame characteristics for tube 1 at flow rate of 1.8 mL/h.

Following the previous procedure, Xu et al. investigated the diffusion combustion characteristics of liquid bio-butanol at micro scales under an electric field [19]. The electric field was imposed to analyze its impact on the flame structure and investigate the quenching phenomenon, in order to improve the stability and efficiency of liquid fuel micro flames, which are known to become unstable and tend to extinguish easily. They observed that the electric field has significant impacts on the micro scale combustion (see Figure 9). Increasing the DC voltage, the quenching flow rate decreased initially and after reaching its trough value, it started to gradually increase. The flame height, however, showed an opposite behavior: increasing at first and then decreasing. Moreover, the flame reached its maximum height faster in the smaller tube. It was also observed that increasing the fuel flow rate results in an almost linear increase of the highest temperature of flame. In this work, the numerical results were in a good agreement with the experimental measurements. The maximum deviation observed in this work was for the flame height (9.13%). In the case of the maximal temperature, the deviations were less than 2.24%. It is noted that the combustion process in this work was modeled by a simplified two-step reaction mechanism. It was however, one of the first studies to model the gasification process of a liquid fuel by calculating the heat change of the liquid droplets.

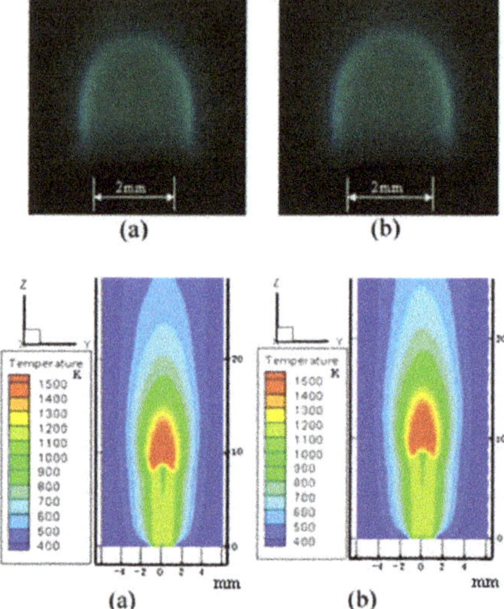

Figure 9. Experimental and numerical study of the effects of electric field on micro-combustion using liquid biobutanol by Xu et al. [19]. Experimental micro flame structure (**Top**) and Flame temperature field of numerical simulation (**Bottom**), with the flow rate of 0.9 mL/h for tube 1. (**a**) 0 V; (**b**) 4000 V.

To understand the limits of spherical hydrogen diffusion flames, Lecoustre et al. performed a study of one-dimensional spherical laminar diffusion flames, using adiabatic porous burners of various diameters [20]. The combustion reaction was modelled with an emphasis on kinetics of extinction, using detailed H_2/O_2 chemical mechanism with 9 species and 28 reactions, and transport properties with updated light component diffusivities. They tested two configurations: (1) normal (H_2 flowing into quiescent air) and (2) inverse (air flowing into quiescent H_2). The flames were simulated using a steady-state laminar flame code with detailed chemistry and transport, initially developed by Sandia National Laboratories for premixed flames [21], and modified to include diffusion flames. For equivalent heat release rates, the inverse flames showed to be smaller and hotter than normal flames. For example, at large flow rates, the peak temperature of inverse flames was 290 K higher than the adiabatic flame temperature of the normal flames, which is attributed to the low Lewis number of the ambient gas for inverse flames fueled by hydrogen. Similar to quenching limits measured in normal gravity on tube burners, the lower quenching limits of normal flames were predicted with a lower bound of 0.25 W, but in case of inverse flames the lower heat release rate increased to 1.38 W. The minimum temperatures at the quenching limits were 1070 K and 903 K for normal and inverse flames, respectively. Some of the aforementioned results are graphically presented in Figure 10.

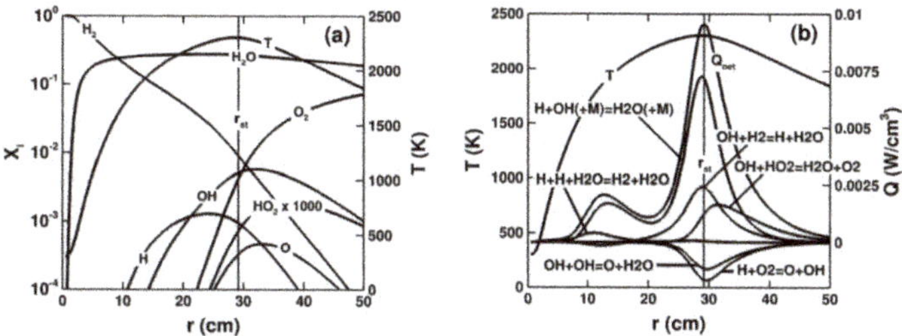

Figure 10. Modeled quenching limits of hydrogen diffusion flames by Lecoustre et al. [20]. Flame structure (**a**) and local heat release rates (**b**) for a large H2/air normal flame on a 5 mm radius burner.

Hossain and Nakamura conducted a numerical investigation to study the thermal and chemical structures formed in the hydrogen-air jet flames in a micro burner, where the excess heat recirculation through the burner wall has the potential to stabilize the miniaturized jet diffusion flame [22]. In their numerical simulations, they considered an axis-symmetric 2D case, where different burner materials and fuel ejecting velocities were tested to understand their roles on the flame structure inside the combustion chamber. Figure 11 presents a sample of the results obtained from these simulations. The thermal structure of the micro-jet flame under different burner materials (titanium and aluminum) showed that the overall flame shape and height were almost identical. They however behaved differently near the burner port. Here, due to the low conductivity of the materials, the heat recirculation from the flame toward the burner is small and therefore the burner tip and its vicinity were continuously heated up. The incoming fuel was also preheated, which increased the maximum flame temperature slightly. Decreasing the fuel velocity resulted in pushing the stoichiometric mixture fraction towards the burner surface, i.e., the flame height decreased, which was also evident in the OH distribution. It is important to note that even at extremely low Reynolds numbers, the effective usage of the burner wall properties and heat transfer can stabilize a significantly small flame. They also showed that enriching the fuel mixture can modify the flame structure and expand the low stability limit.

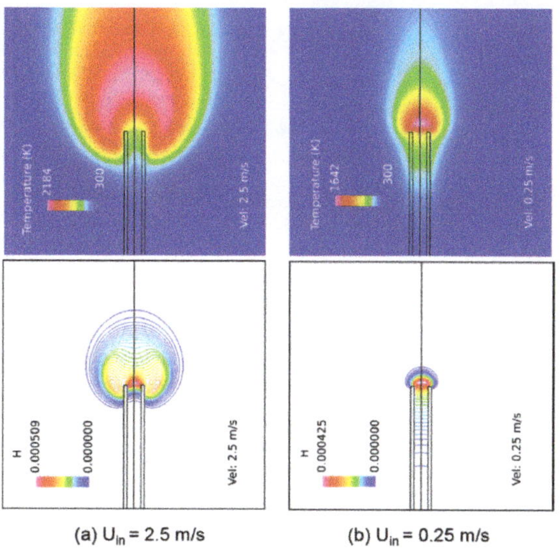

Figure 11. Numerical study of hydrogen-air jet flames in a micro burner by Hossain and Nakamura [22]. 2D temperature and mass fraction of H-atom distributions for different fuel ejecting velocities ((**a**) U_{in} = 2.5 m/s and (**b**) U_{in} = 0.25 m/s) over titanium burner.

In order to investigate the wall thermal interactions with the flame, Zhang et al. conducted a numerical and experimental research to study the combustion characteristics of the non-premixed micro-jet hydrogen flame in an air co-flow, while considering the thermal interactions through the solid walls of the micro tube [23]. The experimental results showed how the flame height and shape, and correspondingly the OH-radical distributions would change with fuel flow velocity. Such results were well captured by the numerical simulations (see Figure 12). The evolution of the flame temperature field with the velocity variations was analyzed by post processing the numerical simulations results. Th flame temperature was shown to reach its maximum of 2300 K for the fuel flow velocity range of 5–25 m/s and to decrease smoothly in the fuel velocity range of 1–5 m/s. The fuel velocity could not go below 0.08 m/s, where the flame extinction occurred. The flame shape and structure were also found to behave differently at fuel flows with high, intermediate and low velocities. The impacts of thermal interactions between flames and the solid walls were different depending on the fuel velocity. For example, for higher fuel velocities, the external surface of the tube had a positive effect on the flame stability, while this effect was negative at lower velocities as the walls absorbed heat from the flame. Figure 12 compares these impacts for different fuel flow velocities. Numerical simulation methodology that was adopted in this work was similar to that of Hossain and Nakamura [22], with the addition of the heat interactions between the fluid and solid by coupling the boundary conditions.

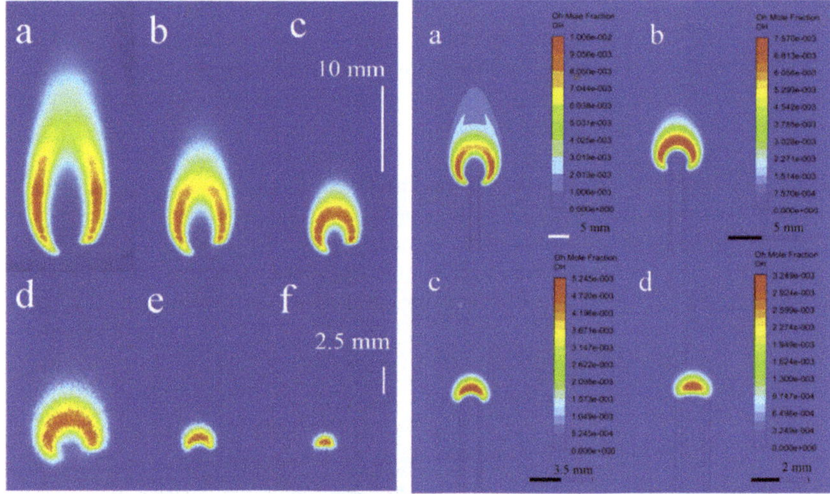

Figure 12. Experimental and numerical study of the thermal interaction with solid micro tube in non-premixed hydrogen jet flames by Zhang et al. [23]. **Left:** OH-PLIF images of hydrogen micro-jet flames at different fuel flow velocities: (**a**) 10 m/s; (**b**) 5 m/s; (**c**) 2.5 m/s; (**d**) 1 m/s; (**e**) 0.35 m/s; (**f**) 0.17 m/s; **Right:** Computational OH mole fractions of hydrogen micro-jet flames at different fuel flow velocities: (**a**) 2.5 m/s; (**b**) 1 m/s; (**c**) 0.35 m/s; (**d**) 0.17 m/s.

Ayed et al. took advantage of the latest advances in numerical investigation of micro-combustion devices and conducted an interactive experimental and numerical study to optimize a micro-mix hydrogen combustion in a test burner and achieve minimum NOx emissions [24]. In this work, authors studied the implications of the turbulent flow field on the NOx emissions by applying different combustion models and characterizing the turbulent micro-mix flame structures. In order to reduce the NOx emissions, they used a cross-flow mixing of air and hydrogen, where gaseous hydrogen is injected perpendicularly into an air cross-flow and reacts in multiple miniaturized diffusion-type flames. The simulations were performed by the commercial CFD tool STAR-CCM+ [25]. In their numerical simulation approach, they used the k-epsilon turbulent model to predict the flow field. To characterize the hydrogen combustion process, they used and compared the following models: (1)

EDM (Eddy Dissipation) model, (2) ED-FR (Eddy Dissipation–Finite Rate) model, and (3) EDC (Eddy Dissipation Concept) model. The calculation of reaction rates and species consumption/formation rates were obtained by implementing a detailed hydrogen reaction mechanism including 19 reversible reactions and 9 species. In the wall treatment of the RANS model, the local dimensionless wall distance, which depends on the wall distance, the friction velocity and the kinematic viscosity was utilized. The geometric model to perform the numerical simulations was derived from the burner configuration APU 030 ED 6.7 [26]. Comparisons between the findings from these three models and the experimental results illustrated qualitative agreement with the experimental findings (see Figure 13); however, the predictions from the EDC model that was closest to the measured values, had 21.5% error. In general, the numerical simulations were able to capture the micro-mix flame anchoring and structure for the operation parameters considered in this work. However, further investigations will be necessary to increase the precision of the numerical results involving diffusion flames, especially at micro scales.

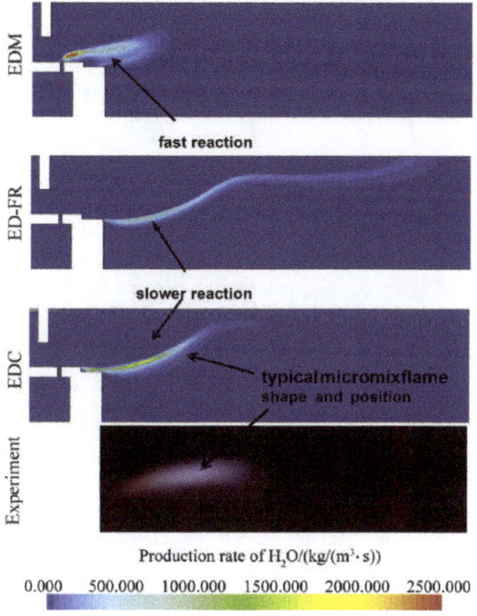

Figure 13. Experimental and numerical study of the dry-low-NOx hydrogen micro-mix combustion chamber by Ayed et al. [24]. Calculated reaction rates in comparison with visual flame structure.

Most recently, Gao et al. investigated the combustion characteristics of micro-jet diffusion flames with a mixture of hydrogen and methane fuels and examined the stabilization mechanisms for different fuel mixture compositions [27]. They reported that flames with hydrogen as the base fuel would always attach to the burner, as opposed to flames with methane as the base fuel. Comparing the dominant intermediate-consumption steps in both methane and hydrogen, those of hydrogen flame were found to have significantly lower activation energies. In addition to the comparisons between hydrogen- and methane-based flames, the impacts of fuel mixing were studied as well, where the combustion characteristics of different fuel mixture ($H_2 + CH_4$) compositions were analyzed and compared (see Figure 14). They reported that the fraction of hydrogen in the fuel mixture has a controlling impact on the burner flame temperature, which was found to be attributed to the reaction rate of $H + O_2 + M \rightarrow HO_2 + M$ as it is suppressed by reactions involving intermediate species such as CH_3 and CH_2O in methane consumption. With that, they suggested that the combination of methane and hydrogen can be considered a good alternative to prevent the potential thermal damages to the chamber.

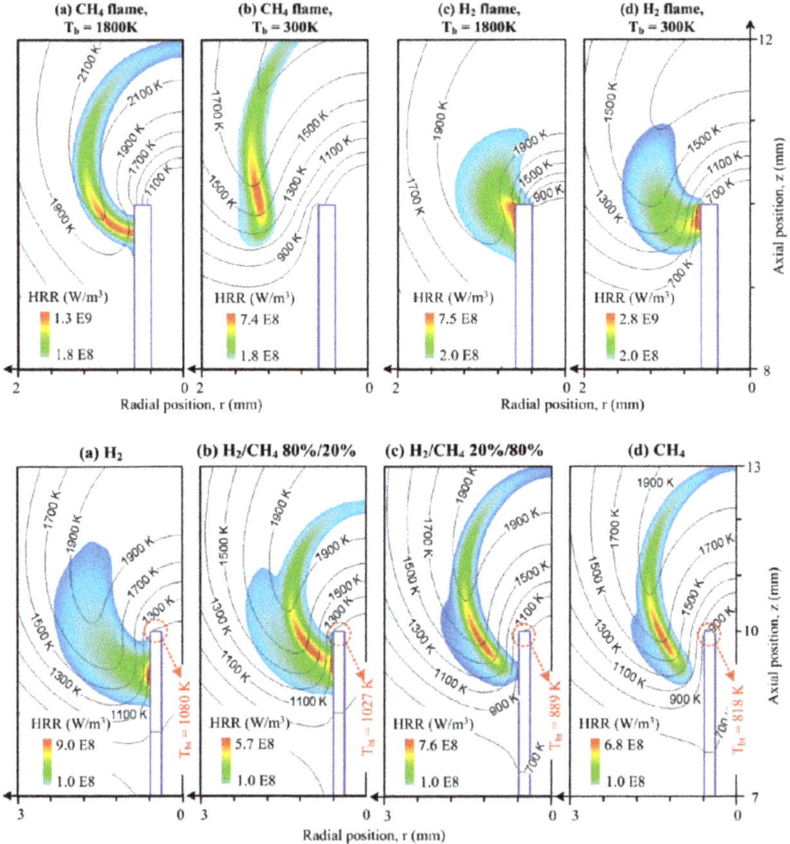

Figure 14. Numerical study of micro-jet hydrogen/methane diffusion flames by Gao et al. [27]. **Top**: Computed temperature contours and heat release rate (HRR) isopleths for CH_4 flames at fuel jet velocity of 0.4 m/s for (**a**) $T_b = 1800$ K (**b**) $T_b = 300$ K, and for H_2 flames at 1.0 m/s for (**c**) $Tb = 1800$ K (**d**) $T_b = 300$ K; **Bottom**: Computed temperature contours and HRR isopleths at the thermal conductive burner condition (the thermal conductivity of burner wall is 16 W/m-K) for (**a**) H_2 flame at 2.0 m/s, (**b**) H_2/CH_4 80%/20% flame at 1.2 m/s, (**c**) H_2/CH_4 20%/80% flame at 0.6 m/s, (**d**) CH_4 flame at 0.5 m/s. Here the flame temperatures range between 2028 K and 2034 K.

Gao et al. continued their investigations by conducting a numerical study of micro-jet diffusion flames using different fuels for several operating conditions by varying the inlet jet velocities and burners proprieties. The objective of this study was to prove the importance of the heat-recirculation impact on the flame stabilization at near-extinction conditions [28]. They used different fuels such as hydrogen, methane and dimethyl ether (DME), and employed a chemical kinetics mechanism with 17 species and 58 reactions. The first part of the study consisted of the analysis of the heat recirculation, Q_{re}, and the heat losses through the burner surfaces, Q_{loss}, by varying the thermal conductivities, k_b, from 1 to 100 W/m-K, and the wall thicknesses, c_b, from 0.2 to 0.8 mm. To evaluate the global thermal effect on the burner, an effective excess enthalpy was defined, $H = (Q_{re} - Q_{loss})/M$, where M is the mass flow rate (kg/s). They reported that the ratio between Q_{re} and Q_{loss} increases monotonically as the jet velocity increases. This results in a critical fuel velocity, where Q_{re} overcomes Q_{loss}, i.e., the heat recirculation can shadow the impact of the heat losses. Correspondingly, as k_b and c_b decrease at near extinction conditions, the flame temperature increases, i.e., the extinction limit for a specific fuel jet velocity can be extended. Figure 15 (left) represents the effective excess enthalpy, *H*, and the maximum flame temperatures of a micro-jet methane combustion, for different burners, over fuel jet velocities ranging from 0.1 m/s to 3.2 m/s. It is evident that the heat-recirculation can assist the combustion regime for positive values of *H*. In second part of the work, simulations were performed for the same burner proprieties, k_b = 6 W/(m-K) and c_b = 0.2 mm, and using different fuels to analyze the impact of the effective excess enthalpy, *H*. Results of this analysis are presented in Figure 15 (right). The investiagotrs concluded that for fuels with Schmidt number (Sc) > 1.0, such as DME (Sc = 1.2), the flame is lifted, resulting in a negligible flame–burner interaction (H ≈ 0). Such behavior was the opposite for fuels with Sc < 1.0, such as methane (Sc = 0.7) and hydrogen (Sc = 0.2), where the chemical kinetics is dominant in the flame base structure, and higher values of *H* are expected.

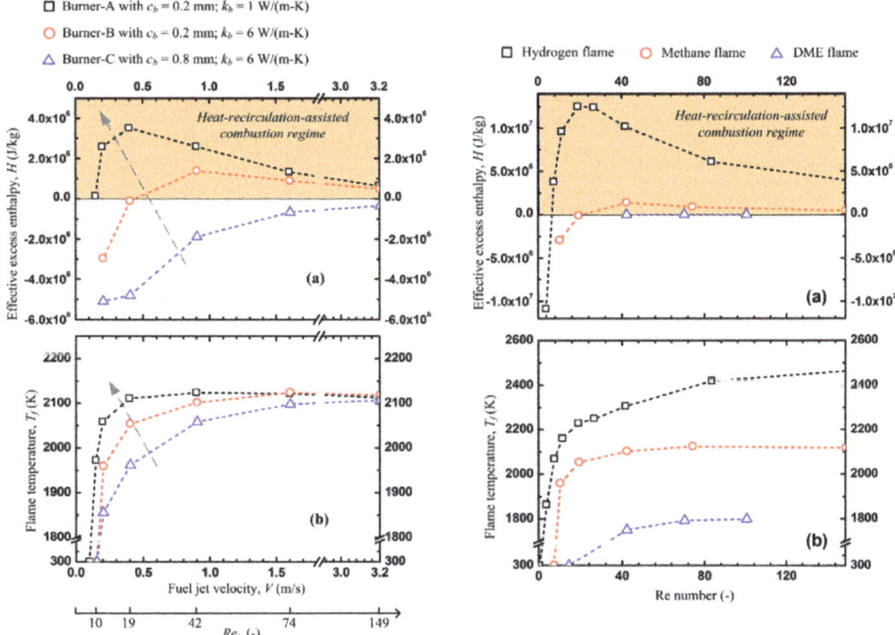

Figure 15. Numerical study on heat-recirculation assisted combustion for small scale jet diffusion flames at near-extinction condition by Gao et al. [28]. **Left**: Effective excess-enthalpies (**a**) and flame temperatures (**b**) over fuel jet velocities range from the extinction limits to 3.2 m/s for different burners: Burner-A with c_b = 0.2 mm and k_b = 1 W/(m-K), Burner-B with c_b = 0.2 mm and k_b = 6 W/(m-K) and Burner-C with c_b = 0.8 mm and k_b = 6 W/(m-K); **Right**: Effective excess-enthalpies (**a**) and flame temperatures (**b**) as a function of Re number for H$_2$, CH$_4$, and DME flames. The burner wall thickness is and the thermal conductivity is c_b = 0.2 mm and k_b = 6 W/(m-K).

Li et al. investigated the mixing performance and the diffusion combustion characteristics in a 2D planar micro-combustor using hydrogen and air separated by a plate [7]. This study was conducted through numerical simulations and consisted of various boundary conditions and burner geometries. Analyzing the mixing performance, flame stability limits and the combustion efficiency, they confirmed that the separating plate improved the mixing performance for lower inlet velocities and channel heights (see Figure 16 (left)), which is attributed to the extended residence time and shortened diffusion distance. They also reported that due to the improved mixing process, the combustion efficiency increased by reducing the combustor height, for the same inlet air velocities (see Figure 16 (right)).

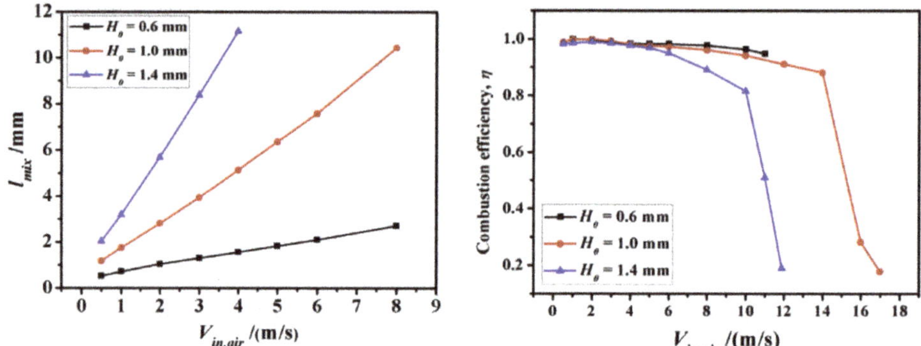

Figure 16. Numerical investigation on mixing performance and diffusion combustion characteristics of H_2 and air in planar micro-combustor by Li et al. [7]. (**Left**): Required mixing distance versus inlet air velocity for micro-combustors with different heights; (**Right**): Variations in combustion efficiency versus inlet air velocity in three micro-combustors with different heights.

In a most recent paper on this topic, Kang et al. [29] conducted a numerical study to reproduce the "flame-street" phenomenon described in the experimental work of Miesse et al. [30]. The "flame-street" is a phenomenon observed in diffusion flames, and is used to describe a separated flame into a discrete series of flame segments. Typically "flame-street" phenomenon has been reported in experimental studies, but this work by Kang et al. seeks to understand its characteristics with the help of numerical simulations. Using a reacting flow solver on the open-source software OpenFOAM [31], and with a chemical reaction mechanism of 21 species and 84 reactions, they simulated a 3D geometry channel with 30 mm in length, 5 mm in width and 0.75 mm in height. They used methane and oxygen as fuel and oxidizer, with inlet flow rates of 100 sccm and 200 sccm, respectively. To consider the heat external losses through the wall, a conjugate heat transfer model was considered, with wall material proprieties identical to the experimental work. The flame-street behavior consisting of a bibrachial, presented a long diffusion tail, a weaker fuel-lean premixed branch, and consecutive downstream "new moon"-like flamelets each with a fuel-lean and a fuel-rich premixed branch (see Figure 17 (left)), where Z_{st} is the fuel mass fraction for an unburnt fuel/oxidizer mixture at stoichiometric conditions. Also, the effects of wall thermal conditions and species mass diffusion on the formation of the flame-street were analyzed, and verified that the flame-street structure was formed at a moderate wall heat loss range, between 800–1100 K wall temperatures (see Figure 17 (right)).

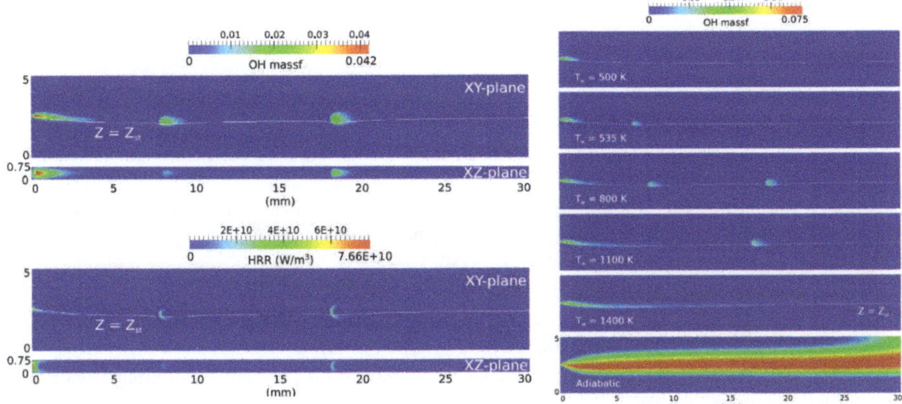

Figure 17. Numerical investigation on the thermo-chemical structures of methane-oxygen diffusion flame-streets in a microchannel by Kang et al. [29]. (**Left**): Contours of OH mass fractions and heat release rate (HRR) on the sliced XY- and XZ-plane along the channel centerline; (**Right**): Effects of wall thermal conditions on flame behaviors.

4. Conclusions

The increasing attentions to the combustion-based micro power generation systems, mainly due to the high energy density of hydrocarbon fuels, has persuaded researchers to expand their efforts to investigate the combustion characteristics at micro scales and seek ways to achieve stabilized combustion and optimize the power generation process in such devices. This work summarized the pertinent investigations revolved around non-premixed (diffusion) combustion regime, which is most commonly employed in practical systems. With the complexities and expenses associated to the experimental analyses, the application of numerical simulations to model combustion at micro scales, especially diffusion flames which are the focus of this work, has been widely considered and shown to provide promising insights. The review in this paper confirmed the strength of the numerical simulations, to both qualitatively and quantitatively predict the flame behavior if used properly. Using different numerical approaches over the past couple of decades, diffusion combustion at micro scale has been extensively studied and characterized. Such case are found to be very sensitive to the operating conditions, i.e., the flame characteristics, stability conditions, and flammability limits can be altered by varying different parameters or the system configuration. For example, conjugate heat transfer or heat recirculation were found to have strong impacts on the flammability limit. Various studies showed that changing the fuel type, fuel/air composition, inlet velocity, burner geometry or material, etc. can change the flame characteristics such flame location, flame height, flame temperature and flame stability considerably. This review of the studies on the numerical investigations and approaches in diffusion combustion characteristics at micro scales presents a pathway for researchers in this field to more efficiently move forward with their research.

Author Contributions: Conceptualization, P.R.R., M.A. and A.M.A.; methodology, P.R.R., M.A. and A.M.A.; investigation, P.R.R., M.A. and A.M.A.; writing—original draft preparation, P.R.R.; writing—review and editing, P.R.R., M.A. and A.M.A.; funding acquisition, P.R.R., M.A. and A.M.A.

Funding: This research received external funding by FAPESP through Project No. 2015/26842-3. A.M. Afonso acknowledges the funding by FCT under Project No. PTDC/EMS-ENE/3362/2014.

Acknowledgments: P. R. Resende acknowledges the funding by FAPESP through Project No. 2015/26842-3. A.M. Afonso acknowledges the funding by FCT under Project No. PTDC/EMS-ENE/3362/2014. M. Ayoobi acknowledges the support from Wayne State University for this publication.

Conflicts of Interest: The authors declare no conflict of interest.

References

1. Ju, Y.; Maruta, K. Microscale combustion: Technology development and fundamental research. *Prog. Energy Combust. Sci.* **2011**, *37*, 669–715. [CrossRef]
2. Nakamura, Y.; Gao, J.; Matsuoka, T. Progress in small-scale combustion. *J. Therm. Sci. Technol.* **2017**, *12*, JTST0001. [CrossRef]
3. Norton, D.G.; Vlachos, D.G. Combustion characteristics and flame stability at the microscale: A CFD study of premixed methane/air mixtures. *Chem. Eng. Sci.* **2003**, *58*, 4871–4882. [CrossRef]
4. Ayoobi, M.; Schoegl, I. Numerical analysis of flame instabilities in narrow channels: Laminar premixed methane/air combustion. *Int. J. Spray Combust. Dyn.* **2017**, *9*, 155–171. [CrossRef]
5. Cova, A.; Resende, P.R.; Cuoci, A.; Ayoobi, M.; Afonso, A.M.; Pinho, C.T. Numerical Studies of Premixed and Diffusion Meso/Micro-Scale Flames. *Energy Procedia* **2017**, *120*, 673–680. [CrossRef]
6. Resende, P.R.; Afonso, A.; Pinho, C.; Ayoobi, M. Impacts of Dilution on Hydrogen Combustion Characteristics and NOx Emissions. *J. Heat Transf.* **2018**, *141*, 012003. [CrossRef]
7. Li, L.; Yuan, Z.; Xiang, Y.; Fan, A. Numerical investigation on mixing performance and diffusion combustion characteristics of H2 and air in planar micro-combustor. *Int. J. Hydrog. Energy* **2018**, *43*, 12491–12498. [CrossRef]
8. Cheng, T.S.; Wu, C.Y.; Chen, C.P.; Li, Y.H.; Chao, Y.C.; Yuan, T.; Leu, T.S. Detailed measurement and assessment of laminar hydrogen jet diffusion flames. *Combust. Flame* **2006**, *146*, 268–282. [CrossRef]
9. Cuoci, A.; Frassoldati, A.; Faravelli, T.; Ranzi, E. A computational tool for the detailed kinetic modeling of laminar flames: Application to C2H4/CH4 coflow flames. *Combust. Flame* **2013**, *160*, 870–886. [CrossRef]
10. Nakamura, Y.; Ban, H.; Saito, K.; Takeno, T. Structures of Micro (Millimeter Size) Diffusion Flames. In Proceedings of the Third International Symposium on Scale Modeling, Nagoya, Japan, 10–13 September 2000; Volume ISSM3-E7.
11. Nakamura, Y.; Yamashita, H.; Saito, K. A numerical study on extinction behaviour of laminar micro-diffusion flames. *Combust. Theory Model.* **2006**, *10*, 927–938. [CrossRef]
12. Cheng, T.S.; Chao, Y.C.; Wu, C.Y.; Li, Y.H.; Nakamura, Y.; Lee, K.Y.; Yuan, T.; Leu, T.S. Experimental and numerical investigation of microscale hydrogen diffusion flames. *Proc. Combust. Inst.* **2005**, *30*, 2489–2497. [CrossRef]
13. Chen, C.P.; Chao, Y.C.; Cheng, T.S.; Chen, G.B.; Wu, C.Y. Structure and stabilization mechanism of a microjet methane diffusion flame near extinction. *Proc. Combust. Inst.* **2007**, *31*, 3301–3308. [CrossRef]
14. CFD-ACE+. 2018. Available online: https://www.esi-group.com/software-solutions/virtual-environment/cfd-multiphysics/ace-suite/cfd-ace (accessed on 20 June 2019).
15. Badra, J.; Masri, A. Design of a numerical micro-combustor for diffusion flames. In Proceedings of the Seventh Mediterranean Combustion Symposium (MCS-7), Sardinia, Italy, 11–15 September 2011.
16. ANSYS-Fluent. 2018. Available online: https://www.ansys.com/products/fluids/ansys-fluent (accessed on 20 June 2019).
17. Li, J.; Huang, J.; Zhao, D.; Zhao, J.; Yan, M.; Wang, N. Diffusion combustion of liquid heptane in a small tube with and without heat recirculating. *Combust. Sci. Technol.* **2012**, *184*, 1591–1607. [CrossRef]
18. Xu, T.; Gao, X.N.; Yang, J.; Gan, Y.H.; Yang, Z.L.; Zhang, Z.G. Experimental and Numerical Simulation Study of the Microscale Laminar Flow Diffusion Combustion of Liquid Ethanol. *Ind. Eng. Chem. Res.* **2013**, *52*, 8021–8027. [CrossRef]
19. Xu, T.; Chen, Q.; Zhang, B.; Lu, S.; Mo, D.; Zhang, Z.; Gao, X. Effects of electric field on microscale flame properties of biobutanol fuel. *Sci. Rep.* **2016**, *6*, 32938. [CrossRef] [PubMed]
20. Lecoustre, V.R.; Sunderland, P.B.; Chao, B.H.; Axelbaum, R.L. Modeled quenching limits of spherical hydrogen diffusion flames. *Proc. Combust. Inst.* **2013**, *34*, 887–894. [CrossRef]
21. Kee, R.J.; Grcar, J.F.; Smooke, M.D.; Miller, J.A.; Meeks, E. *Premix: A FORTRAN Program for Modeling Steady Laminar One-Dimensional Premixed Flames*; Technical Report No. SAND85-8240; Sandia National Laboratories: Albuquerque, NM, USA, 1987.
22. Hossain, A.; Nakamura, Y. Thermal and chemical structures formed in the micro burner of miniaturized hydrogen-air jet flames. *Proc. Combust. Inst.* **2015**, *35*, 3413–3420. [CrossRef]

23. Zhang, J.; Li, X.; Yang, H.; Jiang, L.; Wang, X.; Zhao, D. Study on the combustion characteristics of non-premixed hydrogen micro-jet flame and the thermal interaction with solid micro tube. *Int. J. Hydrogen Energy* **2017**, *42*, 3853–3862. [CrossRef]
24. Ayed, A.H.; Kusterer, K.; Funke, H.H.W.; Keinz, J.; Striegan, C.; Bohn, D. Experimental and numerical investigations of the dry-low-NOx hydrogen micromix combustion chamber of an industrial gas turbine. *Propuls. Power Res.* **2015**, *4*, 123–131. [CrossRef]
25. STAR-CCM+. 2018. Available online: https://mdx.plm.automation.siemens.com/star-ccm-plus (accessed on 20 June 2019).
26. Funke, H.; Recker, E.; Börner, S.; Bosschaerts, W. LES of jets-in-cross-flow and application to the micromix hydrogen combustion. In Proceedings of the 19th International Symposium on Air Breathing Engine, Montreal, QC, Canada, 7–11 September 2009; American Institute of Aeronautics and Astronautics: Montreal, QC, Canada, 2009.
27. Gao, J.; Hossain, A.; Nakamura, Y. Flame base structures of micro-jet hydrogen/methane diffusion flames. *Proc. Combust. Inst.* **2017**, *36*, 4209–4216. [CrossRef]
28. Gao, J.; Hossain, A.; Matsuoka, T.; Nakamura, Y. A numerical study on heat-recirculation assisted combustion for small scale jet diffusion flames at near-extinction condition. *Combust. Flame* **2017**, *178*, 182–194. [CrossRef]
29. Kang, X.; Sun, B.; Wang, J.; Wang, Y. A numerical investigation on the thermo-chemical structures of methane-oxygen diffusion flame-streets in a microchannel. *Combust. Flame* **2019**, *206*, 266–281. [CrossRef]
30. Miesse, C.; Masel, R.; Short, M.; Shannon, M. Experimental observations of methane–oxygen diffusion flame structure in a sub-millimetre microburner. *Combust. Theory Model.* **2005**, *9*, 77–92. [CrossRef]
31. OpenFoam. 2019. Available online: https://www.openfoam.com (accessed on 20 June 2019).

Article

Mechanism of Combustion Noise Influenced by Pilot Injection in PPCI Diesel Engines

Jingtao Du, Ximing Chen, Long Liu *, Dai Liu and Xiuzhen Ma

College of Power and Energy Engineering, Harbin Engineering University, Harbin 150001, China; dujingtao@hrbeu.edu.cn (J.D.); Chenximing@hrbeu.edu.cn (X.C.); dailiu@hrbeu.edu.cn (D.L.); maxiuzhen@hrbeu.edu.cn (X.M.)
* Correspondence: liulong@hrbeu.edu.cn; Tel.: +86-451-8251-8036

Received: 14 April 2019; Accepted: 2 May 2019; Published: 7 May 2019

Abstract: Pilot injection combined with exhaust gas recirculation (EGR) is usually utilized to realize the partially premixed compression ignition (PPCI) mode in diesel engines, which enables the simultaneous decrease of nitrogen oxide and soot emissions to satisfy emission regulations. Moreover, the ignition delay of main injection combustion can also be shortened by pilot injection, and then combustion noise is reduced. Nevertheless, the mechanisms of pilot injection impacts on combustion noise are not completely understood. As such, it is hard to optimize pilot injection parameters to minimize combustion noise. Therefore, experiments were conducted on a four-stroke single-cylinder diesel engine with different pilot injection strategies and 20% EGR as part of an investigation into this relationship. Firstly, the combustion noise was analyzed by cylinder pressure levels (CPLs). Then, the stationary wavelet transforms (SWTs) and stationary wavelet packet transform (SWPT) were employed to decompose in-cylinder pressures at different scales, and thus the combustion noise generated by pilot and main combustion was investigated in both the time and frequency domain. The results show that pilot injection is dominant in the high frequency segment of combustion noise, and main injection has a major impact on combustion noise in the low and mid frequency segment. Finally, the effects of various pilot injection parameters on suppressing combustion noise were analyzed in detail.

Keywords: PPCI diesel engine; pilot injection; combustion noise; stationary wavelet packet transform; stationary wavelet transform

---

1. Introduction

Recently, the requirement of continuous mitigation of particulate matter (PM) and nitrogen oxide (NOx) emissions has become a significant challenge regarding the improvement of combustion in diesel engines. Advanced combustion concepts, such as low-temperature combustion (LTC) and premixed charge compression ignition (PCCI) [1–3], have arisen as a solution for the reduction of emissions, but some limitations must still be faced. That is, these advanced combustion concepts cannot be applied for high load operating cases because the fast fuel-burning and heat release of premixed combustion will deteriorate combustion noise. However, the application of the pilot-main injection strategy to the PCCI combustion mode can reduce combustion noise by shortening the ignition delay (ID) of main injection, which is introduced as partially premixed compression ignition (PPCI) [4,5]. In this way, the reduction of combustion noise by pilot injection has the potential to extend the PPCI mode to the full operating condition. Additionally, higher combustion efficiency, and lower CO_2 and NOx emissions can be obtained using pilot injection in the PPC mode compared with conventional diesel combustion [6–8].

Moreover, the advantages of pilot injection have gained increasing attention, particularly of researchers, regarding the detailed exploration of the effects of pilot injection on combustion noise. It has already been shown that increasing the fuel quantity of the pilot injection or advancing pilot injection timing will result in a progressive decrease of combustion noise [9,10]. However, as reported by Wang Ping et al. [11], the combustion noise can be minimized in the conditions of the moderate pilot injection quantity and dwell. In addition, Alok Warey et al. [12] determined that the minimum combustion noise can be obtained when the energizing dwell between the main and pilot injections decreases below 200 μs. Therefore, throughout these previous combustion noise investigations, it is observed that the conclusions drawn with respect to the pilot injection parameters required to suppress combustion noise are inconsistent. This is probably due to differences of the experimental conditions, but what is foremost is that the influences of pilot injection on combustion noise cannot be interpreted clearly. Meanwhile, the evaluation of combustion noise for situations of varying pilot injection strategies is particularly challenging.

In order to determine the detailed relationship between pilot injection and combustion noise, a robust method is required. Over the last 20 years, numerous methods have been investigated for combustion noise evaluation and analysis. Among these methods, the maximum of the in-cylinder pressure rise rate (MPRR) has been the most intuitive tool for combustion noise assessment [13–17]. In addition, researchers have also demonstrated the existence of an overall relationship between the combustion noise and peak of heat release rate (HRR) [18–20]. It is worth noting that pressure spectrum analysis was of utmost importance in combustion noise assessment. Payri et al. proposed a novel combustion noise assessment method on the basis of in-cylinder pressure spectrum decomposition [14,21,22]. Besides, cylinder pressure levels (CPLs) have been widely used in combustion noise analysis [13,19,23,24], and in our previous work, CPLs was synthesized in several frequency segments to investigate the effects of pilot injection on combustion noise [25]. Unfortunately, the CPLs or synthesized CPLs can reveal the overall noise level only in the frequency domain, and the lack of time-domain information limits combustion noise analysis. However, with the application of signal processing techniques to in-cylinder pressure signals, wavelet transforms (WT) and time-frequency analysis have been proposed to further study the relationship between combustion noise and in-cylinder pressure [23,26–28]. Considering the limitation of CPLs, WT-based methods are regarded as a powerful tool for decomposition of the in-cylinder pressure signal in both the frequency and time domain, which is suitable for use to determine the combustion noise generated by pilot and main combustion in this study. As a classical method in WT, DWT (discrete wavelet transforms) has several drawbacks essentially [29,30]. SWT (stationary wavelet transforms) is proposed to overcome the disadvantages of DWT [31], which could ascertain the signal information in the frequency domain with high precision. Additionally, for the sake of refining the signal information in the high-frequency segment, SWPTs (stationary wavelet packet transforms) are also used as for subtler multiresolution analysis. Therefore, the two robust signal processing methods, namely SWT and SWPT, are utilized to decompose the in-cylinder pressure signal in this work. Besides, the sub-signals of SWT and SWPT are divided into two phases, namely the pilot and main so that the influences of the main and pilot combustion on combustion noise are analyzed individually in different frequency ranges.

Therefore, in this paper, the mechanisms of pilot injection impacts on combustion noise were analyzed on the basis of the experimental data from a single-cylinder direct injection diesel engine operated in high load. The experimental data was measured under a range of pilot injection parameters with an engine speed of 1500 rpm, injection pressure of 125 MPa, indicated mean effective pressure of 1.01 MPa, and exhaust gas recirculation ratio of 20%. At first, the combustion noise was analyzed by CPL. Then, in order to analyze the combustion noise more separately, SWT and SWPT were employed to decompose the in-cylinder pressure in different scales, and the sub-signals were divided into two phases: pilot and main.

2. Experimental Setup

Figure 1 reveals the outline of the experimental system. The engine was a four-stroke diesel engine with water cooling with a common-rail electronic injection system. Table 1 illustrates the fundamental specifications of the engine, and the experimental conditions are listed in Table 2. The experiments were conducted at a fixed speed of 1500 rpm and the indicated mean effective pressure (IMEP) was kept at 1.01 MPa, which is high-load operation conditions for the test engine. The intake pressure and intake temperature were 1.68 bar and 30 °C, respectively. Furthermore, in these experiments, the injection pressure was set at 125 MPa. The swirl ratio and main spray injection timing were fixed at 2.0 and 1° ATDC, respectively. A piezoelectric pressure transducer (Kistler 6052A, Winterthur, Switzerland) was used for the in-cylinder pressure measurement. Fifty cycles of the averaged pressure were utilized of to calculate the HRR.

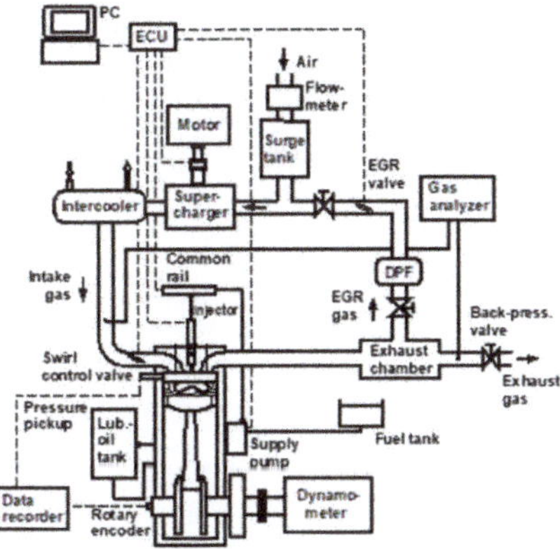

Figure 1. Experimental setup.

Table 1. Specifications of the test engine.

Type	Single-Cylinder, Direct-Injection, Water-Cooled Diesel Engine
Bore × Stroke	85 mm × 96.9 mm
Displacement	550 cc
Compression ratio	16.3
Combustion chamber	Reentrant type (Cavity diameter: 51.6 mm)
Injection system	Common-rail system with a solenoid injector (Max. pressure: 180 MPa) 0.125 mm × 7 hole nozzle (Spray angle: 156°)
Supercharging	External supercharging
EGR system	Low-pressure loop EGR

Table 2. Engine operating parameters.

Injection pressure	125 MPa
Total injection quantity	32 mm^3/stroke
Pilot injection quantity (q_{pilot})	2, 4, 6, 8 mm^3/stroke
Pilot injection timing (θ_{pilot})	−9, −14, −19, −24° ATDC
Main injection timing (θ_{main})	1° ATDC
EGR rate	20%
Swirl ratio	2.0

3. Results and Discussion

As discussed in the following paragraphs, the in-cylinder pressures of the PPCI diesel engine were processed by several methods. Additionally, parametric investigation of the pilot injection quantity and the start of pilot injection were assessed for combustion noise analysis.

3.1. Analysis of In-Cylinder Pressure and Heat Release Rate

The in-cylinder pressures and HRR for the pilot mass of 2, 4, 6, and 8 mm^3 with the pilot injection timing varied from −9 to −24° ATDC in steps of 5 degree (deg) are indicated in Figure 2a–d. It is obvious that as the pilot injection mass becomes larger, the in-cylinder pressure becomes higher, as well as the HRR peaking during pilot injection combustion. This leads to higher in-cylinder temperature before the main spray, and the heat release of premixed combustion during the main spray becomes gentler. Therefore, the first HRR peak during main injection combustion drops as the pilot injection quantity increases. Meanwhile, as the pilot injection mass is increasing and the total quantity of injection fuel is fixed, the decrease of the main injection mass causes a lower HRR peak of the diffusion combustion during main spray combustion (second HRR peak for main combustion). Moreover, with the crank angle of the pilot injection advanced, the HRR of the pilot injection combustion decreases, yet the main spray combustion achieves a high value of HRR.

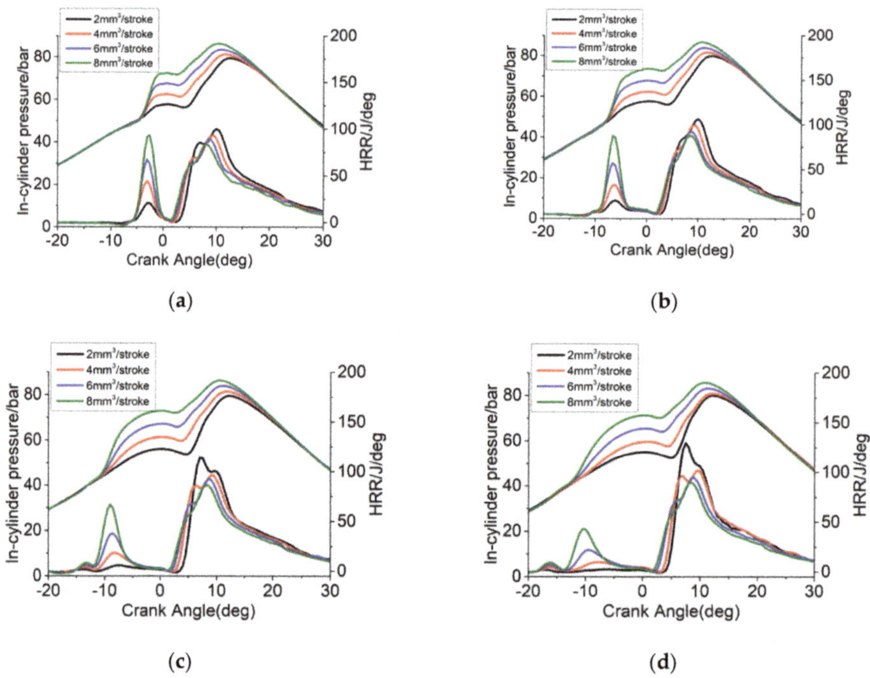

Figure 2. In-cylinder pressure and heat release rates. (**a**) Pilot injection timing = −9° ATDC; (**b**) pilot injection timing = −14° ATDC; (**c**) pilot injection timing = −19° ATDC; (**d**) pilot injection timing = −24° ATDC.

According to the characteristics of HRR, the variations of the in-cylinder pressure rise rate (PRR) are shown in Figure 3a–d. A larger pilot mass results in a higher maxima pressure rise rate (MPRR) during pilot spray combustion and a lower MPRR for main injection combustion. A smaller MPRR for pilot combustion and a larger MPRR for main combustion is observed with earlier pilot injection timing. Particularly, the PRR of main and pilot spray combustion show contrary variations as the amount of pilot injection fuel is changed, so that it is hard to evaluate the combustion noise.

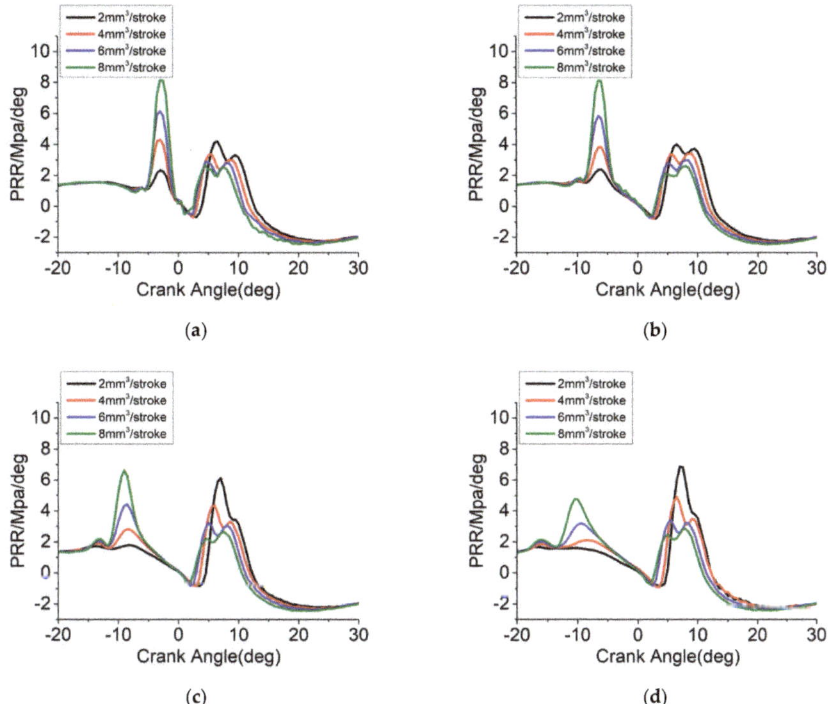

Figure 3. In-cylinder pressure rise rates. (**a**) Pilot injection timing = −9°ATDC; (**b**) pilot injection timing = −14° ATDC; (**c**) pilot injection timing = −19° ATDC; (**d**) pilot injection timing = −24° ATDC.

3.2. Combustion Noise Analysis by Cylinder Pressure Levels

As known, frequency domain information is suitable for the assessment of combustion noise. In this section, fast Fourier transforms (FFT) was applied on measured in-cylinder pressure, so that the in-cylinder pressure data could be transformed to the frequency domain. Then, the sound pressure levels (SPLs) were expressed as cylinder pressure levels (CPLs), which was formulated as follows:

$$CPL = 20 \log(P/P_0),\tag{1}$$

where P_0 and P are the reference sound pressure and in-cylinder pressure, respectively. P_0 was set as 20 µPa in this paper.

The results of CPLs in the situations of pilot injection timing varied from −9 to −24° ATDC and are shown in Figure 4a–d. For the case that the pilot injection timing is −24° ATDC, it is indicated that the increase of pilot injection mass generally results in a drop of the CPL amplitudes. Therefore, in this situation, the evaluation of the combustion noise level lies on the main injection combustion. Nevertheless, when the pilot injection timing comes to −9° ATDC, the variation of the CPLs is relatively complex. That is, in the medium frequency range of 200 to 1000 Hz, a larger amount of pilot injection causes lower CPLs with intense amplitude fluctuations. However, on the contrary, the tendency in the high-frequency segment (higher than 1000 Hz) is that a larger pilot quantity results in higher amplitudes of CPLs. Actually, the combustion excitation imposed on the piston increases the in-cylinder pressure and power output, and is primarily responsible for the CPLs in the mid-frequency segment

(200–1000 Hz). Hence, it can be inferred that the CPLs in the mid frequency segment are mainly influenced by main spray combustion. Besides, as illustrated in Figure 3a, the MPRR of pilot injection occurs in approximately −2° ATDC when the pilot injection timing is −9° ATDC. So, the most violent combustion is induced near the top death center (TDC), causing remarkably high frequency pressure oscillations and radiated noise emission. Accordingly, in the high frequency segment (higher than 1000 Hz), the pilot injection combustion may principally influence the CPLs.

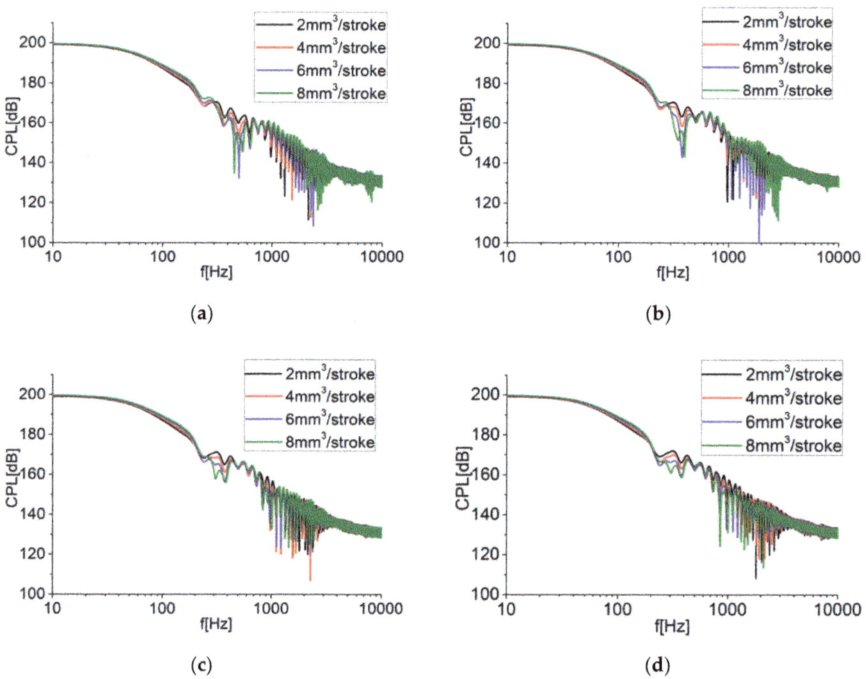

Figure 4. Cylinder pressure levels. (**a**) Pilot injection timing = −9° ATDC; (**b**) pilot injection timing = −14° ATDC; (**c**) pilot injection timing = −19° ATDC; (**d**) pilot injection timing = −24° ATDC.

3.3. SWT of In-Cylinder Pressure Signals

Although CPLs can illustrate the impacts of the pilot and main injection on combustion noise, the analysis proceeds only based on the overall combustion noise level and the effects of the pilot and main injection on combustion noise have not been extracted separately. Therefore, a signal decomposition technique is required in this section. DWT has been proven to be an effective tool for decomposing signals. An approximation and a detailed sub-signal are obtained by the decomposition of a single step DWT, which contains the low and high-frequency components of the original signal. However, the procedures of up and down sampling in DWT lead to aliases of the sub-signals [29]. Therefore, the wavelet energy of the sub-signals, which is regarded as an important quantifier for signal intensity [28], cannot be compared precisely. However, in order to address the issues of DWT, a powerful wavelet transform algorithm called SWT is carried out. Instead of the upsampling and downsampling of the signal in DWT, upsampling of the filter coefficients is employed in SWT, and consequently the translation-invariance is accomplished. Thus, each level of SWT contains the same number of samples as the input. Figure 5 shows the tree structures of five-layer SWT. However, SWT can only decompose approximations, namely low and mid-frequency segments. Therefore, SWT is only suitable for the analysis of combustion noise in low and mid frequency ranges.

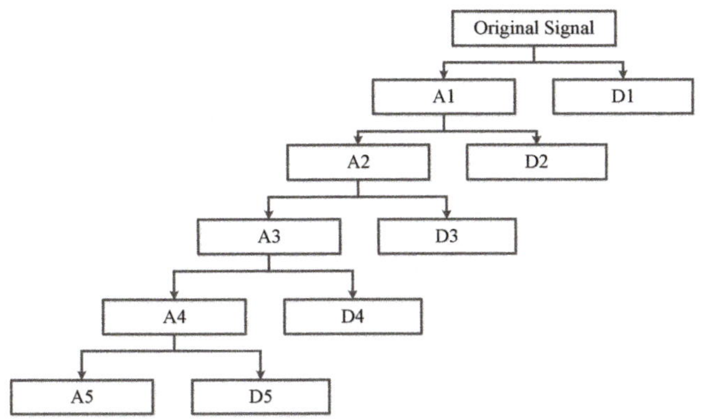

Figure 5. Five-layer SWT.

In this research, using Daubechies 10 as the wavelet, in-cylinder pressure was decomposed into five layers by SWT, including an approximation (A5) and five details (D1, D2, D3, D4, D5). Table 3 shows the frequency band of each sub-signal. The sub-signals, D5 and D4, were selected to investigate the combustion noise in the low and mid frequency segments. Nevertheless, the influences of the pilot and main combustion had to be separated. Thus, the sub-signals were divided into two phases: Pilot and main, as presented in Figure 6. The quantifier of combustion noise intensity was the wavelet signal energy, which is defined as:

$$E_{i,n} = \sum_{k-1}^{j} |x_{n,k}|^2,\qquad(2)$$

where integer n represents a sub-signal serial number, $x_{n,k}$ is the SWT coefficient, integer j is the coefficients number, integer k is the index of coefficients, and integer i is the layer number. Hence, the combustion noise of the pilot and main combustion could be evaluated in the low and mid frequency segments.

Table 3. Frequency band of each sub-signal in stationary wavelet transforms.

Sub-Signals	Frequency Band (Hz)
A5	0–281.25
D5	281.25–562.5
D4	562.5–1125
D3	1125–2250
D2	2250–4500
D1	4500–9000

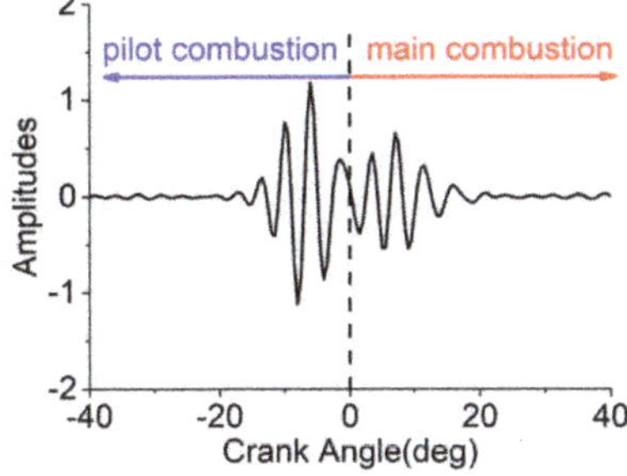

Figure 6. Pilot and main phases of an arbitrary sub-signal.

Figure 7a–d present the combustion noise energy of D4 and D5 during the pilot combustion phase for the condition of various pilot injection timing and pilot injection quantity. As can be seen, the values of combustion noise energy in the low-frequency segment (D5) initially drop, and subsequently rise with the increasing pilot injection quantity. It can be interpreted that a larger amount of pilot injection fuel causes more violent pilot combustion, and partial energy in the low-frequency range may transfer to the high-frequency segment, which will be discussed in the next section. However, when the pilot mass is extremely large (8 mm^3), the combustion noise energy level for all the frequency ranges is largely increased. In addition, advancing the pilot injection timing will also enlarge the combustion noise energy generated by pilot combustion in the low-frequency domain. That is because an extension of the heat release duration of pilot combustion will smooth the process of pilot combustion, which mitigates the mid and high frequency combustion noise energy, as revealed by D4 in Figure 7. Hence, the energy of combustion fluctuation is mainly concentrated on the low-frequency segment. Moreover, for the mid-frequency range, a larger pilot quantity results in higher combustion noise energy. That can be explained by the fact that the PRR plays a key role in the mid-frequency combustion noise [22], and PRR rises as the quantity of pilot injection fuel increases (as illustrated in Figure 3), which causes higher mid-frequency combustion noise energy.

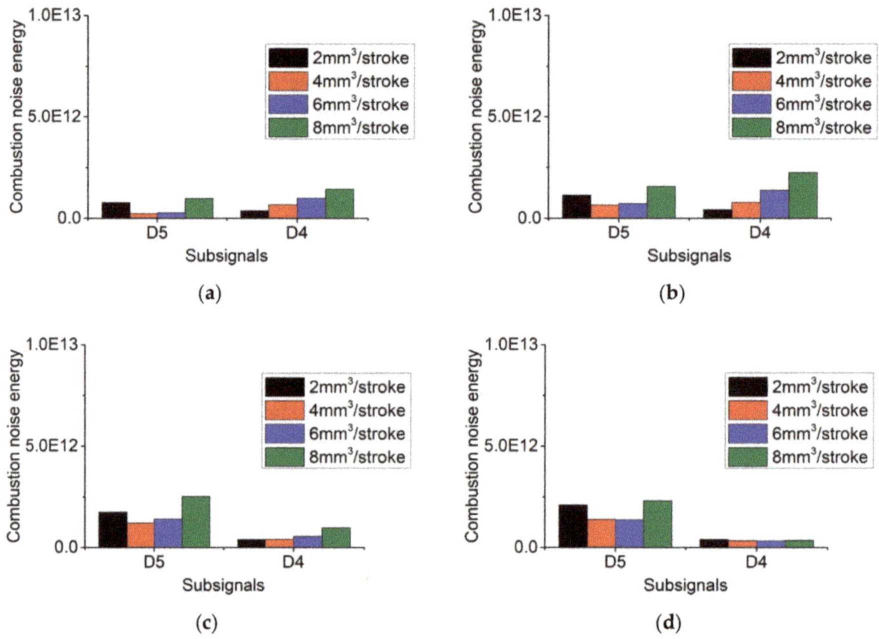

Figure 7. Pilot combustion noise energy of D4 and D5. (**a**) Pilot injection timing = −9° ATDC; (**b**) pilot injection timing = −14° ATDC; (**c**) pilot injection timing = −19° ATDC; (**d**) pilot injection timing = −24° ATDC.

The main combustion noise energy of D4 and D5 is also revealed in Figure 8a–d. The comparisons between various pilot injection parameters indicate clearly that a larger pilot injection quantity (lower main injection quantity) causes lower combustion noise energy during main combustion in both the low and mid-frequency domains. Specifically, in the mid-frequency range, with the pilot injection timing advanced, the main combustion noise energy of the small pilot quantity condition (2 and 4 mm^3) rises as well, but that changes slightly for the large pilot injection timing condition (6 and 8 mm^3). These trends are consistent with the PRR in Figure 3: The PRR during the main combustion phase for the pilot injection quantity of 2 and 4 mm^3 rises greatly as the pilot injection timing advances, however, the changes of the PRR for the large pilot injection quantity are not obvious. In addition, from the comparison with Figures 7 and 8, it can be seen that main spray combustion generates more combustion noise energy than pilot injection for all the pilot injection timings in both the low and mid-frequency ranges. Thus, main injection has a dominant impact on low and mid-frequency combustion noise.

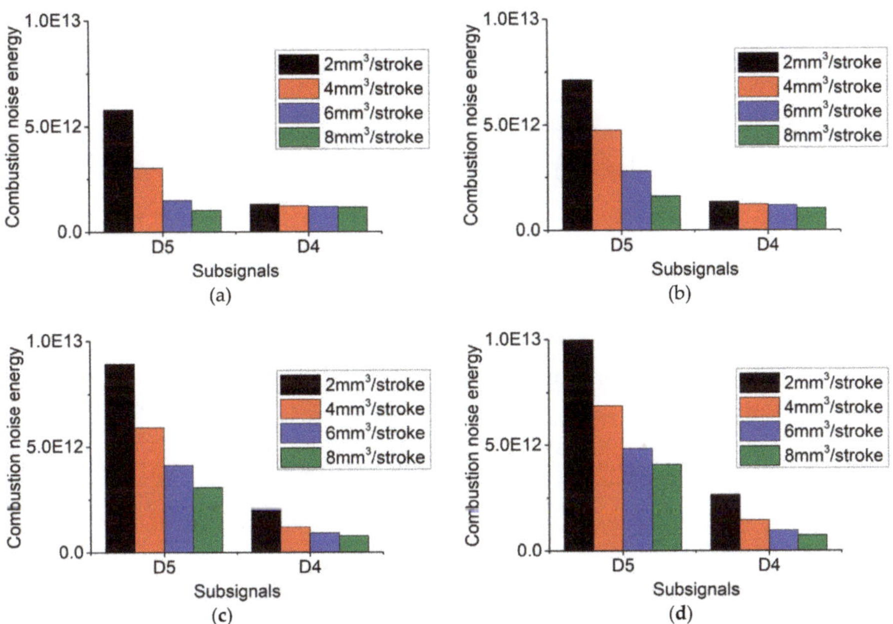

Figure 8. Main combustion noise energy of D4 and D5. (**a**) Pilot injection timing = −9° ATDC; (**b**) pilot injection timing = −14° ATDC; (**c**) pilot injection timing = −19° ATDC; (**d**) pilot injection timing = −24° ATDC.

The main combustion noise energy of D4 and D5 is also revealed in Figure 8a–d. The comparisons between various pilot injection parameters indicate clearly that a larger pilot injection quantity (lower main injection quantity) causes lower combustion noise energy during main combustion in both the low and mid-frequency domains. Specifically, in the mid-frequency range, with the pilot injection timing advancing, the main combustion noise energy of the small pilot quantity condition (2 and 4 mm^3) rises as well, but that changes slightly for the large pilot injection timing condition (6 and 8 mm^3). These trends are consistent with the PRR in Figure 3: The PRR during the main combustion phase for the pilot injection quantity of 2 and 4 mm^3 rises greatly with the pilot injection timing advancing, however, the changes of the PRR for the large pilot injection quantity are not obvious. In addition, from the comparison with Figures 7 and 8, it can be seen that main spray combustion generates more combustion noise energy than pilot injection for all the pilot injection timings in both the low and mid-frequency ranges. Thus, main injection has a dominant impact on low and mid-frequency combustion noise.

3.4. SWPT of In-Cylinder Pressure Signals

From the previous section, the influences of the pilot and main combustion on low and mid-frequency combustion noise were analyzed separately. However, combustion noise in the high-frequency domain plays a key role in radiated noise as well, which should be investigated in detail. According to Figure 5, SWT could only decompose approximations, from which it is hard to obtain the information for high-frequency components. Nevertheless, SWPT can simultaneously separate the approximations and details of the signal, which would be expected to refine the combustion noise energy in the high-frequency segment. Therefore, with Daubechies 10 as the wavelet, SWPT was performed and the in-cylinder pressure signal was decomposed into three layers. Figure 9 shows the tree structure of the three-layer SWPT. Here, $T_{3,n}$ symbolizes sub-signal n in the third layer. The frequency range of each sub-signal is listed in Table 4. The sub-signal, $T_{3,0}$, was not used to investigate high-frequency combustion noise, due to the fact that it only contains information about the low and mid frequency segment.

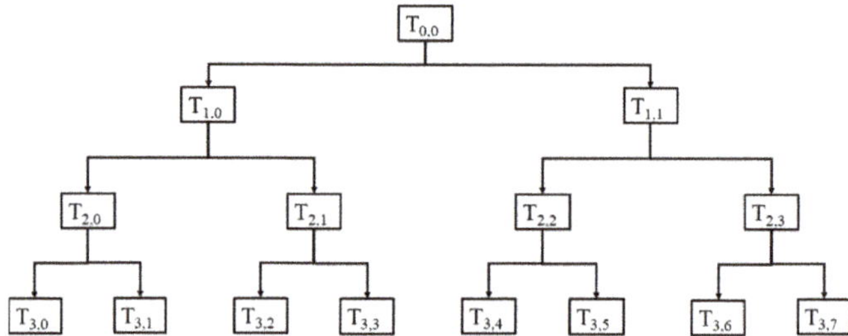

Figure 9. Tree structure of the three-layer stationary wavelet packet transforms.

Table 4. Frequency band of each sub-signal in the SWPT.

Sub-Signals	Frequency Band (Hz)
$T_{3,0}$	0–1125
$T_{3,1}$	1125–2250
$T_{3,2}$	2250–3375
$T_{3,3}$	3375–4500
$T_{3,4}$	4500–5625
$T_{3,5}$	5625–6750
$T_{3,6}$	6750–7875
$T_{3,7}$	7875–9000

In this section, the sub-signals are also divided into two phases just like the previous section, and the pilot and main combustion noise energy in the high-frequency domain were both calculated using the same computational method (Equation (2)). The results are shown in Figures 10 and 11. Obviously, pilot injection influences the combustion noise energy of the sub-signals, $T_{3,1}$ and $T_{3,3}$, whose frequency bands are 1125–2250 Hz and 3375–4500 Hz, respectively. Furthermore, a larger pilot injection quantity results in higher combustion noise energy in general. Additionally, the combustion noise energy level drops with the pilot injection timing advancing, which corresponds to the inference in Section 3.3. Additionally, Figure 11 depicts the main combustion noise energy in the high-frequency domain. It is worthy to note that the main injection only affects the sub-signal, $T_{3,1}$, corresponding to the frequency segment of 1125–2250 Hz. When the pilot injection mass increases, the combustion noise energy of the main injection is progressively reduced. Besides, earlier pilot injection timing causes a larger main combustion noise energy only in the case with a pilot injection quantity of 2 mm^3, and significant changes are not observed for conditions of other pilot masses, which reveals that the influence of the pilot injection timing on combustion noise energy during main combustion is quite small.

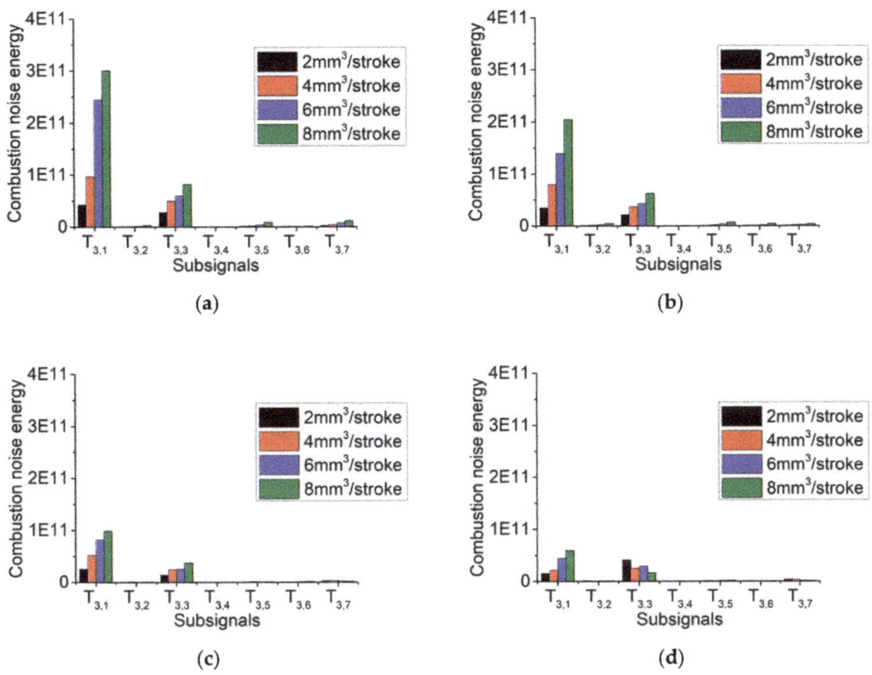

Figure 10. Pilot combustion noise of seven sub-signals. (**a**) Pilot injection timing = −9° ATDC; (**b**) pilot injection timing = −14° ATDC; (**c**) pilot injection timing = −19° ATDC; (**d**) pilot injection timing = −24° ATDC.

Furthermore, for the purpose of determining the significant factor that influences high-frequency combustion noise, comparisons between pilot and main combustion noise energy, namely Figures 10 and 11, were carried out. In the condition of retarded pilot injection timing, just like −9° ATDC, although the pilot mass (2 mm^3) is relatively smaller than the amount of main injection fuel (30 mm^3), the combustion noise energy of pilot injection is nearly equal to that of main injection. In addition, the pilot combustion noise energy is significantly larger than the main combustion noise energy when the pilot quantity increases over 2 mm^3. Considering the results of the comparisons, with the retarded pilot injection timing, the pilot combustion shows remarkable effects on the high-frequency combustion noise. Nevertheless, for the early crank angle of pilot injection, only in the situation of a small pilot mass does the combustion noise energy of main injection dominate. Hence, pilot injection mainly affects the combustion noise in the high-frequency domain.

In addition, the frequency components of the in-cylinder pressure spectrum above 1000 Hz were correlated with the noise radiated [32]. Accordingly, it is feasible to optimize the pilot injection parameters for the mitigation of combustion noise on the basis of the high frequency combustion noise energy analysis. The overall combustion noise energy (pilot and main combustion) of the high frequency segment was calculated, which is shown in Figure 12. It can be seen that a large pilot injection quantity (6, 8 mm^3) should not be combined with late pilot injection timing (−9, −14° ATDC), which leads to excessive combustion noise energy during pilot combustion. Similarly, it is also unreasonable to combine the small pilot injection quantity (2 mm^3) with early pilot injection timing (−24° ATDC), which enlarges the main combustion noise energy. Instead, aiming at suppressing combustion noise energy, It is suitable to select a large pilot mass (6, 8 mm^3) with advanced pilot injection timing (−19, 24° ATDC) and a small pilot quantity (2 mm^3) with retarded pilot injection timing (−9, −14° ATDC).

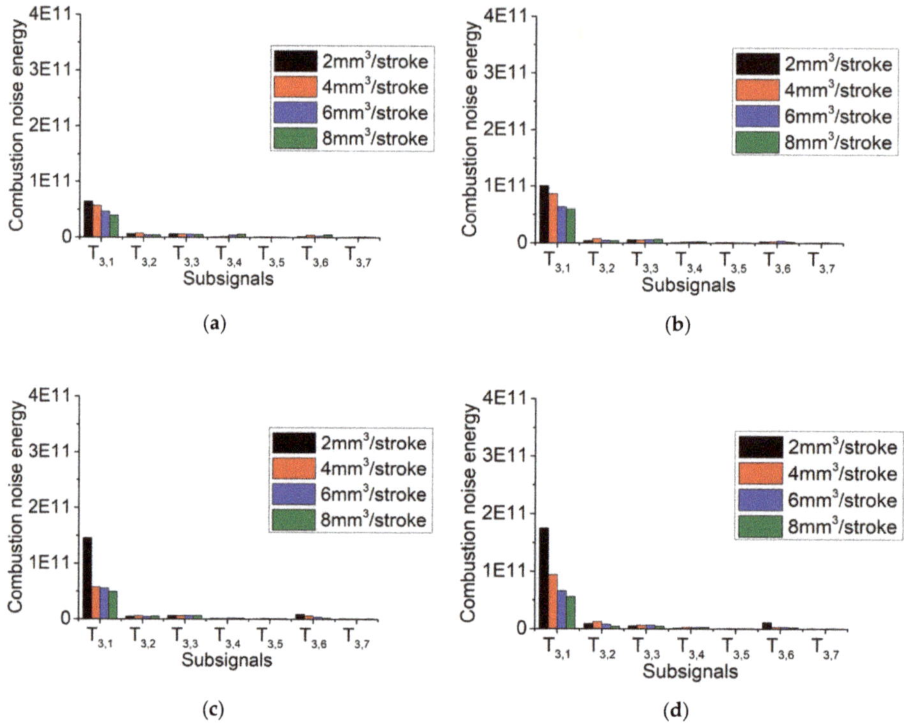

Figure 11. Main combustion noise energy of seven sub-signals. (**a**) Pilot injection timing = −9° ATDC; (**b**) pilot injection timing = −14° ATDC; (**c**) pilot injection timing = −19° ATDC; (**d**) pilot injection timing = −24° ATDC.

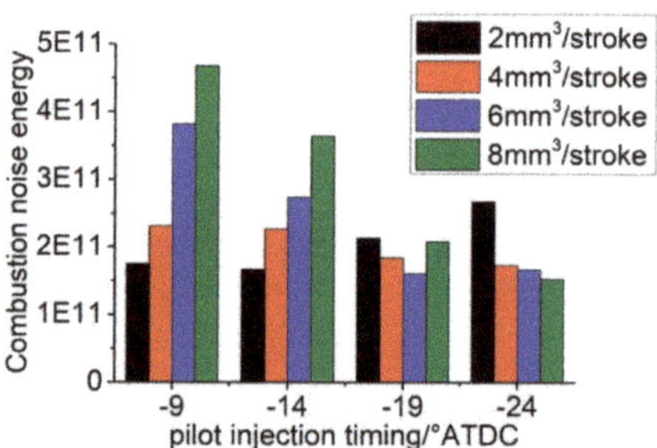

Figure 12. Overall combustion noise energy.

4. Conclusions

In the present study, the influences of pilot injection on combustion noise were analyzed using in-cylinder pressure data measured in a single-cylinder diesel engine. The mechanism of pilot and main injection parameters' effects on combustion noise was captured by means of CPLs, SWT, and SWPT. By means of the analysis and results provided above, conclusions can be stated as follows:

(1) Generally, main injection is the significant factor that affects combustion noise in low and mid frequency segments, and an increase of the pilot mass results in mitigation of combustion noise. Pilot injection plays an essential role in high frequency combustion noise, and larger pilot injection quantities lead to higher combustion noise in the high-frequency domain.

(2) Advancing pilot injection timing will attenuate pilot combustion noise energy in mid and high frequency domains, but will strengthen that in the low-frequency range. Additionally, an earlier pilot injection timing also leads to higher main combustion noise energy in almost the whole frequency spectrum.

(3) From the results of SWPT, pilot and main injection combustion impacts on high-frequency combustion noise are concentrated in the frequency band of 1125–2250 Hz and 3375–4500 Hz, and 1125–2250 Hz, respectively.

(4) According to the view of overall combustion noise energy, a large pilot mass with advanced pilot injection timing and small pilot quantity with retarded pilot injection timing is considered as the optimal pilot injection strategy for the mitigation of combustion noise.

Author Contributions: L.L. proposed the primary research concept. J.D. developed the algorithm and preformed combustion noise analysis. X.C. did the programing work, analyzed the calculated results, and wrote this paper. D.L. performed the experiments and processed the data. X.M. managed the research project.

Funding: This research was funded by the National Natural Science Foundation of China (grant number 51509051), and Marine Power Research and Development Program (grant number DE0302)

Acknowledgments: The experiments presented here had been performed in the Combustion and Power Engineering Laboratory at Kyoto University.

Conflicts of Interest: The author declares no conflict of interest.

References

1. Seamoon, Y.; Changhee, L. Exhaust Gas Characteristics According to the Injection Conditions in Diesel and DME Engines. *Appl. Sci.* **2019**, *9*, 647.

2. Keeler, B.; Shayler, P.J. *Constraints on Fuel Injection and EGR Strategies for Diesel PCCI-Type Combustion*; SAE Technical Paper; SAE International: Troy, MI, USA, 2008.

3. Kanda, T.; Hakozaki, T.; Uchimoto, T.; Hatano, J.; Kitayama, N.; Sono, H. *PCCI Operation with Early Injection of Conventional Diesel Fuel*; SAE Technical Paper; SAE International: Troy, MI, USA, 2005; pp. 584–593.

4. Ge, J.C.; Yoon, S.K.; Kim, M.S.; Choi, N.J. Application of Canola Oil Biodiesel/Diesel Blends in a Common Rail Diesel Engine. *Appl. Sci.* **2017**, *7*, 34. [CrossRef]

5. Ojeda, W.; Zoldak, P.; Espinosa, R.; Kumar, R. *Development of a Fuel Injection Strategy for Diesel LTC*; SAE Technical Paper; SAE International: Troy, MI, USA, 2008.

6. Belgiorno, G.; Dimitrakopoulos, N.; Di Blasio, G.; Beatrice, C.; Tunestål, P.; Tunér, M. Effect of the engine calibration parameters on gasoline partially premixed combustion performance and emissions compared to conventional diesel combustion in a light-duty Euro 6 engine. *Appl. Energy* **2018**, *228*, 2221–2234. [CrossRef]

7. Belgiorno, G.; Dimitrakopoulos, N.; Di Blasio, G.; Beatrice, C.; Tuner, M.; Tunestal, P. *Parametric Analysis of the Effect of Pilot Quantity, Combustion Phasing and EGR on Efficiencies of a Gasoline PPC Light-Duty Engine*; SAE Technical Paper; SAE International: Troy, MI, USA, 2017.

8. D'Ambrosio, S.; Gaia, F.; Iemmolo, D.; Mancarella, A.; Salamone, N.; Vitolo, R.; Hardy, G. *Performance and Emission Comparison between a Conventional Euro VI Diesel Engine and an Optimized PCCI Version and Effect of EGR Cooler Fouling on PCCI Combustion*; SAE Technical Paper; SAE International: Troy, MI, USA, 2018.

9. Torregrosa, A.J.; Broatch, A. Sensitivity of combustion noise and NOx and soot emissions to pilot injection in PCCI Diesel engines. *Appl. Energy* **2013**, *104*, 149–157. [CrossRef]

10. Bo, Y.; Long, W.; Le, N.; Ke, Z. Effects of pilot injection timing on the combustion noise and particle emissions of a diesel/natural gas dual-fuel engine at low load. *Appl. Therm. Energy* **2016**, *102*, 822–828.

11. Wang, P.; Song, X.; Xue, D.; Zhou, H.; Ma, X. *Effect of Combustion Process on DI Diesel Engine Combustion Noise*; SAE International: Troy, MI, USA, 2007.

12. Alok, W.; Francesco, P.; Richard, P.; Alberto, V.; Stephen, B.; Kan, Z.; Paul, C. *Miles. Experimental and Numerical Investigations of Close-Coupled Pilot Injections to Reduce Combustion Noise in a Small-Bore Diesel Engine*; SAE Technical Paper; SAE International: Troy, MI, USA, 2010.

13. Broatch, A.; Margot, X.; Novella, R.; Gomez-Soriano, J. Combustion noise analysis of partially premixed combustion concept using gasoline fuel in a 2-stroke engine. *Energy* **2016**, *107*, 612–624. [CrossRef]

14. Payri, F.; Broatch, A.; Margot, X.; Monelletta, L. Sound quality assessment of diesel combustion noise using in-cylinder pressure components. *Meas. Sci. Technol.* **2009**, *20*, 1–12. [CrossRef]

15. Broatch, A.; Margot, X.; Novella, R.; Gomez-Soriano, J. Impact of the injector design on the combustion noise of gasoline partially premixed combustion in a 2-stroke engine. *Appl. Therm. Eng.* **2017**, *119*, 530–540. [CrossRef]

16. Torregrosa, A.J.; Broatch, A.; Novella, R.; Mónico, L.F. Suitability analysis of advanced diesel combustion concepts for emissions and noise control. *Energy* **2011**, *36*, 825–838. [CrossRef]

17. Giakoumis, E.G.; Rakopoulos, D.C.; Rakopoulos, C.D. Combustion noise radiation during dynamic diesel engine operation including effects of various biofuel blends: A review. *Renew. Sustain. Energy Rev.* **2016**, *54*, 1099–1113. [CrossRef]

18. Russell, M.F.; Haworth, R. *Combustion Noise from High Speed Direct Injection Diesel Engines*; SAE Technical Paper; SAE International: Troy, MI, USA, 1985.

19. Zhang, Q.; Hao, Z.; Zheng, X.; Yang, W. Characteristics and effect factors of pressure oscillation in multi-injection DI diesel engine at high-load conditions. *Appl. Energy* **2017**, *195*, 52–66. [CrossRef]

20. Russell, M.F.; Cavanagh, E.J. *Establishing a Target for Control of Diesel Combustion Noise*; SAE Technical Paper; SAE International: Troy, MI, USA, 1979.

21. Payri, F.; Broatch, A.; Tormos, B.; Marant, V. New methodology for in-cylinder pressure analysis in direct injection diesel engines-application to combustion noise. *Meas. Sci. Technol.* **2005**, *16*, 540–547. [CrossRef]

22. Torregrosa, A.J.; Broatch, A.; Martín, J.; Monelletta, L. Combustion noise level assessment in direct injection Diesel engines by means of in-cylinder pressure components. *Meas. Sci. Technol.* **2007**, *18*, 2131–2142. [CrossRef]

23. Bhat, C.S.; Meckl, P.H.; Bolton, J.S.; Abraham, J. Influence of fuel injection parameters on combustion-induced noise in a small diesel engine. *Int. J. Eng. Res.* **2012**, *13*, 130–146. [CrossRef]

24. Hao, Z.; Zhang, Q.; Zheng, X.; Yang, W. Prediction of the overall noise level of DI diesel engine with calculated in cylinder pressure. *Meas. Sci. Eng.* **2017**, *39*, 2421–2432. [CrossRef]

25. Liu, L.; Fei, H.; Du, J. Analysis of pilot injection effects on combustion noise in PPCI diesel engines. In Proceedings of the ASME 2016 Internal Combustion Engine Division Fall Technical Conference, Greenville, SC, USA, 9–12 October 2016.

26. Qiao, X.; Hou, J.; Wang, Z. Knock Investigation of a Direct Injection-Homogeneous Charge Compression Ignition Engine Fueled with Dimethyl Ether and Liquefied Petroleum Gas. *Energy Fuels* **2009**, *23*, 2006–2012. [CrossRef]

27. Desantes, J.M.; Torregrosa, A.J.; Broatch, A. *Wavelet Transform Applied to Combustion Noise Analysis in High-Speed DI Diesel Engines*; SAE Technical Paper; SAE International: Troy, MI, USA, 2001.

28. Hou, J.; Qiao, X.; Wang, Z.; Liu, W.; Huang, Z. Characterization of knocking combustion in HCCI DME engine using wavelet packet transform. *Appl. Energy* **2010**, *87*, 1239–1246. [CrossRef]

29. Bao, W.; Zhou, R.; Yang, J.; Yu, D.; Li, N. Anti-aliasing lifting scheme for mechanical vibration fault feature extraction. *Mech. Syst. Signal Process.* **2009**, *23*, 1458–1473. [CrossRef]

30. Yang, J.; Park, S.T. An anti-aliasing algorithm for discrete wavelet transform. *Mech. Syst. Signal Process.* **2003**, *17*, 945–954. [CrossRef]

31. Christen, U.; Vantine, K.; Chevalier, A.; Moraal, P.; Scholl, D. A wavelet-based combustion noise meter. In Proceedings of the IEEE International Symposium on Intelligent Control, Munich, Germany, 4–6 October 2006; pp. 533–538.

32. Dernotte, J.; Dec, J.; Ji, C. *Investigation of the Sources of Combustion Noise in HCCI Engines*; SAE Technical Paper; SAE International: Troy, MI, USA, 2014.

Article

Effect of Different Acoustic Parameters on NOx Emissions of Partially Premixed Flame

Kai Deng, Mingxiao Wang, Zhongliang Shen, Yanjun Hu and Yingjie Zhong *

Institute of Energy and Power Engineering, Zhejiang University of Technology, Hangzhou 310014, China; dkai@zjut.edu.cn (K.D.); wmxgood@126.com (M.W.); superman_panda@163.com (Z.S.); huyanjun@zjut.edu.cn (Y.H.)
* Correspondence: zhong_yingjie@zjut.edu.cn; Tel.: +86-0571-88320650

Received: 9 February 2019; Accepted: 1 April 2019; Published: 10 April 2019

Featured Application: The results of the research can be used in the design of low NOx pulsating burners to determine the corresponding frequency and amplitude.

Abstract: The effects of acoustic frequency (f)/0–400 Hz and amplitude (A)/0–1400 Pa on nitrogen oxides (NOx) emissions of a partially premixed flame were investigated experimentally. The mechanism of NOx emissions was analyzed by the evolution of the vortex, which was shown by particle image velocimetry (PIV). From the relationship of NOx emission index (EINOx) and acoustic parameters, it was concluded that a critical frequency (f_c) from 170 Hz to 190 Hz appeared. When the frequency was less than f_c, EINOx decreased linearly with an increase in amplitude. The flame length became shorter, which led to a decrease in the global residence time, and hence, a reduction in reaction time for NOx. However, a direct proportional relationship between EINOx and amplitude was not found when the frequency was larger than f_c. Based on PIV particle scattering images, with an increase of the acoustic frequency, the effects of the acoustic field on the flame base became less significant, but the flame length and reaction space of NOx were gradually increased.

Keywords: acoustic; frequency; NOx emission index; flame length; global residence time

---

1. Introduction

There is a growing interest in alternative technologies for energy generation applications, which can range from the emerging of hybrid solar receiver–combustor (HRC) technologies [1,2] to simpler fuel blending approaches, such as the co-firing of biomass feed stocks with coal for power generation purposes [3,4], because of increasing pressure to conserve ever diminishing fossil fuel resources and to reduce pollutant emissions. Excessive emissions of nitrogen oxide (NOx) from power generation applications is a severe problem that can lead to increased occurrences of acid rain and photochemical smog pollution. This has also led to research and investment into low NOx combustion technology [5–7], such as pulsed combustion technology. Pulse combustion technology is one of the highly combustion efficient and low NOx combustion technologies that have been demonstrated in engineering application. Pulse/pulsating combustion refers to the unstable combustion process under the effects of sound field or pulsating flow, while the steady-state variables such as temperature, pressure, air velocity, and heat release rate fluctuate periodically with time in the combustion zone. This phenomenon about heat excited acoustic oscillations was firstly observed by Rijke, who discovered that strong acoustic oscillations occurred when a heated metallic grid was positioned in the lower half of a vertical tube opened at both ends [8]. Rayleigh stated that the sound could be excited in a closed volume only if the pressure oscillation is in phase with the heat-release oscillation [9]. Based on Rayleigh's criterion, different kinds of pulse combustors, including the Rijke, Hemholtz, Schmidt pulse combustor, were developed and widely used in power and heat generation devices, including boilers, water heaters, industrial furnaces, and pulsating dryers for a long time [10–14].

Previous studies showed that the reductions in NOx could occur for lean premixed flames under acoustic excitation [13–17]. Lean premixed flames are more sensitive to flow perturbations and tend to induce self-excited thermoacoustic oscillations. However, in terms of NOx reduction, the effects of acoustic excitation on partially premixed flame and diffusion flame could potentially be more effective [18–21], as the lean premixed flame inherently has very low NOx emissions. Partially premixed flames have dual flame dynamics characteristic of lean premixed flame and diffusion flame. Therefore, it has engineering and academic value to investigate partially premixed flame, which is also widely used in boilers and industrial furnaces.

Understanding the characteristics of partially premixed combustion systems under acoustic excitation, in terms of NOx emissions, is still limited because the complexity involved in the interaction between upstream disturbances and reacting flow fields [22]. Kim et al. proposed that the reduction of flame length was the main reason for the decrease of the NOx emission index (EINOx), by investigating flames under acoustic frequencies of 256.9 Hz, 400 Hz, 770.6 Hz, and 514 Hz [23]. In conclusion, interaction between acoustic excitation and flame were demonstrated to lead to reduced NOx emission for premixed flame and diffusion flame [24,25], but the effects of acoustic parameters including frequency and amplitude on NOx emissions for partially premixed flame were rarely reported. Recently, Shen [26], Deng [27,28], and Zhong [29] focused on the NOx emission of partially premixed flame in Rijke tubes with an acoustic frequency of 110 Hz, and observed that the reduction of high temperature area and the decrease of residence time were major causes for NOx reduction. However, the exact relationships between amplitude, frequency, and EINOx have not been obtained. Present work focuses on the effects of acoustic frequency (0–400 Hz) and amplitude (0–1000 Pa) on NOx emissions with partially methane/air premixed flame. The correlation of the NOx emission results to the changes in flame properties, such as flame length, flame shape, and global residence time, in response to variations in the controlling parameters was analyzed.

2. Materials and Methods

2.1. Experimental Setup

The experimental system consisted of the combustor and a sound generating device, as seen in Figure 1. It is an atmospheric, partially premixed combustor, consisting of an annular jet burner with a coflow tube and an optically accessible quartz combustion chamber. The improved Burke–Schumann burner, which was referenced in Hassan et al. [21] and Illingworth et al. [30], was designed. This Bunsen burner was installed in the center of a coflow tube (1600 mm $\times$ Φ140 mm) with two rectangular-shaped quartz windows (90 mm $\times$ 400 mm). The burner had an inner diameter, d_F, of 18.0 mm and a lip thickness of 2.0 mm. The inner diameter of the coflow tube, d_C, was 150 mm. The coflow air jet was injected parallel to the central jet of the methane mixture to provide enough oxygen for the combustor. The coflow air was maintained at a sufficiently low velocity, less than 0.1 m/s, so as not to disturb the flow field in the combustion area. Porous medium plates were set in the coflow tube to ensure a uniform airflow. A loud speaker (A0208A, JBL, Northridge, CA, USA) with a signal generator (AFG-2225, GW-INSTEK, Taiwan, China) and amplifier (E-470, Accuphase, Yokohama, Japan) were placed at the bottom of the combustor as a forced acoustic source to provide adjustable acoustic frequencies and amplitudes (see Figure 1).

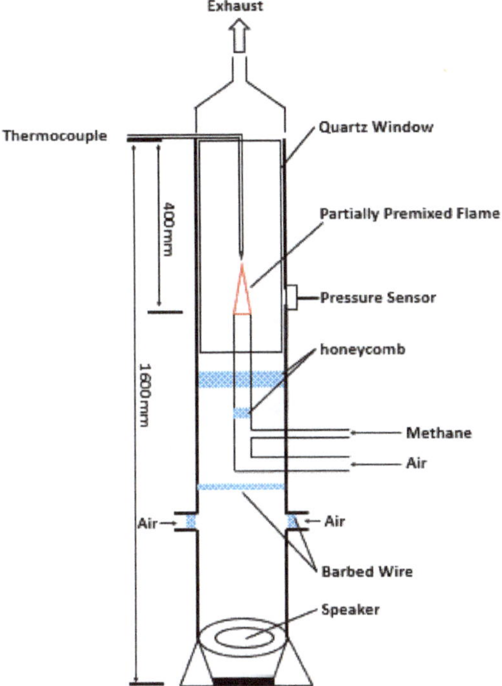

Figure 1. Experimental system.

2.2. Instrumentation and Acquisition

A pressure sensor (CYG1409T, Shuang Qiao, Kunshan, China), which was installed at the flame location and perpendicular to the cylinder's axis, was employed to measure the acoustic frequency and amplitude in the cylinder. The pressure sensor had a measuring range of 0–3000 Pa and a measuring accuracy of 1%. The results were obtained after fast Fourier transform (FFT) by a data acquisition system (PXI-1033, NI-DAQ).

As shown in Figure 2, the flame length was measured by processing the images taken by a high-speed digital camera (MVC1000DAF-GE1000, Microview, Beijing, China). A digital single-lens reflex (DSLR) camera (Canon, EOS 550d, Tokyo, Japan) was used to observe flame structures and flame surface wrinkling. The exposure time for the high-speed camera was set at 1/1000 s. To visualize the flow field, a particle image velocimetry (PIV) system (Dantec Dynamics, Copenhagen, Denmark) was utilized to describe the flame response under acoustic excitation. The PIV system consists of a YAG laser generator with double 200 mJ pulses at 532 nm and a Dantec Dynamics Flow Sense 4 M digital camera with 2048 × 2048 pixels. Al_2O_3 seeding particles about 2.5 μm in diameter were used to seed the coflow air tube.

A 2D cross-correlation FFT algorithm incorporating a multi-pass approach was used to analyze corresponding particle image pairs. The initial pass of the routine analyzed the images using 32 by 32 pixel interrogation regions, and the final pass used interrogation regions of 16 by 16 pixels. A 3-point Gaussian peak search algorithm was used to calculate sub-pixel displacement and a 50% Nyquist overlap of the regions was incorporated.

Phase-resolved PIV measurement was triggered by the signal from the pressure sensor. A compromise on the number of PIV image pairs was made for the phase-resolved PIV experiments. This was because the sampling rate was low due to phase jitter and a longer duration of the experiments would exacerbate the problem associated with deposition of seed particle on the enclosure wall. The

100 image pairs collected were deemed sufficient to produce stable mean velocities for examining the evolution of the flow field structures of partially premixed flame.

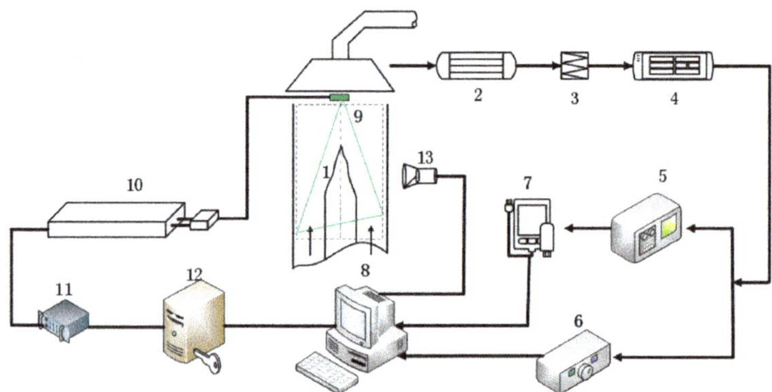

Figure 2. Schematic diagram of the experimental setup used for measurements (1. Flame; 2. Insulation pipe; 3. Filter; 4. Preprocessor; 5. Photon flue gas analyzer; 6. NOx Analyzer; 7. Data Acquisition; 8. Computer; 9. Laser; 10. Laser generator; 11. Synchronizer; 12. Controller; 13. Camera).

The flame length was obtained by flame area image processing [31–33]. The flame contour and flame length were obtained by selecting the threshold value in the flame area image, which was captured by high-speed camera. The average flame length was calculated based on every 2000 photos. An electrochemical NOx sensor (CLD64, ECO Physics, Duernten, Switzerland) was located after the flue gas collection device, and the collection device used was made with reference to Costa's work but with slight modifications for adapting the size of the combustor [34]. Another infrared fuel gas analyzer (Photon, Madur, Vienna, Austria) was used to measure components of the exhaust gas. Measurement precision of components are shown in Table 1.

Table 1. Measurement range and measurement accuracy of the gas analyzer for different gas components.

Component	Measurement Instrument	Measurement Range	Measurement Accuracy
O_2		0–25%	±0.2% or 2% rel
CO_2	MADUR PHOTON infrared gas	0–25%	±0.1% or 3% rel
CO	analyzer	0–20,000 ppm	±3 ppm or 3% rel
CH_4		0–5%	±0.1% or 3% rel
NO	ECO Physics CLD60	0–0.5 to 0–100 ppm	2 ppb
NO_2	electrochemical NOx analyzer	0–0.5 to 0–100 ppm	2 ppb

The NOx emission index (EINOx) was employed to characterize NOx emissions, which were defined as NOx per unit mass fuel. Assuming that all the carbons in the fuel are completely converted into CO_2 and CO, EINOx is defined as:

$$\text{EINOx} = \left(\frac{x_{NOx}}{x_{CO} + x_{CO_2}}\right) \cdot \left(\frac{mMW_{NOx}}{MW_F}\right). \tag{1}$$

Here, x_i is the mole fraction, m is the number of carbon atoms in the fuel, and MW_i is molar mass.

According to the measurement error transfer formula, the arithmetic average error of EINOx is:

$$\Delta\text{EINOx} = \left(\frac{\Delta x_{NOx}}{x_{CO} + x_{CO_2}} + \left(-\frac{\Delta x_{CO} \cdot x_{NOx}}{(x_{CO} + x_{CO_2})^2}\right) + \left(-\frac{\Delta x_{CO_2} \cdot x_{NOx}}{(x_{CO} + x_{CO_2})^2}\right)\right)\left(\frac{xMW_{NOx}}{MW_F}\right). \tag{2}$$

The EINOx deviation was 0.001–0.005 g/kg, which was less than 1% of experimental values, which means the experimental data was accurate.

2.3. Experimental Conditions

The primary equivalence ratios (Φ) of 2 and 3 were selected for comparison because the partially premixed flame would not flash back or blow off under acoustic excitation at these equivalence ratios. Methane and air flow rates were controlled by the mass flow controller (Sheng Ye, SY-93) with precision of approximately 1%. Experimental conditions are summarized in Table 2.

Table 2. Experimental conditions.

Experimental Conditions	CH$_4$ (mL/min)	Air (mL/min)	Φ	Re	Coflow Air (mL/min)	f (Hz)	A (Pa)
Case 1	1000	5000	2	422.1			
Case 2	1500	5000	3	476.8	10,000	0–400	0–1400
Case 3	1500	7500	2	663.1			

The mean velocities of premixed fuel in the combustor were 0.393 m/s, 0.5895 m/s, and 0.426 m/s corresponding to case 1, case 2, and case 3, respectively. The average velocity of coflow air was 0.00964 m/s; therefore, the influence of the coflow air on the flame structures and flame surface wrinkling could be ignored. The Reynolds number ranged from 422.1 to 663.1, which belonged to laminar combustion without acoustic excitation. For laminar flame, the relationship between NOx and the flow field induced by acoustic excitation was more clearly. Comparison of NOx emission characteristics with equivalence ratios and Reynolds numbers were conducted.

The ranges of acoustic frequency and acoustic amplitude were selected between 0–400 Hz and 0–1400 Pa. Firstly, pulsed combustors tend to operate at low frequencies, usually around 100 Hz [17]. Secondly, based on the experimental results of the premixed flame under acoustic excitation, Bourehla [35] found the flame shape changed significantly within the frequency range of 100 Hz to 220 Hz. In addition, the flame structure also changed significantly following an increase in the amplitude. Furthermore, the acoustic response of the speaker had a good behavior between 0–400 Hz and 0–1400 Pa. The maximum of acoustic amplitude was 1400 Pa, and pressure amplitude p'_n could be converted to velocity amplitude u'_n by conversion relations [36,37]:

$$p'_n / u'_n = \rho_m c. \tag{3}$$

Herein n is acoustic mode; ρ_m is density (=0.946 kg/m^3); c is sonic speed (=387.23 m/s). As a result, the maximum of acoustic velocity amplitude was 3.8218 m/s.

2.4. Acoustic Characteristics of Experimental System

Acoustic amplitude and frequency were adjustable in this experiment. The measured results of 180 Hz with 579.04 Pa acoustic are shown in Figure 3; no obvious signal noise was apparant. Therefore, reliability of the acoustic excitation was verified.

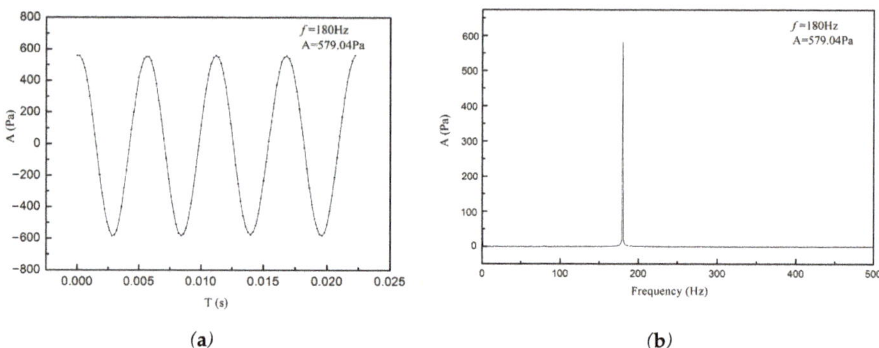

Figure 3. Acoustic pressure of the experimental system; (**a**) Time domain; (**b**) Frequency domain

3. Results and Discussion

3.1. The Relationship between EINOx with Acoustic Amplitude

The NOx emissions for partially premixed flame under acoustic excitation are plotted in Figure 4. The linear relationship between acoustic amplitude and EINOx were presented at a fixed frequency. Under three different premixed gas parameters, the lines displayed a negative slope when the frequency was less than 180 Hz, while a positive slope appeared when the frequency was more than 190 Hz. Therefore, a critical frequency (f_c) was defined between 170 Hz and 190 Hz, which was consistent with Delabroy's results from a liquid boiler [20].

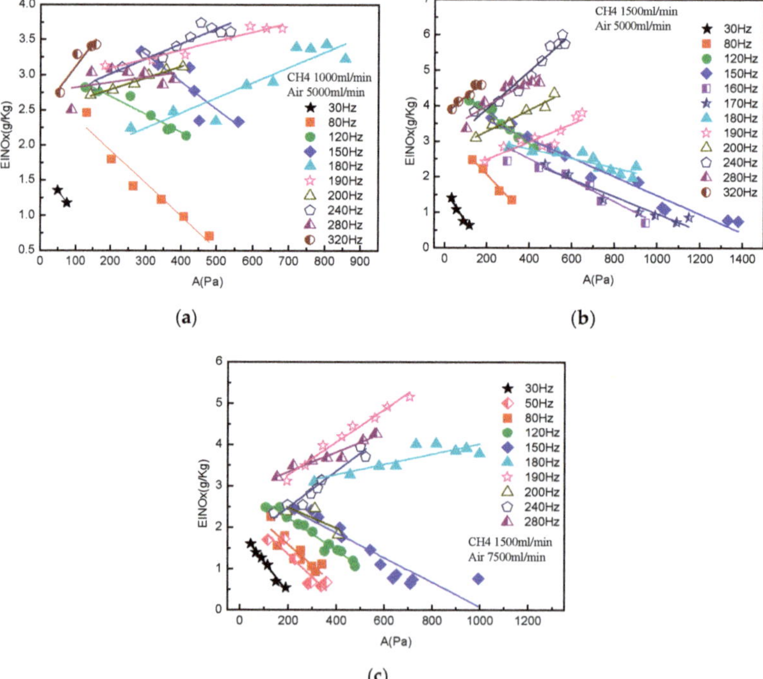

Figure 4. The relationship between acoustic parameters and NOx emission index (EINOx) for partially premixed flame; (**a**) Case 1; (**b**) Case 2; (**c**) Case 3.

EINOx and amplitude had linear relationships in each case. EINOx reduced 15% to 30% with the increase of amplitude by every 100 Pa when frequency was less than f_c. Furthermore, the slope in this region ranged from −0.009 to −0.003. In contrast, the linear relationships were reversed when the frequency was greater than f_c; the positive slopes were also observed to range from 0.001 to 0.006. In a word, EINOx decreased with the increase of acoustic amplitudes while frequency was less than f_c, and increased while the frequency was greater than f_c.

3.2. The Relationship between EINOx and Acoustic Frequency

The effects of acoustic frequency on NOx emissions are shown in Figure 5. There was a clear linear relationship between EINOx and frequency when frequency was less than f_c, as seen in Figure 5. The positive slope of the linear relationship between EINOx and acoustic frequency ranges from 0.001 to 0.003. However, the relationship between the frequency and EINOx is not clear when $f > f_c$. This is related to the flame response to acoustic excitation at different frequencies.

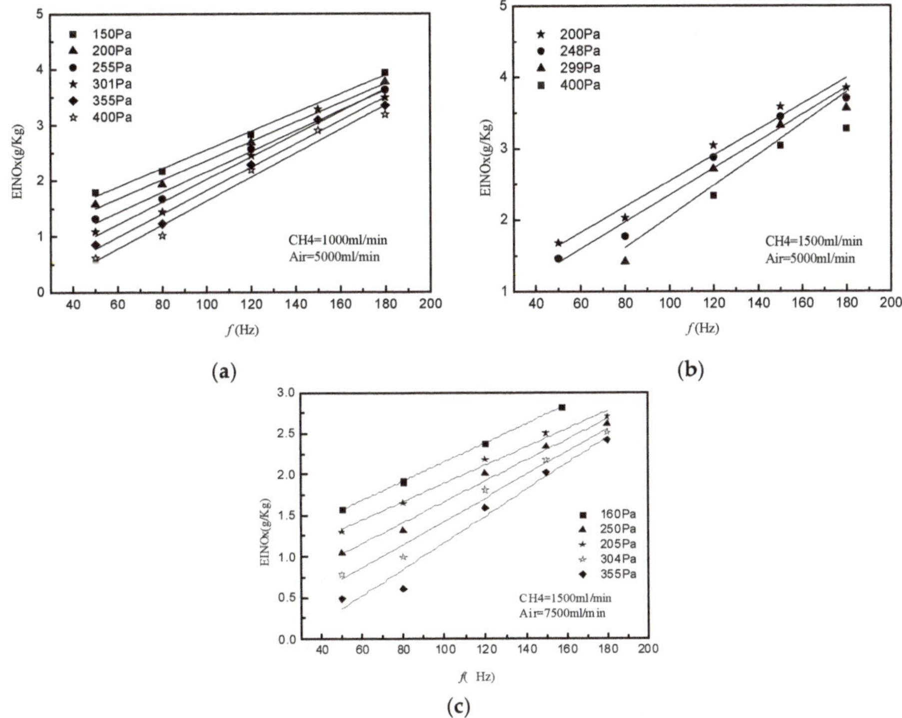

Figure 5. NOx emission characteristics with acoustic amplitudes (0–180 Hz) (**a**) Case 1; (**b**) Case 2; (**c**) Case 3.

3.3. Flame Length and Flame Surface Wrinkling

The flame images were taken by a Canon CMOS (Complementary Metal Oxide Semiconductor) camera to show the flame structures under the different frequencies. In Figure 6, the images are illustrated for case two of partially premixed flames with a range of acoustic frequency from 80 Hz to 280 Hz based on time-average images. Flame surface wrinkles were quite different with acoustic frequencies at a fixed acoustic amplitude of 345 ± 10 Pa. The effects of acoustic forcing on flame wrinkling at 80 Hz to 150 Hz were much stronger than 200 Hz to 280 Hz, which was in good agreement with the results in the change of NOx emissions with acoustic frequency.

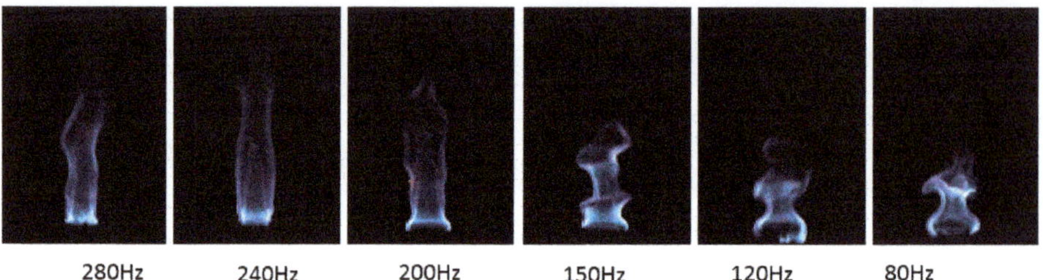

Figure 6. Flame surface wrinkling under different frequencies (345 ± 10 Pa).

Flame length (L_f) can quantitatively describe the relationship between acoustic parameters and EINOx for jet flame. Compared with Figures 4 and 7, this was consistent with the relationship between acoustic amplitude and NOx. Flame length decreased visibly with a gradually increase of acoustic amplitude when the frequency was less than 170 Hz. There was also a good linear correlation between the flame length and the acoustic amplitude when the frequency was less than 170 Hz. However, there was no obvious regularity between flame length and acoustic amplitude when the frequency was larger than 170 Hz.

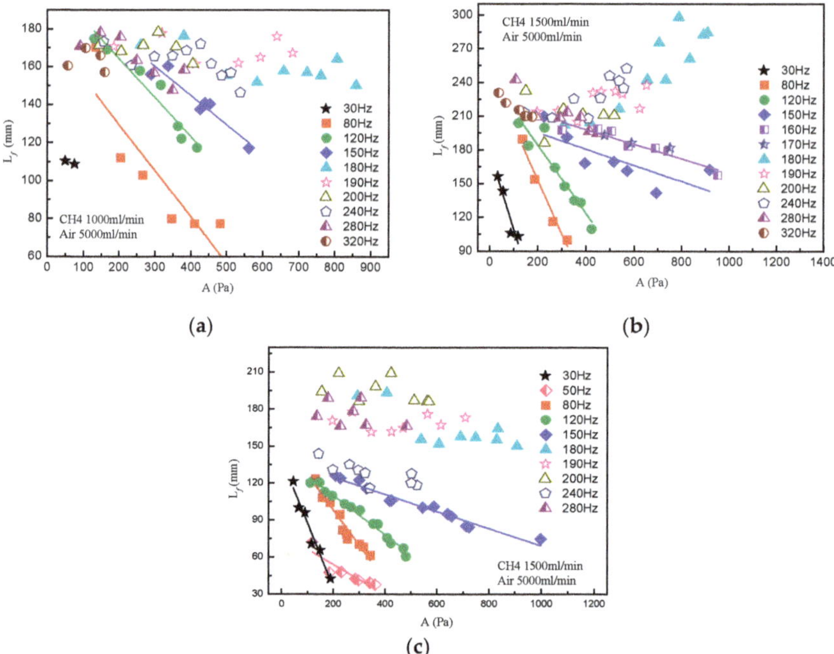

Figure 7. Flame length characteristics with different acoustic parameters; (**a**) Case 1; (**b**) Case 2. (**c**) Case 3.

3.4. Mechanism for Interaction between EINOx and Vortex

To visualize the interaction between flame and vortex, Mie scattering images and velocity fields were obtained by the PIV system to display vortex structure in the partially premixed flame with acoustic parameters of 110 Hz and 321 Pa for case two. In Figure 8, the white circular scattering image indicated that a vortex shape was generated near the exit of the nozzle by acoustic excitation. The vortex grew, left the flame root, and then moved to the downstream zone of the partially premixed flame, which could be observed from the velocity contour. It was thought that the periodic acoustic excitation enhanced the mixing of fuel and air, and reduced the flame length. Research results showed that when the flame length reduced, the reaction region of NOx shortened, reaction rate strengthened, and NOx emission decreased [23,38].

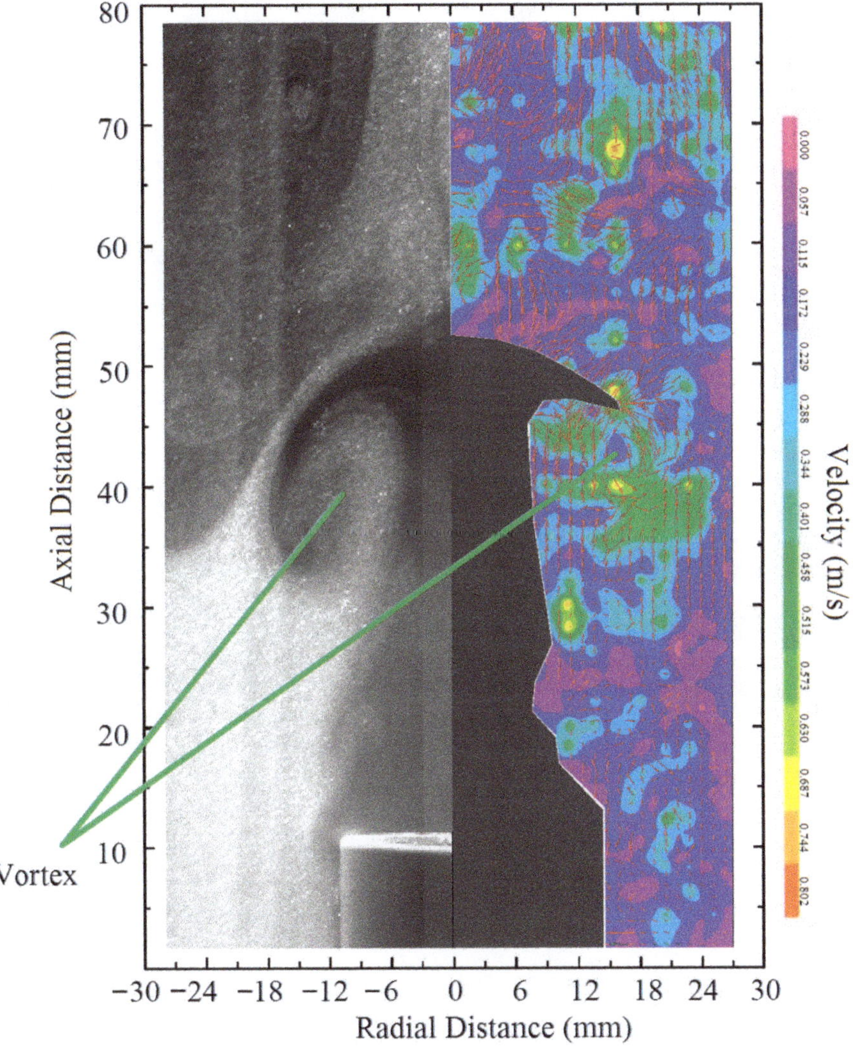

Figure 8. Vortex structure in partially premixed flame under acoustic excitation.

The interaction between vortex and flame was the important mechanism for reduction of NOx in partially premixed flame with acoustic excitation. Figure 9 shows the interaction of flame–vortex under the acoustic frequency range from 80 Hz to 280 Hz in a fixed amplitude at 241 ± 10 Pa. It can be observed that the intensity and position of the vortex showed different characteristics with acoustic frequency. The vortex changed from upstream to downstream with an increase in frequency from 80 Hz to 200 Hz at the same phase. Vortex diameter became larger and the vortex rolled up with surrounding air to the downstream when the acoustic frequency increased. The flame–vortex interaction caused the flame base to broaden when the frequency was less than f_c (f_c^-), although this broadening effect did not reappear when the frequency was greater than f_c (f_c^+). This corresponded to an increase of NOx emissions induced by the weakness of the mixing effect. This was the reason that EINOx for f_c^- was much less than for f_c^+.

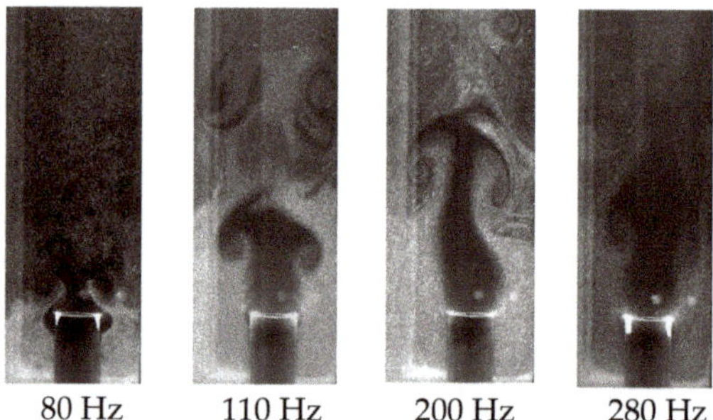

Figure 9. Mie scattering images at different frequency.

As Figure 10 shows, the vortex moved upstream and enhanced the interaction at the base of the flame with an increase of amplitude at 110 Hz, but the influence of amplitude on the length of flame was not obvious at 280 Hz. These changes in flame structure with amplitude variation at different frequencies were well matched with changes of flame length shown in Figure 6. Therefore, NOx emission and flame length both shortened observably with increased acoustic amplitude at 110 Hz; however, it was opposite when the frequency was at 280 Hz.

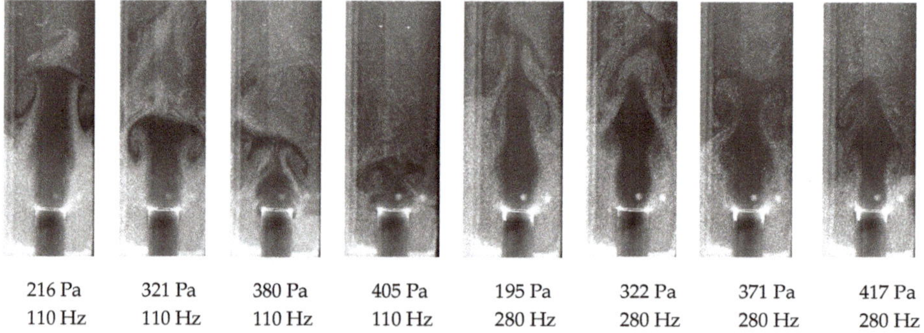

Figure 10. Mie scattering images at different acoustic parameters.

4. Conclusions

The NOx emission of partially premixed flames under acoustic parameters of 0–400 Hz/0–1400 Pa were investigated experimentally. The relationship between frequency, amplitude, and EINOx was discussed based on flame length, flame shape, and vortex. The main conclusions are as follows:

(1) The critical frequency f_c, $f_c \in [170, 190]$, occurs in the changing EINOx of partially premixed flame with the increase of amplitude and frequency. A similar phenomenon is also observed in the change of flame length with acoustic parameters. EINOx decreases with the increase of amplitude when $f < f_c$; there is a linear relationship between EINOx and amplitude, and the slope range is from -0.009 to -0.003; EINOx increases with the increase of frequency in a fixed amplitude, and there is also a linear relationship between EINOx and frequency, where the slope range is from 0.01 to 0.03. However, EINOx increases with an increase of amplitude when $f > f_c$, and there is a linear relationship between EINOx and amplitude, where the slope ranges from 0.001 to 0.006.

(2) The flame length decreases with the increase of amplitude when $f < f_c$; the reason is that an increasing acoustic effect makes the flame wrinkle and deform, and the mixing effect of the flow field is enhanced. The flame length increases with the increase of amplitude when $f > f_c$; the reason is that the influence of acoustics on the flame becomes weakened in this frequency range; flame diameter is decreased, and fuel gas velocity is increased with the increase of amplitude. This is the dominant factor for the decrease of EINOx.

(3) Interaction between flame and vortex influences the NOx emission. When acoustic amplitude increases at $f > f_c$, the vortex moves to the upstream zone of the partially premixed flame, there is enhanced interaction at the flame base, and shortened flame length; therefore, NOx emissions decrease.

Author Contributions: Conceptualization, M.W. and Z.S.; methodology, Z.S.; software, Z.S.; validation, M.W., Z.S., and K.D.; formal analysis, M.W.; investigation, Z.S.; resources, K.D.; data curation, M.W. and Z.S.; writing—original draft preparation, M.W. and Z.S.; writing—review and editing, K.D.; visualization, Y.H.; supervision, K.D.; project administration, Y.Z.; funding acquisition, K.D.

Funding: This research was supported by Zhejiang Provincial Natural Science Foundation of China under Grant No. LY18E060012, National Natural Science Foundation of China under Grant No. 51106139, and the China Scholarship Council under Grant No. 201708330386.

Acknowledgments: Thanks to the reviewers and editor for manuscript improvements. This research was supported by Zhejiang Provincial Natural Science Foundation of China under Grant No. LY18E060012, National Natural Science Foundation of China under Grant No. 51106139, and the China Scholarship Council under Grant No. 201708330386.

Conflicts of Interest: The authors declare no conflict of interest.

References

1. Medwell, P.R.; Nathan, G.J.; Chan, Q.N.; Alwahabi, Z.T.; Dally, B.B. The influence on the soot distribution within a laminar flame of radiation at fluxes of relevance to concentrated solar radiation. *Combust. Flame* **2011**, *158*, 1814–1824. [CrossRef]

2. Wang, C.; Chan, Q.N.; Kook, S.; Hawkes, E.R.; Lee, J.; Medwell, P.R. External irradiation effect on the growth and evolution of in-flame soot species. *Carbon* **2016**, *102*, 161–171. [CrossRef]

3. Chen, C.; Chan, Q.N.; Medwell, P.R.; Yeoh, G.H. Co-combustion characteristics and kinetics of microalgae Chlorella vulgaris and coal through TGA. *Combust. Sci. Technol.* **2018**, 1–20. [CrossRef]

4. Chen, C.; Chen, F.; Cheng, Z.; Chan, Q.N.; Kook, S.; Yeoh, G.H. Emissions characteristics of NOx and SO_2 in the combustion of microalgae biomass using a tube furnace. *J. Energy Inst.* **2016**, *90*, 806–812. [CrossRef]

5. Glassman, I.; Yetter, R.A.; Glumac, N.G. *Combustion*, 4th ed.; Academic press: Manhattan, NY, USA, 2014; pp. 154–196.

6. Wang, C.; Wang, P.; Zhao, L.; Du, Y.; Che, D. Experimental Study on NOx Reduction in Oxy-fuel Combustion Using Synthetic Coals with Pyridinic or Pyrrolic Nitrogen. *Appl. Sci.* **2018**, *8*, 2499. [CrossRef]

7. Cho, I.; Lee, Y.; Lee, J. Investigation on the Effects of Internal EGR by Variable Exhaust Valve Actuation with Post Injection on Auto-ignited Combustion and Emission Performance. *Appl. Sci.* **2018**, *8*, 597. [CrossRef]

8. Raun, R.L.; Beckstead, M.W.; Finlinson, J.C.; Brooks, K.P. A review of Rijke tubes, Rijke burners and related devices. *Prog. Energy Combust. Sci.* **1993**, *19*, 313–364. [CrossRef]

9. Rayleigh, J.W.S. The explanation of certain acoustic phenomena. *Nature* **1878**, *18*, 319–321. [CrossRef]

10. Putnam, A.A.; Belles, F.E.; Kentfield, J.A.C. Pulse combustion. *Prog. Energy Combust. Sci.* **1986**, *12*, 43–79. [CrossRef]

11. Zinn, B.T. *Pulsating Combustion*, 4th ed.; Academic press: New York, NY, USA, 1986; pp. 113–181.

12. Drogue, S.; Breininger, S.; Ruiz, R. Minimization of NOx emissions with improved oxy-fuel combustion: Controlled pulsated combustion. *Ceram. Eng. Sci. Proc.* **1994**, *15*, 147–158.

13. McQuay, M.Q.; Dubey, R.K.; Nazeer, W.A. An experimental study on the impact of acoustics and spray quality on the emissions of CO and NO from an ethanol spray flame. *Fuel* **1998**, *77*, 425–435. [CrossRef]

14. Keller, J.O.; Hongo, I. Pulse Combustion—The Mechanisms of NOx Production. *Combust. Flame* **1990**, *80*, 219–237. [CrossRef]

15. Michel, Y.; Belles, F.E. Effects of Flue-Gas Recirculation on NOx Production and Performance of Pulse Combustion Hot-Water Boilers. *Combust. Sci. Technol.* **1993**, *94*, 447–468. [CrossRef]

16. Au-Yeung, H.W.; Garner, C.P.; Hanby, V.I. An experimental study of the effects of combustion frequency and pressure amplitude on the NO emissions from pulse combustors. *J. Inst. Energy* **1998**, *71*, 204–208.
17. Poppe, C.; Sivasegaram, S.; Whitelaw, J.H. Control of NOx emissions in confined flames by oscillations. *Combust. Flame* **1998**, *113*, 13–26. [CrossRef]
18. Hardalupas, Y.; Selbach, A. Imposed oscillations and non-premixed flames. *Prog. Energy Combust. Sci.* **2002**, *28*, 75–104. [CrossRef]
19. Chao, Y.C.; Huang, Y.W.; Wu, D.C. Feasibility of controlling NOx emissions from a jet flame by acoustic excitation. *Combust. Sci. Technol.* **2000**, *158*, 461–484. [CrossRef]
20. Delabroy, O.; Lacas, F.; Poinsot, T.; Candel, S.; Hoffmann, T.; Hermann, J.; Vortmeyer, D. A study of NOx reduction by acoustic excitation in a liquid fueled burner. *Combust. Sci. Technol.* **1996**, *119*, 397–408. [CrossRef]
21. Hassan, M.I.; Wu, T.W.; Saito, K. A combination effect of reburn, post-flame air and acoustic excitation on NOx reduction. *Fuel* **2013**, *108*, 231–237. [CrossRef]
22. Kim, K.T.; Lee, J.G.; Quay, B.D.; Santavicca, D.A. Response of partially premixed flames to acoustic velocity and equivalence ratio perturbations. *Combust. Flame* **2010**, *157*, 1731–1744. [CrossRef]
23. Kim, M.; Choi, Y.; Oh, J.; Yoon, Y. Flame–vortex interaction and mixing behaviors of turbulent non-premixed jet flames under acoustic forcing. *Combust. Flame* **2009**, *156*, 2252–2263. [CrossRef]
24. Chung, D.H.; Lin, T.H.; Hou, S.S. Flame synthesis of carbon nano-onions enhanced by acoustic modulation. *Nanotechnology* **2010**, *21*, 435604. [CrossRef]
25. Kim, K.T.; Hochgreb, S. The nonlinear heat release response of stratified lean-premixed flames to acoustic velocity oscillations. *Combust. Flame* **2011**, *158*, 2482–2499. [CrossRef]
26. Shen, Z.; Deng, K.; Lu, B.; Zhong, Y. Effects of Equivalence Ratio and Velocity on NOx Emissions in Methane Partially Premixed Flames with Acoustic Excitation. *Proc. CSEE* **2013**, *33*, 54–60.
27. Deng, K.; Zhao, Y.; Fang, D.; Zhong, Y. NOx Emission from Methane Jet Diffusion Flames at Different Coaxial Air Flow Rates. *J. Chin. Soc. Power Eng.* **2010**, *11*, 874–882.
28. Deng, K.; Zhong, Y.; Li, H. Experimental study on NOx emission from methane self-excited pulsating combustion. *J. Chin. Soc. Power Eng.* **2010**, *30*, 528–535.
29. Zhong, Y.; Deng, K.; Li, H.; Li, W.; Wu, M. Experimental Study of NOx Emission in Partially Premixed Flame under Acoustic Forcing. *J. Eng. Thermophys.* **2011**, *32*, 1609–1612.
30. Illingworth, S.J.; Waugh, I.C.; Juniper, M.P. Finding thermoacoustic limit cycles for a ducted Burke-Schumann flame. *Proc. Combust. Inst.* **2013**, *34*, 911–920. [CrossRef]
31. Meunier, P.; Costa, M.; Carvalho, M.G. The formation and destruction of NO in turbulent propane diffusion flames. *Fuel* **1998**, *77*, 1705–1714. [CrossRef]
32. Morcos, V.H.; Abdel-Rahim, Y.M. Parametric study of flame length characteristics in straight and swirl light-fuel oil burners. *Fuel* **1999**, *78*, 979–985. [CrossRef]
33. Cheng, T.S.; Chao, Y.C.; Wu, D.C. Effects of partial premixing on pollutant emissions in swirling methane jet flames. *Combust. Flame* **2001**, *125*, 865–878. [CrossRef]
34. Costa, M.; Parente, C.; Santos, A. Nitrogen oxides emissions from buoyancy and momentum controlled turbulent methane jet diffusion flames. *Exp. Therm. Fluid Sci.* **2004**, *28*, 729–734. [CrossRef]
35. Bourehla, A.; Baillot, F. Appearance and Stability of a Laminar Conical Premixed Flame Subjected to an Acoustic Perturbation. *Combust. Flame* **1998**, *114*, 303–318. [CrossRef]
36. Dah-You, M.A.A. Simple theory of Rijke tube oscillation. *Acta Acust.* **2001**, *26*, 289–294.
37. Dah-You, M.A.A. Exact solution of the Rijke tube equation. *Acta Acust.* **2002**, *27*, 288.
38. Oh, J.; Heo, P.; Yoon, Y. Acoustic excitation effect on NOx reduction and flame stability in a lifted non-premixed turbulent hydrogen jet with coaxial air. *Int. J. Hydrogen Energy* **2009**, *34*, 7851–7861. [CrossRef]

MDPI

St. Alban-Anlage 66

4052 Basel

Switzerland

Tel. +41 61 683 77 34

Fax +41 61 302 89 18

www.mdpi.com

Applied Sciences Editorial Office

E-mail: applsci@mdpi.com

www.mdpi.com/journal/applsci

www.ingramcontent.com/pod-product-compliance
Lightning Source LLC
LaVergne TN
LVHW071506180726
843512LV00014B/1031